D0984740

DNA STRUCTURE AND FUNCTION

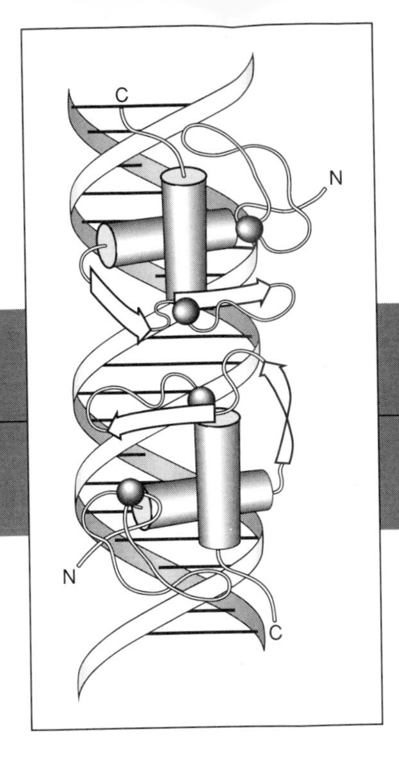

DNA STRUCTURE AND FUNCTION

Richard R. Sinden

Albert B. Alkek Institute of Biosciences and Technology
Center for Genome Research
Department of Biochemistry and Biophysics
Texas A&M University
Houston, Texas

ACADEMIC PRESS
San Diego New York Boston
London Sydney Tokyo Toronto

Front cover: Visualization of the TATA-box binding protein (TBP) and its associated proteins that form a preinitiation complex in all eukaryotes for transcribing DNA to messenger RNA. Acrylic painting in collaboration with Dr. S. K. Burley, Howard Hughes Medical Institute, Rockefeller University. Illustration copyright by Irving Geis.

This book is printed on acid-free paper. ♾

Academic Press, Inc.
A Division of Harcourt Brace & Company
525 B Street, Suite 1900, San Diego, California 92101-4495

United Kingdom Edition published by
Academic Press Limited
24-28 Oval Road, London NW1 7DX

Library of Congress Cataloging-in-Publication Data

Sinden, Richard R.
DNA structure and function / Richard R. Sinden.
p. cm.
Includes bibliographical references and index.
ISBN 0-12-645750-6
1. DNA. 2. Molecular genetics. I. Title.
QP624.S56 1994
574.87'3282--dc20 94-10464
CIP

PRINTED IN THE UNITED STATES OF AMERICA
96 97 98 99 EB 9 8 7 6 5 4 3 2

*To my father and mother for their genetic contribution;
to my wife, Jane for sharing her DNA; and to my children,
David and Laura, may you treat the DNA of your ancestors
with care and respect.*

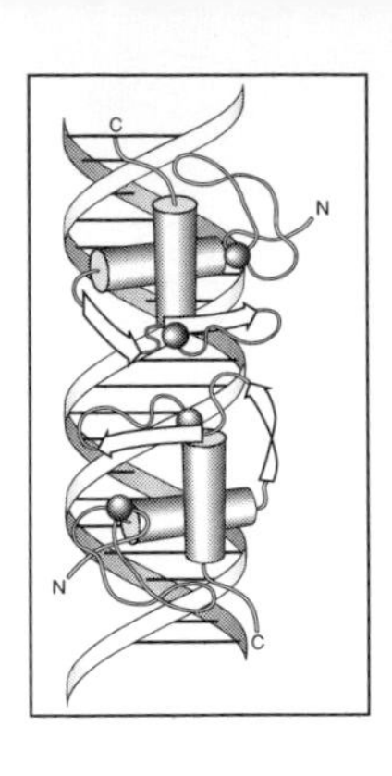

Contents

CHAPTER 3

DNA Supercoiling

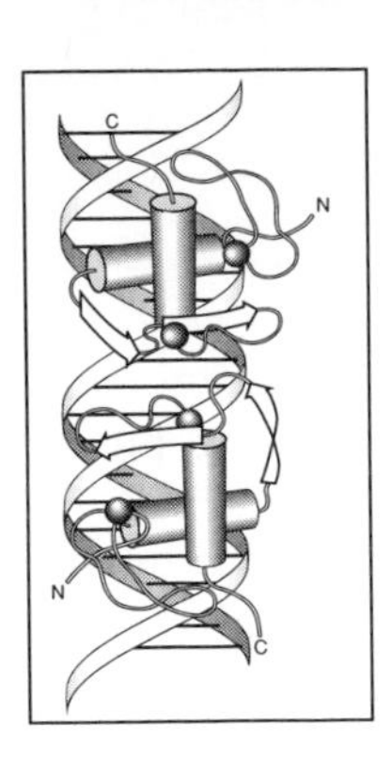

Preface

For years after description of the right–handed DNA double helix by Watson and Crick, DNA was viewed by many as a uniform molecule. With the genetic information encoded as a linear array of triplet codons, it seemed that the key to understanding the regulation of gene expression or the processes of replication, repair, and recombination would likely lie in the interaction of specific proteins with regions of defined base sequences within uniform DNA molecules. However, within the past 15 years our understanding of the complex nature of DNA structure has grown considerably. Defined, ordered sequence DNA (dosDNA) including inverted repeats, mirror repeats, direct repeats, and homopurine·homopyrimidine elements can form a number of alternative DNA structures. Phased A tracts lead to stable DNA bending, inverted repeats can form cruciform structures, alternating purine·pyrimidine sequences can form left-handed Z–DNA, and homopurine·homopyrimidine regions with mirror repeat symmetry can form intramolecular triplex structures. Other dosDNA sequences include A + T-rich regions that can exist as stably unwound regions at origins of DNA replication and at matrix attachment regions, and guanine-rich regions at telomeres that can form triplex and quadruplex structures. The list of alternative conformations of DNA that can form in sequences found in the human genome, and in the genomes of other organisms, will continue to grow.

Although enormous progress has been made in elucidating the structures formed by dosDNA elements, relatively little is definitively known about the biology of alternative DNA conformations. The human genome is littered with homopurine·homopyrimidine elements and alternating purine pyrimidine tracts that can form triplex structures and left-handed Z-DNA, respectively. Probably every human gene that has been sequenced has one or more dosDNA elements that could participate in the formation of an alternative DNA conformation. Why is the human genome so full of dosDNA elements if they are not biologically important? Could these elements simply be the results of errors in DNA polymerization or the products of unequal genetic recombination events that lead to genetic expansion? Are dosDNA elements integral parts of the sophisticated, elaborate interactive dance between the DNA and the many proteins that leads to the coordinated and developmental expression responsible for development from sperm and egg to adult?

This book is intended to serve as a source of information about the many structures of DNA that can form in dosDNA elements. It should be useful for graduate students, advanced undergraduates, and all scientists interested in a survey of the structures of DNA and the possibilities for the involvement of DNA in biological reactions. The first chapter describes the basics of DNA organization— the bases, the base pairs, the B–DNA helix, properties of the B-DNA double helix— and surveys various chemicals and enzymes that react with DNA. The next chapter on DNA bending provides a detailed example of how the primary sequence of DNA directs a defined shape to the DNA double helix. This chapter also begins to introducethe experimental rationale and procedures used to study DNA structure. Specific experiments are presented in "Details of Selected Experiments" sections at the end of several chapters or experiments (or concepts beyond the scope of the general text) are presented in boxes. Boxed sections add another level of sophistication to an appreciation for the details of DNA structure or the details of structural studies. Chapter 3 presents a simplified description of DNA supercoiling and its biological significance. The next three chapters (Chapters 4–6) discuss three major alternative helical forms of DNA that have been studied extensively in the last 15 years: cruciforms, Z–DNA, and intramolecular triplex DNA. Chapter 7 contains a brief description of other non–B–DNA alternative forms of DNA. The list of structures presented in Chapter 7 is necessarily incomplete, since new DNA structures are continually being described. Although an attempt is made to present the structures of alternative DNA conformations and possibilities for their involvement in biology, the reader should remember that this is an intense area of research. In 5 years we may have a much clearer understanding of alternative DNA structures and their role in replication, repair, recombination, and the regulation of gene expression. Chapter 8 presents basic principles of the interactions between DNA and proteins. This field represents one of the most active with great progress being made in solving the X–ray crystal structures of

DNA–protein cocrystals. Chapter 9 briefly discusses the significance of the organization of DNA into chromosomes.

I am deeply indebted to the many scientists who have shared their ideas with me, both in written and verbal form. I greatly appreciate the efforts of Iain L. Cartwright, James E. Dahlberg, Maxim Frank-Kamenetskii Fred Gimble, Myron Goodman, Paul J. Hagerman, James Hu, Terumi Kohwi-Shigematsu, David M. J. Lilley, Donal S. Luse, and Miriam Ziegler for reading selected chapters of this book. I thank Richard Gumport and William Scovell for reading the entire manuscript. I greatly appreciate the contributions of Jan Klysik, David Ussery and especially Karl Drlica for very careful reading of multiple drafts of the entire manuscript. I thank Vladimir Potaman for carefully proofreading the page proofs. I am also very grateful for the patience and persistence of artist Patti Restle of Calyx Studio, Cincinnati, Ohio (formerly with the Medical Illustration Department at the University of Cincinnati College of Medicine, Cincinnati, Ohio). Patti drew all the figures for this book, either *de novo* or from my modification of drawings from the literature. I also thank Beverly Domingue for assistance with preparation and proofreading of the manuscript.

Richard R. Sinden

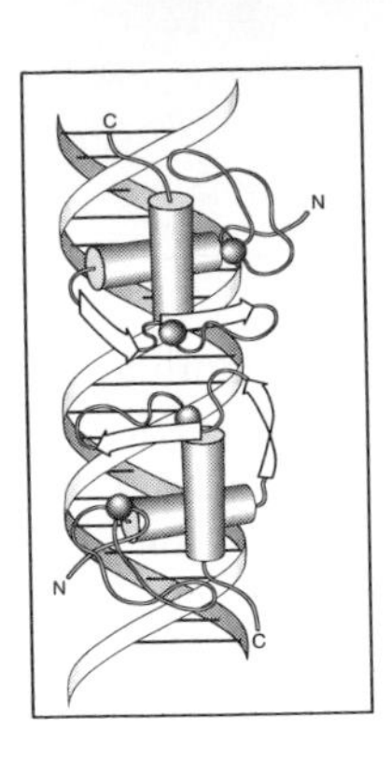

Foreword

The past fifteen years have produced an immense refinement of our ideas about DNA structure and the interactions with proteins. Up to this point, DNA was known as a straight, right-handed double helix composed of two strands held together through complementary Watson–Crick base pairing. While most of this remains correct for the bulk of our knowledge of DNA, the work done over the past few years has demonstrated that almost none of these points is immutable. Within the context of the "standard" B-DNA duplex, systematic, sequence-dependent structural variation occurs, particularly in the way in which sequential base pairs see each other. These properties become especially important when the deformation of DNA is required, as when it bends around proteins such as a histone core. On a larger scale, DNA can become left–handed and can adopt three- and four-stranded conformations. The trajectory of the axis may be systematically deformed, as shown in the curved structure adopted by phased oligoadenine tracts. Base pairing can be broken or rearranged to form helical junctions and cruciform structures, and alternative base–base interactions such as Hoogsteen pairing and the guanine tetrad are also important.

This new appreciation of the wealth of conformations available to DNA is the result of a number of advances in techniques. Some of these can fairly be placed at the high-tech end of the scale of laboratory techniques, while some are rather less so. Probably the single most important contribution has come from the organic chemists, who have provided methods that enable us to synthesize chemically virtually any sequence of oligonucleotides up to about 100 bases in length and sufficiently pure for the most demanding of physical techniques. This has enabled single-crystal X-ray studies to generate an immense structural resource which has been extended to solution by multidimensional NMR spectroscopic methods. These standard structural methods are now being supplemented by new approaches, such as cryoelectron microscopy and fluorescence resonance energy transfer, that generate larger-scale structural information. A second powerful approach involves the application of the methods of molecular genetics to the study of DNA structure. The ability to clone any sequence into multicopy plasmids and then study them under the conditions of negative supercoiling has opened an entirely new world on DNA conformational flexibility. This combination of physical chemistry and molecular biology provides powerful structural, thermodynamic, and kinetic information for studying local DNA structure. Third, enzyme and chemical probing approaches have also been very important in the study of local DNA structure. The reactivity or accessibility of certain positions within a structure may be probed using chemicals such as dimethyl sulpfhate or osmium tetroxide, or the effects of modifications introduced at the time of synthesis may be studied.

Topology expands the structural repertoire of DNA considerably. Negative supercoiling provides a way of trapping within the structure of circular DNA molecules large amounts of free energy that can be used to stabilize otherwise improbable DNA structures. This is of considerable biological importance, and subtle interplay can occur. For example, since promoter function requires changes in DNA winding, many promoters are sensitive to the prevailing level of superhelical stress in the template. Yet, we now know that the action of transcription can itself generate local supercoiling effects in some circumstances. These two effects can be coupled together in a complex manner.

Perhaps the most important aspects of DNA structural variation are likely to be found in the mechanics of molecular recognition and manipulation by proteins. Even for proteins whose main function is just to bind a specific DNA sequence and repress transcription, distortion of DNA is almost the norm, with either local bending or twisting accompanying binding. Some proteins are required to manipulate the DNA structure to carry out their function. Take the initiation of transcription as an example: in the eubacteria the cAMP-dependent activator CAP bends its cognate sequence by about 90°, while in eukaryotes the TATA box-binding protein TBP introduces a massive distortion into the DNA, both bending and opening the minor groove. Similarly, proteins involved in site-specific recombination generate precise wrapping of DNA to juxtapose specific sequences for splicing reactions, while certain classes of nuclease and other proteins recognize the geometry of branched DNA structures in a highly selective manner.

Just because a given sequence can adopt a certain structure in the test tube, this is not a guarantee that it will occur inside the cell, and a major goal in this area is an elucidation of the biological role of DNA structural variability. There is no doubt that DNA does possess an immense conformational flexibility that can be exploited by the topology or in interactions with proteins. Doubtless the next fifteen years will generate many more examples of this, and hopefully some more surprises.

David M. J. Lilley

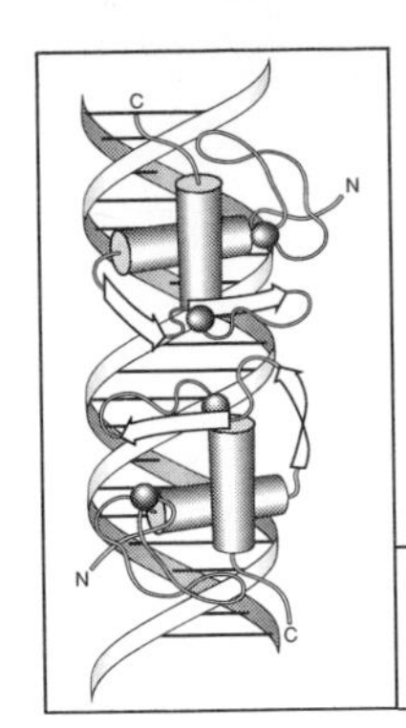

CHAPTER

Introduction to the Structure, Properties, and Reactions of DNA

A. Introduction

The classical view of the DNA (deoxyribonucleic acid) double helix described by Watson and Crick in 1953 is an artifact of textbooks, but for many years most scientists accepted the idea that DNA was structurally a very uniform molecule (Figure 1.1). However, DNA does not exist as the monotonously uniform helix, as will become evident in the following chapters.

The importance and significance of DNA being the genetic material can be appreciated from an understanding of the double-helical structure of DNA. The critical feature of DNA is its linear order of the four nucleotides. With four simple bits of information (the four bases), the DNA encodes all the information necessary for development from sperm and egg to a life form as complex as the one reading this book. The cell is the result of incredible design and engineering. With immense pleasure and frequent awe and amazement we, as scientists, unravel the processes of DNA replication, repair of DNA damage, reorganization of DNA by genetic recombination, regulation of gene expression for transcription into messenger RNA (mRNA), and translation of mRNA into protein.

The structure of DNA physically protects the all-important atoms of the bases from chemical modification by the environment. The hydrogen bonding

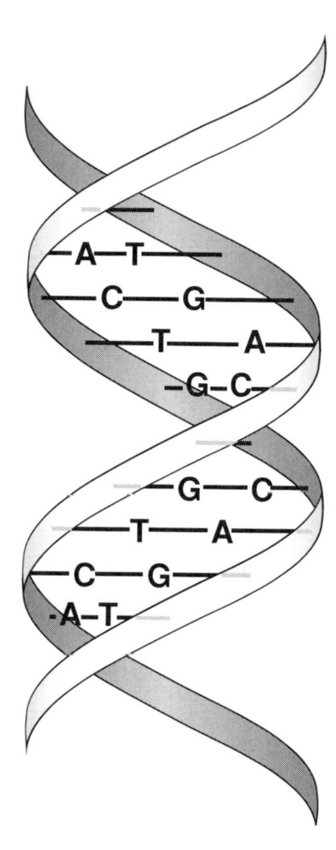

Figure 1.1 Watson–Crick double helix. The model for DNA described by Watson and Crick in 1953 is a right-handed helix, meaning that the strands rotate in a clockwise configuration as they move away from you, as you look down the length of the double helix. The bases are located on the inside of the helix and are shown by letters A, T, C, and G. The phosphate backbone is represented by the ribbons on the outside of the structure.

surface of the individual bases is on the inside, at the center of the double helix. The phosphate backbone is on the outside of the helix. Such design, however, produces an outer surface that is monotonous and uniform. Many proteins that recognize and bind to the DNA in a sequence-independent manner have evolved. These proteins recognize general structural features of the DNA, such as the grooves. For certain processes such as replication and transcription, the DNA must unwind to provide access to the genetic information buried within the center of the double helix. Still other processes involve recognition of particular DNA sequences buried inside a monotonous helix. This is a formidable problem since a sequence-specific DNA binding protein (such as a repressor) must probe down inside the grooves of the helix.

A sophisticated view of the DNA double helix has come into focus through analysis of diffraction patterns of DNA fibers and resolution of crystal structures by X-ray crystallography. More recently, two-dimensional and three-dimensional nuclear magnetic resonance (NMR) data of DNA in solution have provided three-dimensional coordinates for the positions of individual atoms in DNA. The picture that emerges is one of an extremely variable helical structure. The structure of the DNA helix is not at all uniform and monotonous. Moreover, the DNA is a dynamic molecule that can undergo a wide variety of rearrangements in its secondary structure. In fact, many years will pass before we understand enough about the rules that govern the subtleties of DNA structure to be able to accurately predict the structure from the linear array of bases. This book presents our current understanding of the subtleties of the helical forms of DNA and the myriad alternative or non-B-DNA forms that DNA can adopt.

B. The Structure of Nucleic Acids

To understand the structure of B-form DNA and numerous structural variations in the DNA helix, it is important to have an appreciation of the individual components of DNA. DNA is composed of aromatic bases (a purine or pyrimidine ring), ribose sugars, and phosphate groups. The many variations in the structures of the bases and the sugars, and in the structural relationship of the base to the sugar, give rise to differences in the helical structure of DNA.

1. Bases

a. Purines: Adenine and Guanine

Two different heterocyclic aromatic bases with a purine ring (composed of carbon and nitrogen) are found in DNA. The numbering system for the purine ring is shown in Figure 1.2. The two common purine bases found in DNA, *adenine* and *guanine,* are also shown in Figure 1.2. These are synthesized in cells *de novo* in multistep biochemical reactions with the base being built on a phosphorylated ribose sugar molecule. The last common intermediate in their synthesis is inosine. Adenine has an amino group (-NH_2) on the C6 position of the purine ring (carbon at position 6 of the purine ring). Guanine has an amino group at the C2 position and a carbonyl group at the C6 position.

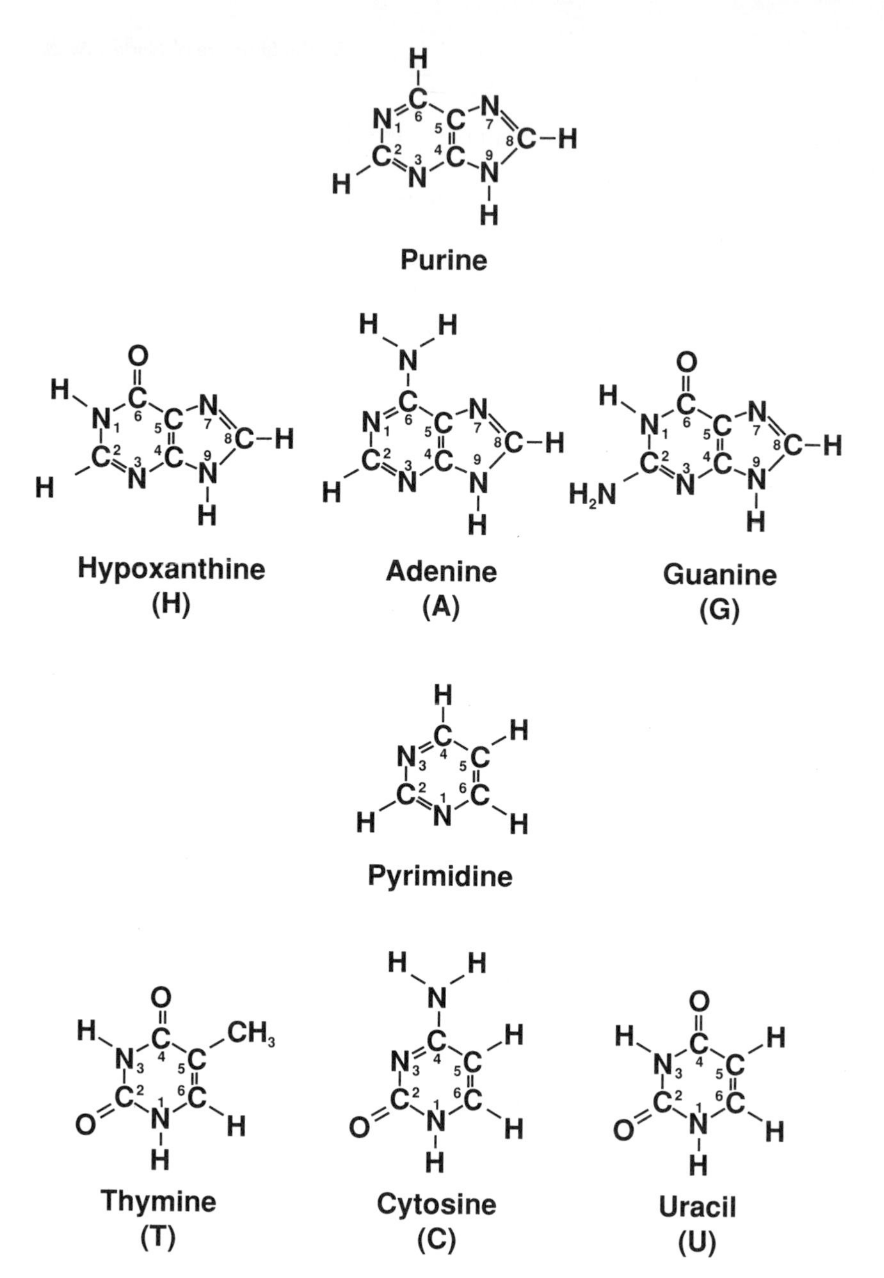

Figure 1.2 Structures of purine and pyrimidine bases. (*Top*) The purine ring structure is composed of 6-member and 5-member aromatic rings. The numbering systems for the rings are indicated on the purine ring and on the structures of hypoxanthine (H), adenine (A), and guanine (G). (Hypoxanthine, although not a common, natural component of DNA, is used *in vitro*. The nucleoside of hypoxanthine is called inosine.) (*Bottom*) The pyrimidine ring structure is a 6-member aromatic ring. The numbering system is indicated on the structures of thymine (T), cytosine (C), and uracil (U).

b. Pyrimidines: Thymine, Cytosine, and Uracil

The two pyrimidine bases commonly found in DNA are *thymine* and *cytosine*. These are also synthesized in cells *de novo* in multistep reactions. The structures of the basic six-member ring, thymine, and cytosine are shown in Figure 1.2. Thymine contains a methyl group at the C5 position with carbonyl groups at the C4 and C2 positions. Cytosine contains a hydrogen atom at the C5 position and an amino group at C4.

Uracil is similar to thymine but lacks the methyl group at the C5 position (Figure 1.2). Uracil is not usually found in DNA. It is a component of ribonucleic acid (RNA) in which it is utilized in place of thymine as one of the pyrimidines. RNA also differs from DNA in the structure of the sugar moiety, as described later.

c. Purines and Pyrimidines as Informational Molecules

The purines and pyrimidines are well suited to their roles as the informational molecules of the cell. The differential placement of hydrogen bond donor and acceptor groups gives the bases the unique structural identity that allows them to serve as the genetic information. The hydrogen atoms of amino groups provide hydrogen bond donors, whereas the carbonyl oxygens and ring nitrogens provide hydrogen bond acceptors. The aromatic nature of the rings means that they are rigid planar molecules. This flatness is important in the organization of bases within the helix, since it allows the bases to stack uniformly within the helix. As described subsequently, this stacking helps protect the chemical identity of the bases.

2. Sugars

a. Ribose Sugar Is Found in RNA

The source of the ribose (or deoxyribose) sugar in the biochemical synthesis of purines and pyrimidines is 5-phosphoribosyl pyrophosphate (PRPP), which is derived from α-D-ribose-5-phosphate. The purine and pyrimidine rings are both synthesized on the β-D-ribose ring (Figure 1.3). β-D-Ribose is a 5-carbon sugar with a hydroxyl group (-OH) on each carbon. Carbons 1 and 4 are joined into a five-member ring through the C4 hydroxyl oxygen. This ribose sugar is found in all RNA molecules.

b. Deoxyribose Sugar Is Found in DNA

In DNA a slightly different sugar, β-D-2-deoxyribose, is found. This is a derivative of β-D-ribose in which the hydroxyl (–OH) at the 2′ position is replaced by a hydrogen (–H). Biochemically this is done by the enzyme ribonucleotide reductase which converts all ribonucleoside diphosphates (or occasionally triphosphates) in a chemical reduction reaction from 2′ OH to 2′ H.

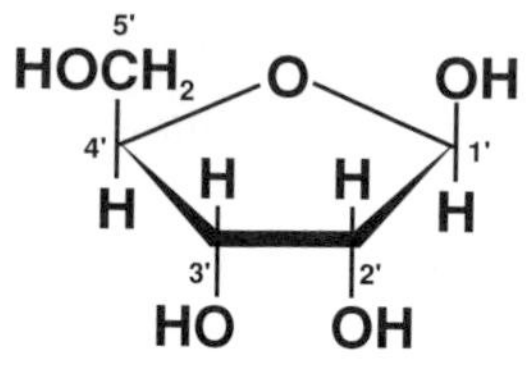

β-D-ribose

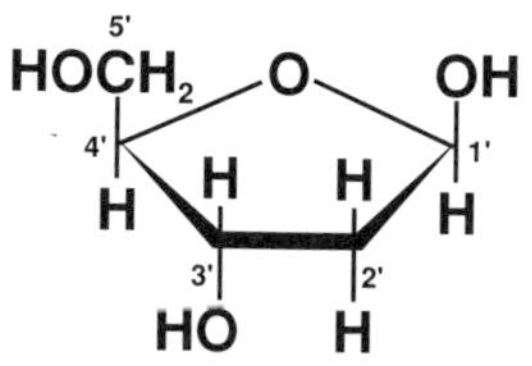

β-D-2-Deoxyribose

Figure 1.3 Structures of β-D-ribose and β-D-2-deoxyribose. (*Top*) β-D-Ribose is the ribose sugar found in RNA molecules. (*Bottom*) β-D-2-Deoxyribose, which is found in DNA, contains a hydrogen rather than a hydroxyl at the 2′ position. The positions of the carbon atoms in the ribose ring are numbered with primes, for example, 2′, to distinguish these atoms from those of the bases.

The sugar moiety of DNA is one of the more flexible and dynamic parts of the molecule. Figure 1.4 shows the structures of the common sugar conformations that are found in the various forms of DNA. The sugar ring structure is easy to envision if one thinks of an envelope. In the *envelope* form, the four carbons form a plane at the corners of the body of the envelope. The oxygen is at the position representing the top of the envelope flap. The oxygen can be bent out of the plane of the body of the envelope. Twisting the C2′ and C3′ carbons relative to the other atoms results in various *twist* forms of the sugar ring. To form the C2′ endo form of the ribose sugar, C2′ twists up from the plane of the four carbons. To form the C3′ endo, C3′ twists down out of the plane of the four carbons.

Dideoxyribonucleotides are used in DNA sequencing reactions. A 2′,3′ dideoxyribonucleotide has hydrogen atoms at both the 2′ and 3′ positions (see Figure 1.5). When incorporated into a DNA chain, the dideoxyribonucleotide blocks further polymerization, since there is no 3′ OH to which another base can be added. This will become apparent from the ensuing discussion of the phosphodiester bond and polynucleotides. DNA sequencing strategies are discussed later in this chapter.

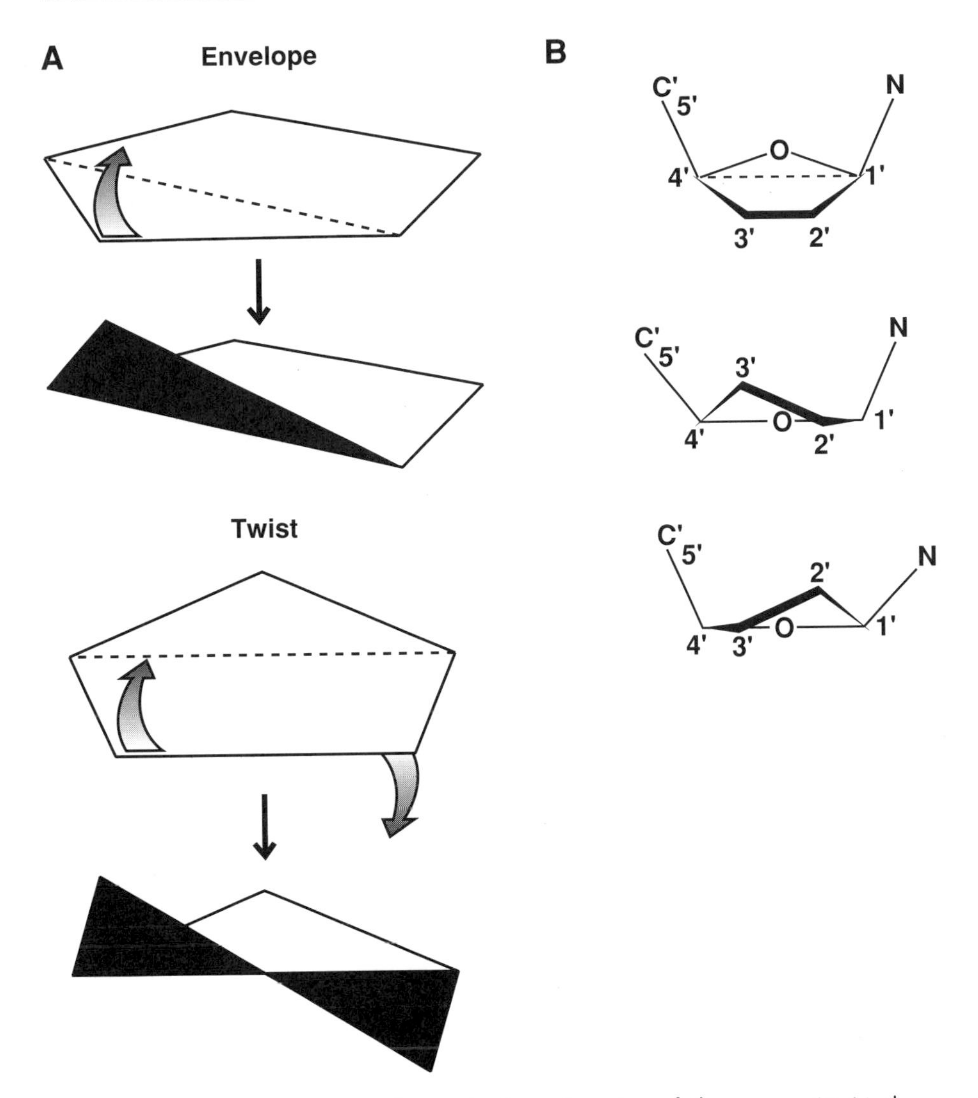

Figure 1.4 Sugar puckers. (A) Envelope and twist configurations of ribose sugars. (*Top*) In the envelope conformation the four carbons make up the body of the envelope, the tip of the flap is the oxygen molecule. Note that the flap of the envelope can be folded up or down. (*Bottom*) In the twist conformation the 2′ and 3′ carbons can be bent up or down making two different twist conformations. Only one conformation is shown. (B) (*Top*) A representation of an envelope conformation of a ribose sugar. (*Middle*) A representation of the C3′ endo conformation of the ribose sugar. (*Bottom*) A representation of the C2′ endo ribose sugar conformation.

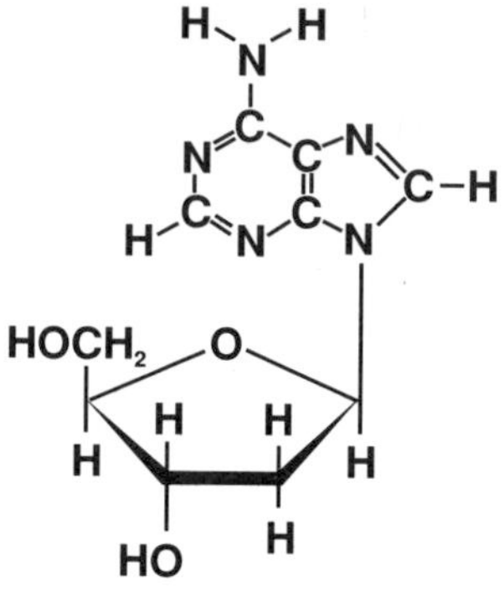

Deoxyadenosine (a nucleoside)

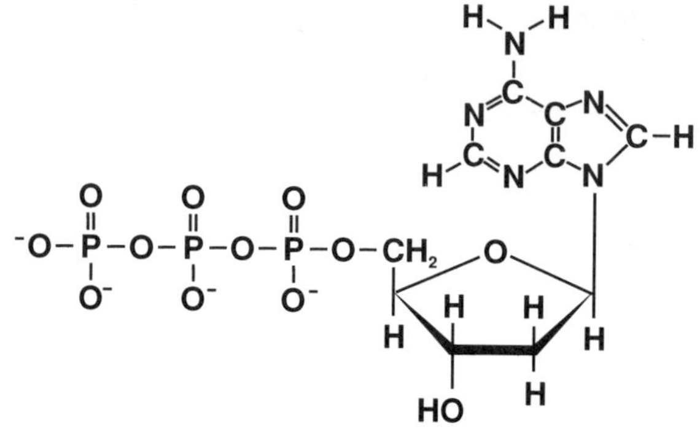

Deoxyadenosine 5' -triphosphate (dATP) (a nucleotide)

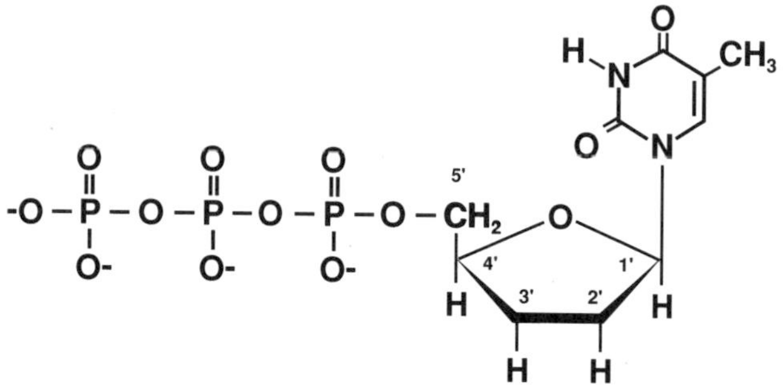

2', 3'-Dideoxythymidine triphosphate (ddTTP)

Figure 1.5 Structure of nucleosides and nucleotides. Deoxyadenosine (a nucleoside) (*left*) is composed of the base adenine linked through a glycosidic bond to the C1′ position of a 2′ deoxyribose sugar. Deoxyadenosine 5′-triphosphate (dATP, a nucleotide) is composed of the base adenine bound to a deoxyribose sugar containing a triphosphate group on the 5′ carbon. In the structure of 2′,3′-dideoxythymidine triphosphate (ddTTP), there is no hydroxyl group on either the 2′ or the 3′ position of the ribose. Since all DNA polymerases require a 3′ OH for addition of the next deoxyribonucleotide, in the absence of a hydroxyl group at this position further chain elongation cannot occur. Dideoxynucleotides are used for DNA sequencing reactions.

3. Nucleosides and Nucleotides

Nucleosides—adenosine, guanosine, thymidine, and cytidine—are the terms given to the combination of base and sugar. The term *nucleotide* refers to the base, sugar, and phosphate group. The structure of adenosine triphosphate is shown in Figure 1.5, in which the phosphate group is attached to the 5′ carbon of the ribose. The nucleotide can have one, two, or three phosphate groups designated α, β, and γ for the first, second, and third, respectively (Figure 1.5). For example, adenosine 5′-triphosphate, abbreviated ATP, contains three phosphate groups on the 5′ carbon of adenosine. A phosphate group can also be attached to the 3′ carbon of ribose rather than the 5′ carbon.

The structure of 2′,3′-dideoxythymidine triphosphate (ddTTP) is also shown in Figure 1.5. This nucleotide is utilized in the chemical sequencing of DNA, as described in Section H,2.

The nucleotides are found in DNA, RNA, and various energy carriers such as nicotinamide adenine dinucleotide (NAD^+) and flavin adenine dinucleotide (FAD). Since a great deal of potential chemical energy is found in the β–γ pyrophosphate bond, ATP and guanosine 5′-triphosphate (GTP) are energy carriers in the cell. The ribonucleotides are synthesized as monophosphates that must be converted to diphosphates and then to triphosphates before being incorporated into RNA. To make deoxyribonucleotides for incorporation into DNA, all ribonucleoside diphosphates are reduced to deoxyribonucleotides by a single enzyme in cells: ribonucleotide reductase.

The bond between the sugar and the base is called the *glycosidic* bond. This bond is said to be in the β (up) configuration with respect to the ribose sugar, in contrast to the α (down) position of the hydrogen. The base is free to rotate around the glycosidic bond. The two standard conformations of the base around the glycosidic bond are *syn* and *anti.* The *anti* conformation reflects the relative spatial orientation of the base and sugar as found in most conformations of DNA, for example, B-form DNA. The *syn* conformation is found (in conjunction with a different sugar pucker) in Z-form DNA (described in Chapter 5).

4. The Phosphodiester Bond

In DNA and RNA the individual nucleotides are joined by a 3′–5′ *phosphodiester* bond. In the short polynucleotide shown in Figure 1.6, the nucleotides are joined from the 3′ sugar carbon of one nucleotide, through the phosphate, to the 5′ sugar carbon of the adjacent nucleotide. (This is termed the 3′–5′ phosphodiester bond.) This bond is formed during the biochemical synthesis of DNA by the enzyme DNA polymerase. (The bond is the same in RNA which is formed by RNA polymerase.) To extend the polynucleotide shown in Figure 1.6, an additional base would form a phosphodiester bond through the phosphate attached to the incoming deoxyribonucleoside triphosphate to the 3′ position of the polynucleotide chain. The β–γ pyrophosphate group is split off and hydrolyzed into individual phosphate molecules. This makes the reaction thermodynamically favorable.

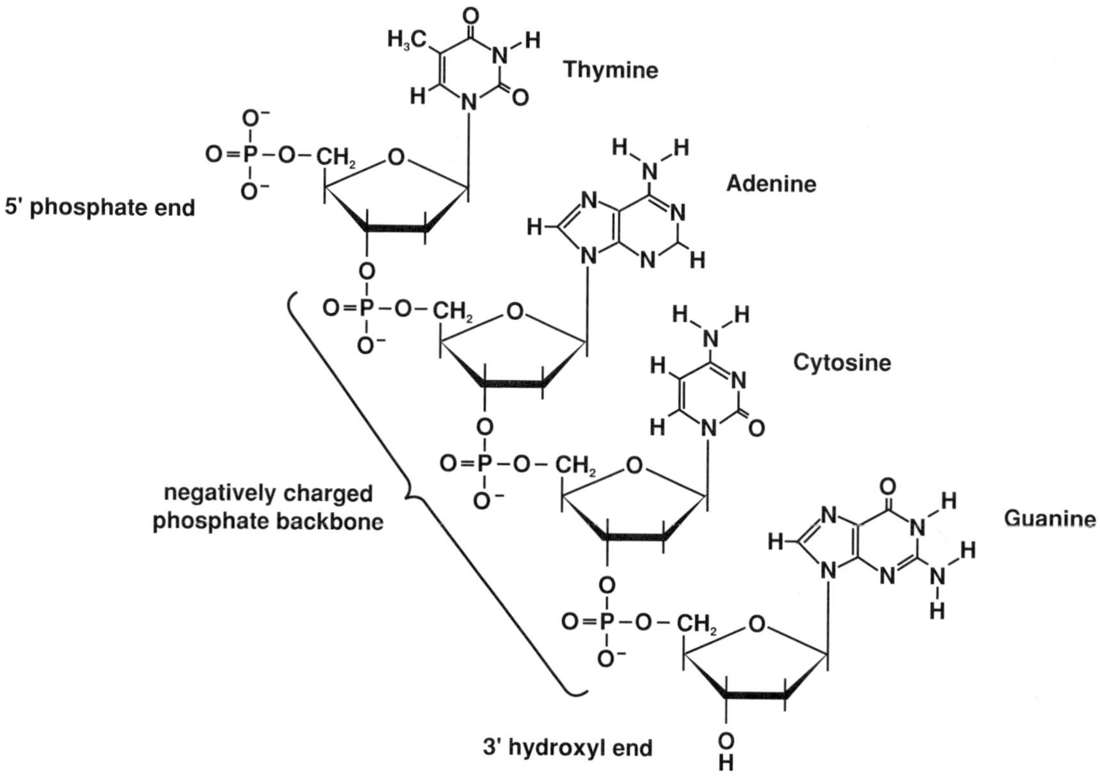

Thymine

Adenine

Cytosine

Guanine

5' phosphate end

negatively charged phosphate backbone

3' hydroxyl end

Figure 1.6 A single strand of DNA. This chain of DNA is composed of the bases thymine, adenine, cytosine, and guanine. Note that the nucleotides are linked through the phosphate groups connected between the 5′ carbon and the 3′ carbon of adjacent deoxyribose sugar molecules. The chain of DNA has a negatively charged phosphate backbone. The DNA chain has two chemically distinct ends. At the 5′ end (*top*) a phosphate group is attached to the 5′ carbon of the deoxyribose sugar. The 3′ end (*bottom*) has a hydroxyl group on the 3′ carbon. This sequence of DNA would be written TACG since sequences are written in the 5′ → 3′ direction by convention.

An important point to be made about the structure of a polynucleotide is that it has *two distinct ends* called 5′ and 3′. Chemically and biologically these ends are quite distinct. This property gives each DNA strand *polarity*. Frequently, DNA exists with a hydroxyl group at the 3′ ends (3′ OH) and a single phosphate group at the 5′ ends (5′ PO_4). The processes of DNA replication and transcription occur by the addition of a 5′ nucleoside triphosphate onto the 3′ hydroxyl group of the terminal nucleotide of the polynucleotide. Nucleases, enzymes that cut DNA, can cleave the phosphodiester bond in one of two places: on the 3′ or 5′ side of the phosphate. A cut made on the 3′ side of the phosphate leaves a 3′ OH and a 5′ PO_4, whereas a cut made on the 5′ side of the phosphate would produce a 5′ OH and a 3′ PO_4.

C. The Structure of Double-Stranded DNA

Watson and Crick first described the structure of the DNA double helix in 1953. A representation of their model is shown in Figure 1.1. Duplex DNA is a right-handed helix formed by two individual DNA strands aligned in an *antiparallel* fashion. This means that one strand is oriented in the 5′→3′ direction and the other in the 3′→5′ direction. The two strands are held together by *hydrogen bonds* between individual bases. The bases are *stacked* near the center of the cylindrical helix. The base stacking provides considerable stability to the double helix. The sugar and phosphate groups are on the outside of the helix and form a "backbone" for the helix. There are about 10 base pairs (bp) per turn of the double helix.

Two important pieces of information were critical for the development of this structure, one of which was "Chargaff's Rules." In the early 1950s, Chargaff pointed out that the amount of adenine always equalled the amount of thymine and the amount of guanine always equalled the amount of cytosine (Chargaff, 1951; Zamenhof *et al.*, 1952). This was true for DNA purified from a wide variety of organisms, and true regardless of the total G + C (or A + T) content. This condition is met by having two strands of DNA in which the bases are hydrogen bonded with strict complementary base pairing. Specifically, A only pairs with T (A·T) and G only pairs with C (G · C). (See Table 1.1 for conventions in designating polynucleotide chains and base pairs.) The second piece of information came from X-ray diffraction patterns of DNA fibers (Wilkins and Randall, 1953; Wilkins *et al.*, 1953) which showed that the geometric shape of DNA is a right-handed helix. Watson and Crick used this information to deduce a model for the structure of DNA.

The double-helical model of DNA seems simple and straightforward. However, the elucidation of this structure was not trivial. Watson and Crick had no way of knowing that DNA was necessarily composed of two strands, that these strands were antiparallel, or that the bases were paired exclusively A · T and G · C. In addition, because of tautomerization and ionization, Watson and Crick were not even sure of the chemical form of the bases. This is a very important point because tautomerization and ionization change the electronic configuration of the bases, which changes their base-pairing properties. Moreover, as discussed subsequently, there are many different ways to form hydrogen bonds between two bases (see Figure 1.11).

Table 1.1
Conventions for Writing DNA Sequences

Sequence polarity	
The sequence of DNA is written left to right in the 5′ to 3′ direction; thus, the sequence GGAATTCC refers to 5′ GGAATTCC 3′	
Base pairs	
Base pairs are designated by a dot	
A · T[a] or G · C	
Oligonucleotides	
Oligonucleotides are designated by a dash or written simply as two successive letters	
Dinucleotides	A-T, G-C or AT, GC
Trinucleotides	G-A-T or GAT
Repeating oligonucleotides	
Repeating units within polymers are designated by poly (x) with x as the repeating base(s)	
Mononucleotide	poly(A)
Dinucleotide	poly(AT)[b]
Trinucleotide	poly(GAT)
Double-stranded repeating polymers	
The polymers that are base paired (indicated with a dot) are written with 5′ to 3′ polarity	
Mononucleotide	poly(A) · poly(T)[c]
Dinucleotide	poly(AT) · poly(AT)
Trinucleotide	poly(GAT) · poly(ATC)

[a]To distinguish DNA from RNA, a *d* to indicate deoxy could be included, e.g., dA · T or dG · dC. When it is clear that DNA rather than RNA is being discussed, the *d* is frequently omitted.
[b]Could be written poly(dA-dT).
[c]Could be written poly(dA) · poly(dT).

1. Hydrogen Bonding and Base Stacking Hold the DNA Double Helix Together

A hydrogen bond is a short, noncovalent, directional interaction between a covalently bound H atom (donor) that has some degree of positive charge (due to attachment to a nitrogen or oxygen atom) and a negatively charged acceptor atom. The negative charge (acceptor) is provided by electrons on a carbonyl oxygen (-C=O) or the lone pair electrons on nitrogen (N:). In the DNA double helix, the N and O atoms involved in hydrogen bonding are separated by 2.82–2.92 Å. (An angstrom is 10^{-10}m.) In the A · T base pair are two hydrogen bonds, as shown in Figure 1.7, that are separated by 2.82 and 2.91 Å (Seeman *et al.*, 1976b). In the G · C base pair three hydrogen bonds are separated by 2.84–2.92 Å (Rosenberg *et al.*, 1976). Typically, hydrogen bonds are weak, having only 3–7 kcal/mol with a dis-

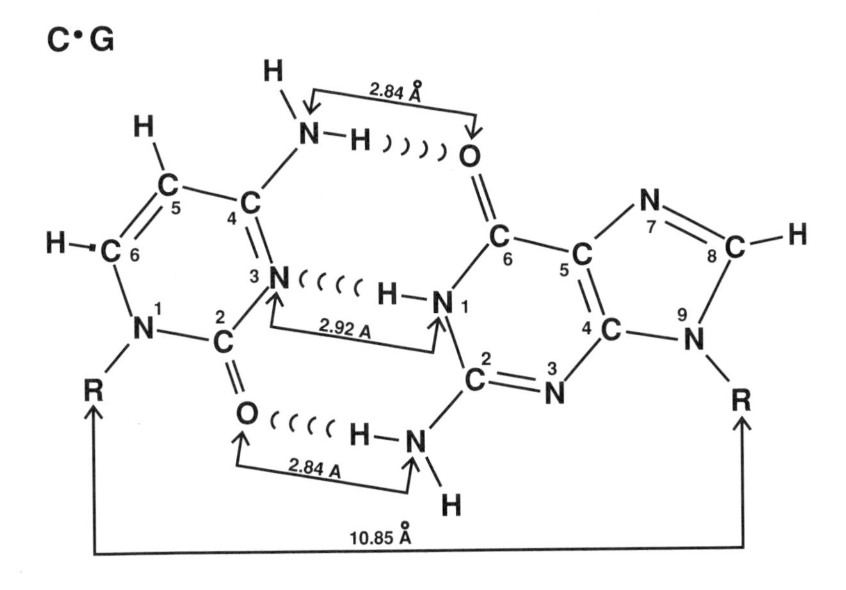

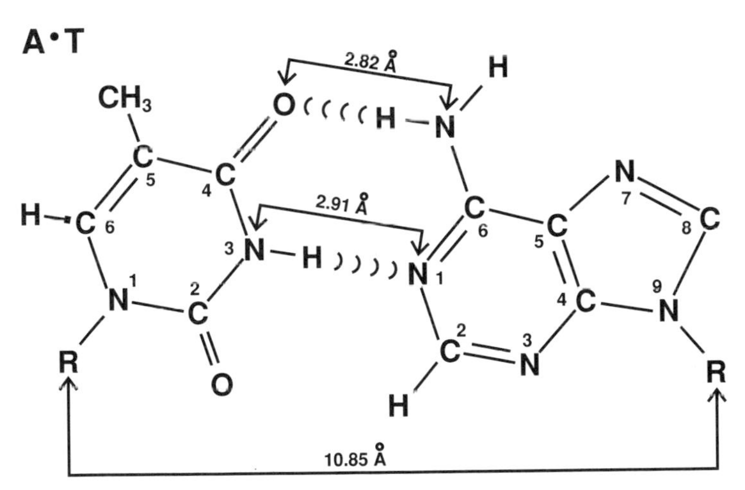

Figure 1.7 Watson–Crick base pairs. The interatom hydrogen bond distances and distances between the C1′ positions of the ribose sugars are indicated. The curved lines represent the hydrogen bonds. The curves are in the direction of the hydrogen bond acceptor (N or O atoms). Figure modified with permission from Arnott *et al.* (1965).

tance between C, N, or O atoms of 2.6–3.1 Å. In contrast, covalent bonds have 80–100 kcal/mol. Although hydrogen bonds are directional, they can be deformed by stretching and bending. In DNA, the hydrogen bonds have 2–3 kcal/mol. This is weaker than most hydrogen bonds and is due to geometric constraints within the double helix.

Since the aromatic bases are planar, they can stack nicely on one another. Hydrophobic interactions and Van der Waals forces are involved in the stacking interaction, which is estimated to be 4–15 kcal/mol per dinucleotide. Van der Waals interactions involve dipole–dipole interactions and London dispersion interactions (transient dipole interactions). Note that stacking provides energies of stabilization similar to those provided by hydrogen bonding (considering a dinucleotide). An analogy commonly used to illustrate base stacking in DNA is that of a stack of coins. The position of a single coin is stabilized in the stack by the coins above and below it.

Differences between the characteristics of base stacking and hydrogen bonding energies contribute to the heterogeneity of the DNA helix structure. The overall energy of hydrogen bonding depends predominantly on base composition; that is, all A · T and C · G base pairs have relatively the same geometry and strength of hydrogen bonding. On the other hand, base stacking energies depend on the sequence of the DNA. For example a 5′ GT dinucleotide stack has a stacking energy of 10.51 kcal/mol, whereas a 5′ TG stack has an energy of 6.78 kcal/mol. Base stacking energies are listed in Table 1.2.

Once the DNA double helix is formed, it is remarkably stable. The individual interactions stabilizing the helix are weak, but the sum of all interactions makes a very stable helix. This cooperative type of stability will be discussed in more detail in Section F,2.

Table 1.2
Base Pair Stacking Energies

Dinucleotide base pairs	Stacking energies (kcal/mol/stacked pair)[a]
(GC) · (GC)	−14.59
(AC) · (GT)	−10.51
(TC) · (GA)	− 9.81
(CG) · (CG)	− 9.69
(GG) · (CC)	− 8.26
(AT) · (AT)	− 6.57
(TG) · (CA)	− 6.57
(AG) · (CT)	− 6.78
(AA) · (TT)	− 5.37
(TA) · (TA)	− 3.82

[a]Data from Ornstein *et al.* (1978).

2. Wobble, Tautomerization, and Ionization Cause "Deviant" Base Pairs

a. Wobble and Other "Deviant" Base Pairs

The classic Watson–Crick base-pairing scheme is only one of several. There are many other ways in which two bases can be held together by hydrogen bonding. Several of these are shown in Figure 1.8. Reverse Watson–Crick base pairs are formed when one nucleotide rotates 180° with respect to the complementary nucleotide. In this case, the glycosidic bonds (and phosphate sugar backbones) are in a trans rather than a cis orientation. Because of symmetry in the hydrogen bonding potential of T at the C2–N3–C4 positions it can rotate on the N3–C6 axis to form a reverse Watson–Crick A · T base pair. This type of base pairing is found in parallel DNA discussed in Chapter 7. Hoogsteen base pairs utilize the C6–N7 face of the purine for hydrogen bonding with the Watson–Crick (N3–C4) face of the pyrimidine (Hoogsteen, 1963). A characteristic feature of Hoogsteen base pairing is that the N7 position of the purine is base-paired, altering the chemical reactivity of this position. A reverse Hoogsteen base pair involves flipping one of the bases 180° with respect to the other.

There are other ways to hydrogen bond two bases in which the spatial relationship of the bases is different from that found in the Watson–Crick (or reverse Watson–Crick base pair. These are *anti–syn* base pairs and wobble base pairs. The glycosidic bond, as discussed above, is typically in the *anti* configuration. An *anti–syn* base pair can form between two purines in which one purine is in the typical *anti* form and one in the unusual *syn* configuration. A G(*anti*) · A(*syn*) pair is shown in Figure 1.8 and an A(ionized-*anti*) · G(*syn*) base pair is shown in Figure 1.11. Bases can also form wobble pairs, in which the position of one base relative to its complement is shifted within the flat plane of the aromatic rings (and hydrogen bonds). A wobble G · T base pair is shown in Figure 1.8. As discussed subsequently, in addition to ionized mispairs, 5-bromodeoxyuridine (5-BrdUrd) and 2-aminopurine (2-AP) can form wobble base pairs (Sowers *et al.*, 1986,1989).

Mispairing errors during DNA replication that lead to mutations do not occur at the high frequency one might expect given the possibilities for tautomeric shifts, ionizations, and wobble base pairings because bacterial and animal cells have evolved elaborate mechanisms to deal with base mispairing. The processes of DNA replication and DNA repair are designed to correct these naturally occurring errors. DNA polymerases have proofreading functions to survey the bases they have just incorporated. If an incorrect nucleotide has been incorporated, the polymerase backs up and a nuclease activity of the polymerase removes the incorrect nucleotide. Various other DNA repair systems correct mismatches missed by DNA polymerase. These systems

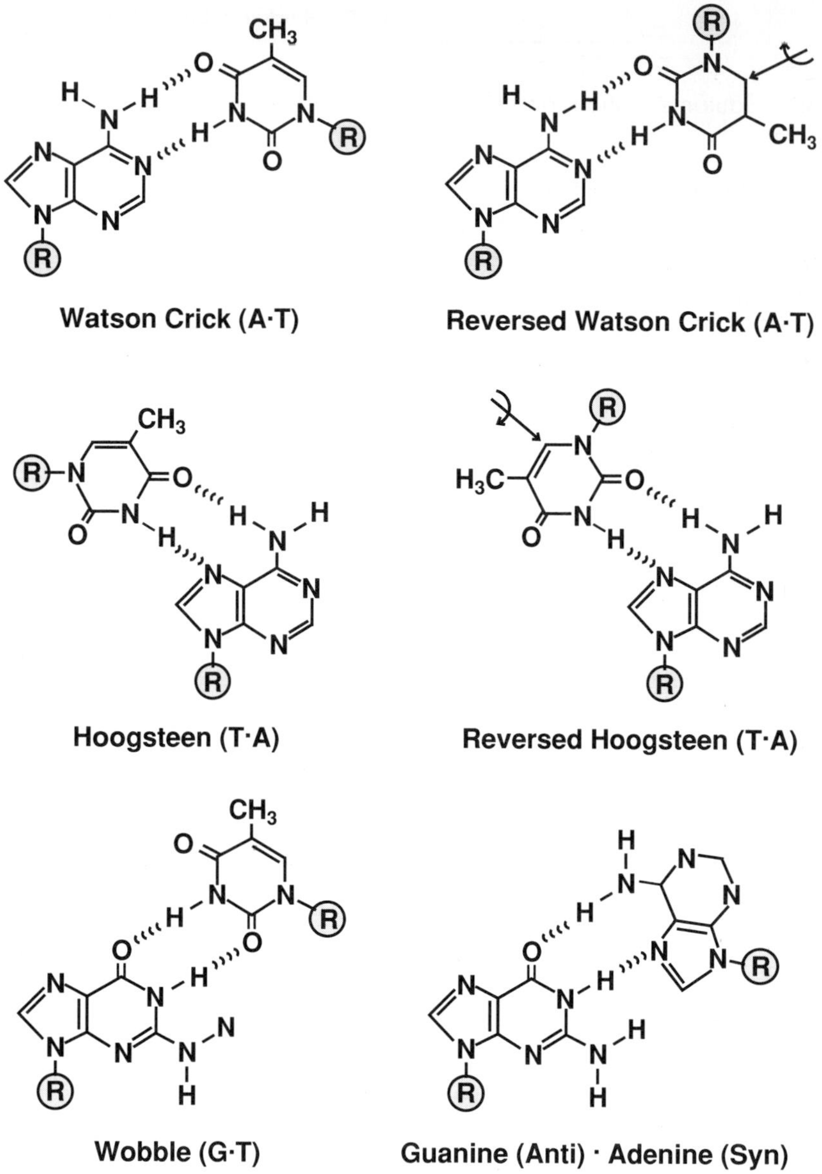

Watson Crick (A·T)

Reversed Watson Crick (A·T)

Hoogsteen (T·A)

Reversed Hoogsteen (T·A)

Wobble (G·T)

Guanine (Anti) · Adenine (Syn)

Figure 1.8 Base pairing schemes. The top pair of A·T base pairs represent the traditional Watson–Crick configuration (*left*) and a *reversed* Watson–Crick pair, with a 180° rotation of the pyrimidine base (*right*). The center two T·A pairs represent Hoogsteen base pairs. In a standard Hoogsteen base pair the pyrimidine uses its Watson–Crick surface to pair with the N1, C6, N7 side of the purine base. A 180° rotation of the pyrimidine results in formation of a *reversed* Hoogsteen base pair. The bottom left structure shows a wobble G·T base pair. The pyrimidine has been shifted up the length of about one C–C (or C–N) bond. The structure on the bottom right shows a G(anti) · A(syn) base pair with the glycosidic bond of the adenine in the syn configuration.

reduce mutation rates in microbial organisms to 0.0033 mutation per genome per replication (Drake, 1991).

b. Tautomerization

The representations shown in Figures 1.2 and 1.7 are only *one* possible chemical form of the purine and pyrimidine bases. Figure 1.9 shows the classic keto–enol, amino–imino alternative chemical forms of the bases. For example, the C6 keto (C=O) position of guanine can undergo a *tautomerization* to an enol form (with a -OH at the C6 position). For this tautomerization to occur, the double bond must shift from the carbonyl group to the nitrogen–carbon bond in the ring. In a similar fashion, an amino nitrogen (-NH_2) can undergo a transition to an imino form (=NH). The imino or enol forms of the bases each have two isomeric forms that can exist, depending on the position of the hydrogen relative to the lone pair electrons (see the isomers shown in Figure 1.9). Tautomerization is extremely significant, *because it can reverse the polarity of hydrogen bonding sites.*

The bases are in chemical equilibrium between the two alternative tautomeric forms. The equilibrium favors the keto and amino forms by a ratio of about 10^4 to 1. Because the polarity of hydrogen bonding is reversed on tautomerization, rare *mispairs* can form. Enol-G will pair with T, keto-T will pair with G, imino-A will pair with C, and imino-C will pair with A as shown in Figure 1.9B. Tautomerization should create mispairs in the DNA at a frequency of 10^{-4} during the synthesis of DNA. However, there is little, if any, evidence suggesting that keto–enol and amino–imino tautomerizations occur during the synthesis of DNA *in vivo*. Mispairs involving ionizations and wobble base pairs may occur more frequently in DNA.

c. Ionized Base Pairs

In addition to keto–enol tautomerizations, bases can exist in ionized forms that also change their hydrogen bonding properties. This was first realized with several "classic" base analogs, 5-BrdUrd and 2-AP, that were believed to cause mutations by undergoing keto–enol tautomerizations. The base analog 5-BrdUrd, which contains a bromine atom at the C5 position of uracil, is a base analog of thymine (the bromine atom of 5-BrdUrd and methyl group of T are about the same size). The incorporation of 5-BrdUrd into DNA leads to a rather high frequency of a type of mutation called a transition mutation, in which a Py · Pu base pair mutates to the other Py · Pu base pair. For example, 5-BrdUrd · A is changed to a C · G base pair, which is effectively a T · A→C · G transition. The classic explanation for the high mutation frequency induced by 5-BrdUrd is that this base analog has a high frequency of keto–enol tautomerization. [See an elegant paper by Topol and Fresco (1976) for more details on base mispairing and its relationship to mu-

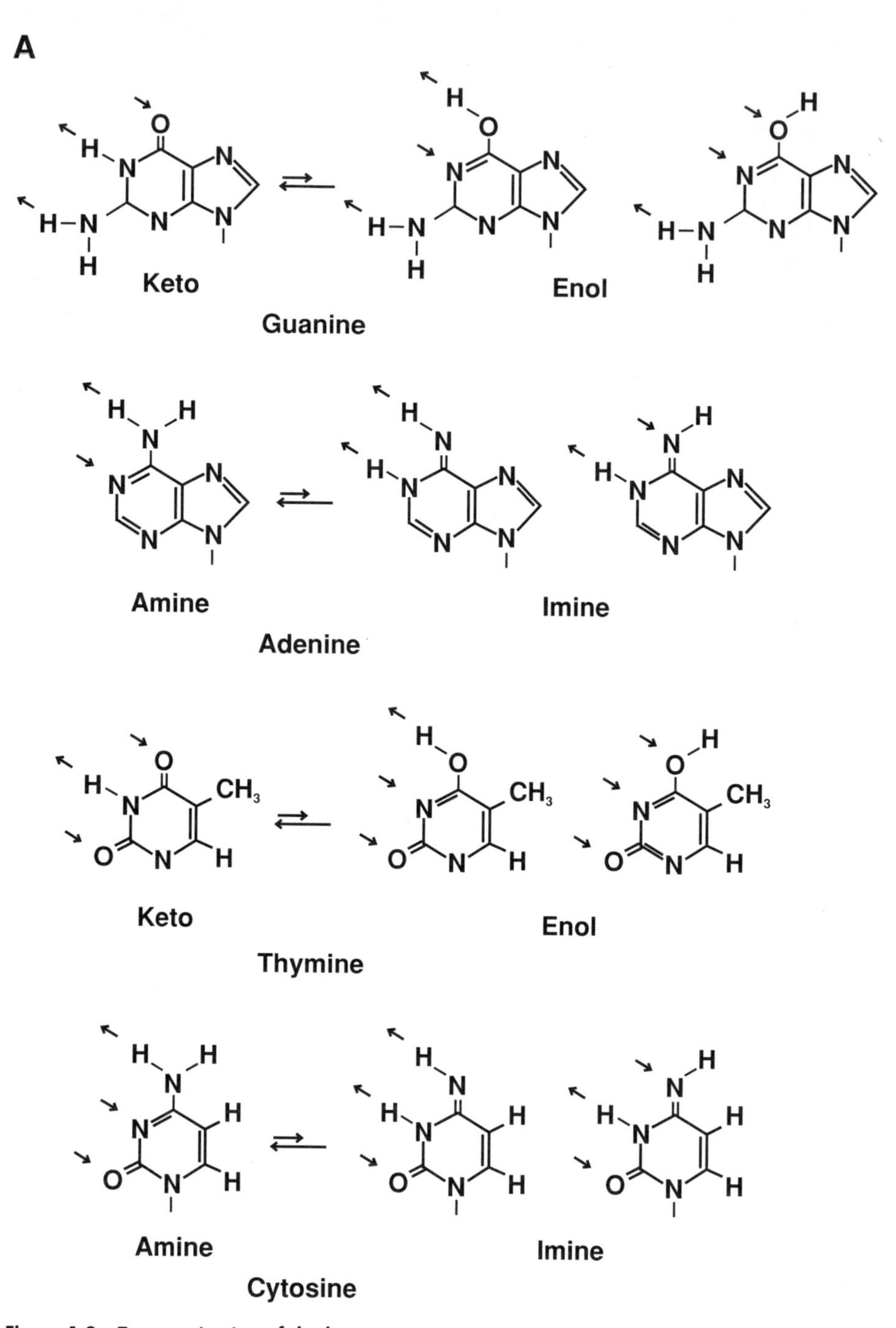

Figure 1.9 Tautomerization of the bases. (A) Base tautomerization. The bases can exist in different chemical isomeric forms. The top panel shows guanine which normally exists in the keto configuration (right) in which there is a carbonyl group at a C6 position of the ring. The hydrogen bonding properties of the bases are indicated by the arrows. The arrows pointing away from the hydrogens represent hydrogen bonding donors. The arrows pointing toward the negative centers

B

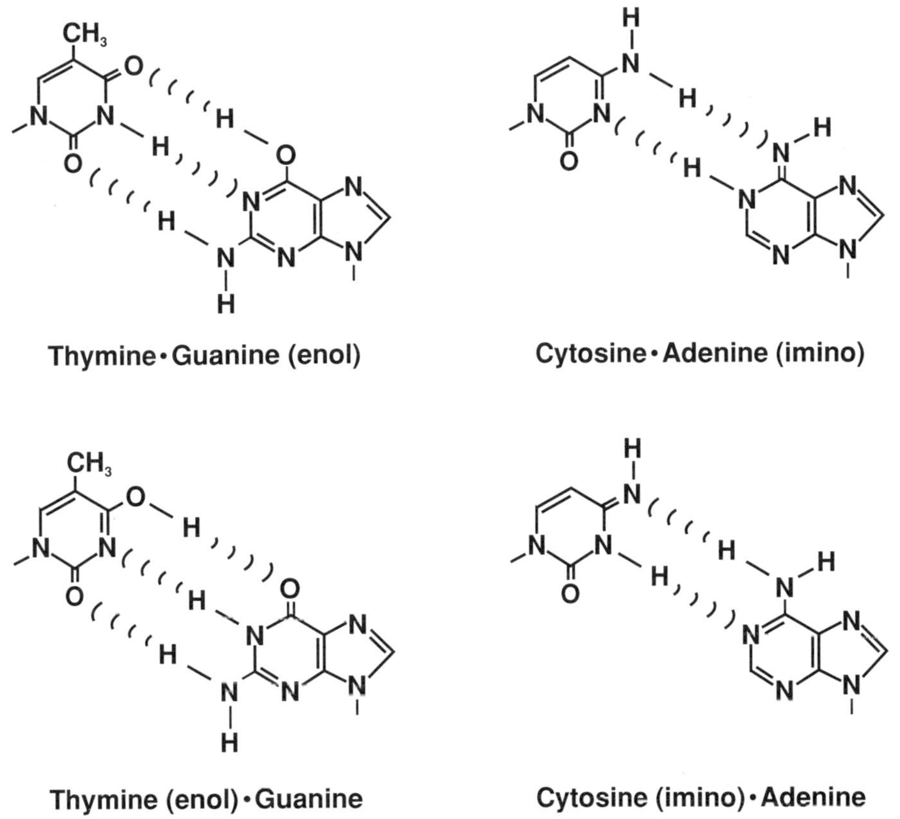

Thymine•Guanine (enol)

Cytosine•Adenine (imino)

Thymine (enol)•Guanine

Cytosine (imino)•Adenine

(the unpaired electrons of the oxygen and nitrogen) represent hydrogen bonding acceptors. The enol configuration can exist rarely, at a frequency of about 10^{-4}. Here a hydrogen moves from the N1 position to the oxygen at the C6 position making a hydroxyl group. In the two possible configurations of the enol form of the guanine, the hydrogen on the oxygen is pointed to the left or to the right. Note the change in the electronic hydrogen bonding properties between the keto and enol forms. The direction of the arrows for hydrogen bonding potentials are reversed at the N1 position and the C6 position for the left-most enol form. For the right-most enol form, the arrows denoting hydrogen bonding potential are both pointing in toward the nitrogen and the oxygen. The second panel shows the tautomeric amine–imine structures of adenine. The typical amine form is shown on the left and the rare tautomeric imine isomers are shown on the right. The amine form has an amino group on the C6 position, whereas the imine forms have an imino group at this position. The hydrogen from the amino group moves to the N1 position of the ring on formation of the imino group. This event necessarily reverses hydrogen bonding properties of the N1 position of the ring. This position may contribute hydrogen bonding in either direction depending on the location of the hydrogen in the imino form. The third panel shows the keto–enol tautomeric forms of thymine and the fourth panel shows the amine–imine forms of cytosine. (B) Base pairing of tautomeric isomers. The rare tautomeric forms of the bases can be involved in alternative base pairing schemes. The top panel shows a mispair between thymine and the enol form of guanine (*left*) and a mispair between cytosine and the imino form of adenine (*right*). The bottom panel shows mispairs between the enol form of thymine and guanine (*left*) and the imino form of cytosine and adenine (*right*). For nearly 30 years researchers believed that when such mispairs form during DNA replication they can lead to transition mutations. However, the ionizations shown in Figure 1.11 may be more frequent sources of mutations.

A

Keto

Ionized

5 - Bromodeoxyuracil

Ionized

2 - Aminopurine

B

5-BrdUrd • Adenine

5-BrdUrd (ionized) • Guanine

Thymine • 2-Aminopurine

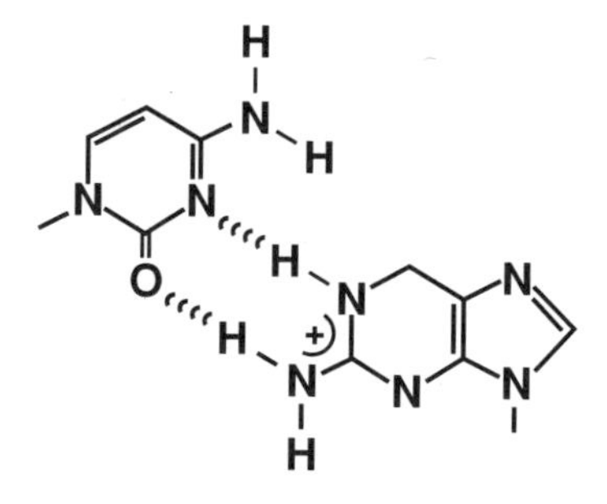

Cytosine • 2-Aminopurine (ionized)

tation.] However, recent experimental evidence has challenged this idea. Goodman and colleagues have obtained experimental evidence that 5-BrdUrd can form a 5-BrdUrd · G base pair when 5-BrdUrd exists in an *ionized* form (Sowers *et al.*, 1989). In this situation, the hydroxyl (-OH) at C4 ionizes and acquires a negative charge on loss of the hydrogen, a reaction favored under basic pH conditions. Similarly, 2-AP, believed to mispair because of amino–imino tautomerizations, has been shown to form ionized and wobble base pairs (Sowers *et al.*, 1986). Recent evidence has shown that DNA polymerase can incorporate ionized base pairs into DNA (Yu *et al.*, 1993). The ionized forms of 5-BrdUrd and 2-AP are shown in Figure 1.10, as are the 5-BrdUrd(ionized) · G base pair and the 2-AP (ionized) · T base pair.

The natural bases can also undergo ionizations, as shown in Figure 1.11. When this occurs, a number of non-Watson–Crick base pairs can arise. Adenine is prone to protonation, a condition favored by low pH (acidic conditions), which can lead to the formation of an $A^+ \cdot C$ base pair. The $A^+ \cdot C$ requires a wobble pairing scheme that is discussed earlier. Cytosine is also very prone to protonation. Protonation of C can lead to a $C^+ \cdot G$ base pair. This, however, requires a Hoogsteen pairing scheme that is discussed earlier. On loss of a proton, the ionized form of thymine can form a $T^- \cdot G$ base pair. In certain sequence contexts protonation can occur very readily at near neutral pH.

3. Helix Parameters

The structure of DNA can be described by a number of parameters that define the helix (Dickerson *et al.*, 1989). Saenger (1984) describes the structure of numerous forms of DNA in great detail. It is useful to become familiar with these terms since this will facilitate understanding the structural variations in DNA shape and structure to be described later. Some of parameters are illustrated in Figures 1.12 and 1.13.

Helix sense refers to the helical rotation of the double helix. The structure described by Watson and Crick is a right-handed (clockwise) helix. Most helical forms of DNA are right-handed. Left-handed DNA, called Z-DNA, will be discussed in Chapter 5.

Figure 1.10 Ionization and base pairing schemes of 5-bromodeoxyuridine (BrdUrd) and 2-aminopurine (AP). (A) 5-BrdUrd and 2-AP are in equilibrium with the ionized forms shown on the right. Ionization of 5-BrdUrd involves dissociation of a hydrogen from the N3 position of the pyrimidine ring. Ionization of 2-AP involves the association of a hydrogen at the N1 position of the purine ring. (B) 5-BrdUrd · A and 2-AP · T form normal Watson–Crick pairs and are thus classified as *base analogs* of, respectively, T and A. The 5-BrdUrd · G and 2-AP · C mispairs involve the ionized forms of these base analogs.

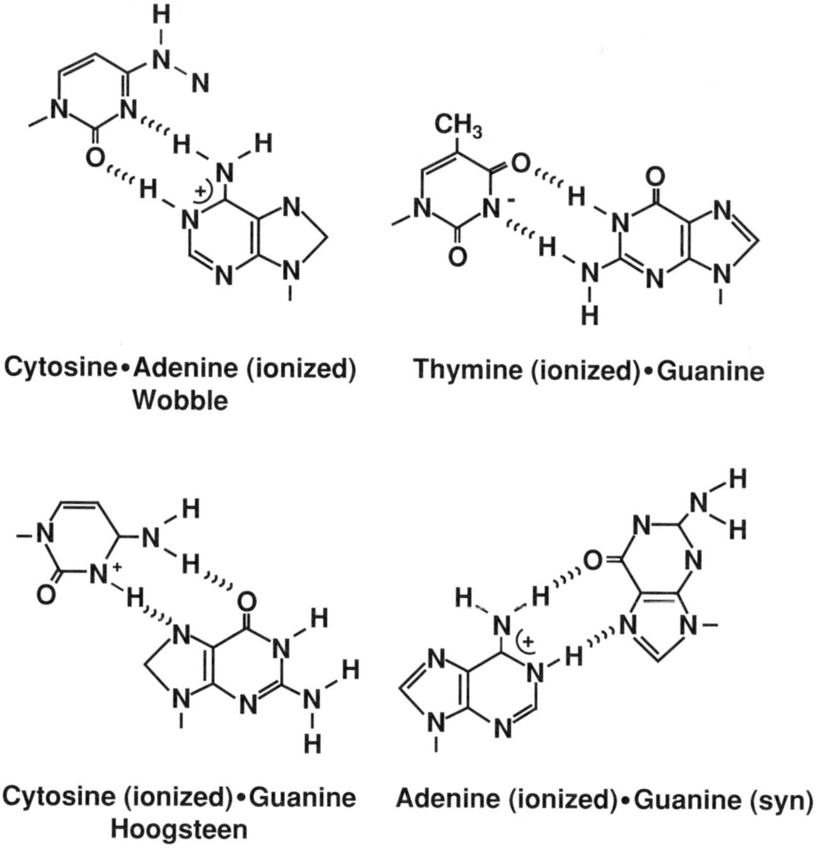

Cytosine•Adenine (ionized) Wobble

Thymine (ionized)•Guanine

Cytosine (ionized)•Guanine Hoogsteen

Adenine (ionized)•Guanine (syn)

Figure 1.11 Base pairs involving ionized forms of bases. Several examples of alternative base pairing schemes involving the ionized forms of bases are presented. The C · A(ionized) base pair involves wobble pairing (*top left*). The C(ionized) · G base pair involves Hoogsteen base pairing (*bottom left*) and the A(ionized) · G(syn) base pair involves hydrogen bonding between an A in the anti position and a G in the syn configuration (*bottom right*). The T (ionized) · G base pair forms with Watson–Crick hydrogen bonding surfaces (*top right*).

Residues per turn refers to the number of base pairs in one helical turn of DNA, that is, the number of bases needed to complete one 360° rotation. The structure described by Watson and Crick, "textbook B-form" DNA, contains 10 bp per turn. DNA in solution contains 10.4–10.5 bp per turn, although this value can vary considerably as a function of base composition.

Axial rise is the distance between adjacent planar bases in the DNA double helix. In textbook B-form DNA there are about 3.4 Å between adjacent base pairs.

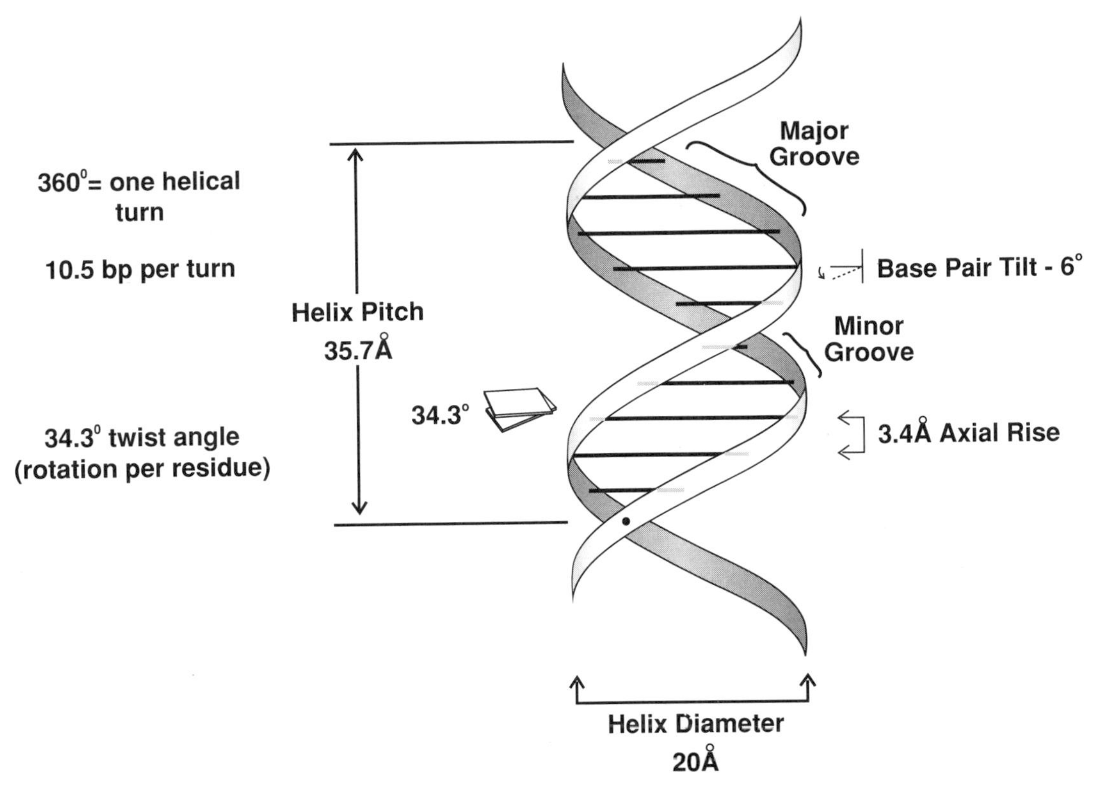

Figure 1.12 The DNA double helix in solution: structural parameters. The Watson–Crick double helix is composed of about 10.5 base pairs per helical turn. Since 360° constitutes one helical turn, there would be a 34.3° twist angle or rotation per residue between adjacent base pairs (see Table 1.4). The helix pitch or length per helical turn is 35.7 Å. The axial rise or distance between two planar base pairs is 3.4 Å. The base pair tilt or deviation from the horizontal plane of the bases is about -6°. The helix diameter or the width of the helix is about 20 Å. Note the positions of the minor groove and the major groove.

Helix pitch is the length of one complete helical turn of DNA. In textbook B-form DNA, one helical turn of 10 bp is completed in 34 Å.

Base pair tilt refers to the angle of the planar bases with respect to the helical axis. The tilt angle is measured by considering the angle made by a line drawn through the two hydrogen bonded bases relative to a line drawn perpendicular to the helix axis. This is illustrated in Figures 1.12, 1.13, and 1.14. A base pair that was perfectly flat, that is, perpendicular to the helix axis, would have a tilt angle of 0°. In B-form DNA the bases are tilted by only −6°. In A-form DNA the base pairs are significantly tilted at an angle of 20° (see Figure 1.16).

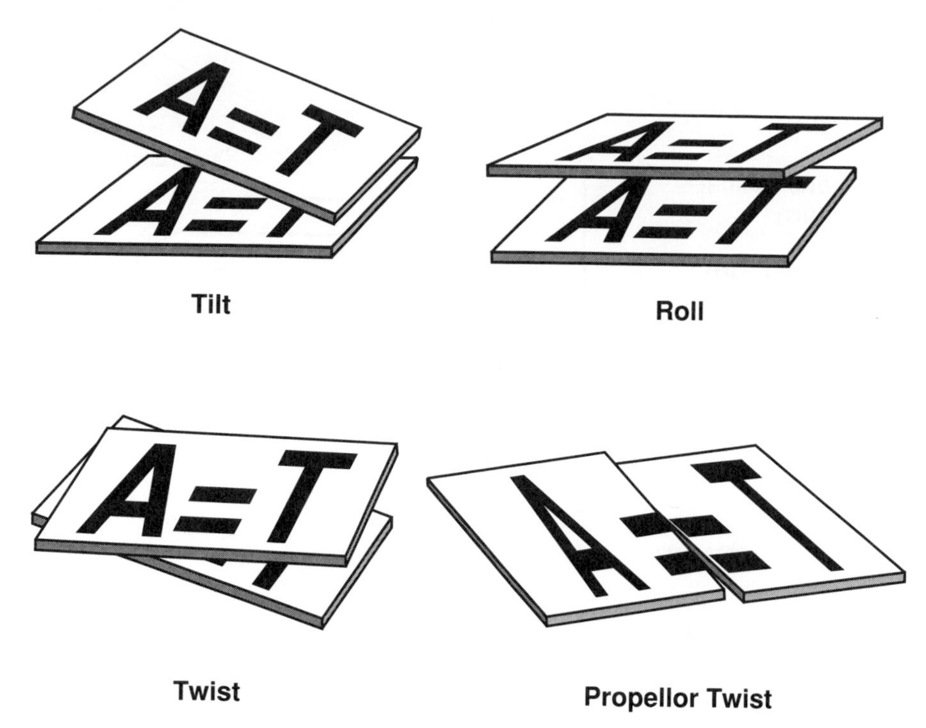

Figure 1.13 Tilt, roll, twist, and propeller twist. Tilt refers to the angle between two planes of base pairs. It is an angle of deflection in the direction of hydrogen bonds between two base pairs. The tilt angle opens or closes toward the phosphate backbone. Roll angle refers to the angle of deflection of two planar base pairs in a direction perpendicular to the direction of the hydrogen bonds between the base pairs. Twist refers to the rotation of one base pair with respect to another. Propeller twist refers to the angle of roll of one base relative to the other within a pair of hydrogen bonded bases.

Base pair roll refers to the angle of deflection of a base pair with respect to the helix axis along a line drawn between two adjacent base pairs relative to a line drawn perpendicular to the helix axis. This is also shown in Figure 1.13. Compare this to tilt in which the line angle is measured along a line drawn through the base pair. The roll and tilt angles are offset by 90°. Tilt angles and roll angles will be discussed in more detail in Chapter 2.

Propeller twist, illustrated in Figure 1.13, refers to the angle between the planes of two paired bases. A base pair is rarely a perfect flat plane with each aromatic base in the same plane. Rather, each base has a slightly different roll angle with respect to the other base. This makes the two bases look like an airplane propeller.

Diameter of the helix refers to the width in Å across the helix. B-DNA has a diameter of 20 Å.

Rotation per residue, or *twist angle,* sometimes designated *h,* refers to the angle between two adjacent base pairs. Consider the angle between lines drawn through two adjacent base pairs. In textbook B-form DNA with 10 bp in one 360° helix turn of DNA, the rotation per residue is 36°. For B-form DNA in solution with 10.5 bp per turn, $h = 34.3°$.

4. B-Form DNA

The structure of B-form DNA, the most common form, was originally deduced from X-ray diffraction analysis of the sodium salt of DNA fibers at 92% relative humidity (Langridge *et al.,* 1960a,b). B-Form DNA is pictured in Figure 1.12, where various helix parameters and features are indicated. A molecular model is shown in Figure 1.14. There are about 10.5 bp per right-handed helical turn in B-DNA (helix parameters are listed in Table 1.3). The form of the ribose sugar is C2′ endo. The term B-form DNA will be used to refer to the right-handed helical form commonly found for DNA in solution.

A dominant feature of B-form DNA is the presence of two distinct grooves, a major and a minor groove, shown in Figures 1.12 and 1.15. These two grooves obviously provide very distinct surfaces with which proteins can interact. As discussed in Chapter 8, different DNA binding proteins have domains that interact with either the major or the minor groove. Certain chemicals and drugs can interact specifically with either the major or the minor groove. Different functional groups on the purine and pyrimidine bases are accessible from the major or the minor groove (Figure 1.15). The Watson–Crick hydrogen bonding surfaces are not available to solvent or pro teins, since the functional groups involved in hydrogen bonding are interacting with each other (in complementary base pairs) at the center of the double helix. The Hoogsteen hydrogen bonding surface of purines is accessible through the major groove. This is evident by looking down the axis of the double helix as shown in the representation in Figure 1.14. In this projection, the stacked base pairs form a central core surrounded by the phosphate backbone. The center of the helix is a relatively chemically inert place to store genetic information.

5. A-Form DNA

A-Form DNA was originally identified by X-ray diffraction analysis of DNA fibers at 75% relative humidity (Fuller *et al.,* 1965). The structure of A-DNA is shown in Figure 1.16 and the helix parameters are listed in Table 1.3. The grooves are not as deep as in B-DNA, and the bases are much more tilted

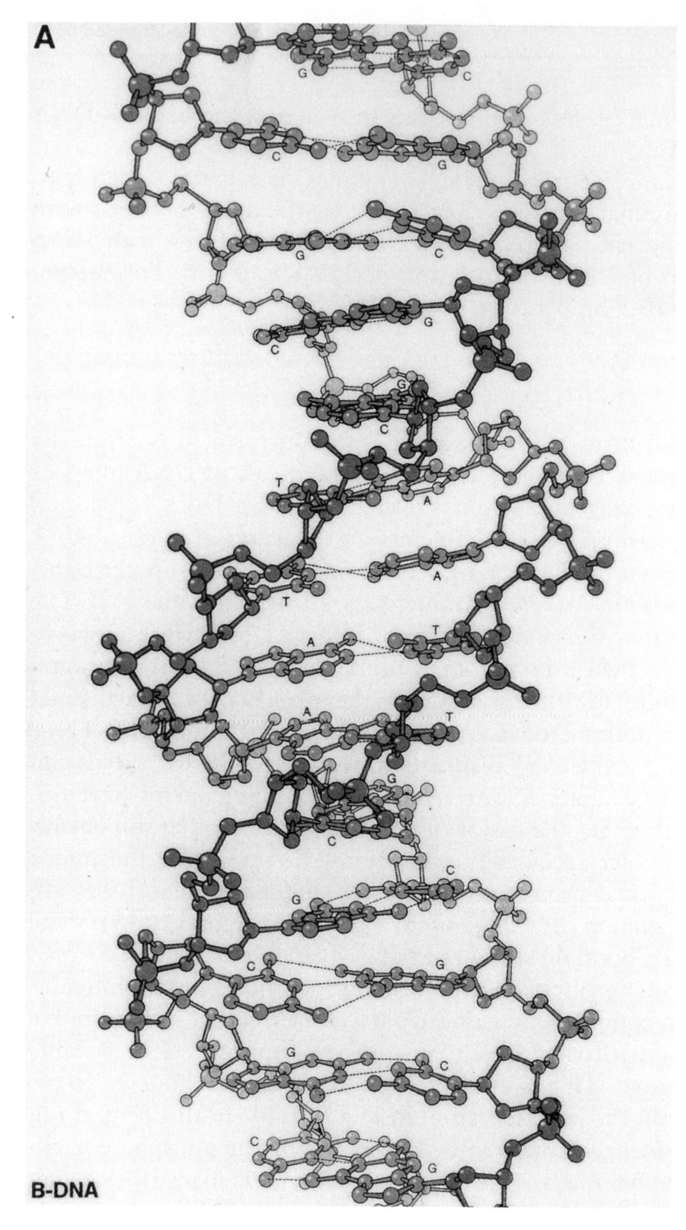

Figure 1.14 B-DNA helix. (A) In this model of the B-DNA helix the phosphate backbones can be seen as smooth right-handed coils on the outside of the helix. This view looks into a minor groove at the center of the model. Major grooves are seen above and below the minor groove. (Copyright by Irving Geis.)

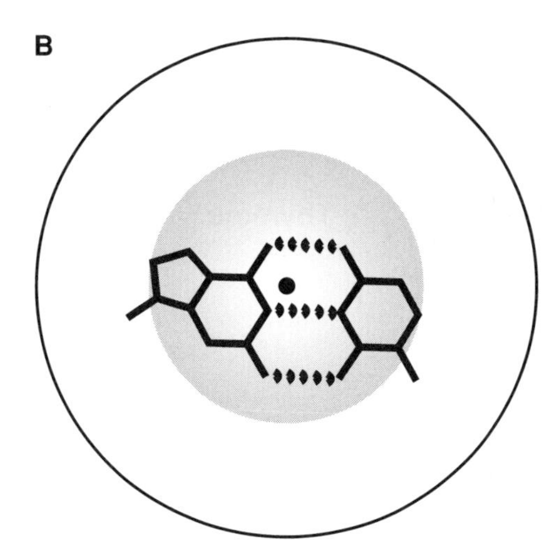

Figure 1.14 *Continued.* (B) In the B-DNA helix the hydrogen bonded base pairs are stacked near the center of the helix. The center of the helix passes nearly symmetrically through the Watson–Crick hydrogen bonds.

Table 1.3
Helix Parameters

Parameter	A-DNA	B-DNA	Z-DNA
Helix sense	Right	Right	Left
Residue per turn	11	10 (10.5)[a]	12
Axial rise (Å)	2.55	3.4	3.7
Helix pitch (°)	28	34	45
Base pair tilt (°)	20	−6	7
Rotation per residue (°)	33	36 (34.3)[a]	−30
Diameter of helix (Å)	23	20	18
Glycosidic bond configuration			
dA, dT, dC	anti	anti	anti
dG	anti	anti	syn
Sugar pucker			
dA, dT, dC	C3′ endo	C2′ endo	C2′ endo
dG	C3′ endo	C2′ endo	C3′ endo
Intrastrand phosphate–phosphate distance (Å)			
dA, dT, dC	5.9	7.0	7.0
dG	5.9	7.0	5.9

[a]Values in parentheses are the residues per turn and rotation per residue for B-form DNA as it exists in solution of physiological ionic strength. Other values are taken from X-ray diffraction data.

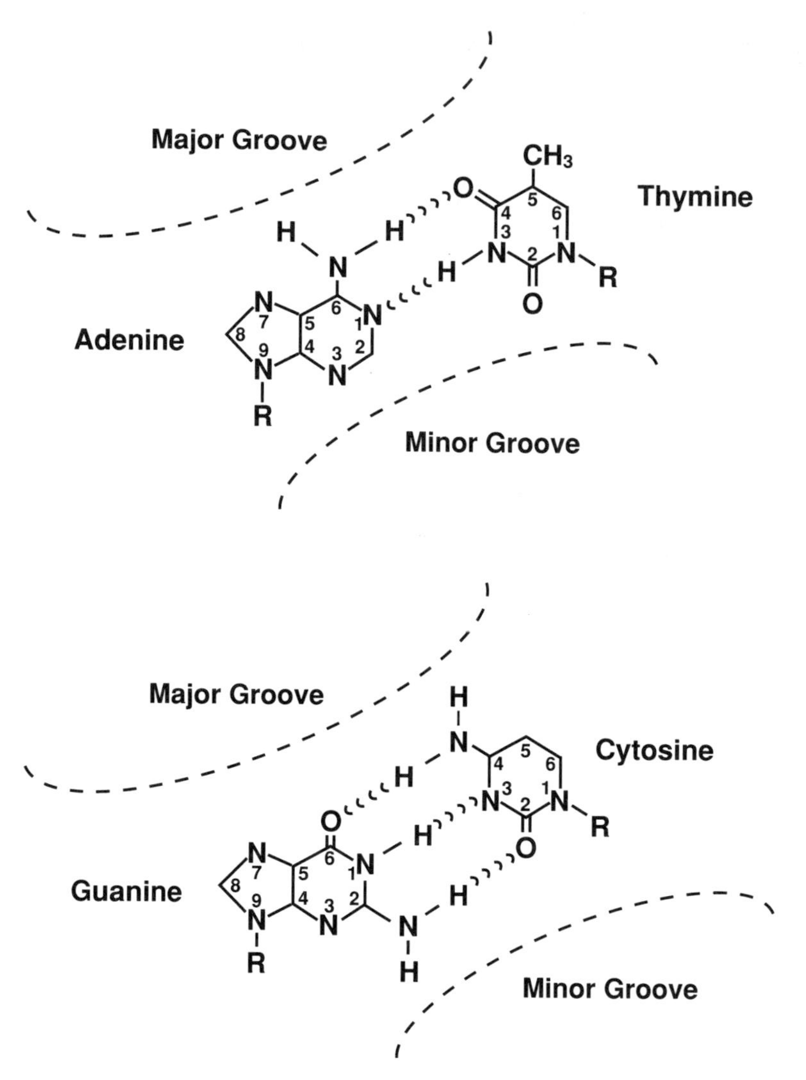

Figure 1.15 Accessibility of functional groups in DNA. The relative positions of the major groove and the minor groove are indicated with respect to an adenine · thymine and a guanine · cytosine base pair. The O6 position of guanine and the N7 positions of guanine and adenine represent accessible positions within the major groove of DNA that can be attacked by electrophiles.

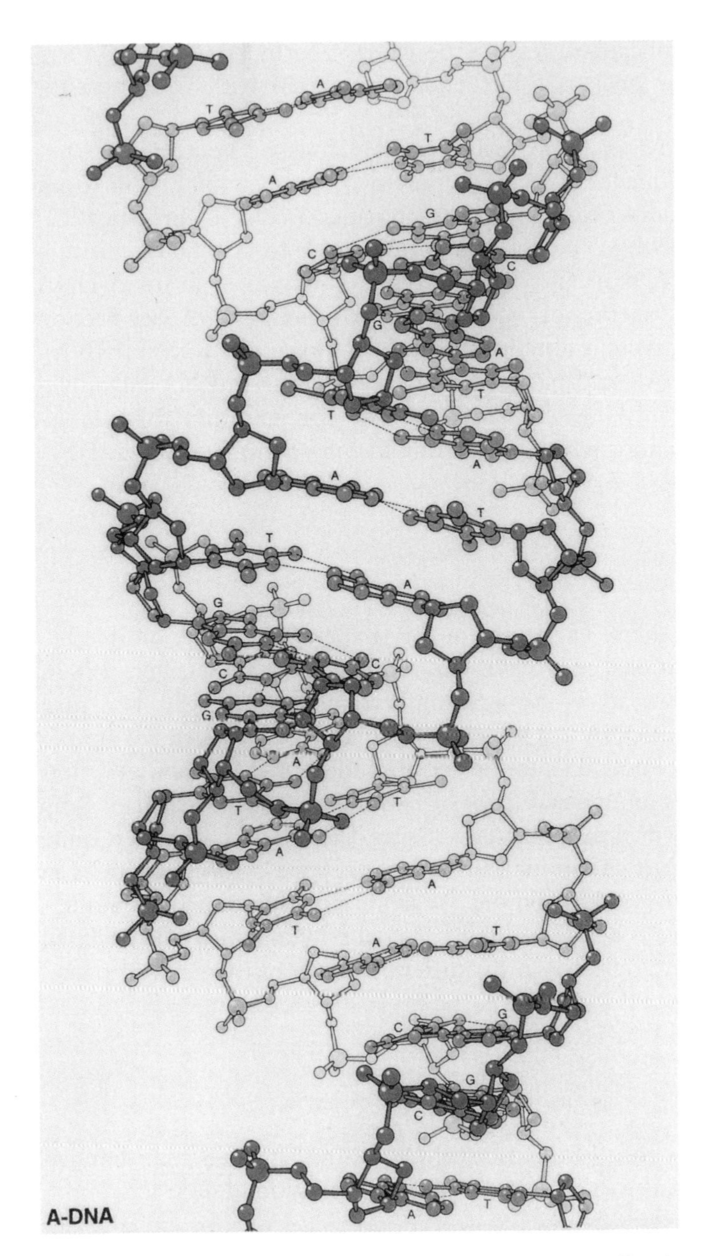

Figure 1.16 A-DNA helix. The structural model for A-DNA differs from that for B-DNA in several ways. First, the bases are tilted significantly with respect to the helix axis. Second, the grooves are not as pronounced or as deep as in B-DNA. Third, the bases are not stacked at the center but are more toward the outside of the helix. (Copyright by Irving Geis.)

(to 20°). Another significant difference between A-form DNA and B-form DNA is that the sugar pucker is C3′ endo (compared with C2′ endo for B-DNA).

Does A-DNA exist in biological systems? Runs of homopurine · homopyrimidine DNA sequence [poly(dG) · poly(dC), for example] seem to set up an A-like helix, as determined by characteristic circular dichroism (CD) spectra (Fairall *et al.*, 1989). Therefore, it is reasonable to assume that within a generally B-like DNA molecule, specific regions may exist in an A-DNA form. This would be a function of sequence composition of DNA (see Section E). RNA frequently exists in a double-helical form in transfer RNAs (tRNA), ribosomal RNAs (rRNA), and parts of messenger RNAs (mRNA). Double-stranded RNA forms an A-like helix. The ribose configuration for double-stranded RNA is C3′ endo, which is a distinguishing feature of the A-DNA helix.

6. C-, D-, and T-DNA

Numerous other subtle variations in the shape of the DNA double helix, in specialized situations, may have biological relevance (Saenger, 1984). C-DNA forms in fibers at 57–66% humidity and has 9.3 bp per turn (Marvin *et al.*, 1961). D-DNA is a structure with a helix repeat of 8.5 bp per turn. Runs of poly(dA) · poly(dT) are believed to adopt a D-like helix (Arnott *et al.*, 1974). T-DNA is an unusual form of DNA with a helix repeat of 8 bp per turn. T-DNA is purified from bacteriophages T2, T4, and T6 and is quite different from most DNA. As shown in Figure 1.17, the cytosine residues in T-DNA contain a hydroxymethyl on the 5′ carbon. In addition to this modification, a glucose residue is added to the hydroxymethyl, making glucosylated DNA. These changes are reflected in the different shape of the helix of bacteriophage T4 DNA.

7. Z-DNA

Z-DNA is a left-handed helix that is very different from right-handed DNA forms. Z-DNA can form in alternating purine–pyrimidine tracts under certain conditions, including high salt, the presence of certain divalent cations, or DNA supercoiling. Compared with B-DNA, there are major structural differences in the sugar pucker, rotations about the glycosidic bond, and orientation of base pairs within the helix (Table 1.3). The structure of Z-DNA is presented in detail in Chapter 5.

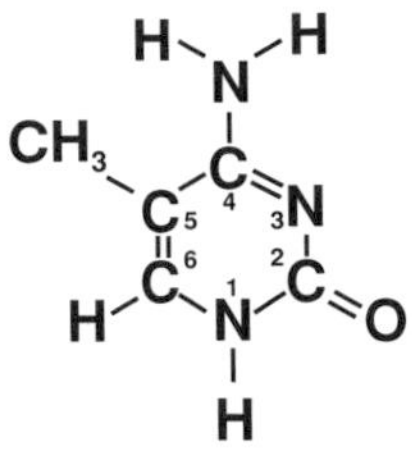

5-Methylcytosine

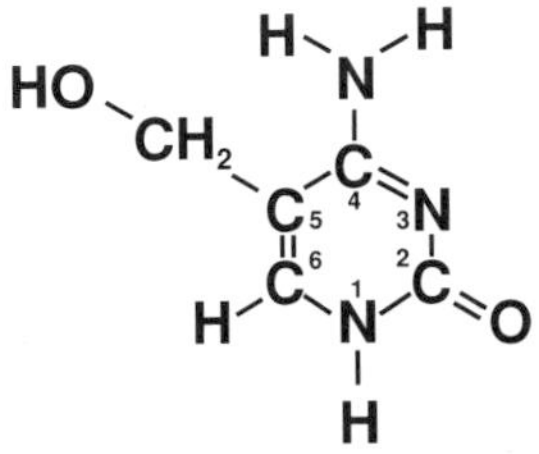

5-Hydroxymethylcytosine

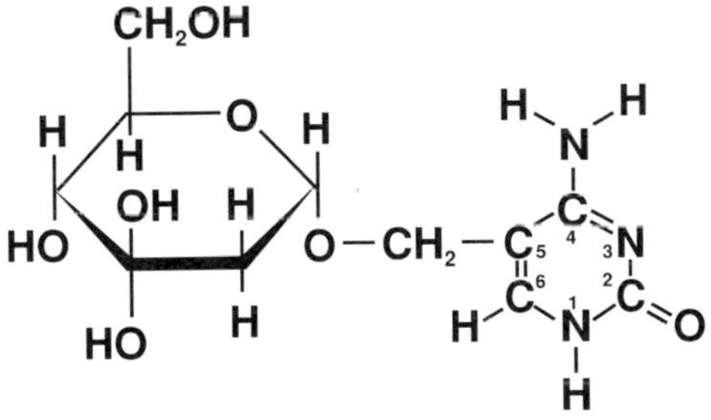

α-Glucosylated derivative of 5-Hydroxymethylcytosine

Figure 1.17 Modifications to DNA. 5-Methylcytosine is commonly found in eukaryotic DNA. Methylation, which occurs at CpG sequences, has been associated with inhibition of gene expression. In some inactive genes all cytosines in CpG sequences are methylated. DNA from bacteriophage T4 contains 5-hydroxymethylcytosine rather than cytosine. In addition, many of the 5-hydroxymethyl positions are glucosylated. The structure of an α-glucosylated derivative of 5-hydroxymethylcytosine is shown. These modifications to bacteriophage T4 DNA cause a helical structure termed T-DNA that is quite different from B-form DNA.

D. The Biological Significance of Double Strandedness

It is significant biologically that the genetic information exists as a double-stranded B-form DNA molecule. First, the two complementary strands provide templates that can be copied by DNA polymerase, producing two exact copies of the genetic information. Second, the double-stranded structure is also of critical importance when either strand is damaged by genotoxic chemicals or ionizing or ultraviolet irradiation. By having two complementary copies of the genetic information, the undamaged strand can serve as a template for repair of the damaged strand. Third, the B-form helix is designed to protect the chemical identity of the genetic information. Hydrogen bonding, base stacking interactions, and hydration of the helix stabilize and chemically insulate the Watson–Crick informational coding surfaces from the environment. Female germ-line cells in humans are produced early in development and may exist more than 40 years before fertilization. The accumulation of damages to the genetic information in germ-line cells could not be tolerated. The stability of DNA, aided by its packaging and condensation with proteins, ensures that the genetic information is chemically inert.

E. Sequence-Dependent Variation in the Shape of the DNA

A canonical textbook B-form helix is not likely to exist in living organisms because the actual shape of a region of DNA will depend on its base composition, the local sequence environment on either side of the region, and the composition of the cellular milieu. Evidence for local variation in helix parameters has come from the X-ray crystallographic analysis of DNA (Dickerson and Drew, 1981; Dickerson, 1983). Analysis of the crystal structure of the 12-bp sequence 5′CGCGAATTCGCG3′ showed that the twist angle between adjacent base pairs varied considerably, from 32° to 45°. In addition, not all identical dinucleotides had the same twist angle, which means that adjacent bases or flanking sequence can influence the shape of a dinucleotide. The *average* twist angles for all possible dinucleotide combinations are shown in Table 1.4. Since individual twist angles are different, the actual helical repeat and therefore the exact shape of a 10-bp helical turn of DNA can vary considerably. Since each base pair of a helical turn could be any of the four bases (considering one strand) and since there are nine dinucleotides

Table 1.4
The 10 Twist Angles of B-DNA

Dinucleotide	Twist Angle (h)[a]
(AA) · (TT)	35.6 ± 0.1
(AC) · (GT)	34.4 ± 1.3
(AG) · (CT)	27.7 ± 1.5
(AT) · (AT)	31.5 ± 1.1
(CA) · (TG)	34.5 ± 0.9
(CC) · (GG)	33.7 ± 0.1
(CG) · (CG)	29.8 ± 1.1
(GA) · (TC)	36.9 ± 0.9
(GC) · (GC)	40.0 ± 1.2
(TA) · (TA)	36.0 ± 1.0

[a]Data from Kabsch *et al.* (1982).

per helical turn, there are 262,144 (4^9) possible variations in helix structure for a 10-bp piece of DNA. This calculation considers only the dinucleotide twist angles shown in Table 1.4. This number actually represents a minimum, since the helix parameters associated with a particular dinucleotide may vary depending on flanking base sequence and base composition.

Certain polymeric regions of single base runs in one strand—poly(dA) · poly(dT) or poly(dG) · poly(dC), for example—can adopt an unusual helical form that is quite different from typical B-form DNA. In these cases, only after a certain length of the repeating nucleotide is reached will a different helical shape be formed. Poly(dA) · poly(dT) regions of DNA adopt an unusual structure called *heteronomous DNA* (Arnott *et al.,* 1983). Heteronomous DNA has a helix repeat of 10 bp per turn. A very unusual feature characteristic of heteronomous DNA is that the deoxyribose sugar in the dA strand is C3′ endo whereas the deoxyribose in the dT strand is C2′ endo. The helical changes associated with phased runs of As that are responsible for DNA bending constitute another example of sequence-directed variation in the DNA helix. Bent DNA is discussed in detail in chapter 2. A third example is the formation of A-form-like tracts of DNA in runs of poly(dG) · poly(dC). Short pieces of DNA with runs of G give CD spectra like those of A-DNA (Fairall *et al.,* 1989). A local region of DNA containing runs of guanines may be A-DNA like, making this region structurally quite different from flanking B-form DNA. Runs of dG · dC longer than 20 bp can form triple strands and four-stranded structures as discussed in Chapters 6 and 7.

F. Physical Properties of Double-Stranded DNA

1. Ultraviolet Absorption Spectra of DNA

DNA absorbs ultraviolet (UV) light in a band centered around 260 nm. The absorption profile of DNA (Figure 1.18) can be used as a measure of the concentration and purity of a DNA sample. The concentration of DNA in solution is determined by measuring the absorbance at 260 nm, A_{260}. When $A_{260} = 1$, the concentration of the DNA is about 50 μg/ml (the actual value depends on base composition since the bases differ in extinction coefficient). This is calculated from Beer's law, $A = \epsilon lc$ (or $c = A/\epsilon l$) where A is the absorbance, l is the width of the light path of the cuvette in cm (usually 1 cm), ϵ is the extinction coefficient, and c is the concentration of the DNA in mol/L. The extinction coefficient of DNA is ϵ = 6600 L/mol cm. The A_{260}/A_{280} ratio is frequently used as a measure of the purity of DNA (proteins absorb UV light maximally at 280 nm). A ratio of 1.8–2.0 is generally an indication of a pure preparation of DNA. A lower ratio is an indication of protein contamination. Single-stranded DNA has a 40% increase in absorbance relative to double-stranded DNA; generally $A_{260} = 1$ is equivalent to about 36 μg/ml for single-stranded DNA. The increase in absorbance on heating DNA, called the hyperchromic shift, has been used as a simple assay for denaturation.

2. Denaturation and Renaturation

The double-helical structure of DNA is remarkably stable. This stability is derived from two chemical forces, hydrogen bonding and base stacking interactions, as discussed in Section C,1. In addition, the helix is solvated or covered with water molecules which form a "shell of hydration" around the DNA. To melt the two strands or *denature* the DNA, all these stabilizing forces must be overcome.

The two strands of DNA come apart readily on incubation at pH > 12 or pH < 2 due to ionization of the bases (Figure 1.19). As discussed earlier, ionization results in a change in the hydrogen bond donor/acceptor properties of the bases, which will disrupt the normal A · T and C · G Watson–Crick hydrogen bonds. In addition, the shell of hydration surrounding the DNA is disrupted at very high or low pH, destabilizing base stacking. Acid treatment of DNA leads to depyrimidation and depurination, the loss of bases by cleavage of the glycosidic bond. The phosphodiester bond is more susceptible to hydrolysis at these abasic sites. Thus, strong acid will degrade DNA. Since single strands of DNA are relatively stable in alkali, denaturation is usually accomplished by alkali treatment.

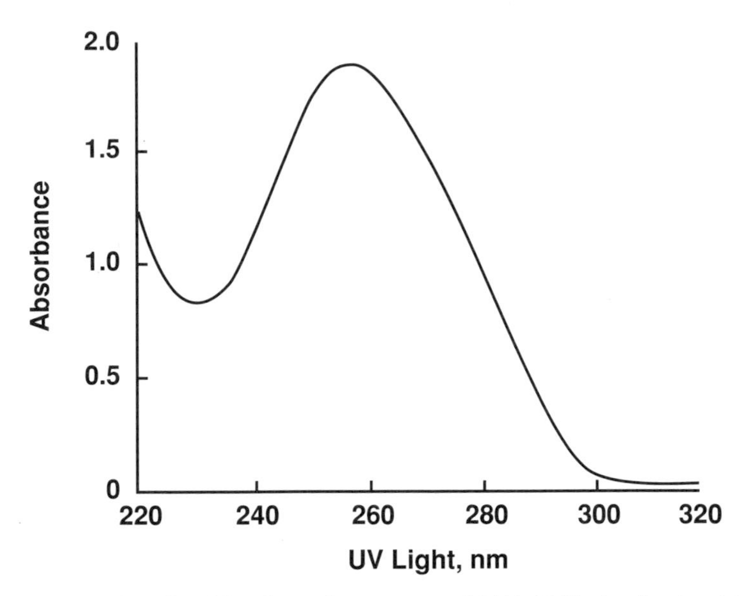

Figure 1.18 Ultraviolet absorption spectrum of DNA. DNA absorbs ultraviolet light in the range of 240 to 280 nm. The maximum absorption occurs at about 260 nm; the exact value is dependent on base composition. A typical UV absorption profile is shown here. The aromatic ring structures of the bases are responsible for absorbing ultraviolet light. Typically an absorbance of 1 at 260 nm is equivalent to a concentration of about 50 μg/ml DNA. One measure of DNA purity is the A_{260}/A_{280} ratio. This ratio is used because proteins, a frequent contaminant of DNA preparations, absorb at 280 nm. An A_{260}/A_{280} ratio of 1.8–2 is an indication of a pure DNA preparation.

Increasing the temperature of DNA destabilizes the double helix, resulting in the separation of the two strands. Heat both disrupts the hydrogen bonds and destroys the shell of hydration of DNA leading to a loss of forces holding the two strands together. Experimentally, there is a linear relationship between the G + C content and the melting temperature of DNA (Marmur and Doty, 1962). The temperature at which 50% of a DNA sample is melted is called the melting temperature or T_m. Moreover, there is a linear relationship between the T_m of DNA and the calculated stacking energies (listed in Table 1.2). This suggests that the thermal stability of DNA is a function of base stacking, not only hydrogen bonding (see Saenger, 1984, for more discussion).

DNA denaturation can be measured or monitored in many different ways. One method involves measurement of a characteristic increase in the absorbance at 260 nm called *hyperchromicity*, which results from the unstacking of the bases. Another method employs enzymes specific for single-stranded DNA (such as S1 nuclease). Selective binding reactivity of single- or

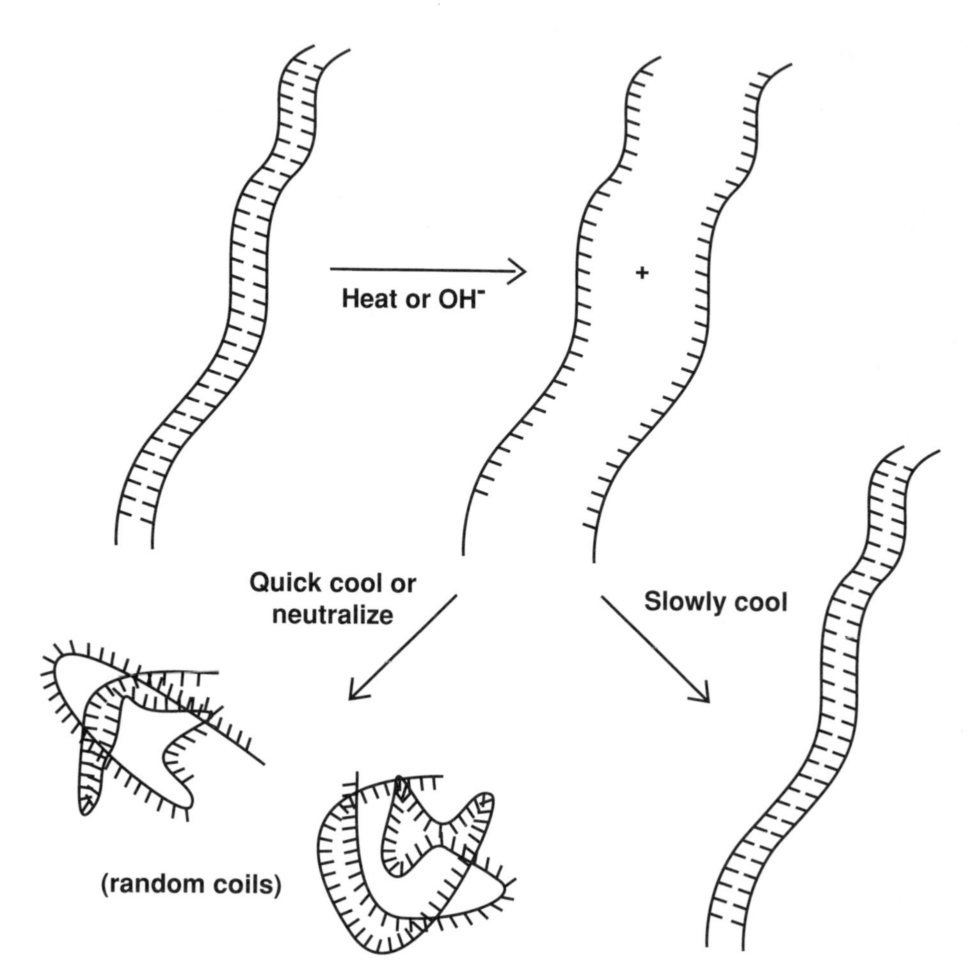

Figure 1.19 Denaturation and renaturation of DNA. The double helical configuration of DNA can be denatured by heat, alkali, or acid. In this process the two strands separate as base stacking and hydrogen bonding interactions are disrupted. If one quickly removes the heat or neutralizes the DNA solution (quick cool or neutralize), the DNA will collapse into a compact random coil in which some bases are hydrogen bonded. If a denatured solution of DNA is slowly cooled, the two single strands can reform a paired double helical molecule. The process requires a *nucleation event* in which a region of complementary bases on opposite strands finds each other, comes into register, and begins to form a hydrogen bonded double helix with stacked base pairs. Once nucleation has occurred, the rest of the DNA molecule rapidly renatures.

double-stranded DNAs with various surfaces can also be used. For example, hydroxylapatite, a $CaPO_4$ precipitate, will bind double-stranded DNA but not single-stranded DNA at certain phosphate concentrations.

Rapid removal of denaturing conditions, such as quickly cooling a heat-denatured DNA sample, results in the collapse of single strands into an unordered random coil (Figure 1.19). In this configuration the two strands cannot reform a double helix. However, if DNA is slowly cooled, interstrand nucleation can occur in which small complementary regions on the two opposite DNA strands come together and form a short double-helical region. Once nucleation has occurred, the rest of the double helix *renatures* very rapidly. The rate limiting step in renaturation is the nucleation event.

Negative DNA supercoiling, which is discussed in Chapter 3, can also drive the melting of a region of DNA. Supercoiled DNA contains a great deal of free energy that can be used to transiently or stably melt local regions of supercoiled DNA. Not surprisingly, these are usually regions of DNA rich in A + T.

3. DNA Hybridization

DNA hybridization involves the formation of a double-stranded nucleic acid, either a DNA double helix or an RNA–DNA duplex. Typically this involves use of a single-stranded radioactive probe sequence, prepared from all or part of a cloned gene, or a chemically synthesized oligonucleotide. Hybridization analysis makes use of the unique linear array of bases, which in one strand specifies a unique complementary strand. For example, a 15- to 21-bp region of DNA encoding 5–7 amino acids of a protein may occur nowhere else in a human genome. Thus, a 15- to 21-nucleotide (nt) single DNA strand will only hybridize to a unique complementary single region of the entire 3 billion bases in human DNA.

Experimentally, hybridization is accomplished by denaturing the DNA to be analyzed, adding a labeled hybridization probe (a piece of cloned DNA, chemically synthesized DNA, or even RNA), and then incubating the reaction below the T_m of the DNA. Because this is a bimolecular reaction, the concentration of the DNAs and the time of hybridization will affect the extent of hybridization. Moreover, the stability of a DNA duplex is dependent on the ionic strength of the solvent and on temperature. The appropriate temperature for hybridization depends on the G + C content of DNA and, for short pieces of DNA, on length. The temperature should be slightly below the T_m of the duplex to allow the stable duplex to form while melting out hybridization products containing several mismatched bases. This set of conditions is referred to as hybridization *stringency*. Assays for the extent of hybridization make use of the techniques available for selective detection of single- and double-stranded DNA (usually binding to hydroxylapatite or binding to nitrocellulose).

DNA hybridization provides an extremely powerful tool in molecular biology. Hybridization allows the identification and cloning of specific genes, analysis of levels of mRNA in cells, analysis of the copy number of sequences in the genome, and DNA fingerprinting, among other applications.

G. Chemicals That React with DNA

Many carcinogens and mutagens, present in the air we breathe and the water we drink, cause a very large number of chemical modifications to the DNA. Chemical damage can lead to mutations by interfering with the fidelity of DNA replication (i.e., by changing the base-pairing properties of bases or by errors introduced during DNA repair processes). The specificity of reactivity of individual bases with certain chemicals has provided extremely powerful probes of DNA structure as well as providing methods for the chemical sequencing of DNA (Section H,1). This section will limit a discussion of chemicals to those that are particularly useful in molecular genetic research in the areas of DNA structure analysis and gene expression. Many good reviews and books are available on DNA chemistry and chemical modification of DNA (Brown, 1974; Waring, 1981; Singer and Kusmierek, 1982; Paleček, 1991).

1. Nucleophiles

Nucleophiles (which means nucleus-loving) are electron-rich reagents with a slight negative charge. These reagents are reactive with carbons that have a slightly positive charge that are adjacent to a ring nitrogen (Figure 1.20A). These positions are the C4 and C6 positions of thymidine and cytidine and the C2, C6, and C8 positions of adenosine and guanosine. Two nucleophiles important in the context of this and following chapters are hydrazine and hydroxylamine.

a. Hydrazine

Hydrazine ($H_2N{-}NH_2$) is widely used to modify DNA. One reaction of hydrazine (Figure 1.20 B) involves the amino nitrogen on the C4 position of cytosine. Hydrazine displaces the amino group forming N4 aminocytosine. As discussed subsequently, hydrazine is used in chemical sequencing reactions. Its usefulness in DNA sequencing relies on a different reaction leading to opening of the pyrimidine ring. (This reaction with thymine is shown in Figure 1.23.)

b. Hydroxylamine

Hydroxylamine (H_2NOH) is a very strong nucleophile that hydroxylates amino groups, predominantly those on A and C (Figure 1.20C). The formation of a hydroxylamino group in place of the amino group can lead to a change in the base-pairing properties of the hydroxylamino derivatives of A and C. Hydroxylamine is a powerful mutagen because it disrupts the normal hydrogen bonding properties of the modified bases, preventing A and C from pairing with T and G, respectively.

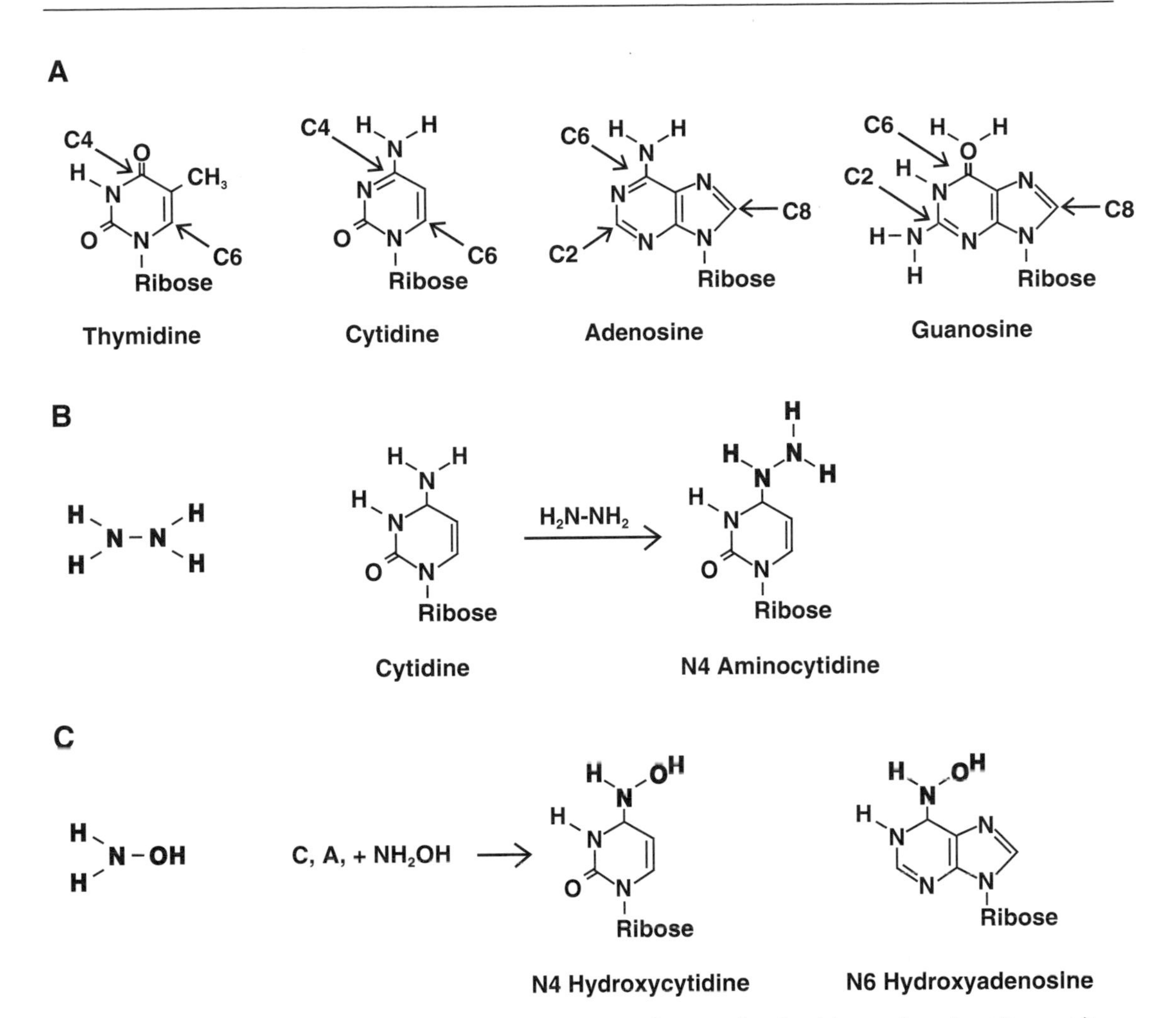

Figure 1.20 Site of nucleophilic attack. (A) The sites of nucleophilic attack on thymidine, cytidine, adenosine, and guanosine are indicated by the arrows. Purines (G and A) are reactive at the C6, C2, and C8 positions. Pyrimidines (T and C) are reactive at the C4 and C6 positions. (B) Hydrazine reacts with the C4 position of cytidine forming N4 aminocytidine. The hydrazine atoms are shown in bold. (C) Hydroxylamine reacts with the N4 position of cytidine or the N6 position of adenosine forming N4 hydroxycytidine and N6 hydroxyadenosine, respectively. The hydroxylamine atoms are shown in bold.

2. Electrophiles

Electrophiles (electron-loving reagents) seek an unbonded electron pair. The ring nitrogens and exocyclic nitrogens and oxygens on the purine and pyrimidine rings are good sites for electrophilic attack (Figure 1.21A). For example, thymine can react at N3 or at the oxygens at the C2 and C4 positions; cytosine can react at the N3, N4, or O2 positions; guanine at the N1, N3, N7, or O6 positions; and adenine at the N1, N3, N7, or N6 positions.

a. Alkylating Agents

Alkylating agents, including ethylmethylsulfonate (EMS), ethylnitrosourea (ENU), *N*-methyl-*N*′-nitro-*N*-nitrosoguanidine (NMNG), and mitomycin C (not shown) are commonly used as chemical mutagens (Figure 1.21B). These chemicals are mutagenic because, depending on the site of alkylation, they can alter the keto–enol equilibrium. In addition, the normal Watson–Crick hydrogen bonding can be physically affected by a bulky alkyl adduct on the base. As shown in Figure 1.21A, many positions of attack are those that can be involved in Watson–Crick or Hoogsteen hydrogen bonding interactions.

b. Dimethylsulfate

Dimethylsulfate (DMS) and its reaction product with guanine are shown in Figure 1.21B. DMS preferentially methylates the N7 position of guanine. Because DMS is a small molecule, it can easily penetrate the major groove of DNA and reach the nitrogen at the N7 ring position. Treatment of DMS-modified DNA with piperidine then leads to cleavage of the phosphodiester bond at the site of methylation. This chemical reaction is one of the Maxam–Gilbert chemical sequencing reactions for DNA described in Section H,1. (See box entitled "Footprinting a Protein on DNA".)

"Footprinting" a Protein on DNA

DMS is used very commonly to "footprint" DNA, that is, when a protein is bound to a specific region of DNA, the N7 positions of guanine are protected from chemical modification by DMS. Following modifi-cation, on chemical cleavage by piperidine treatment, cleavage will occur at all modified guanines in the DNA except at those protected, or footprinted, by the protein. By end-labeling a DNA molecule and introducing one break on average per molecule, the pattern of bands observed on a DNA sequencing gel corresponds to every position where DMS-modified guanine occurs in the DNA.

c. Bifunctional Alkylating Agents

Bifunctional alkylating agents are those with two reactive groups. The structures of nitrogen and sulfur mustard (mustard gas) are shown in Figure 1.21B. These bifunctional reagents can covalently react with two different bases in DNA, forming *intrastrand* cross-links between two bases in the same DNA strand or *interstrand* cross-links that covalently link the two complementary strands of DNA.

d. Chloroacetaldehyde and Bromoacetaldehyde

Chloroacetaldehyde (CAA) and bromoaceteldehyde (BAA) react at the Watson–Crick hydrogen bonding surface, predominantly with cytosines and adenines and occasionally with guanines. Since the reactive groups are Watson–Crick hydrogen bonding sites, these chemicals are specific for single-stranded DNA. The structures and reaction products of CAA and BAA are shown in Figure 1.21C. The site of chemical modification can be identified by chemical cleavage of CAA- or BAA-modified DNA with hydrazine, DMS, or formic acid followed by treatment with piperidine. (The rationale for this cleavage reaction is identical to that described for the Maxam–Gilbert chemical sequencing reactions.) Regions of double-stranded DNA containing CAA- or BAA-modified bases are also susceptible to digestion with S1 nuclease, since modification at the Watson–Crick surfaces precludes formation of a normal double strand of DNA and gives modified DNA some single-stranded character.

e. Diethylpyrocarbonate

Diethylpyrocarbonate (DEPC) is shown in Figure 1.21C, as is the product of its reaction with DNA. DEPC reacts with the N7 positions of adenine and guanine. Since the N7 position is relatively protected from DEPC in normal B-DNA, DEPC only reacts efficiently with purines in single-stranded DNA or when the base exists in the *syn* configuration (as occurs in Z-DNA).

f. Osmium tetroxide

Osmium tetroxide (OsO_4) and its reactions with thymine are shown in Figure 1.21C. OsO_4 reacts strongly with thymine and to a lesser degree with cytosine. The 5,6 double bond is reactive, forming C5 and C6 osmate ester bonds with OsO_4. This double bond would clearly be most reactive when DNA is single stranded. Consequently, OsO_4 is a sensitive chemical probe of cruciforms loops, B–Z junctions, and triplex–duplex junctions, each of which has single-stranded character. The reactivity of OsO_4 is greatly increased in the presence of a ligand (pyridine). Moreover, in the absence of a ligand, reaction products are different than with a ligand.

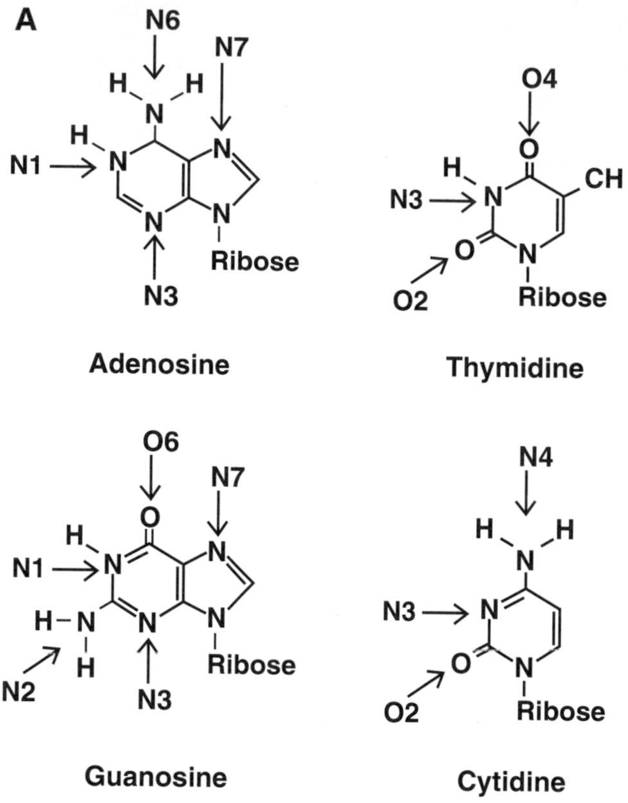

A

Adenosine

Thymidine

Guanosine

Cytidine

Figure 1.21 Sites of electrophilic attack. (A) The sites of electrophilic attack on adenosine, thymine, guanine, and cytidine are indicated by arrows. These sites represent ring nitrogens, amino nitrogens, and carbonyl oxygens. For example, the N1, N3, N7, and N6 positions of adenosine and the N1, N2, N3, O6, and N7 positions of guanosine are available to electrophilic attack. The O2, N3, and O4 positions of thymidine and the O2, N3, and N4 positions of cytidine are reactive. (B) A number of alkylating reagents and, in some cases, their reaction products with DNA are shown. Dimethyl sulfate (DMS) reacts at the N7 position of guanine forming N7 methylguanosine. The alkylating reagents methylmethane sulfonate (MMS), ethylmethane sulfonate (EMS), ethylnitrosourea (ENU), and *N*-methyl-*N'*-nitro-*N*-nitrosoguanidine (NMNG) are common mutagens. Nitrogen and sulfur mustards are bifunctional alkylating reagents and can form monoadducts or intrastrand cross-links with DNA. Chloroacetaldehyde and bromoacetaldehyde react preferentially with C and A. Diethylpyrocarbonate and osmium tetroxide react preferentially with G and T, respectively. Nitrous acid reacts at the N4 position of cytidine causing a deamination reaction in which uridine is formed.

B

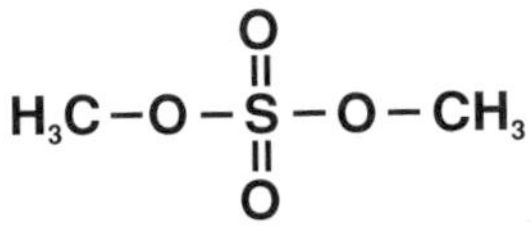

O
||
$H_3C-O-S-O-CH_3$
||
O

Dimethyl Sulfate (DMS)

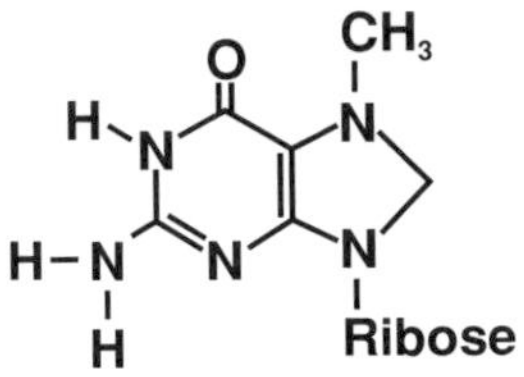

CH_3
O
H
N
N
H−N
N
N
H
Ribose

N7 Methylguanosine

O
||
$H_3C-S-O-CH_3$
||
O

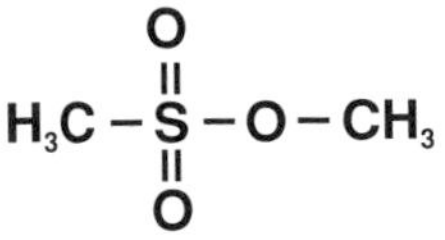

Methyl Methane Sulfonate (MMS)

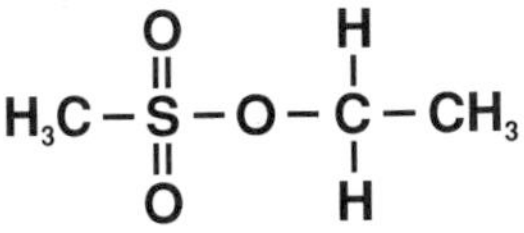

O H
$H_3C-S-O-C-CH_3$
O H

Ethyl Methane Sulfonate (EMS)

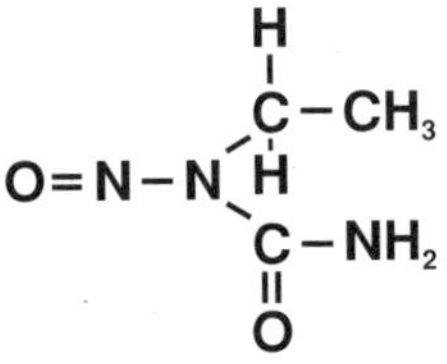

H
$C-CH_3$
O=N−N H
$C-NH_2$
||
O

Ethyl Nitrosourea (ENU)

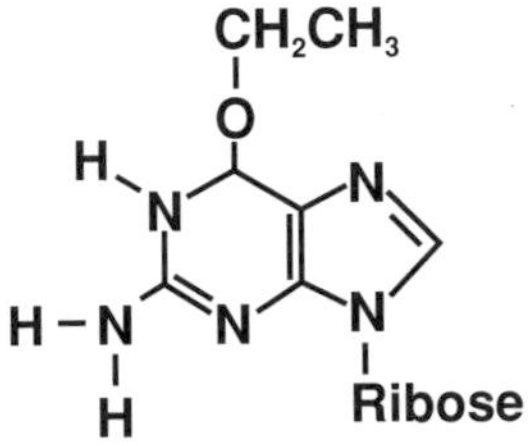

CH_2CH_3
O
H
N
N
H−N
N
N
H
Ribose

O6 Ethylguanosine

H
$N-NO_2$
N−N=C
N−N=O
CH_3

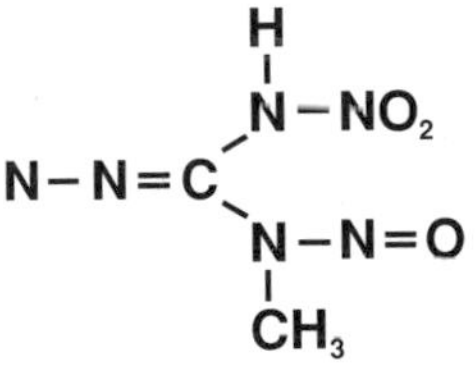

N-Methyl-N' Nitro-N-Nitrosoguanidine

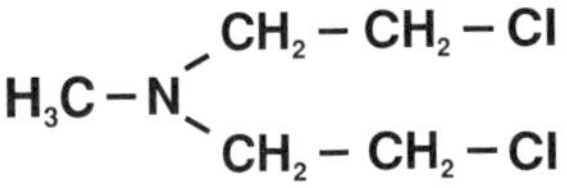

CH_2-CH_2-Cl
H_3C-N
CH_2-CH_2-Cl

Nitrogen Mustard

CH_3
$CH_2-CH_2-N-CH_2-CH_2-OH$
O
H
N
N
H−N
N
N
H
Ribose

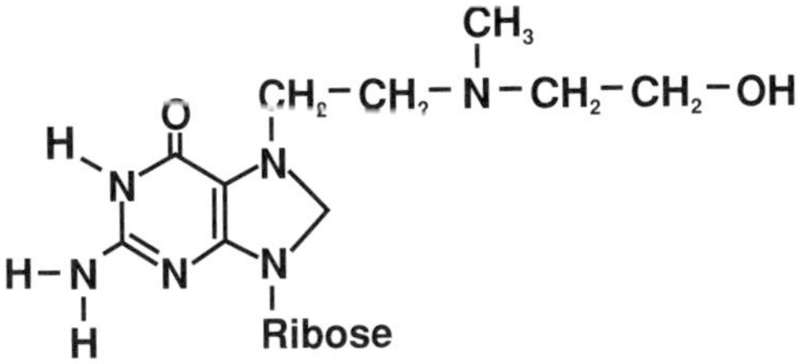

Nitrogen Mustard Monoadduct

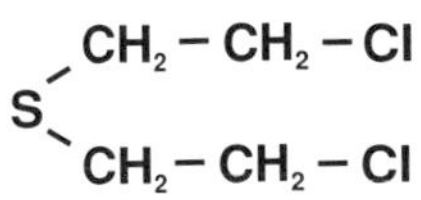

CH_2-CH_2-Cl
S
CH_2-CH_2-Cl

Sulfur Mustard

$CH_2-CH_2-S-CH_2-CH_2$
Cl
H
O
N
N
H−N
N
N
H
Ribose
O
N
N
H
N
N
N−H
Ribose
H

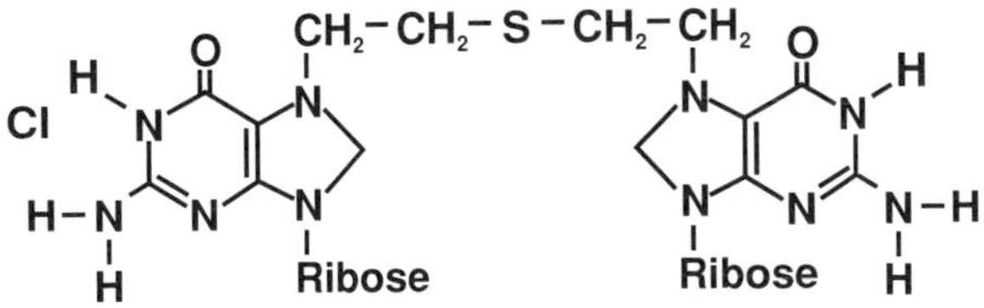

Sulfur Mustard Crosslink

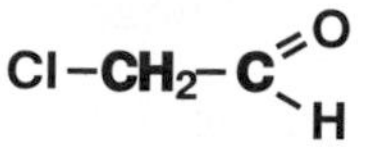

$Cl-CH_2-CHO$

Chloroacetaldehyde

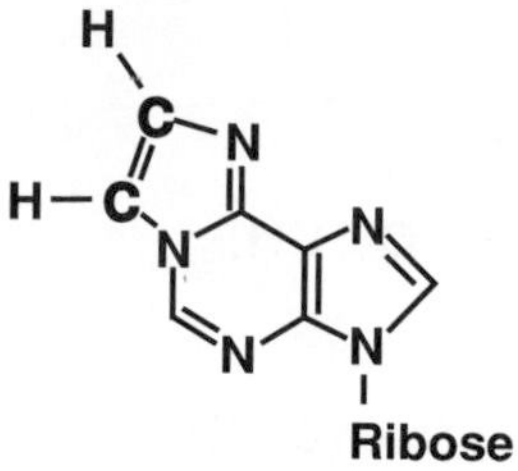

N1, N6 Etheno Adenosine

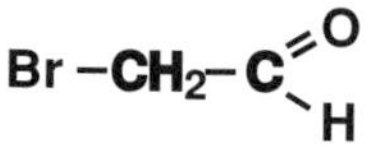

$Br-CH_2-CHO$

Bromoacetaldehyde

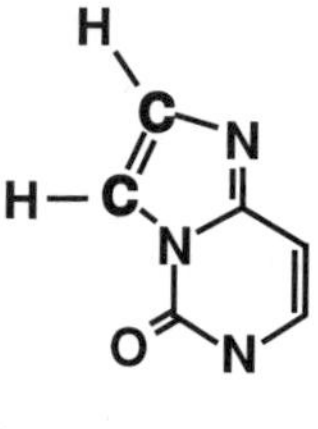

N3, N4 Etheno Cytosine

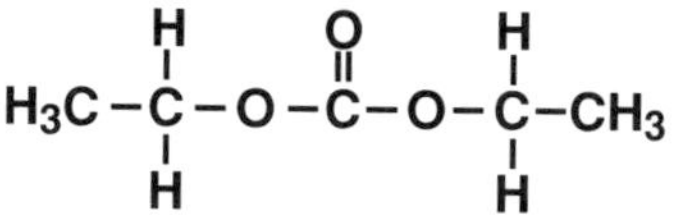

$H_3C-CH_2-O-CO-O-CH_2-CH_3$

Diethylpyrocarbonate

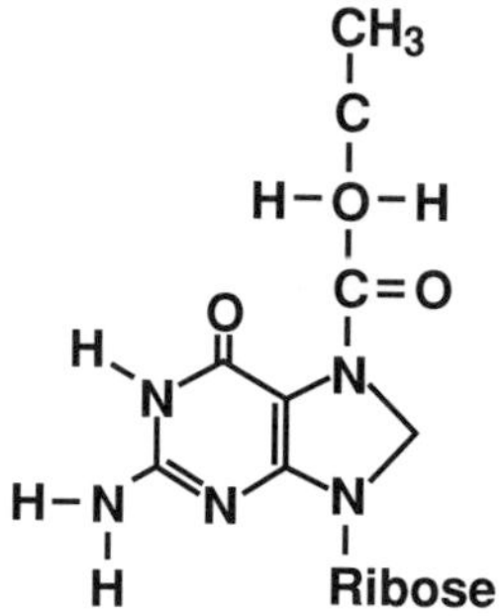

N7 Carboxyethyl Guanosine

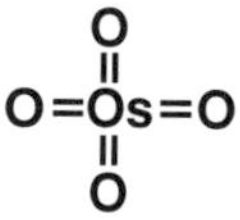

$O{=}Os({=}O)_2{=}O$

Osmium Tetroxide

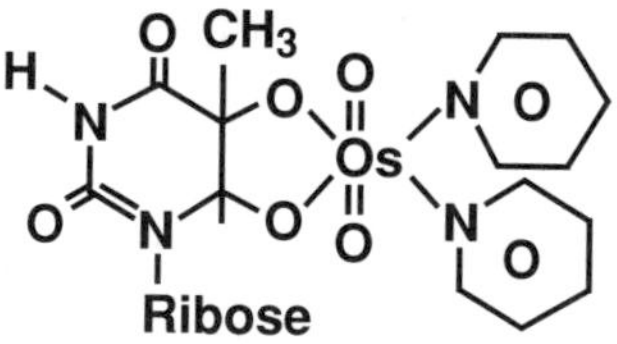

Osmium Tetroxide, Pyridine - Thymidine

$O{=}N-O-H$

Nitrous Acid

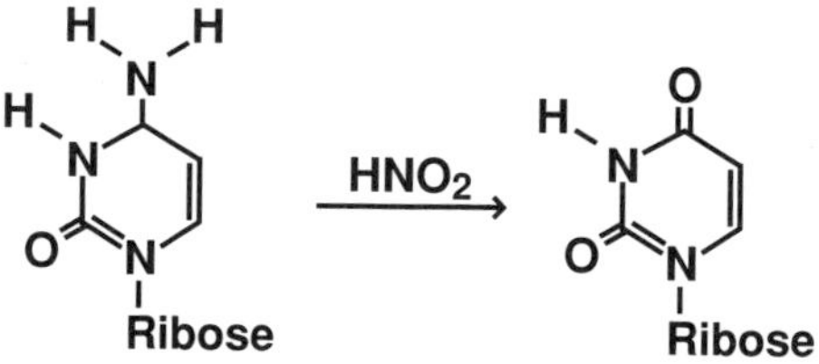

Cytidine $\xrightarrow{HNO_2}$ Uridine

Figure 1.21 *Continued.*

g. Nitrous acid

Nitrous acid (HNO_2) reacts with amino groups on cytosine, adenine, or guanine in an oxidative deamination reaction (Figure 1.21C). The deamination, leading to the formation of a carbonyl group, results in the conversion of cytosine to uracil, adenine to hypoxanthine, and guanine to xanthine. Deamination of C and A leads to a reversal of the polarity of the hydrogen bonding surfaces. As a consequence of deamination, U will pair with A and hypoxanthine will pair with C. This mispairing can result in the introduction of mutations during DNA replication. Xanthine, like guanine, will pair with cytosine; consequently, the deamination of guanine to form xanthine does not create a mispair.

Note that deamination reactions occur spontaneously at an alarmingly high rate. About 10^4 deaminations of C occur per human cell per day. Although DNA repair mechanisms have evolved to deal with this aspect of DNA chemistry, mutations do occur frequently at potential sites of spontaneous or acid-induced deamination.

3. DNA Cross-Linking Reagents

A number of chemicals with two reactive sites can form cross-links in DNA. These cross-links can be *intrastrand* or *interstrand* in nature. An intrastrand cross-link is one in which two bases in the same strand are covalently linked together. Cisplatin, a drug used to treat several cancers, and diepoxybutane are two examples of compounds that bind to adjacent bases in one strand of double-stranded DNA forming an intrastrand cross-link. Mitomycin C and nitrogen and sulfur mustards (Figure 1.21B) are bifunctional alkylating agents that can form either interstrand or intrastrand cross-links. Psoralen derivatives are compounds that form interstrand cross-links in DNA. Psoralens have been extremely useful as *in vivo* probes of DNA supercoiling and various alternative DNA conformations as discussed in Chapters 4–6.

a. Reaction of 4,5´,8-Trimethylpsoralen with DNA

4,5´,8-Trimethylpsoralen is a planar, aromatic compound that binds to DNA by sliding (intercalating) between two adjacent stacked base pairs of DNA (Figure 1.22). On absorption of 360-nm light, psoralen photobinds primarily to pyrimidine bases. Psoralen photobinds to only one strand of the DNA forming *monoadducts* on absorption of one photon of light. If a pyrimidine is present in an adjacent base pair in the opposite strand, a second photoreactive bond can photobind forming interstrand cross-links. Psoralen exhibits the preference 5´ TA > 5´ AT > 5´ TG >> 5´ GT in its rate of cross-linking various dinucleotides. Although 5´ TA is the preferred cross-

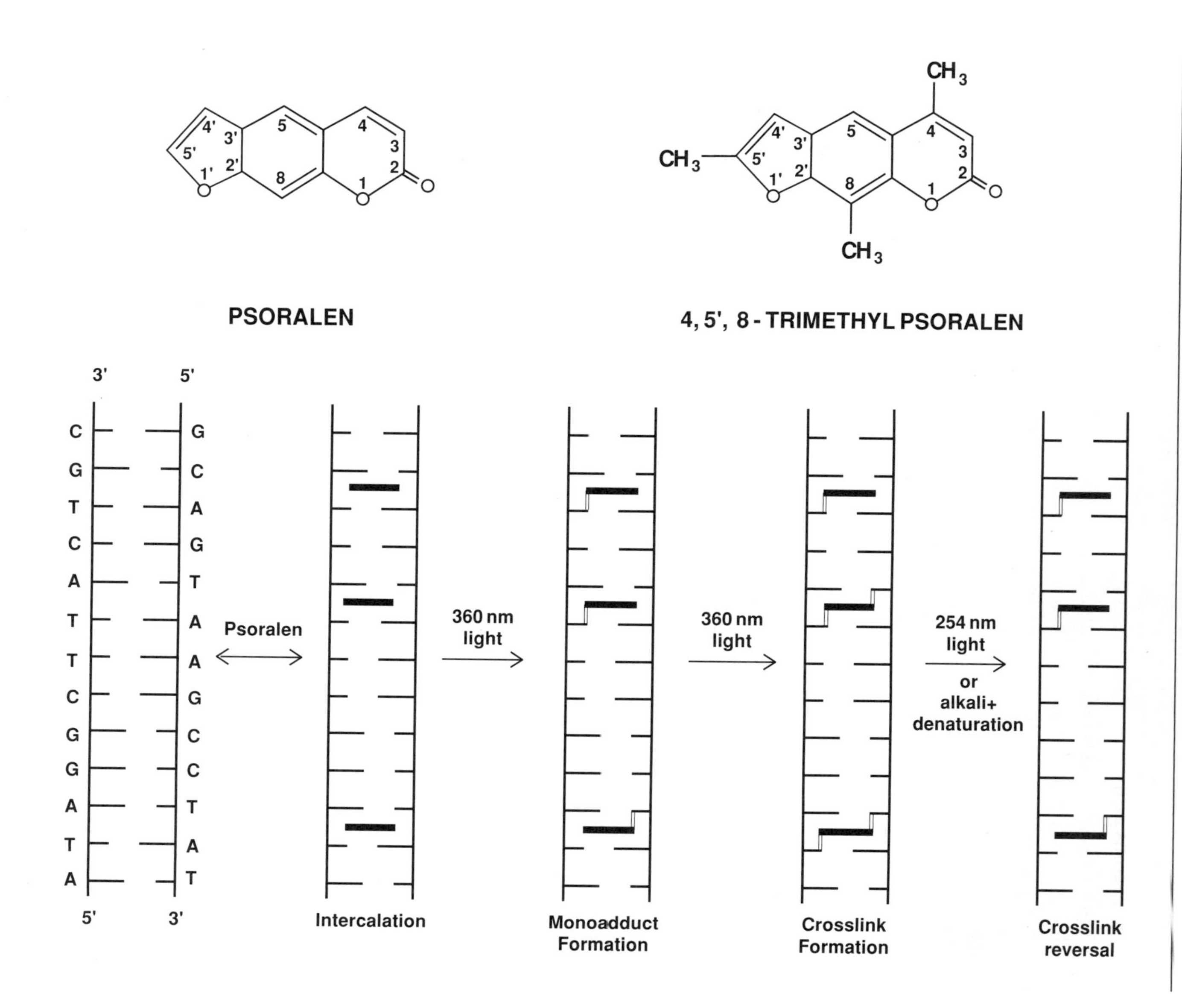
PSORALEN
4, 5', 8 - TRIMETHYL PSORALEN
CH3
3' 5'
CGTCATTCGGATA
GCAGTAAGCCTAT
5' 3'
Psoralen
360 nm light
360 nm light
254 nm light
or
alkali+ denaturation
Intercalation
Monoadduct Formation
Crosslink Formation
Crosslink reversal

linking site, not all 5′ TAs react equally because of the heterogeneity of helix structure in different sequence contexts of DNA.

H. DNA Sequencing Strategies

1. Maxam and Gilbert Chemical Sequencing of DNA

The chemical sequencing strategy of Maxam and Gilbert (1980) is based on the principle that different bases have different reactivities to a variety of chemicals. Thus, specific bases can be modified by certain chemicals. Modified bases have a more labile glycosidic bond that can lead to depurination or depyrimidation, creating an apurinic or apyrimidinic site (AP site) where the phosphate backbone can be broken easily. Analysis of the cleavage products on a denaturing acrylamide gel that separates the DNA on the basis of size (with base pair resolution) allows determination of the sequence of the DNA.

Experimentally, DNA is treated with a chemical having a preferential reactivity for one base (or two bases). For example, DMS reacts preferentially with the N7 of guanine (Figure 1.23). On treatment with base, the five-member ring of N7-methylguanine opens. Hydrazine can be used to open the C and T pyrimidine rings. In the presence of a high salt concentration, hydrazine reacts with cytosine. Once the purine or pyrimidine ring opens, piperidine is used to displace the modified base from the deoxyribose. The piperidine on the ribose catalyzes a β-elimination of the phosphates from the ribose. This leads to a specific break in the DNA at the position of the chemically modified base. By initially labeling *only one strand* of the DNA with ^{32}P (or ^{35}S) and treating DNA with the chemical modification/cleavage reactions,

---

Figure 1.22 Psoralen and its reaction with DNA. (*Top*) Psoralen and 4,5′,8-trimethylpsoralen, very useful probes of DNA structure. (*Bottom*) The reactions of psoralen with DNA. Psoralen first intercalates reversibly between various base pairs of DNA. The binding of psoralen is believed to be relatively sequence nonspecific, so psoralen binds freely at most positions throughout the DNA double helix. This binding does not require light. Following absorption of 360-nm light (middle structure), psoralen forms monoadducts in DNA. Monoadducts are formed between the 3,4 pyrone or 4′,5′ furan double bonds of psoralen and the 5,6 double bond in pyrimidine bases. Psoralen reacts preferentially with pyrimidines, especially thymine. On absorption of a second photon of 360-nm light, psoralen can form interstrand cross-links, covalent bonds between the two strands of DNA. The most preferential cross-link site in DNA is the 5′ TA dinucleotide in which the psoralen photobinds to thymines in adjacent base pairs in opposite strands. Interstrand cross-links can be reversed by treatment with 254-nm light, which reverses the cyclobutane ring formed by the fusion of the two double bonds, or by treatment with alkali and elevated temperature.

O
H N N
NH2 N N
Guanine

DMS (1)

O CH3
H N N
+
H2N N N
N7 Methylguanine

OH^- (2)

5' O CH3
H N N C=O
H2N N N-H
O=P-O-CH2 O
O-
O
O-=P-O-
O
3'

O
H N CH3
O N
Thymine

$H_2N\text{-}NH_2$ (5)

O
H N CH3
5' H-N
H2N
O N
O=P-O-CH2 O
O-
H
O
O=P-O-
O
3'

Piperidine
:N

(3)

(6)

O CH3
H N N C=O
H2N N N-H H
H

5'
O
O=P-O-CH2 OH H
O- C=N+
H
O
-O=P-O-
-O
3'

O
H N CH3
H2N H-N
O N
H

5'
O
O=P-O-
O-

(4)

OH H H
H2C=C-C=C-C=N+
H
O-
O=P-O-
O
3'

a series of bands is observed following gel electrophoresis that corresponds to the position of the chemically modified base in DNA. The electrophoretic analysis of DNA sequencing reactions is outlined in Figure 1.24.

2. Sanger Dideoxyribonucleotide Sequencing of DNA

The procedure devised by Sanger and co-workers for sequencing DNA uses a DNA polymerase to synthesize a DNA strand complementary to one of the original strands (Sanger and Coulson, 1975). Since DNA polymerase does not begin synthesis *de novo* on DNA, a primer hybridizing 5′ to the region of interest is required. A labeled deoxyribonucleotide triphosphate [α-^{32}P]dATP, for example) is added to radioactively label the synthesized DNA strand. (Alternatively, the 5′ end of the primer can be labeled using polynucleotide kinase and [γ-^{32}P]ATP.) In the Sanger sequencing procedure during polymerization, the random incorporation of a 2′,3′ *dideoxynucleotide* leads to chain termination since there is no 3′ OH that is required by DNA polymerase for the addition of another nucleotide (Figure 1.5). Therefore, unique size DNA fragments are produced whose lengths correspond to the position of specific bases. Four reaction mixtures are set up to sequence DNA, each one containing three of the deoxyribonucleotides triphosphates (for example dATP, dGTP, and dCTP). The fourth nucleotide (in this case dTTP) is included at a concentration lower than the other three, and the corresponding dideoxyribonucleotide (ddTTP) is added. The concentration of the ddXTP (where X refers to A, T, C, or G) relative to the corresponding dXTP will determine the probability of incorporation of the ddXTP and the frequency of chain termination. For sequencing a short piece of DNA close to the primer, a high concentration of ddXTPs is used. Low concentrations of ddXTPs will allow sequencing of DNA thousands or more bases from the primer. The reaction products are then analyzed by electrophoresis on a denaturing polyacrylamide gel. The Sanger sequencing procedure is outlined in Figure 1.24.

Figure 1.23 Maxam–Gilbert chemical sequencing reactions. Some of the chemical sequencing reactions used to sequence DNA are shown. (1) Guanine is methylated by dimethylsulfate (DMS) forming N7 methylguanine. (2) In the presence of alkali, the guanine ring opens. (3) In the presence of piperidine the open guanine ring leaves the sugar residue, breaking the glycosidic bond. (4) Eventually the phosphodiester bond between the phosphate group and the open sugar ring is broken, leading to a complete scission of the phosphodiester backbone. (5) Hydrazine covalently binds to the C6 and C4 atoms in the thymine ring. (6) This addition of hydrazine opens the thymine ring, making it susceptible to cleavage in the presence of the piperidine.

A

51 61

5' ^{32}P $(N)_{50}$GCGCGGAATTCCGAGA 3'
3'$(N)_{50}$CGCGCCTTAAGGCTCT 5'

Maxam and Gilbert Chemical Cleavage Products

G Cleavage Products (Top Strand)	Modified Base
^{32}P $(N)_{50}$	G - 51
^{32}P $(N)_{50}$GC	G - 53
^{32}P $(N)_{50}$GCGC	G - 55
^{32}P $(N)_{50}$GCGCG	G - 56
^{32}P $(N)_{50}$GCGCGGAATTCC	G - 63
^{32}P $(N)_{50}$GCGCGGAATTCCGA	G - 65
^{32}P $(N)_{50}$GCGCGGAATTCCGAGA 3'	None

T Cleavage Products (Top Strand)	
^{32}P $(N)_{50}$GCGCGGAA	T - 59
^{32}P $(N)_{50}$GCGCGGAAT	T - 60
^{32}P $(N)_{50}$GCGCGGAATTCCGAGA	None

Sanger Dideoxy Sequencing Reaction Products

G Reaction Products

5' $(N)_{50}$ddG
5' $(N)_{50}$GCddG
5' $(N)_{50}$GCGCddG
5' $(N)_{50}$GCGCGddG
5' $(N)_{50}$GCGCGGAATTCCddG
5' $(N)_{50}$GCGCGGAATTCCGAddG

T Reaction Products

5' $(N)_{50}$GCGCGGAAddT
5' $(N)_{50}$GCGCGGAATddT
5' $(N)_{50}$GCGCGGAATTCCGAGA

Figure 1.24 DNA sequencing strategies. (A) A short sequence of DNA with the nucleotide composition from bp 51 to 66 is shown with the 5′ end of the top strand labeled with ^{32}P. Below this sequence are the Maxam and Gilbert chemical cleavage products for modification and cleavage reactions at G and T residues. DNA is treated at a level of chemical modification at which there is, on average, only one modified base per fragment. Cleavage at the modified positions within the popu-

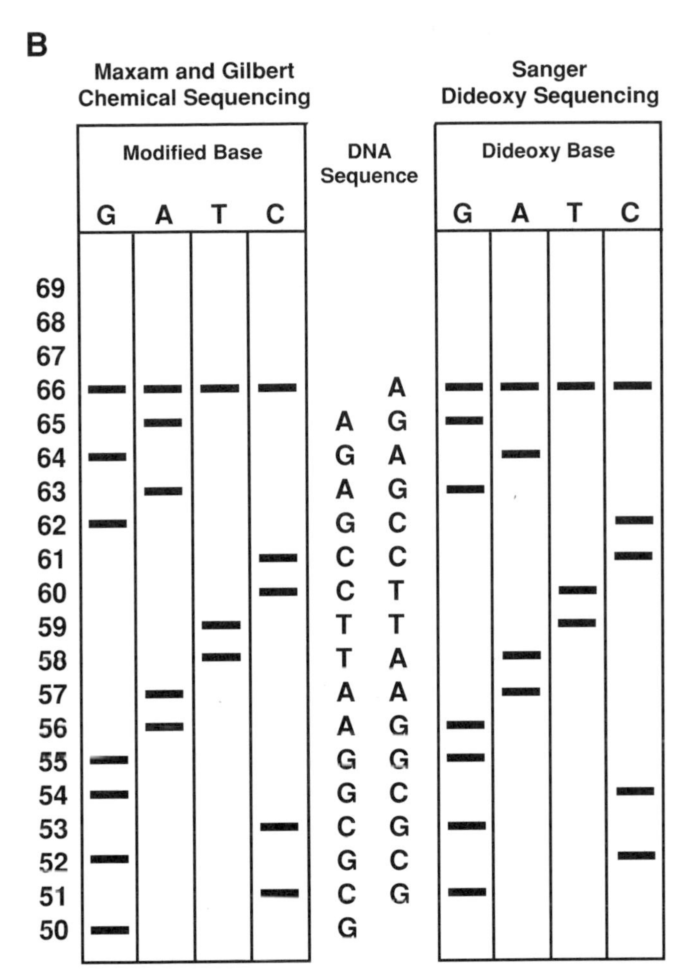

lation will generate a family of variously sized DNA molecules. In the case of guanine modification, the sizes of the fragments are 50, 52, 54, 55, 62, and 64 nucleotides in length (within the region shown), which corresponds to the modification of guanines at positions 51, 53, 55, 56, 63, and 65, respectively. In addition, a full-length unmodified product 66 base pairs long will also be present. This figure also shows Sanger dideoxy sequencing reaction products. The incorporation of a dideoxyribonucleotide (shown in Figure 1.21) leads to the cessation of elongation of the DNA chain. Incorporation of ddG leads to reaction products that are 51, 53, 55, 56, 63, and 65 nucleotides long. (B) The chemical cleavage products or polymerase reaction products are then analyzed by denaturing polyacrylamide gel electrophoresis. A representative DNA sequencing gel pattern comparing the reaction products of the Maxam and Gilbert chemical sequencing and the Sanger dideoxy sequencing reactions is shown. Because of the nature of the cleavage product, the Maxam–Gilbert sequencing products run one base shorter than those of the Sanger dideoxy method.

I. Enzymes That Modify DNA Structure

Many proteins react with DNA. The processes of DNA replication, genetic recombination, DNA repair, restriction/modification, supercoiling, and DNA degradation require breakage and rejoining of phosphodiester bonds, the breakage of the glycosidic bond, the chemical modification of bases, and the denaturation and renaturation of the double helix. This section provides a brief description of selected proteins involved in maintaining, manipulating, and utilizing DNA.

1. DNA Replication

DNA polymerases are responsible for replicating the DNA. They synthesize a new strand of DNA using a preexisting strand as a template. They utilize nucleoside triphosphates adding a 5′ mononucleotide to the 3′ OH end of the nascent DNA chain. Thus, polymerization occurs in the 5′ to 3′ direction. Because of the 5′ to 3′ directionality of the polymerase, one strand of a replication fork, the *leading strand,* can be replicated continuously. The other strand, the *lagging strand,* must be replicated in short "Okazaki" pieces. There are three polymerases in *Escherichia coli* called pol I, pol II, and pol III which are encoded by the *polA, dinA,* and *pol C* (or *dnaE*) genes, respectively. In *E. coli,* pol III is primarily responsible for replication of the chromosome. pol I is involved in joining the short Okazaki pieces. pol I and pol II are involved in DNA repair functions. In eukaryotic cells there are three nuclear polymerases (α, β, and δ) and one mitochondrial polymerase γ. Polymerase α is believed to replicate the leading strand, δ may replicate the lagging strand, and β is involved in DNA repair. These enzymes and the process of DNA replication are described in more detail by Kornberg and Baker (1992).

Primase is a required enzyme because DNA polymerase does not initiate synthesis *de novo* on a single-stranded DNA template. DNA polymerase must have a *primer* on which to add 5′ mononucleotides. The primer, which must be base-paired to the template, can be a short piece of either DNA or RNA with a 3′ OH end. RNA polymerase *can* begin synthetis *de novo* of RNA complementary to a strand of DNA. A specialized RNA polymerase called primase, encoded by *dnaG* in *E. coli,* makes the primer for synthesis by DNA polymerase III.

DNA helicases are proteins that move down the DNA separating the two strands or denaturing the double helix. Helicases require energy from ATP for their activity. Helicases work ahead of DNA polymerase, separating the strands for polymerase. There are helicases with two different polarities, those track along a strand in the 3′ to 5′ direction and those that track from 5′ to 3′.

DNA ligase is an enzyme that forms a covalent phosphodiester bond between a 3′ OH and 5′ PO_4 at the ends of two polynucleotide chains (frequently at a *nick* in one strand of a double helix). Ligase requires energy from ATP to seal the DNA. DNA ligase is used extensively in DNA cloning methodologies.

2. Genetic Recombination

The *RecA protein* from *E. coli* is a multifunctional protein that binds to single-stranded DNA and binds to DNA containing damaged bases. RecA contains a protease activity and is responsible for the catalytic cleavage of several repressors as well as the autocatalytic cleavage of itself. RecA is responsible for catalyzing the process of genetic recombination by promoting the binding and exchange of a third strand.

Resolvase are enzymes that cut four-stranded DNA structures (Holliday recombinational intermediates) into two molecules. They are required for the separation or resolution of two homologous DNA molecules exchanging genetic information. Resolvases are discussed in more detail in Chapter 4. For detailed information on the enzymes involved in genetic recombination, see Cox and Lehman (1987).

3. DNA Repair

Many enzymes recognize and bind to damaged bases, bulges, or kinks formed by mispairs arising from environmental or chemical damage, misincorporation of an incorrect nucleotide during replication, or spontaneous deamination changing the chemical identity (and base-pairing characteristics) of a base. For more information on the processes of DNA repair and the enzymes involved, see an excellent text by Friedberg (1985).

The *UvrABC exonuclease* from *E. coli* recognizes a distortion caused by damage to one strand of DNA. The UvrABC enzyme is required to repair pyrimidine–pyrimidine dimers caused by exposure to 254-nm ultraviolet light. It initiates *excision repair* at many damages in DNA including T–T, T–C, and C–C dimers, psoralen monoadducts and interstrand cross-links, and bases with bulky groups attached. UvrABC exonuclease cuts 4–5 bases 5′ to the damage and 6–7 bp 3′ to the damage. The damaged region of single-stranded DNA is removed (with the aid of the uvrD helicase), and DNA polymerase I then fills the resulting gap.

DNA glycosylases cleave the glycosidic bond in nucleic acids. Specific glycosylases recognize damaged bases (i.e., alkylated bases, uracil resulting from the deamination of cytosine, or thymine dimers) and produce an apurinic or apyrimidinic (AP) site.

AP endonucleases are the enzymes that introduce nicks at AP sites. Individual enzymes can cut on either the 3′ or the 5′ side of the AP site. The nick allows a nuclease to digest the damaged DNA strand containing the AP site, opening up a gap for repair synthesis by DNA polymerase.

4. Transcription and Gene Regulation

RNA polymerase is the enzyme complex that makes an RNA chain from a double-stranded (or single-stranded) DNA template. RNA polymerase in *E. coli* is composed of 2 α subunits, 1 β subunit containing the catalytic activity, 1 β' subunit that has DNA binding activity, and a δ subunit. RNA is synthesized with the same 5′ and 3′ polarity as DNA, in which nucleotides are added onto the 3′ OH end of the chain. In *E. coli* there is one RNA polymerase molecule that synthesizes all types of RNA. The types of RNA are mRNA, which is translated into proteins; rRNA, which is a structural component of ribosomes; and tRNA, which brings amino acids to the ribosome to be incorporated into protein. How is the *E. coli* RNA polymerase regulated so the appropriate levels of the three types of RNA messages are made? *Escherichia coli* has multiple *sigma factors* that provide the specificity to RNA polymerase to encode rRNA, tRNA, or various types of mRNA.

In eukaryotes, in addition to mRNA, tRNA, and rRNA, there are "guide RNAs" used for post-translational editing of mRNA, RNA for incorporation into ribonucleoproteins involved in splicing eukaryotic mRNAs, and RNA for incorporation into certain transcription factors. In eukaryotes specialized RNA polymerases synthesize these various RNA molecules. RNA polymerase I synthesizes large ribosomal RNA, RNA polymerase II synthesizes messenger RNA, and RNA polymerase IIi synthesizes the 5S small ribosomal RNA and tRNAs. These polymerases, like the *E. coli* enzyme, are large multisubunit proteins. All eukaryotic polymerases have two subunits (L and L′) that have similarity with the *E. coli* β and β' subunits. In addition, there are many other subunits associated with these proteins, some of which are specific for the polymerase.

5. Transcription Factors

Precise regulated control of gene expression is essential for proper growth, development, and subsequent maintenance of an organism. In *E. coli,* the catabolite activating protein (CAP), also called the catabolite regulatory protein (CRP), acts as a transcription factor to regulate the level of expression of multiple genes. In eukaryotic cells there are many families of *tran-*

scription factors that bind to regulatory regions of genes. A single gene may require binding of as many as 6–10 individual transcription factors for proper expression throughout the life of an organism.

6. Restriction and Modification

The processes of restriction and modification may act as an "immune system" for bacteria, protecting the organism from infection by foreign DNAs. Type II restriction endonucleases recognize and bind specifically to a unique recognition site that is typically a 4-, 6-, or 8-bp palindrome. After binding they introduce a double-stranded cut at the sequence. Restriction cleavage occurs at one site producing "blunt ends" or, if it occurs in a staggered fashion, "overhanging ends" (see Table 1.5). The blunt ends contain a base pair at both ends produced by the cut, whereas overhanging ends contain a few bases of single-stranded DNA protruding from the end of the duplex molecule. The overhanging end can have a recessed 3′ OH or 5′ PO_4 group. Staggered cuts within a palindromic restriction site produce "sticky" or self-complementary ends.

In a process called *modification,* a cell protects its own DNA from the nuclease action of restriction endonucleases. Specific DNA methylases can methylate A or C residues in the DNA at the restriction recognition site. The corresponding restriction enzyme cannot cut the methylated DNA.

Table 1.5
Restriction Endonucleases/Methylases

Enzyme	Recognition site	Product	
*Eco*RI	5′-GAATTC-3′ 3′ CTTAAG-5′	5′-G 3′-CTTAA	AATTC-3′ G-5′
*Kpn*I	5′-GGTACC-3′ 3′-CCATGG-5′	5′-GGTAC 3′-C	C-3′ CATGG-5′
*Pvu*II	5′-CAGCTG-3′ 3′-GTCGAC-5′	5′-CAG 3′-GTC	CTG-3′ GAC-5′
*Eco*RI methylase	5′-GAATTC-3′ 3′-CTTAAG-5′	Me | 5′-GAATTC-3′ 3′-CTTAAG-5′ | Me	

7. DNA Degradation

Many enzymes, unlike restriction enzymes, degrade DNA in a manner unrelated to sequence. These fall into two categories: *endonucleases* and *exonucleases*. Endonucleases break the phosphodiester bond at any point along the DNA chain. Endonucleases can nick and digest covalently closed circular DNA and can introduce breaks internally within a linear molecule. Exonucleases require a free end at which to initiate digestion. Digestion by exonucleases can occur in the 5′ to 3′ direction and require a terminal 5′ PO_4, or digestion can occur in the 3′ to 5′ direction and require a 3′ OH terminus. Nucleases can be specific for single- or double-stranded DNA.

S1 nuclease, purified from *Aspergillus oryzae,* is one of the most widely used nucleases in molecular biological research. S1 nuclease will digest single strands of DNA (or RNA) as well as single-stranded "tails" or single-stranded loops or gaps in DNA. Under certain conditions it will cut at the site of a mismatched base or at sites of chemical modification where the normal Watson–Crick base pairing has been disrupted. S1 nuclease has been important for analysis of cruciforms, Z-DNA, and intramolecular triplexes.

Deoxyribonuclease I from bovine pancreas digests DNA to small oligonucleotides (about 4 bases in length). It will digest either double- or single-stranded DNA. It is commonly used to introduce random nicks into DNA to permit labeling of DNA with ^{32}P-dNTPs in a DNA polymerization reaction. DNase I is also used in "footprinting" studies where nicks are introduced throughout a DNA molecule *except* where the DNA is protected by a tightly bound protein. Although DNase I does not have a specific nucleotide requirement for cutting, certain positions along the DNA are cut more readily than others.

Exonuclease III, purified from *E. coli,* digests only one strand of double-stranded DNA from the 3′ OH end, releasing 5′ monophosphate nucleotides. The enzyme prefers a recessed 3′ OH and digests poorly from an overhanging 3′ OH end. The enzyme stops when it encounters a protein tightly bound to DNA or a base containing one of many chemical modifications. For example, exoIII has been used to map the position of psoralen adducts in DNA, as described in Chapter 5.

Lambda exonuclease, isolated from *E. coli* infected with bacteriophage λ, is similar to exoIII except that it requires a 5′ PO_4 end of DNA. It prefers to digest from a blunt end of DNA. λ exonuclease has also been used to map the sites of covalent chemical modification of DNA. λ exonuclease will pause, but not necessarily stop, at many chemically modified bases.

Phosphatases remove the 5′ phosphate from nucleotides and from the end of a DNA chain to produce a 5′ OH terminus. Treatment of linear double-stranded DNA molecules with phosphatase removes the natural PO_4 group left by most nucleases.

DNA kinase places a phosphate group on a 5′ OH terminus of a DNA chain. This reaction requires ATP (shown in Figure 1.15), with the γ phosphate transferred to the terminal 5′ OH of the DNA. Following treatment of linear DNA with phosphatase, reaction with DNA kinase and γ-^{32}P-ATP leads to the incorporation of radioactive ^{32}P onto the 5′ ends of each strand of the DNA.

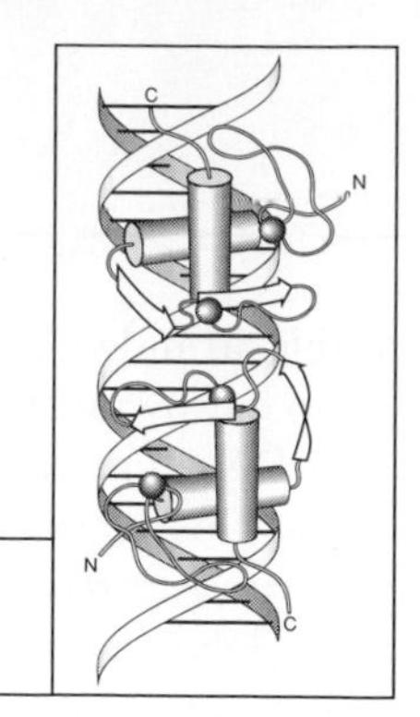

2 CHAPTER

DNA Bending

A. Introduction

In science, once a system or phenomenon is understood, future events related to the phenomenon are predictable. In some respects DNA is a simple molecule. A plasmid DNA of known sequence can be cut into a number of different-sized pieces by a restriction enzyme. Molecules of equal length should exhibit similar flexibility. Pieces of DNA shorter than the persistence length of 150–200 bp will behave as rather stiff rods that cannot be easily bent into a circle (see Hagerman, 1988, for discussion of persistence length). Larger pieces adopt a "random coil" shape in solution. DNA molecules of defined size behave in a very predictable way when run on agarose or acrylamide gels. Shorter molecules migrate more rapidly than larger molecules. In agarose and acrylamide gels there is a linear relationship between the log of the distance migrated and the length of the DNA in base pairs.

Occasionally, a DNA band of known length does not run at its expected position on an acrylamide gel. One of the most striking examples of such anomalous migration was a 414-bp piece of kinetoplast DNA from *Crithidia fasciculata* (Marini *et al.,* 1982,1983). This DNA migrated as if it were twice as long in an acrylamide gel but migrated at its proper position in an agarose gel. The anomalous migration of the kinetoplast DNA on polyacrylamide gels was attributed to the kinetoplast DNA being either stably bent or kinked (Figure 2.1). Although the migration through acrylamide gels is not completely understood in physical terms, the migration is believed to depend on the ability of DNA to "snake" through the gel matrix. A relatively straight piece of DNA that is rather flexible can easily snake through the gel. On the other hand, DNA containing a permanent bend or kink is not as flexible and can get hung up in the gel matrix. (Consider a cooked piece of spaghetti that can snake its way through the gel more easily than a stiff piece that is bent or curved at the center. It will be much more difficult for the stiff piece to move through the gel matrix.) For DNA fragments of equal sizes, a

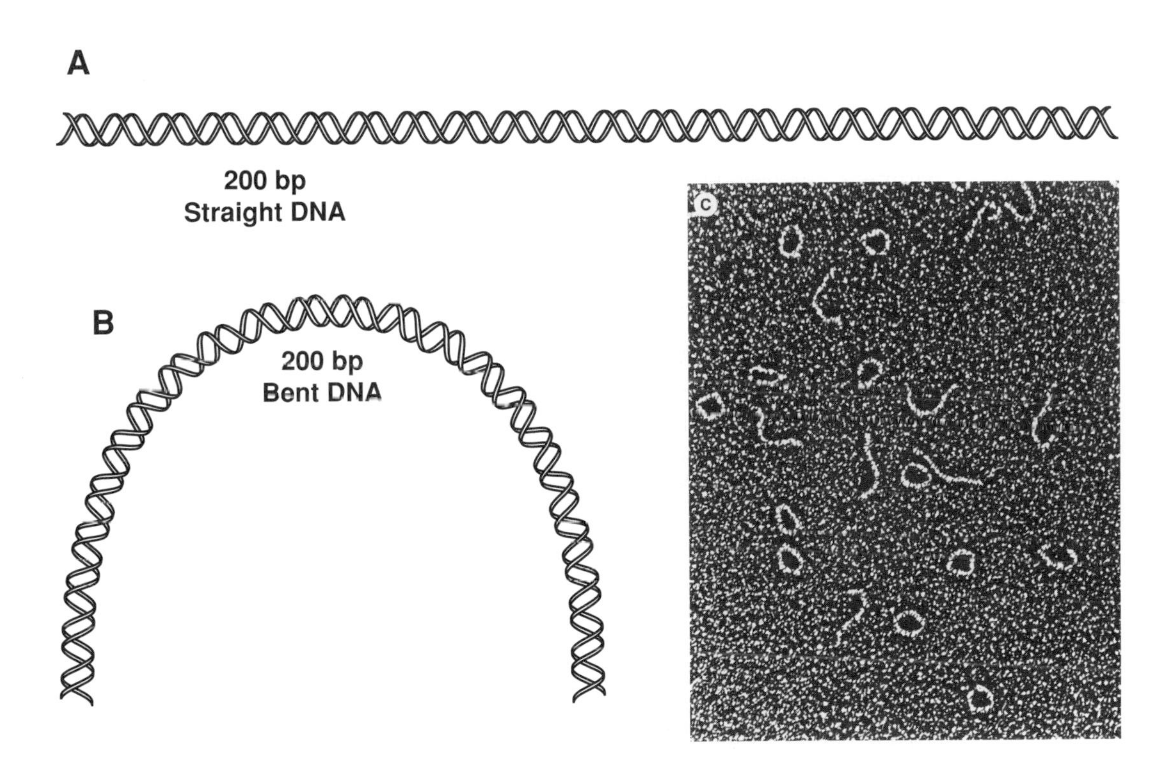

Figure 2.1 Straight DNA and bent DNA. (A) A representation of a 200-bp straight DNA molecule. The persistence length of DNA is 150–200 base pairs. Persistence length is the length of DNA that resists easy deformation, or a length that is not considered very flexible. DNA smaller than its persistence length cannot be easily bent into a circle. (B) A 200-bp piece of bent DNA is shown. Some DNAs, because of their primary base sequence, will adopt a stable bent or curved configuration. (C) The electron micrograph shows a 223-bp fragment of *Crithidia fasciculata* DNA containing 18 phased A_4 to A_6 tracts. Although the fragment is linear and is blunt ended, many molecules appear to be intact circles. Without the phased A tracts, the DNA would appear as linear, relatively straight molecules. Courtesy of Jack D. Griffith.

bent fragment will take more time to snake through the pores or matrix of the gel than a nonbent DNA fragment. (This does not occur in agarose gels because the matrix or pore size is believed to be larger.)

The initial part of this chapter describing DNA bending experiments will be presented from a historical perspective for two reasons. First, it is a convenient way to gradually introduce a number of important concepts related to DNA bending and to introduce other important aspects of DNA structure and function. Second, it provides a small window on "the scientific process." This process involves developing hypotheses and models and then designing experiments to test or disprove the models. In the DNA bending story, the experimental design was frequently simple and the experimental results quite clear. However, the experiments designed to test the various models for the molecular basis of bending did not all support a single unified explanation. The DNA bending experiments provide an example of how the frequent paradox of scientific discovery leads to a deeper understanding of the physical realities of nature. Moreover, incongruities in the bending experiments suggest that the actual molecular mechanisms behind sequence-induced DNA bending are not yet completely understood.

B. The Wedge Model and Junction Model for DNA Bending

1. Sequence Induced Bending of DNA

Wu and Crothers (1984) devised a clever way to map the site of bending in the kinetoplast DNA. They cloned the bent kinetoplast DNA fragment as a dimer of a 241-bp *Hin*dIII DNA fragment. By cutting the dimer with a series of different restriction enzymes that cut only once within each 241-bp fragment, different 241-bp DNA fragments were produced (Figure 2.2). The sequence organization of each fragment would be different from that of the original fragment. Such a set of DNA molecules is called *circularly permuted.* Although all these isomers contain the same base sequence and, for the most part, the same bend in the DNA, they do not migrate at the same rate during electrophoresis in an acrylamide gel.

Circularly permuted molecules do not migrate at the same rate because the bends are at different locations within the molecule. This creates different end-to-end distances. The end-to-end distance for E in Figure 2.2 is nearly as long as that for a straight DNA molecule. The end-to-end distance is shortest when the bend is positioned at the center of the molecule, as in circularly permuted isomer H. For molecules of the same size, the end-to-end distance is important in determining gel mobility. The molecule with the shortest end-to-end distance will migrate most slowly in an acrylamide gel. This pattern was observed by Wu and Crothers (1984) and is illustrated in Figure 2.2C. The distance in migration is plotted as a function of the position of the restriction cutting site used to generate the circularly permuted DNAs. Extending a line from the curves in Figure 2.2C results in a point of intersection that represents the center of the site of bending.

The DNA sequence shown in Figure 2.3 was found at the bend center. There are five runs of 4 or 5 As (A_{4-5}) which, in each case, are preceded by a C and followed by a T ($CA_{4-5}T$) . In addition, the runs of A are phased by 10 bp. The DNA bending hypothesis suggested that the runs of A and the 10-bp phasing were important in bending. Crothers also suggested that bending may occur at a *junction* resulting from the interruption of B-form DNA by an A tract that adopts a non-B-DNA helix. Trifonov proposed a second model for bending, in which a *wedge* angle was associated with the AA dinucleotide, and proposed that bending could be attributed to the summation of the wedge angles of the AA dinucleotides.

2. The Wedge Model for DNA Bending.

The wedge model for DNA bending assumes that the AA dinucleotide contains a "wedge" angle that causes a deflection in the axis of the DNA double helix (Figure 2.4; Trifonov and Sussman, 1980; Ulanovsky *et al.*, 1986). The sum of wedges pointing in the same direction, a condition met by the 10-bp phasing, leads to the bending of DNA. As illustrated in Figure 2.5, the wedge angle can result from a wedge along the tilt axis or a wedge along the roll axis. Calladine and Drew (1992) discuss the physical basis for wedge angles for DNA.

Ulanovsky *et al.* (1986) used measurement of the efficiency of ligation of small DNA molecules into circles to calculate the wedge angle of an AA dinucleotide. As a short piece of DNA with a defined curvature is ligated together into increasingly long polymers, at some length the total angle of curvature will result in the formation of a circle of DNA. By determining the length at which the DNA forms a circle and by knowing the number of AA dinucleotides responsible for the 360° curvature, the individual AA wedge angle can be determined. To calculate the total wedge angle, Trifonov synthesized the following DNA sequence:

```
TCTCTAAAAAATATATAAAAA
          TTTTTATATATTTTTAGAGA
```

This sequence contains two runs of $(A)_5$ that are 10 bp apart. Since the average helix repeat is 10.5 bp, it is necessary to make a 21-bp fragment to ensure that the A tracts will remain in phase when multiple 21-mers are ligated to-

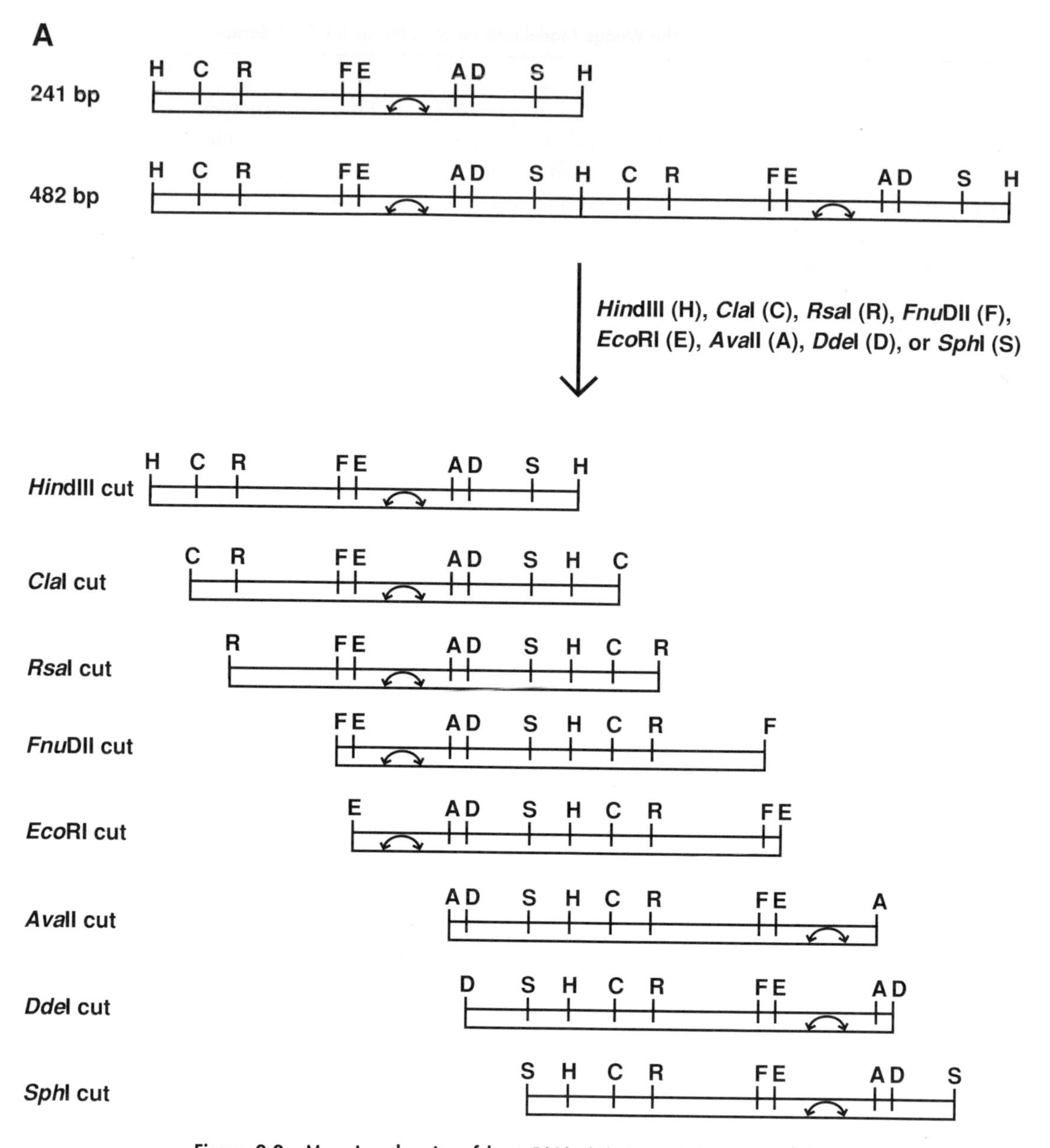

Figure 2.2 Mapping the site of bent DNA. (A) A restriction map of the DNA molecule used to identify the site of bending in DNA from *Crithidia fasciculata.* The bend site is represented by the curved arrow in the box. The box represents a 241-bp piece of DNA. The position of the bend affects the mobility of the DNA through a gel as described in the text. To differentially position this bend within the 241-bp molecule, a family of *circularly permuted* DNA molecules was constructed. These molecules were formed by the head-to-tail ligation of two 241-bp pieces of DNA into a 482-bp molecule. Cutting with a restriction enzyme that cuts once in the fragment will generate a 241-bp piece of DNA in which the location of the bend is dependent on the site of cutting. For example, cutting with *Cla*I positions the bend near the center of the molecule. Cutting with *Eco*RI or *Ava*II posi-

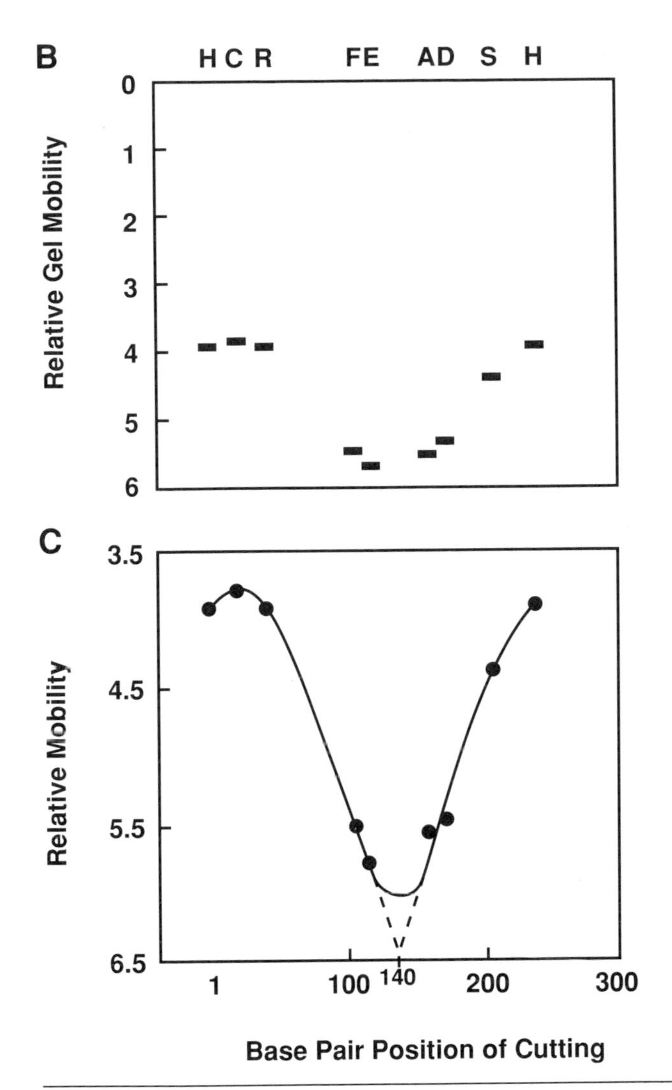

tions the bend very near the opposite ends of the DNA molecule. (B) The relative electrophoretic gel mobilities of these molecules are shown. The box represents an idealized polyacrylamide gel in which the bands represent the DNA after electrophoresis from top to bottom. Bent DNA migrates slowly during gel electrophoresis; DNA that does not contain a bend migrates fastest. The molecule cut with *Cla*I (C) is the most bent whereas the molecule cut with *Eco*RI (E) is the straightest and migrates fastest in the gel. (C) The relative mobility is plotted as a function of the base pair at the position of cutting. The line drawn through the data points intersects the *x* axis at 140 bp, indicating the cutting position at which the DNA is least bent and migrates the most rapidly in the gel. Cutting in the very center of the bend would make the DNA migrate most rapidly. Therefore, this identifies the bend center at position 140 bp in the original 241-bp DNA fragment. Adapted with permission from *Nature* (Wu and Crothers, 1984). Copyright (1984) Macmillan Magazines Limited.

1 11 21 31 41

CCCAAAAATGTCAAAAAATAGGCAAAAAATGCCAAAAATCCCAAC

Figure 2.3 The bend locus in *Crithidia fasciculata* DNA. This DNA sequence is the DNA bending site identified by Wu and Crothers in the experiment described in Figure 2.2. This bending locus contains blocks of five or six As repeated every 10 base pairs. In addition, the As are preceded by a C and followed by a T. Subsequent studies showed that the flanking C and T residues were not required for bending. The requirement for bending is to have multiple tracts of A (> 3 As) that are phased by 10 bp.

Wedge Model

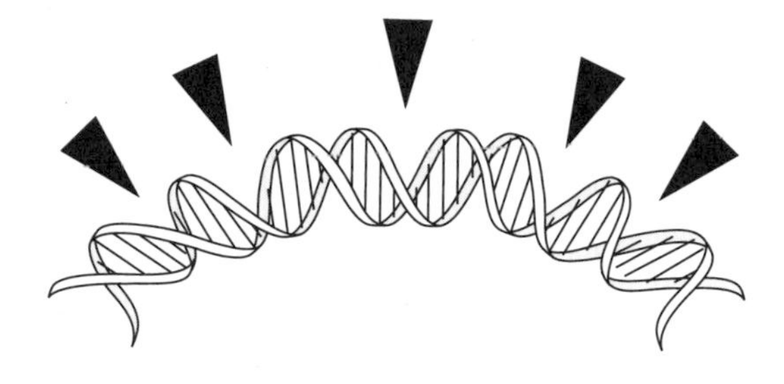

Junction Model

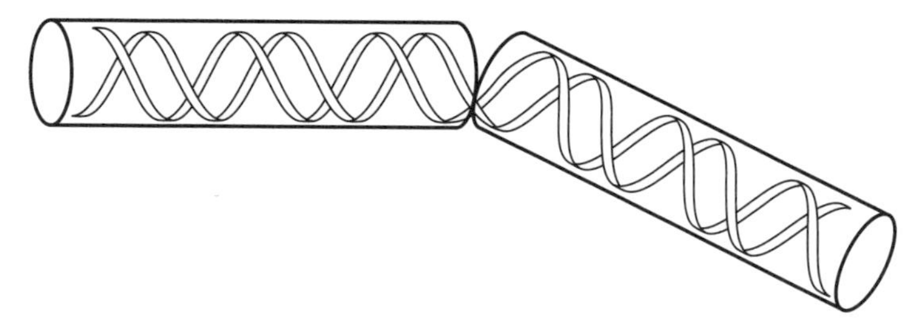

Figure 2.4 The wedge and junction models for DNA bending. The wedge model suggests that bending is the result of driving a wedge between adjacent base pairs at various positions in the DNA. The wedge model predicts a small wedge between each AA dinucleotide in all four A tracts. The total bend will be the sum of all the individual wedges. The junction model suggests that the A tract of DNA forms a slightly different helix than the normal B-DNA helix. A kink or bend is believed to exist at the junction of these two different helices. The sequence shown in Figure 2.3 would contain four junctions, one at each end of the A tract.

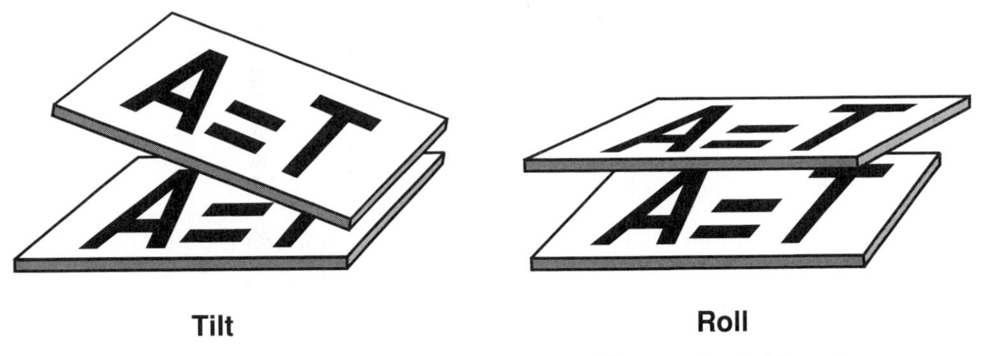

Figure 2.5 Tilt and roll wedges. As discussed in Chapter 1, different angles of deformation are possible when one base stacks on another. One angle is the *tilt angle,* which is in the direction of hydrogen bonding. Another is the *roll angle,* which occurs at 90° to the direction of hydrogen bonding. The tilt angle opens in the direction of the phosphate backbone whereas roll can open toward either the major or the minor groove.

gether. The optimal size for circle formation in polymers of this sequence was about 126 bp (Ulanovsky *et al.*, 1986). This length is well below the persistence length of DNA, which is 150–200 bp. A straight 126-bp piece of DNA will not circularize efficiently. Thus, efficient circularization is an indication of stable 360° curvature. Within the 126-bp molecule there are 66 AA dinucleotides; the sum of their individual wedge angles is responsible for the 360° curvature. Apparently making a few corrections, Trifonov estimated a total wedge angle (of both tilt and roll components) of 8.7° for each AA dinucleotide. This angle probably represents an upper limit of the wedge angle.

3. The Junction Model for DNA Bending

Wu and Crothers (1984) proposed the junction model for DNA bending (Figure 2.4) which suggests that there is a bend at the junction of B-form DNA and a non-B-DNA helix associated with A tracts. The suggestion that the A tracts might adopt a non-B-DNA helix was based, in part, on the knowledge that poly(dA) · poly(dT) can adopt a non-B-DNA helix called heteronomous DNA (Arnott *et al.*, 1983). In addition, modeling studies had suggested that DNA would bend at the junction between A-DNA and B-DNA (Selsing *et al.*, 1979). There are actually two junctions, a 5′ and a 3′ junction, relative to the A-tract helix. The site of bending could be associated with the 3′ end, the 5′ end, or both ends of the A-tract helix. Initially Crothers felt that bending was probably localized at the 3′ end of the A tract.[1] This conclusion was supported by experiments showing greater binding associated with a molecule in which the 3′ ends of A tracts were in phase at 10 bp compared with a molecule in which the 5′ ends of A tracts were in phase (Koo *et al.*, 1986).

4. Tests of the Phasing Hypothesis

An important component of both the junction and the wedge model is 10-bp phasing. For DNA to contain a region of stable curvature and to produce the anomalous gel mobility, the small individual bends must be oriented in the same direction or the same plane (if bends are not all directed in the same plane, the DNA will adopt a conformation that is effectively linear). If bends were placed every 1.5 turns of the DNA helix (15–16 bp), the DNA would be maintained in a zig-zag shape. Indeed, DNA with bends every 15 bp migrates slightly faster than a molecule with out-of-phase bends (or linear DNA) upon electrophoresis in an acrylamide gel. These patterns of curvature are shown in Figure 2.6.

To test the phasing hypothesis, Hagerman (1985) synthesized oligonucleotides of sequence GA_3T_3C, $G_2A_3T_3C_2$, and $G_3A_3T_3C_3$. These oligonucleotides are self-complementary and can hybridize to form duplex DNA. These duplexes can then be ligated into longer polymers. By ligating the individual oligonucleotides, the A_3 block in one DNA strand will be phased at 8-, 10-, and 12-bp intervals. If the 10-bp phasing were important, only the $G_2A_3T_3C_2$ polymer would exhibit anomalous migration on polyacrylamide gel electrophoresis. In Figure 2.7, mobility data are presented by dividing the apparent size in base pairs (determined by electrophoretic mobility) by the actual size of the DNA (determined by counting the ladders on the gel). The bp_{app}/bp_{seq} data points are plotted as a function of size of the DNA. The polymers with A tracts phased at 8 and 12 bp gave a horizontal line indicative of nonbent DNA. The $G_2A_3T_3C_2$ polymer with A tracts phased at 10 bp exhibited an upward slope in the plot of bp_{app}/bp_{seq} until a plateau was reached (at $bp_{app}/bp_{seq} = 2.2$). This pattern of electrophoretic migration is diagnostic for curved or bent DNA, and demonstrates the importance of 10-bp phasing in DNA curvature. A representation of the acrylamide gel for the data in Figure 2.7 is shown in Figure 2.17 at the end of this chapter.

[1]Note that an A tract contains As in one strand and Ts in the other. Obviously it has two 3′ ends and two 5′ ends. When we talk about a 3′ or 5′ end of the A tract we refer to the 3′ or 5′ side relative to the strand containing the As.

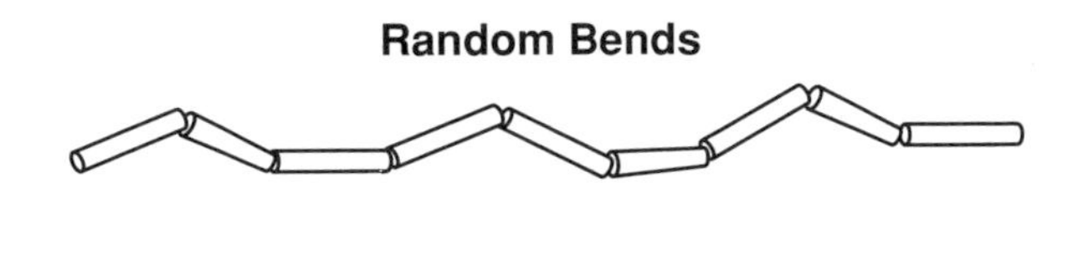

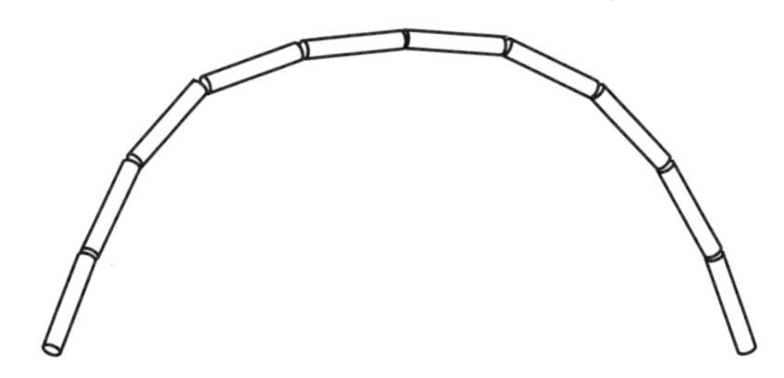

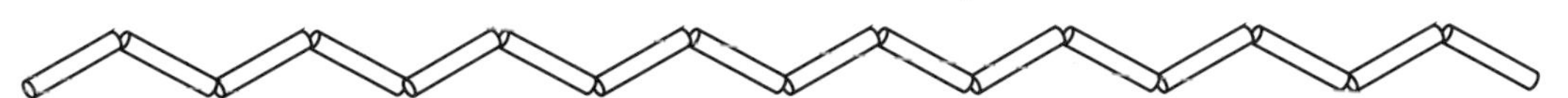

Figure 2.6 The significance of phasing in DNA bending. An A tract will introduce a small bend or deflection of the helix axis in DNA. However, the phenomenon of "bent DNA" or "stably curved DNA" that exhibits anomalous migration on polyacrylamide gel electrophoresis requires a number of small individual bends that are in phase. If bends are random, as shown in the top molecule, the DNA will migrate true to its length in an acrylamide gel. Only when bends are phased by 10.5 bp is the stable curvature shown in the middle structure observed. This DNA migrates anomalously slowly on electrophoresis in an acrylamide gel. If, as shown in the bottom structure, bends are phased by 16 bp, successive bends will be directed alternatively up and down. This creates a zig-zag molecule, which is unusually straight and migrates slightly more rapidly than unbent DNA (or DNA containing random bends) in an acrylamide gel.

5. Sequence Requirements for DNA Bending

Koo, Wu, and Crothers (1986) synthesized a large number of oligonucleotides containing various lengths of A tracts that were phased at different lengths. These oligonucleotides were conceptually similar to those described by Hagerman but were not symmetrical and thus had an A tract in only one strand of the DNA. Polymers with A_{4-9} were bent, with bending being optimal for A_6. The Koo *et al.* polymer with A_3 phased at 10 bp was not significantly bent. (The Hagerman bent sequence with A_3T_3 contained an A_3 tract in both strands that must contribute to bending.) A continuous run of As is

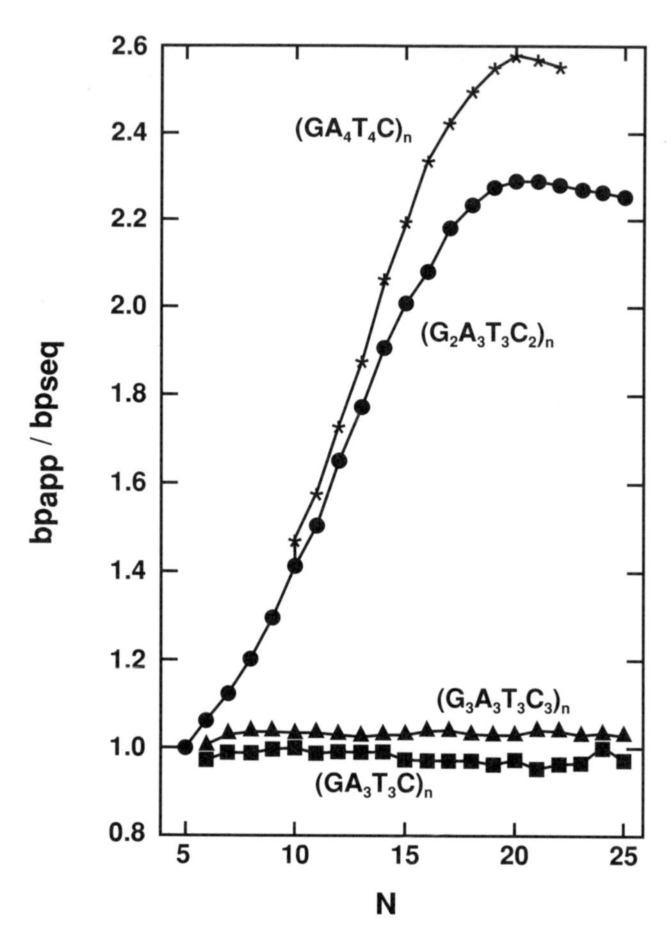

Figure 2.7 The analysis of DNA bending on polyacrylamide gels. Chemically synthesized 8-, 10-, and 12-bp oligonucleotides were ligated into families of molecules. Their mobility on electrophoresis on a polyacrylamide gel was determined, as shown in Figure 2.17. On the *y* axis is plotted the apparent length in base pairs (bp_{app}) divided by the actual length of the sequence (bp_{seq}). On the *x* axis is plotted the number of individual oligonucleotides ligated (*N*). When DNA bends, the apparent length becomes much larger than expected, leading to an increase in (bp_{app}/bp_{seq}). The straight configuration has a ratio of 1. The $(GA_4T_4C)_n$ and $(GGA_3T_3CC)_n$ polymers, represented by the stars and circles, respectively, are bent. Bending is most pronounced when the polymer is ≥ 20 monomers. Although the $(GA_4T_4C)_n$ sequence is not mentioned in the text, it is included as another example of a bent DNA polymer. In contrast, the ($G_3A_3T_3C_3$) and (GT_3A_3C) polymers, represented by triangles and squares, respectively, are not bent. Adapted with permission from Hagerman (1985). Copyright © 1985 by the American Chemical Society.

required for bending, since replacement of the central A in A_5 with C, G, or T destroys the bending. There is no sequence requirement for a particular base 5′ or 3′ to an A tract for bending, although flanking sequences can influence curvature.

DNA sequences that do not contain runs of As can also be bent. The bends observed in DNA lacking phased A tracts are usually not as large as A-tract bends. These sequences have not been as well studied as A-tract-induced bends.

6. Analysis of $(GA_4T_4C)_n$ and $(GT_4A_4C)_n$ Leads to Development of a Refined Wedge Model

The wedge model is a "nearest neighbor model" which assumes that bending can be explained by wedge properties attributed solely to an AA dinucleotide. It assumes, in its simplest form, that there will be no influence on the AA dinucleotide wedge angle from flanking bases or the sequence contexts of DNA. Considering the polymers $(GA_4T_4C)_n$ and $(GT_4A_4C)_n$, one might expect them to migrate with identical rates on electrophoresis since they each contain A_4 tracts that are phased at 10-bp intervals. However, the (GA_4T_4C) polymer is bent but the (GT_4A_4C) polymer is not bent (Hagerman, 1986). This result may seem to rule out the simple basic wedge model proposed by Trifonov.

Ulanovsky and Trifonov (1987) realized that the relationships of phased A tracts in the two strands are actually quite different for (GA_4T_4C) and (GT_4A_4C). Because GA_4T_4C and GT_4A_4C are symmetrical, A tracts in both DNA strands must be considered. Although both strands of a DNA duplex contain phased runs of As, the relationship between the A tracts in the opposite strands in GA_4T_4C compared to GT_4A_4C is different. The difference in A-tract phasing between these two molecules is illustrated in Figure 2.8. In GA_4T_4C, the 3′ ends of the A tracts in opposite strands occur at the same position in the helix axis. In GT_4A_4C, the 5′ ends of the A tracts coincide at the same point along the helix axis.

A key to understanding the difference in the electrophoretic mobility of $(GA_4T_4C)_n$ and $(GT_4A_4C)_n$ lies in the realization that the roll and tilt components of the wedge model are different symmetrically. For an AA dinucleotide, tilt can open either toward the As or toward the Ts. The tilt angle is believed to open toward the As as shown in Figures 2.5 and 2.9. In a DNA helix containing NNNAANNN, the replacement of the As with Ts to form NNNTTNNN results in a 180° reversal of the direction of the tilt angle. The roll angle, however, is unchanged when As are replaced by Ts because the roll angle opens toward the major groove and the positions of the grooves remain the same whether an AA or TT dinucleotide occurs in a helical region.

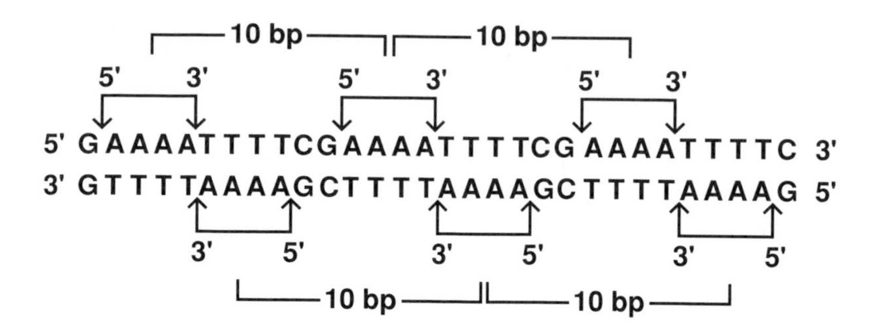

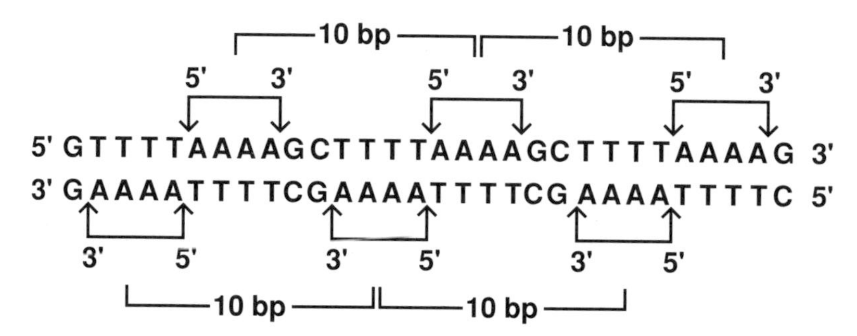

Figure 2.8 Asymmetry of GA_4T_4C and GT_4A_4C. At first glance, the relationship between phased A tracts in $(GA_4T_4C)_n$ and $(GT_4A_4C)_n$ might seem identical. In fact, the A blocks are separated in phase by 10 bp in both strands. In opposite strands, the A tracts are offset by 5 bp. If, however, one considers the phasing of the 5′ and 3′ junctions of the A tract, the sequences are quite different. In GA_4T_4C (*top*), the 3′ ends of the A tracts in the complementary strands are opposite each other. In the GT_4A_4C sequence (*bottom*), the 5′ ends of the A tracts in the complementary strands are opposite each other. If one considers a *simple* wedge model, there would be little difference between the A_4T_4 and T_4A_4 oligonucleotides. However, if one considers a wedge model in which roll and tilt wedges are involved, the reversal of T · A with A · T results in a different direction of tilt but not roll. In this case, the AA tilt wedges reverse direction by 180° with respect to the helix axis. Considering phasing introduced by a junction model, the relationships between the 5′ and 3′ junctions in opposite strands are quite different in the two different molecules.

Assuming that bending is the result of a sum of all vectorial tilt and wedge components for AA dinucleotides in both strands of the DNA, Ulanovsky and Trifonov (1987) calculated that the AA dinucleotide has a tilt of 2.4° opening toward the AA and a roll angle of 8.4° opening toward the major groove (see Figure 2.9).

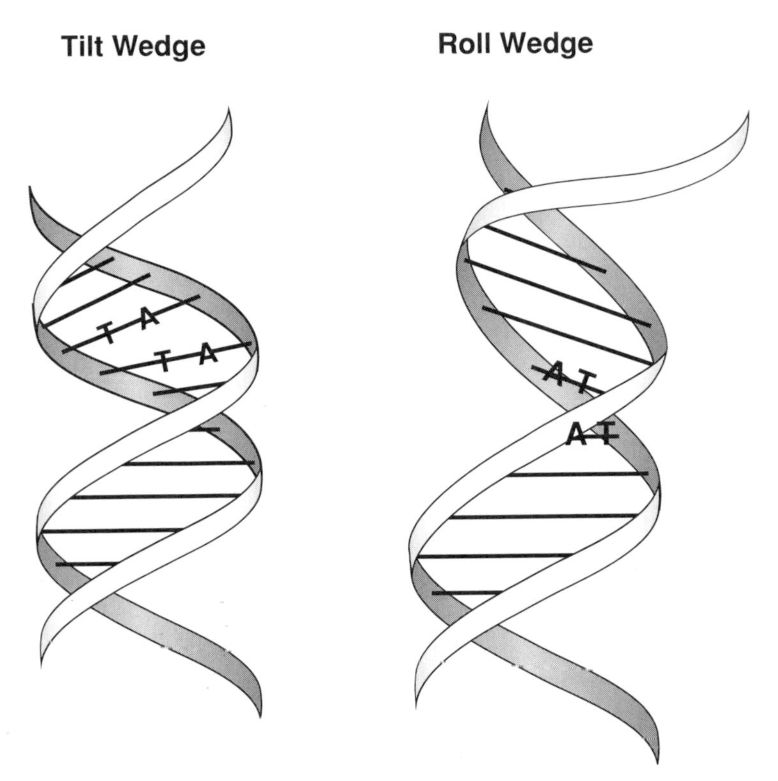

Figure 2.9 Tilt wedge and roll wedge. A tilt angle or a tilt wedge points toward the phosphate backbone, giving a bend in the DNA that points in the direction perpendicular to the phosphate backbone. In contrast, a roll wedge opens toward the major groove, compressing the minor groove.

7. A Refined Junction Model

The simple junction model first proposed by Crothers, in which bending was attributed predominantly to a deflection at the 3′ end of the A tract, did not adequately explain experimental results. A refined wedge model was needed to explain the normal migration behavior of DNA that seemingly should be bent. As illustrated in Figure 2.10, Koo and Crothers (1988) proposed that bending was due to a tilt deflection at both the 3′ and the 5′ junction. At the 5′ junction the tilt was toward the T, whereas at the 3′ junction the tilt was toward the A. In addition, these investigators argued that there was a roll that opened the minor groove only at the 5′ junction.

Differences in bending behavior of $(GA_4T_4C)_n$ and $(GT_4A_4C)_n$ are also predicted by the refined junction model. Since a roll component of bending is

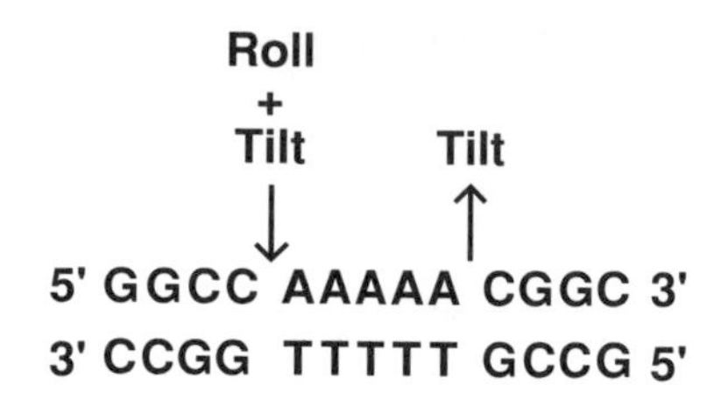

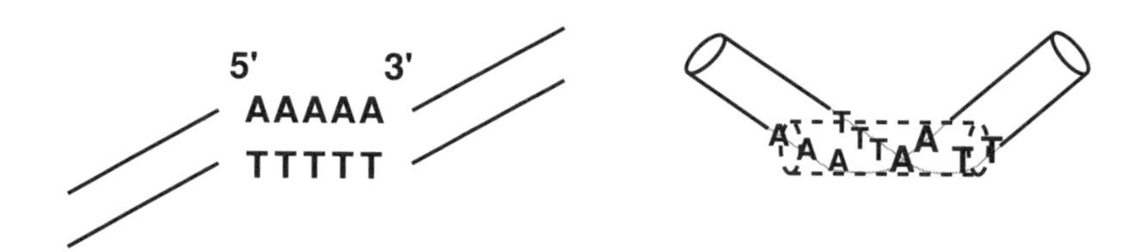

Figure 2.10 Refined junction model. In the refined junction model, there is a tilt toward the T at the 5′ junction of the A tract and a tilt toward the A at the 3′ junction. In addition, there is a roll opening toward the minor groove at the 5′ junction of the A tract. Therefore, the bend is a composite of three short straight helices: the B-form helix 3′ to the A tract, the A tract helix, and the B-form helix 5′ to the A tract. The bends then would occur as a mixture of roll and tilt bends at the 5′ junction of the A tracts and as a tilt bend at the 3′ junction of the A tracts. This relationship is shown in a linear configuration and in a helical configuration.

limited to the 5′ junction of the A tract, this component of bending will be at the same point in the helix axis in GT_4A_4C and at different positions in GA_4T_4C (Figure 2.8). Although it is not intuitively obvious why the A_4T_4 sequence is bent whereas the T_4A_4 sequence is not bent, one should be able to appreciate the symmetry differences in the relationships of the tilt and roll components at the 3′ and 5′ junctions from analysis of Figure 2.8.

8. Does Either the Junction Model or the Wedge Model Adequately Explain DNA Bending?

Neither the pure junction model nor the pure wedge model may provide a completely accurate explanation of the physical basis for DNA bending. Both models reasonably predict the bending of many DNA sequences containing runs of As. However, there are some A-tract-containing non-A-tract-bent DNAs whose bending is not predicted by either model.

Burkhoff and Tullius (1988) obtained results that were inconsistent with both the wedge and the junction model. A hydroxyl radical footprinting technique was used to analyze the width of the minor groove. Hydroxyl radicals are generated by the incubation of an EDTA (ethylenediamine tetraacetic acid) iron (II) complex with hydrogen peroxide. The hydroxyl radical (· OH) removes a hydrogen from the deoxyribose sugar, resulting in a break in the phosphodiester backbone. Since the hydroxyl radical attacks deoxyribose from the minor groove, the reactivity of DNA is dependent on the width of this groove. The hydroxyl radical footprinting pattern for B-DNA showed uniform cutting at each base. Kinetoplast bent DNA showed a reduction in cutting in the A tracts. This result was interpreted as reflecting a narrowing of the minor groove along the A tract. The polymers studied by Hagerman, $(GA_4T_4C)_n$ and $(GT_4A_4C)_n$, showed very different results. The hydroxyl cleavage pattern in the $(GA_4T_4C)_n$ bent polymer showed periodic reduction in the cleavage along the A tracts. There was no reduction in cleavage along A tracts in the $(GT_4A_4C)_n$ polymer. These results suggest that in the $5'A_4T_43'$ sequence the run of As can adopt a helix different than B-form DNA, a helix in which a gradual narrowing of the minor groove occurs. This narrowing does not occur in the A tracts in the $5'T_4A_43'$ sequence.

There is evidence for cooperativity in the formation of the A-tract DNA structure. Leroy *et al.* (1988) presented imino-proton-exchange NMR (nuclear magnetic resonance) experiments to show that the structures of A_nT_n and T_nA_n sequences were quite different. Long proton exchange times for A · T base pairs were associated with longer lengths of A tracts in the A_nT_n but not T_nA_n orientation. The shorter times in T_nA_n oligonucleotides were similar to lifetimes found in B-form DNA. Long proton exchange times were found for all sequences that exhibited anomalous migration on polyacrylamide gels. Nadeau and Crothers (1989) have also confirmed that cooperative structural changes in helix structure occur in a run of As as the length of the tract is increased. Three As do not set up the "A-tract helix" responsible for bending whereas A_4 does begin to adopt a non-B-DNA helix and bend DNA. By the time a length of A_{6-7} is reached, the transition from a B-form to a different helix, an A-tract structure, is complete.

At present there are two models, the wedge model and the junction model, that both explain reasonably well the bent nature of DNA containing phased A tracts. Although these models are based on a different initial premise they both predict the shape of many, but not all, bent molecules. In both models curvature is introduced into the DNA at each A tract by a combination of deviations of the tilt angle and roll angle of bases in the A tracts.

C. Protein-Induced DNA Bending

There are many examples of DNA bending on association with protein. In some cases, the DNA is actually wrapped around a protein core. The wrapping of DNA around the enzyme DNA gyrase, around nucleosomes, and around the bacterial histone-like HU protein are extreme examples. In a less dramatic fashion, many other proteins bind to DNA and introduce a slight bend that is more analogous to the bending introduced by A tracts in DNA. The protein-induced bending of DNA is shown in Figure 2.11.

The catabolite activator protein (CAP) and the *lac* repressor provide two examples of different effects of protein binding to DNA: CAP bends DNA whereas the *lac* repressor does not. In an experiment similar to that described for mapping the bend locus in kinetoplast DNA, Wu and Crothers (1984) demonstrated that CAP binding introduces a bend into DNA. Circularly permuted 203-bp DNA fragments containing the *lac* promoter region were constructed (the *lac* promoter region contains the CAP binding site and the *lac* operator sequence which binds the *lac* repressor). With no CAP bound, the circularly permuted DNAs had nearly identical mobility in the gel in the absence of CAP. CAP was bound to the fragments an the DNA–protein complexes were analyzed by polyacrylamide gel electrophoresis. The mobility of DNA–protein complexes with the CAP binding site located at the center of the DNA was significantly reduced relative to when the CAP binding site was located near the end of the DNA fragment. The X-ray cocrystal structure of the CAP–DNA complex has subsequently shown a bent angle of 90° (Schultz *et al.*, 1991). To demonstrate that the difference in mobility was due to bending and not to the location of protein bound to DNA, the *lac* repressor was also bound to these DNA fragments. There was essentially no difference in the mobility of the circularly permuted DNAs bound with the *lac* repressor. In fact, *lac* repressor DNA complexes ran at nearly the same position as the DNA fragment with CAP bound near the end, despite the fact

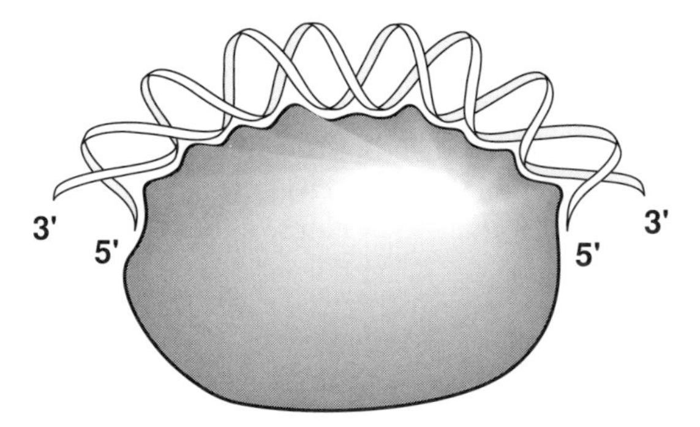

Figure 2.11 DNA bending around proteins. DNA can wrap around a protein as various domains or surfaces of the protein interact with the phosphate backbone or with functional groups in the major (or minor) groove.

that these proteins are of different sizes. This result suggests that the mobility in the gel is a function of the shape of the DNA rather than the size of the protein bound to the DNA.

D. The Biology of DNA Bending

Nature has found many uses for DNA bending. One is the control of access to promoters, the switch regions that turn genes on by bending or looping the DNA. Another is the control of initiation of DNA replication which, since it represents a major commitment for the cell, must be very carefully regulated. A third is site-specific integration of one DNA molecule into another. A fourth is DNA repair. Many chemicals that modify DNA and certain errors of DNA replication introduce bends into DNA. Finally, the organization and compaction of DNA into prokaryotic or eukaryotic cells involves the wrapping of DNA very tightly around DNA-packaging proteins (HU protein and histones). Although the organization of DNA with chromosomal packaging proteins does not involve sequence- or site-specific bends, the tight curvature of the DNA is required for this interaction. This section present examples of DNA bending in biological reactions. Our understanding of the significance of bending and the flexibility of DNA in biology continues to grow.

1. DNA Bending and Gene Regulation

In bacterial cells there are many examples in which DNA bending plays a critical role in the control of gene expression from a promoter. This control may involve the formation of a DNA loop in which the sequences that must be recognized by RNA polymerase are covered by repressor proteins that are bound to DNA stabilizing the loop. Alternatively, the critical RNA polymerase or other DNA binding sequences may be physically unavailable because they exist in a tightly curved loop of DNA that sterically is inaccessible to the protein.

The arabinose operon is one of the best understood examples of the role of bending or looping in gene expression (Schleif, 1992). The arabinose operon encodes three proteins—AraB, AraA, and AraD—that are required for *Escherichia coli* to utilize the sugar arabinose. These genes are not normally expressed in *E. coli*. However, when arabinose is present in the environment, the three Ara proteins are produced. AraC mediates both negative regulation and positive regulation. The regulatory region of the arabinose operon is about 300 bp immediately upstream of the start of transcription (see Figure 2.12). In the arabinose operon, the dimeric repressor protein

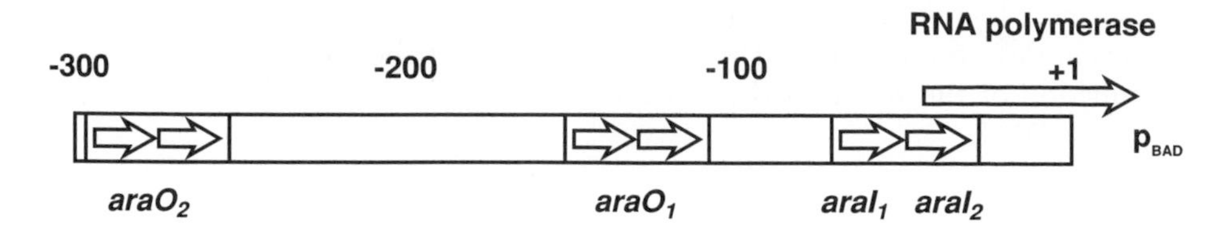
A
RNA polymerase
-300
-200
-100
+1
pBAD
araO2
araO1
araI1
araI2

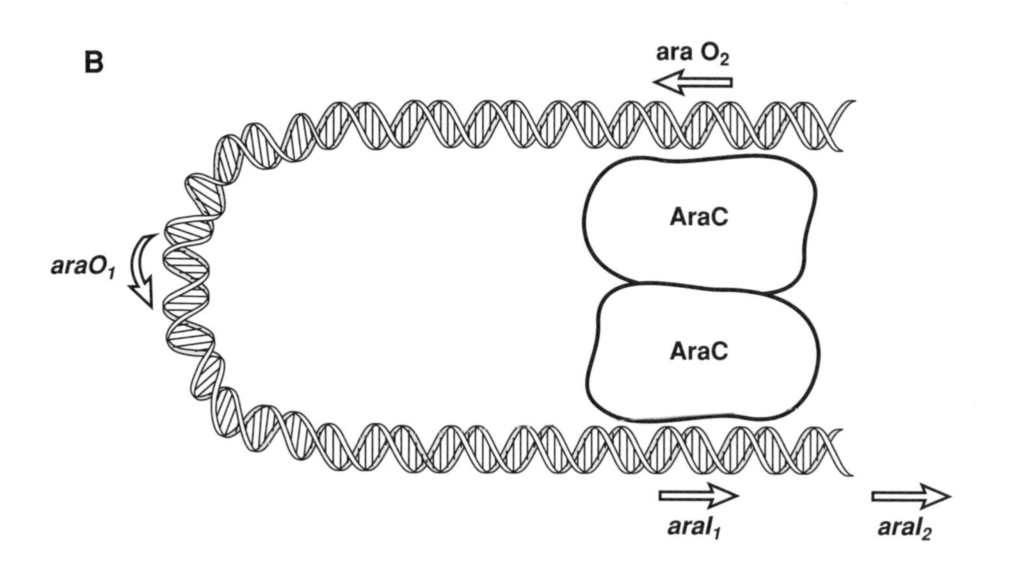
B
ara O2
AraC
AraC
araO1
araI1
araI2

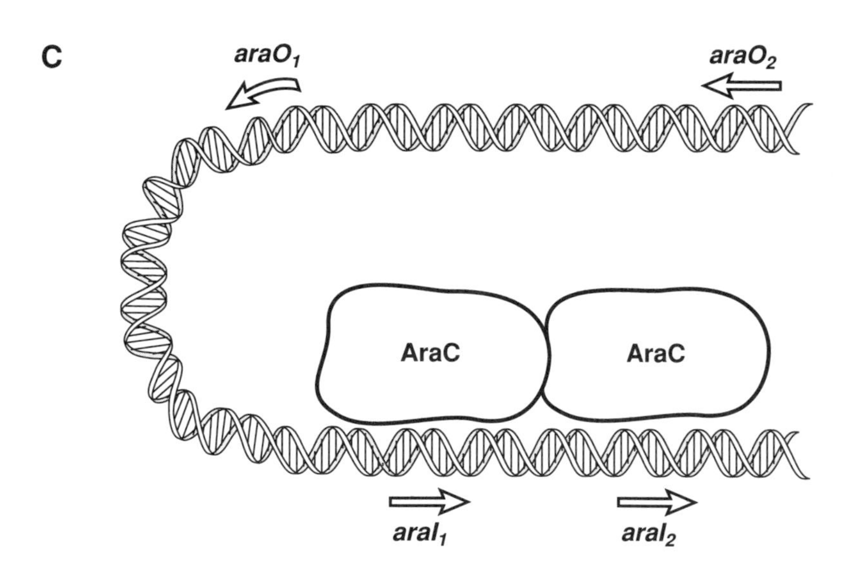
C
araO1
araO2
AraC
AraC
araI1
araI2

AraC binds to three operator sites on DNA: (1) *ara*O_2, which is at position −265 to −294; (2) *ara*O_1, which is at position −110 to 148; and (3) *ara*I, which is at position −4 to −78. The *ara*I site actually contains two separate binding sites *ara*I_1 (position −55 to −78) and *ara*I_2 (position −35 to −4). During repression of the operon, a dimer of AraC binds to the *ara*O_2 and *ara*I operator sites, which are 210 bp apart. For this simultaneous binding to occur, the DNA must form a loop between these two sites as illustrated in Figure 2.12. The formation of this repression loop requires that the DNA be supercoiled. (DNA supercoiling is discussed in detail in the next chapter.) Because the 210-bp length is close to the persistence length of DNA, there is little flexibility in the DNA between the two binding sites. In addition, the two binding sites occur on one "face" or side of the DNA double helix. Assuming a helical repeat of 10.5 bp per turn, binding sites separated by 210 bp are exactly 20 helical turns apart. Bending this DNA will bring the two sites on the same side of the DNA into physical contact. The introduction of 5 bp between *ara*O_2 and *ara*I_1 destroys repression by AraC because it puts the two sites out of phase on different sides of the double helix. Proteins bound to out-of-phase binding sites will not physically interact when the DNA bends. These experiments are explained in Section E2 at the end of this chapter (see Figure 2.18).

To initiate transcription of the arabinose operon, the *ara*I–*ara*O_2 repression loop must be broken. Arabinose, which is utilized by the products of this pathway, acts as an inducer by binding to AraC and inducing a conformational change that causes the release of one subunit of the AraC dimer from the *ara*O_2 site. This subunit of the AraC dimer then binds to the *ara*I_2 site (Figure 2.12). With an AraC dimer bound to *ara*I, the DNA–protein complex recruits RNA polymerase to initiate transcription of the operon.

The CAP protein (cyclic AMP receptor protein), which bends DNA when bound, can also disrupt the repression loop. In the arabinose operon the CAP binding site is located between *ara*O_1 and *ara*I_1. CAP binding may introduce a bend into the DNA that makes the initial repression loop energetically unfavorable (or may destabilize a preformed repression loop).

Figure 2.12 DNA bending in the arabinose operon. (A) The regulatory elements of the arabinose operon include the *ara*O_2 site at the left end of the operon, the *ara*O_1 site at about −120 bp, and an *ara*I_1–*ara*I_2 site at around −30 bp. RNA polymerase begins transcription at bp +1. (B) A repression loop formed by an AraC dimer binding to both *ara*O_2 and *ara*I_1. (C) A transcriptionally active configuration formed by an AraC dimer binding to *ara*I_1 and *ara*I_2. The release by AraC of the *ara*O_2 site is mediated by the binding of the inducer arabinose.

2. DNA Bending and DNA Replication

There are several examples in which natural bends in DNA exist at origins of replication (see Table 2.1). Specific bends can also be introduced by DNA–protein interactions at the origin of DNA replication. These phenomena have been found in both prokaryotes and eukaryotes.

Simian virus 40 (SV40) provides an example in which DNA bending is important in replication. The virus encodes a protein (large T antigen) which, among other things, is essential for DNA replication. Large T antigen binds to a specific site (shown in Table 2.1) at the SV40 origin of replication. The recognition sequence for dimeric binding contains two 5-bp recognition sequences separated by an A tract. This binding site is naturally bent. If the A tract, which does not make contact with the protein, is replaced by "random" bases the origin region is no longer bent and the large T antigen does not bind as tightly (Ryder *et al.,* 1986). This result suggests that a slight bend is required for binding of the large T antigen.

Bacteriophage λ provides another example in which bending is required for DNA replication. Replication from the bacteriophage λ origin requires binding of the O protein to four sites at the origin (the nucleotide sequence of the 19-bp DNA binding site for O protein is shown in Table 2.1). Each of the four binding sites contains two G + C-rich recognition regions separated by a central region containing a run of 3 or 4 As. Moreover, the four binding sites are separated by A tracts. As expected, these A tracts introduce a stable bend into the λ origin (Zahn and Blattner, 1987). Binding of O protein increases the curvature of the origin region at two levels. First, binding of the O protein increases the bending at the short A tract between the two regions of the 19-bp O binding site that make contact with the protein. Second, protein–protein contact between the O proteins folds and organizes the entire origin region into a compact three-dimensional structure called an "O-some."

3. DNA Bending and Site-Specific Integration of Bacteriophage λ

Bacteriophage λ, on infection into *E. coli,* can enter a *lytic* phase in which its DNA is replicated, new viruses are packaged, and the cell bursts releasing new virus particles. At other times, bacteriophage λ can enter a *lysogenic* phase in which the virus slips its genome into the *E. coli* chromosome. This integration into the chromosome is accomplished by *site-specific recombination,* which occurs through the interaction of two specific regions of DNA. These regions are called the *att*P site (the *att*achment site on the *p*hage

Table 2.1
Bend Sites at Replication Origins

Origin	Sequences
Yeast ARS1[a]	111 121 131 141 151 161
	GACAAATGGTGTAAAAGACTCTAACAAAATAGCAAATTTCGTCAAAAATGCTAAGAAATAGGTT
Bacteriophage λ[b]	39031 39041 39051 39061 39071 39081 39091
	GTGCATCCCTCAAAACGAGGGAAAATCCCCTAAAACGAGGGATAAAACATCCCTCAAATTGGGGGAT
	39101 39111 39121 39131 39141 39151
	TGCTATCCCTCAAAACAGGGGGACACAAAAGACACTATTACAAAAGAAAAAAGAAAA
SV40[c]	1 11 21 31
	GCCTCGGCCTCTGCATAAATAAAAAAAATTAGTC

[a]Snyder *et al.* (1986).
[b]Zahn and Blattner (1985).
[c]Dean *et al.* (1987).

chromosome) and the *att*B site (the *att*achment site on the *b*acterial chromosome). Within these *att* sites, specific cuts are made by the bacteriophage λ Int protein and the two molecules are joined together.

The integration host factor (IHF), an *E. coli* protein, is an important component in the integration of the bacteriophage λ chromosome. The binding of IHF to three sites—H1, H2, and H′ on the *att*P sequence—results in the introduction of three 140° bends in the DNA. These bends result in the formation of a specific three-dimensional DNA–protein complex called an "intasome." The *att*B DNA then interacts with this intasome for the cutting and rejoining reactions required for site-specific integration. Bending of the *att*P region by IHF is discussed in more detail in Section E,3 at the end of this chapter.

4. DNA Bending and DNA Repair

When DNA is damaged, mutations can occur. As discussed in Chapter 1, covalent modification of bases can result in a change in their electronic configuration. This will disrupt hydrogen bonding, effectively creating a mispair at the damaged site. If DNA replication occurred using a damaged base in its alternative electronic configuration, a mutation would result. For example, cytosine normally pairs with guanine in its keto configuration. If a chemical modification to guanine forced it into its enol configuration, G would hydrogen bond with T. On the next round of DNA replication the strand containing the T will have undergone a G · C to A · T transition. Thus, it is quite important that cells develop sensitive ways to detect DNA damage and prevent the introduction of mutations in DNA.

Enzymes involved in DNA repair have evolved to sense the shape of the DNA double helix. A bulge resulting from a natural mispair (G · A, for example) or a mispair resulting from DNA damage (C · O6-methyl G, for example) is recognized by various repair enzymes. However, repair enzymes are probably not sensitive to subtle changes in twist or variations in the number of base pairs per turn. Smooth bends in the DNA introduced by phased A tracts or DNA looping by proteins do not seem to trigger attack by DNA repair enzymes.

Several types of damage to DNA introduce bends into DNA. An *intrastrand* cross-link, a covalent link between two bases in the same DNA strand, is one example. Ultraviolet light irradiation at 254 nm results in the formation of pyrimidine dimers in DNA, one type of intrastrand cross-link (Figure 2.13). The pyrimidine dimer introduces a bend of 7–30° into the helical axis of DNA (Husain *et al.*, 1988; Wang and Taylor, 1991). Pyrimidine dimers are recognized by several different repair enzymes. The UvrABC exo-

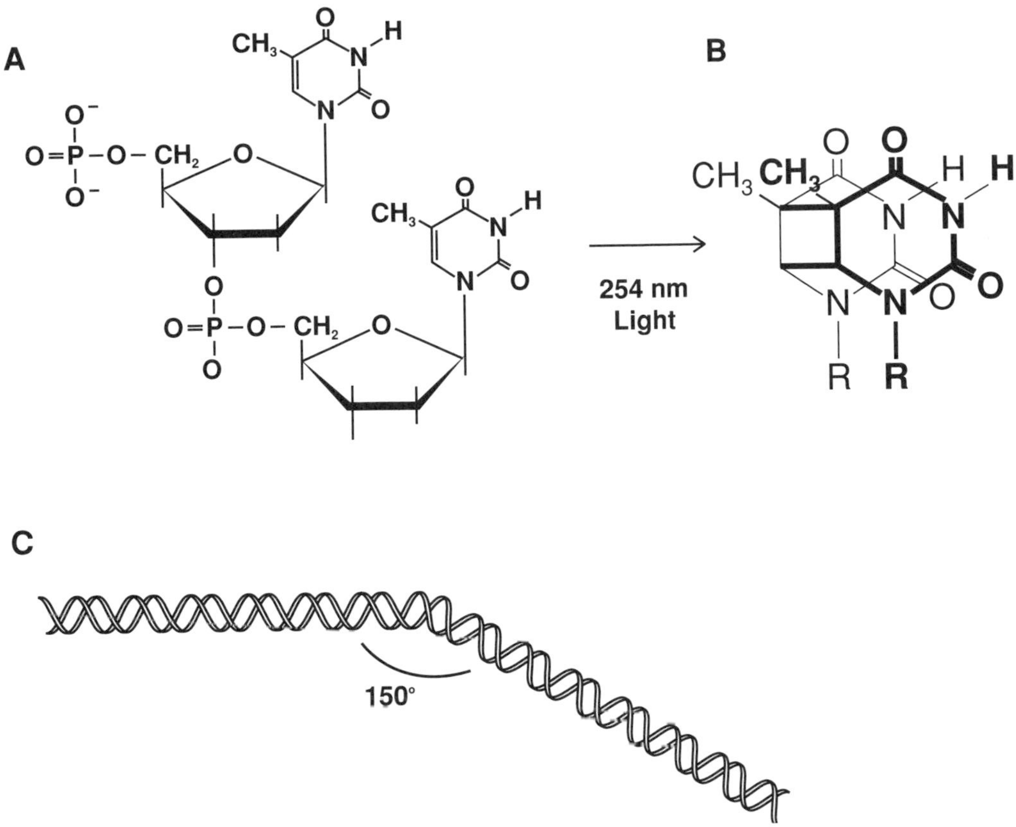

Figure 2.13 **DNA bending introduced by pyrimidine dimers.** (A) Adjacent thymines in one strand of DNA. Dimerization occurs when, on absorption of 254-nm light, the two 5,6 double bonds of adjacent thymines are covalently joined by a cyclobutane ring (B). This event will necessarily remove the 36° twist angle between the two adjacent base pairs and will introduce a kink of about 30° in the DNA (C).

nuclease may recognize the sharply kinked or bent shape of intrastrand cross-linked DNA.

Bulged nucleotides represent another helix anomaly that, if not corrected by DNA repair enzymes, can lead to a mutation. As shown in Figure 2.14, bulged nucleotides occur when there are extra nucleotides in one strand of DNA. During the process of DNA replication, slippage of the template can occur, especially in runs of a single base. Homopolymeric runs are often "hotspots" for naturally occurring addition and deletion mutations. The mechanism by which slipped misalignment can generate additions and dele-

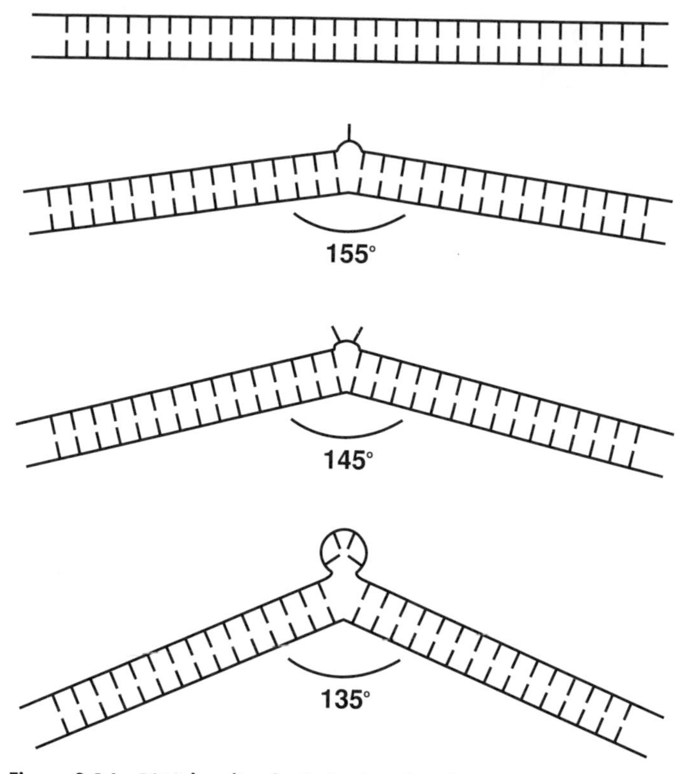

Figure 2.14 DNA bending by bulged nucleotides. A heteroduplex can be formed in which extra nucleotides exist in one strand. A perfectly paired molecule (*top*) is followed by heteroduplexes with 1, 2, and 4 extra nucleotides in the top strand. A single extra nucleotide can exist in an extrahelical structure (as shown) or can be intercalated between two stacked base pairs (as shown in Figure 2.16). A single bulged nucleotide can bend DNA by about 25°. When larger mismatches occur, a loop forms in one strand, and the magnitude of the bend increases.

tions is described in Figure 2.15. A single extra nucleotide in one strand creates only a small perturbation of the structure of the double helix. In some cases, a single extra nucleotide remains "extrahelical" and bulged away from the double helix. In other cases, the extra nucleotide intercalates between two stacked base pairs (Figure 2.16). A single bulged nucleotide can kink the DNA by as much as 25° (Rice and Crothers, 1989). Moreover, as the number of bulged nucleotides increases, the angle of the bend increases (Hsieh and Griffith, 1989). The general feature of DNA kinking introduced by intrastrand cross-links or bulged nucleotides may be the structural feature that is recognized by repair enzymes such as UvrABC nuclease.

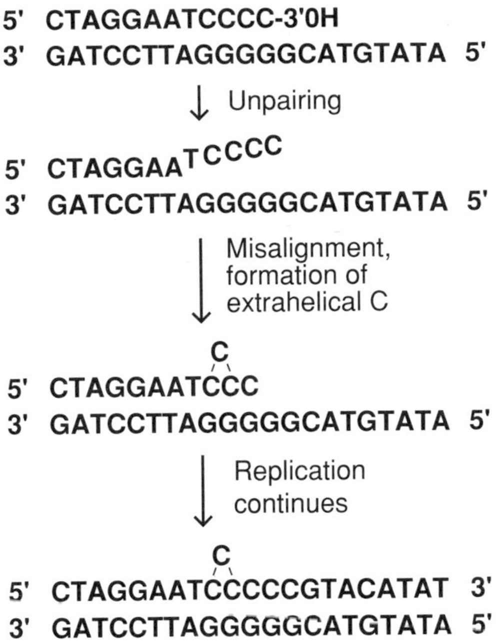

5' CTAGGAATCCCC-3'0H
3' GATCCTTAGGGGGCATGTATA 5'

↓ Unpairing

5' CTAGGAA^{TCCCC}
3' GATCCTTAGGGGGCATGTATA 5'

↓ Misalignment, formation of extrahelical C

C
5' CTAGGAATCCC
3' GATCCTTAGGGGGCATGTATA 5'

↓ Replication continues

C
5' CTAGGAATCCCCCGTACATAT 3'
3' GATCCTTAGGGGGCATGTATA 5'

Figure 2.15 Slipped misalignment during DNA replication. A heteroduplex region, in which extra nucleotides occur in one strand, can result from an error of DNA replication in a process called slipped misalignment. As DNA polymerase replicates several Gs in the template strand, it can dissociate from the DNA allowing the 3′ end of the newly synthesized strand to unpair or "breathe." The 3′ end progeny strand can reform hydrogen bonds but with the Cs and Gs in a different register so one C becomes extrahelical. The end of the progeny strand is said to have "slipped" or misaligned with respect to the template strand. As DNA polymerase rebinds this template and extends the progeny strand, there would be an extra C in the progeny strand. The slipped misalignment may also involve multiple base pairs. In addition, slipped misalignment can occur between direct repeats separated by hundreds of base pairs (see Figure 7.3).

E. Details of Selected Experiments

1. A-Tract Phasing and DNA Bending

To test the idea that DNA bending requires runs of As phased at 10-bp intervals, Hagerman synthesized three different 10-bp oligonucleotides: 5′ GAAATTTC 3′, 5′ GGAAATTTCC, and 5′ GGGAAATTTCCC (alternatively written as $[G_iA_3T_3C_i]$, where $i = 1, 2$, or 3). Since each single strand has inverted repeat symmetry, it is self-complementary. Thus, hybridization of each oligonucleotide (with itself) leads to double-stranded 8-, 10-, and 12-bp molecules:

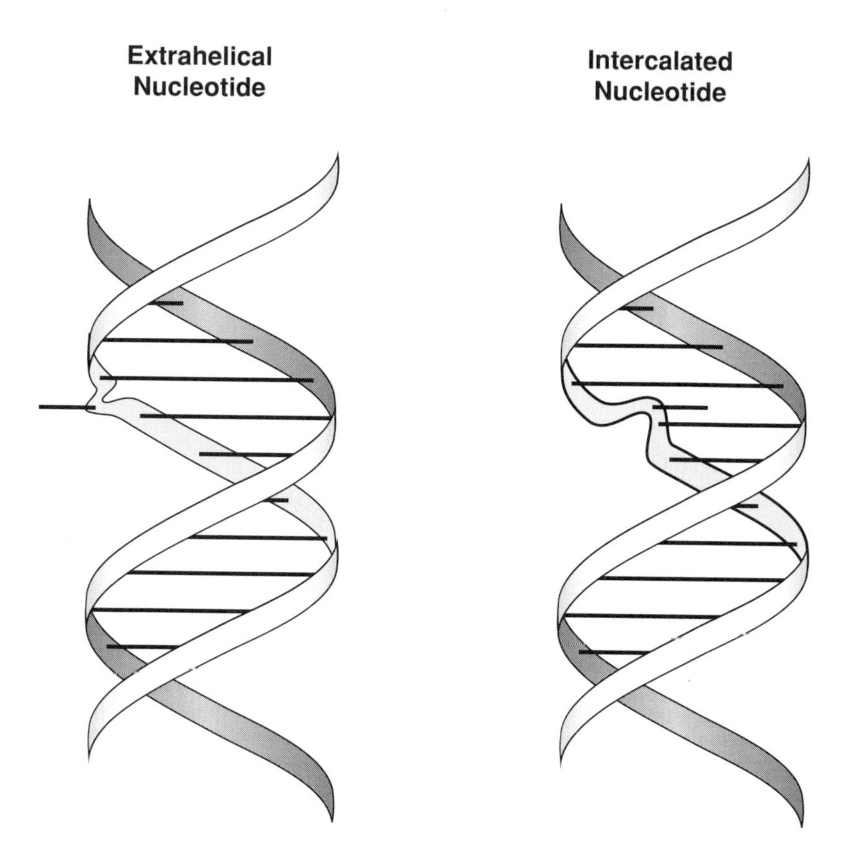

Figure 2.16 Extrahelical and intercalated extra nucleotides. An extra nucleotide in one strand for which no complementary nucleotide exists in the opposite strand can exist in two different configurations. The nucleotide can be extrahelical or extended away from the DNA helix, leading to a distortion of the phosphate backbone at that position and a bending of the double helix. In some sequence contexts, an extra nucleotide can be assimilated into the helix by intercalating the base between two stacked adjacent base pairs. An intercalated nucleotide can have little effect on the overall stability and structure of the double helix.

5′GAAATTTC3′ 5′GGAAATTTCC3′ and 5′GGGAAATTTCCC3′
3′CTTTAAAG5′, 3′CCTTTAAAGG3′, 3′CCCTTTAAAGGG5′

In the presence of DNA ligase (and ATP and Mg^{2+}), the double-stranded oligonucleotides are joined together to form multimers. This creates a family of molecules that, when analyzed by gel electrophoresis, shows a uniform repeating pattern referred to as a "ladder." In this case, the length of each member of the family increases by 8, 10, or 12 bp. For DNA molecules that do not contain a stable bend, the migration through the gel is proportional to the log of the molecular weight. Hagerman used plasmid pBR322 DNA cut with the restriction enzyme *Hae*III as molecular weight markers. Since the sequence of this plasmid has been determined, the exact sizes of the multiple fragments produced by the *Hae*III digestion are known. These DNA fragments, for the most part, migrate true to size on electrophoresis in acrylamide gels. The sizes of these marker bands are indicated to the left of the lanes containing the three experimental samples, designated A, B, and C in Figure 2.17. By examining the bands migrating near the 80- and 89-bp marker bands, the 80-, 88-, 96-, 104-, . . . bp bands for the $(GA_3T_3C)_n$ family (lane A) can be identified. Members of this family run true to size (as straight DNA in the gel). Likewise, the bands for the $(GGGA_3T_3CCC)_n$ family ran as expected for their lengths of 84, 96, 108, 120, . . . bp (lane C). However, the spacing of the bands corresponding to $(GGA_3T_3CC)_n$ in lane B was not that expected for 10 bp. Assigning the band migrating equivalent to the 51-bp band in the marker lane as 50-bp, the next several molecules, which are physically 60, 70, 80, 90, . . . bp in length, migrate as DNAs of about 65, 80, 100, 120, and 145 bp. Multimer number 20 (200 bp) migrates at about 2.3 times its apparent length or the same as a molecule of 460 bp.

The migration of these polymers is plotted in Figure 2.7 as $[bp_{app}/bp_{seq}]$ plotted as a function of length of the multimer. Each of these molecules contained the core sequence GAAATTTC, which consists of bend-inducing A tracts (one in each of the complementary strands of DNA) and identical flanking bases. The distortion in helix structure at each A tract, which leads to the unusually slow migration for the 10-bp family, is present in the 8- and 12-bp families. The only difference between the three molecules is the relative spacing of the A tracts, demonstrating that A tracts must be spaced at 10-bp intervals (not 8 or 12 bp) to observe stably bent DNA.

2. Phasing and DNA Looping

In many cases, the formation of a DNA loop involves two sites on the DNA that are relatively close. If one were to loop a piece of DNA that is less than several hundred base pairs, the two surfaces at the ends of the DNA that would contact one another would constitute a defined area that is restricted by the helical phasing of the DNA. There is not enough flexibility in the rotation of the helix over short distances for all surfaces to have an equal probability of making contact. However, if looping occurs between sites that are thousands of base pairs apart, the helical phasing between the two sites becomes less important because of the greater degree of flexibility inherent in longer pieces of DNA.

The helical phasing of protein binding sites on the DNA can be very important in gene regulation. Experimental data demonstrating the importance

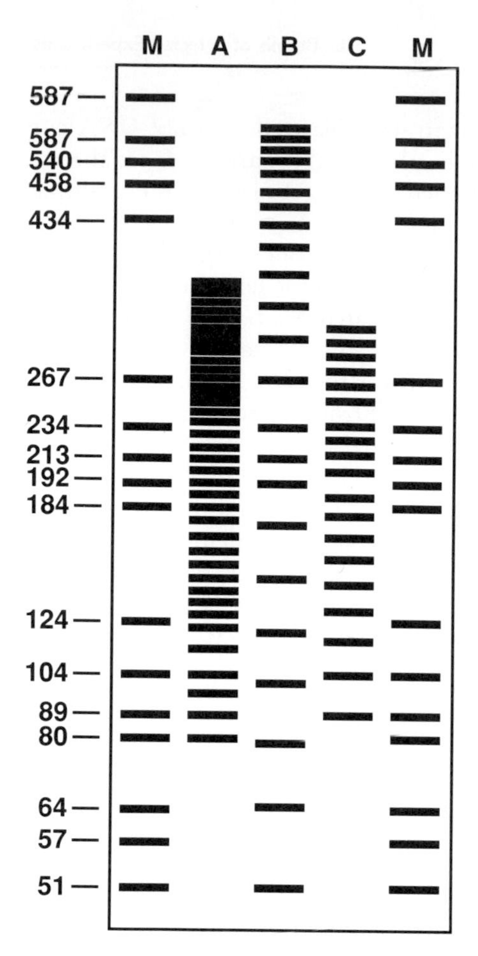

Figure 2.17 DNA bending analyzed by polyacrylamide gel electrophoresis. This representation of a polyacrylamide gel shows one method used to analyze DNA bending. The bands represent the positions of migration of variously sized DNA fragments of different lengths. The outside lanes labeled M represent the positions of marker DNA generated from a *Hae*III restriction digest of plasmid DNA pBR322. The base pair sizes of individual bands are indicated on the left of the figure. Lanes A, B, and C show the positions of migration of three different synthetic oligonucleotides that are ligated into families of varying chain length. Lane A shows molecules resulting from the ligation of the 8-mer GA_3T_3C, lane B represents ligation of the 10-mer $G_2A_3T_3C_2$, and lane C represents ligation of the 12-mer $G_3A_3T_3C_3$. The migrations of molecules in lanes A and C are true to length whereas the bands in lane B do not migrate at their expected positions. For example, the bottommost band at position 51 in lane B represents a 50-bp DNA fragment. The lengths of the bands above this should represent 60, 70, 80, 90, 100, 110, and 120 bp. The 120-bp band migrates as a molecule of near the 192-bp marker, which is an apparent molecular weight much greater than its actual molecular weight. To generate the graph in Figure 2.7, the apparent mobility of 192 bp is divided by the actual length, in this case 120 bp, to generate the (bp_{app}/bp_{seq}) value. The (bp_{app}/bp_{seq}) values are then plotted as a function of actual length of the DNAs. The molecules in B, which are bent, produce a curve with an increasing slope, whereas the molecules shown in lanes A and C produce a horizontal line.

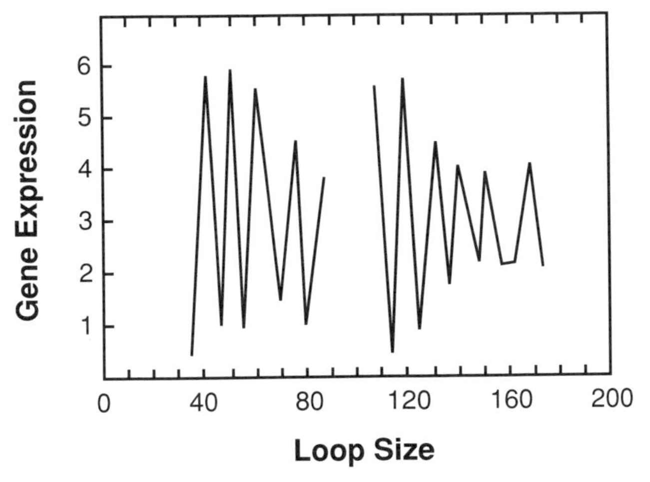

Figure 2.18 Phasing, looping, and gene expression. This figure illustrates the effect of loop size on gene expression in the arabinose operon under conditions of repression (excess AraC and no inducer). The control elements and repression loop of the regulatory region of the arabinose operon are shown in Figure 2.12. The distance between the *araI* and *ara*O_2 sites is critically important for transcription from this promoter. Lee and Schleif (1989) measured the level of gene expression in a number of different constructs in which the length of DNA between *ara*O_2 and *araI* regions varied. As shown here, changing the spacing between these sites by 5 bp had a marked effect on the level of gene expression. When the distance between *araI* and *ara*O_2 was 34 bp there was a very low level of gene expression. When this distance was increased to 38 bp the level of expression increased more than 10-fold. The addition of 5 bp more giving a 43.5-bp spacing, resulted in a 6-fold decrease in the level of gene expression. The addition of complete helical turns allows a high level of gene expression whereas the addition of half a helical turn drastically decreases the level of expression in the arabinose operon. This result demonstrates a 10-bp phasing required for arabinose gene expression that can be explained by examining the structures in Figure 2.12. In Figure 2.12B, the interaction of AraC requires that the two DNA binding sites are on a face of the helix that can come in contact when the DNA bends. If an additional 5 bp were introduced between the *ara*O_2 and *ara*I_1 sites, when the DNA bent one AraC binding site would now be on the opposite site of the DNA helix. Consequently, contact could not be made at the *ara*O_2 site by an AraC dimer bound the first AraC 250 bp away at the *ara*I_1 site. (Data from Lee and Schleif, 1989.)

of helical phasing and looping is shown in Figure 2.18 for the arabinose system. Processes controlled by the interaction of two relatively closely spaced proteins on DNA should exhibit a 10.5-bp periodicity. When in phase, the process should occur readily. When out of phase, the process should not occur or should occur at a slower rate. Data from Lee and Schleif (1989) for transcription from p_{BAD} are shown in Figure 2.18. When the *ara*O_2 and *ara*I sites are moved from 34 to 170 bp apart, a 10.5-bp helical phasing is observed. The addition of half a helical turn (5 bp) relative to optimally oriented sites reduces expression. A periodicity of gene expression due to changing the helical phase relationship of two control sites in the DNA of a promoter has been observed for several other systems including the *lac* operon.

3. Determining the Three-Dimensional Organization of the λ Site-Specific Recombination Complex

The organization of DNA binding sites for Int, IHF, and other proteins in the λ *att*P site is shown in Figure 2.19. Three IHF binding sites are bent by about 140° when IHF is bound. Since it would be difficult to interpret changes in mobility when dealing with three bends in a DNA molecule, Snyder *et al.* (1989) analyzed changes in mobility of DNA containing pairs of IHF bend sites. Two molecules were constructed in which each contained about half the *att*P region. One molecule consisted of the left half of the *att*P site containing IHF sites H1 and H2. The second consisted of the right half of the *att*P site containing H2 and H′.

The Crothers′ laboratory demonstrated that the angle between two bends could be determined by the electrophoretic mobility of DNA. The rationale for this experiment is illustrated in Figure 2.20. Migration is a function of the end-to-end distance of a DNA molecule. In Figure 2.20A, the molecule with two bends on the opposite sides of the helix will migrate more rapidly on a polyacrylamide gel as shown in Figure 2.20B. The molecule with the bends on the same face or side of the DNA double helix will migrate most slowly. Molecules with intermediate angles of rotation at the second bend site will migrate with an intermediate mobility. Therefore, determining the spacing at which electrophoretic mobility is the slowest identifies a molecule in which two bends are in the same direction.

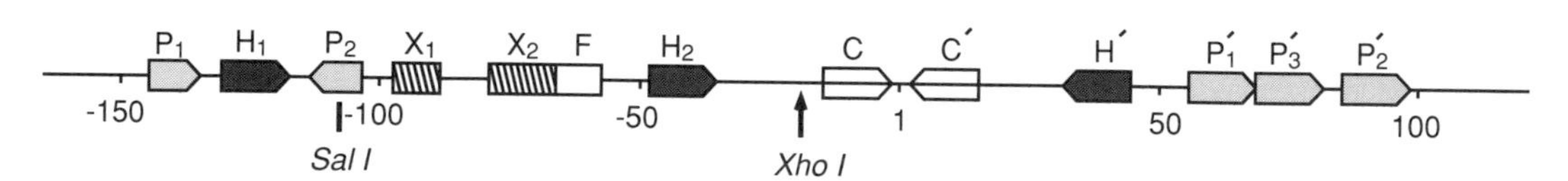

Figure 2.19 Sequence elements of the bacteriophage λ *attP* site. Several different sequence elements occur within the 250-bp bacteriophage λ *attP* site. (This is the site in bacteriophage λ at which integration into the *attB* site in the host chromosome occurs.) The blocks marked H1, H2, and H′ represent sequences where IHF, the integration host factor, binds. The binding of IHF to these sites introduces a bend of 140° at each site. This bending (shown in Figure 2.22) precisely positions other DNA sites within the *attP* region into a three-dimensional configuration in which excision and integration reactions occur. X_1 and X_2 are binding sites for the Xis protein, C is a binding site for the Int protein, and F is a binding site for the Fis protein. Sites labeled "P" represent various transcriptional promoters. The *SalI* and *XhoI* sites used for cloning in the "spacer DNA squences" are indicated.

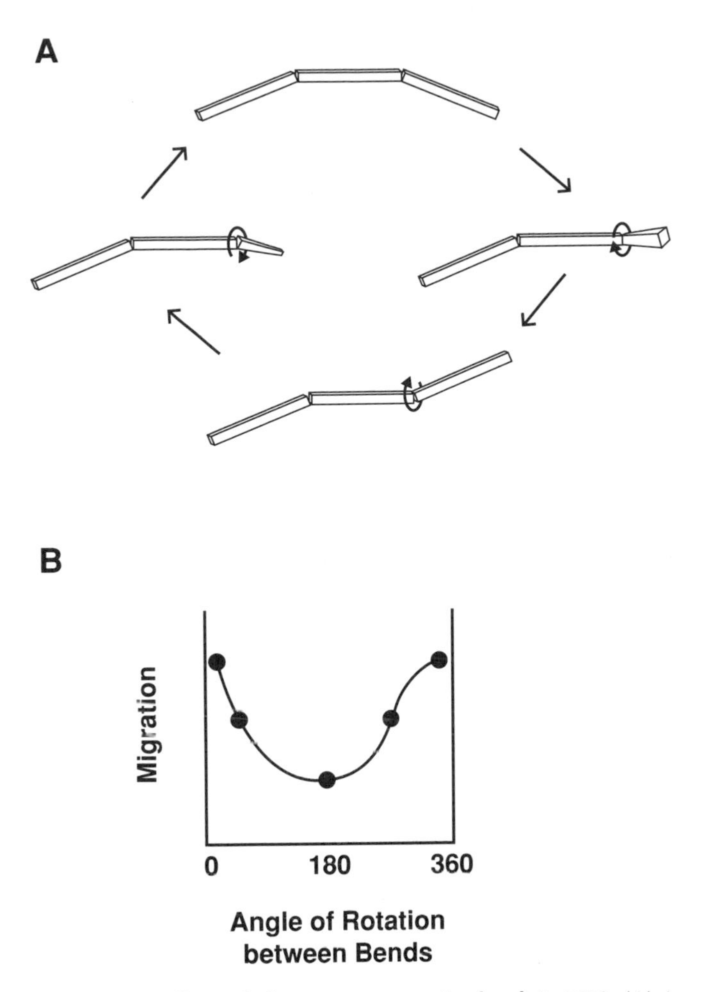

Figure 2.20 Effects of phasing two successive bends in DNA. (A) A molecule that contains two bends. The left-most bend is in the same position in all molecules whereas the second bend is rotated by 90° in each of the four different molecules. The angle of rotation between the two bends significantly influences mobility during gel electrophoresis. In the top molecule, in which the bends are pointing in the same direction in the plane of the page, the bends are *phased;* this molecule migrates anomalously slowly. As the second bend is rotated 90°, 180°, or 270° away from the direction of the first bend, the two bends are no longer in phase and the migration of these molecules increases. The relative angle between two bends can be determined by varying the spacing between two bends. (B) A graph is obtained in which the rate of migration is plotted as a function of the angle between the bends (which can be changed by adding spacer DNA between the two bend sites). The molecule that migrates most slowly identifies the one in which two bends are most closely in phase.

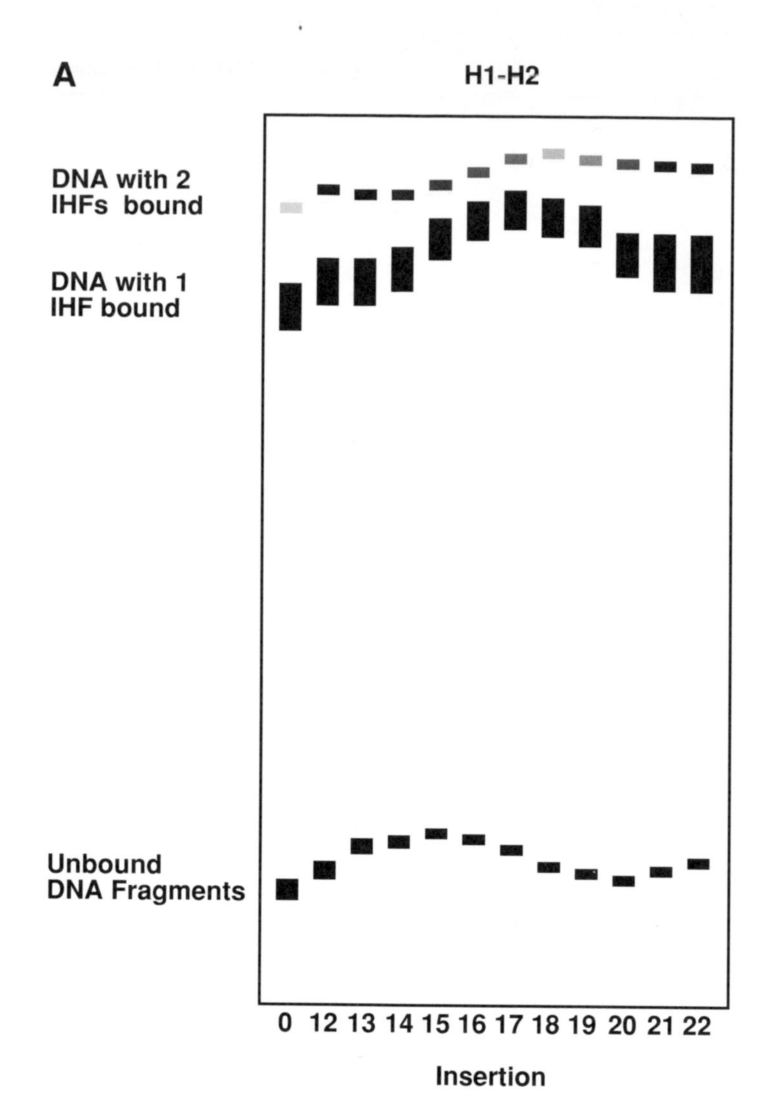

Figure 2.21 Determining the spatial organization of the λ *att*P–IHF complex. (A) An idealized gel, described by Snyder *et al.* (1989), in which IHF protein was added to a DNA fragment containing the left half of the *att*P site, which includes the IHF binding sites H1 and H2. A piece of spacer DNA of 12–22 bp was inserted between the H1 and H2 binding sites, changing the angle of rotation between the two IHF binding sites in increments of about 36°. The position of unbound DNA fragments with no IHF bound is shown at the bottom of the gel. The sinusoidal curve in the migration during gel electrophoresis of the unbound DNA as a function of the spacer length indicates that this DNA is slightly bent. Migration of the molecule is fastest with no insert or with a 20-bp insert. With a 15-bp insert, which changes the angle between the two bends by 514° (or 1.5 helical turns), mobility is slower. (B) The top graph shows an analysis of the relative mobility of DNA containing no IHF bound. In this case the presentation is slightly different than shown in previous figures (2.2, 2.20). Here the relative mobility is plotted as a function of migration so 0.8 represents a molecule that migrates most slowly, whereas a value of 1.2 represent a molecule that migrates most quickly. This periodicity of migration demonstrates that the two natural bends in the DNA are on opposite sides of the site of the linker insertion and are phased by about 180°.

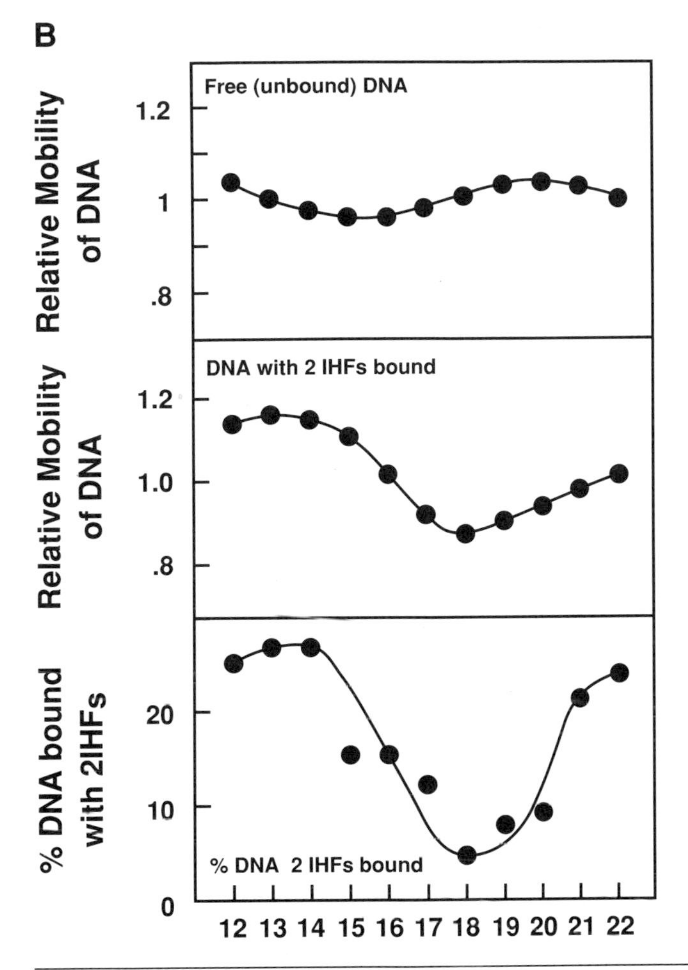

One or two IHF molecules bound to the sequence drastically reduce the electrophoretic mobility of the DNA, predominantly because of the size and geometry of the protein–DNA complex. Two bends flank the linker insertion site, as evidenced by the sinusoidal mobility of the family of molecules. DNA with the single IHF protein bound was most bent with the insertion of 17 bp. The top line in A, which represents a smaller fraction of the total population, shows the mobility of DNA with two IHF molecules bound. The insertion of an 18-bp spacer produces the slowest migrating or most bent doubly bound DNA (center graph). With no insert or with the 18-bp spacer, there was very little of the doubly bound species. With the introduction of an additional 5 bp between IHF bind sites, much more of the doubly bound species was present (bottom graph). One interpretation of this result is that when IHF binds to the H1 and H2 sites in the molecule with the 18-bp spacer, there is steric clash of the DNA sequences flanking the IHF binding sites. From these experiments, the natural angle of bending between the IHF sites was calculated as described in text. Adapted with permission from *Nature* (Snyder *et al.,* 1989). Copyright (1989) Macmillan Magazines Limited.

Snyder *et al.* (1989) inserted "spacers" between the H1 and H2 sites (in the *Sal*I site) and between the H2 and H′ sites (in the *Xho*I site) in each half of the *att*P DNA molecule. The length of the spacers varied in increments of 1 bp from 12 to 22 bp. Thus, these researchers varied the relative angle of rotation of the two IHF bend sites in increments of about 34° (360° ÷ 10.5). Using the set of molecules in which the insert varied from 12 to 22 bp, the phasing of the two IHF binding sites moved through one complete helical turn of DNA. IHF was added to the family of half *att*P sites and the mobility of the DNA–protein complexes was analyzed by gel electrophoresis. Three bands were observed: "naked DNA" or unbound molecules, DNA with a single IHF bound, and DNA with two IHF molecules bound (one to each IHF binding site). The pattern of migration for the H1–H2 family is shown in Figure 2.21A. Several important conclusions can be drawn from these results. First, the *att*P DNA with H1 and H2 sites is naturally bent, as evident by the sinusoidal mobility of the unbound DNA as the spacer length was increased. The molecule with the 15-bp spacer migrated most slowly, indicating that this was the most bent molecule of the set. If the 15-bp spacer made the molecule maximally bent, then the natural *att*P sequence (without the 15-bp insert) would be relatively straight. The natural *att*P sequence must contain two natural bends flanking the *Sal*I site that are out of phase by about 180°. Second, the majority of the DNA migrated as *singly bound molecules*. The migration of the DNA containing one bound IHF molecule was much slower than that of the unbound molecule because the protein bound to the DNA increases the molecular weight of the complex and because the geometry of the molecule changes. The singly bound species also showed a periodicity in gel mobility. The H1–H2 molecule with a 17-bp spacer migrated most slowly, indicating that this molecule was the most bent. Third, the amount of DNA migrating as the *doubly bound species* was reduced relative to the amount of DNA containing a single IHF molecule. In addition, the amount of doubly bound DNA varied as the spacer length changed. With a spacer length of 18 bp, very little doubly bound *att*P DNA was observed. The significance of this observation is discussed subsequently.

From analysis of the mobility and the amounts of the various bands in this experiment (Figure 2.21A), conclusions can be drawn about the directionality of the bends and, therefore, the three-dimensional organization of the *att*P DNA–protein complex. In Figure 2.21B, the relative mobility of unbound DNA, the relative mobility of doubly bound DNA, and the percentage of DNA with two IHFs bound are plotted as a function of the size of the spacer (or insertion). There is an inverse relationship between the extent of bending and the amount of the doubly bound species. The most bent molecule had the least doubly bound species. The doubly bound DNA molecule which was the most bent was the one with the 18-bp spacer, suggesting that

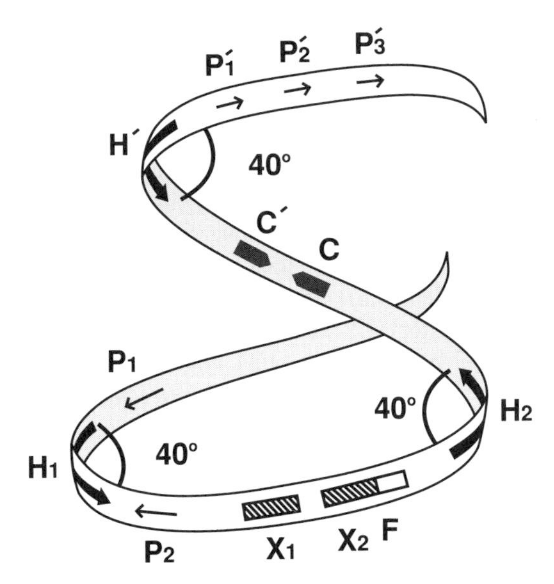

Figure 2.22 Three-dimensional organization of the λ *att*P region. The three-dimensional organization of the λ *att*P region was determined from the experiments described in Figure 2.21. IHF introduces a 140° bend at each of the three binding sites. This bending organizes the DNA into this rather tight spiral configuration. There are numerous proteins bound to the specific binding sites in this DNA molecule, as indicated in Figure 2.19. This large three-dimensional DNA–protein structure is important in the precise alignment of Int nuclease binding sites (c) with the *att*B site on the bacterial chromosome. During integration of bacteriophage λ, the DNA is cut at the Int binding sites (c →← c) between two of the 140° bends, as shown. Ligation of these ends to ends generated from specific cuts made in the bacterial DNA results in the insertion of λ DNA into the *E. coli* chromosome. Adapted with permission from *Nature* (Snyder *et al.*, 1989). Copyright (1989) Macmillan Magazines Limited.

the 18-bp spacer places the two 140° IHF-induced bends on the same face of the double helix. If the two bends are on the same side of the DNA helix, the two "arms" of DNA flanking both bend sites may be in physical contact with each other. This steric clash may be physically and energetically unfavorable. Moreover, this steric constraint may be responsible for the reduced amount of doubly bound species with the 18-bp spacer. Removal of the arms flanking the bend should relieve this steric constraint, allowing IHF to bind to both H1 and H2 sites. The authors reported, as expected, that removal of 97 and 54 bp from the left and right arms, respectively, increased the amount of doubly bound molecules observed.

Using the information from the experiment shown in Figure 2.21 (and similar data for the H2–H′ right half of the *att*P site) the authors calculated the angle between the three bend sites H1, H2, and H′ in the natural sequence. In addition, they proposed the three-dimensional shape for the *att*P site shown in Figure 2.22. From the analysis of the doubly bound DNA molecules, it was evident that an 18-bp spacer placed the two IHF-induced bends on the same face of the helix. Knowing this, the relative angle between these bends in the natural *att*P sequence could be calculated by multiplying the average twist angle (34.3°) by the number of base pairs inserted between the IHF binding sites ($18 \times 34.3° = 617.4°$). Since the helix repeat has a 360° per unit repeat interval (360°, 720°, 1080,° etc.), an introduced angle of 617.4° that restores the bends to a unit repeat interval (720°) means that the natural bends are offset by 102.6° or almost 3 bp worth of twist angle. This allows determination of the three-dimensional path of the DNA in space, as shown in Figure 2.22.

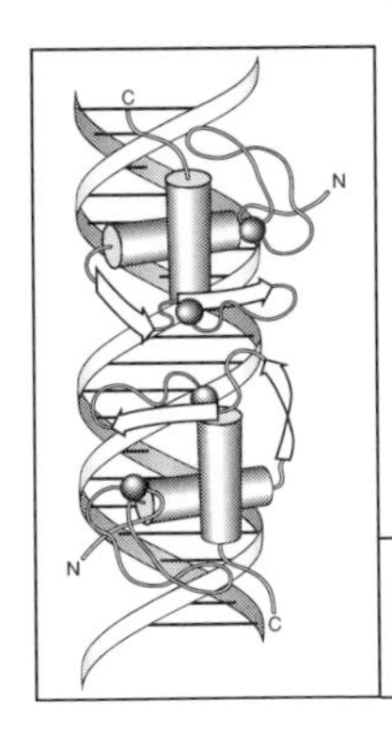

CHAPTER 3

DNA Supercoiling

A. Introduction

Anyone who has twiddled with a rubber band and contemplated the many contorted structures that arise intuitively understands DNA supercoiling. This analogy will be used to illustrate the physical principles of supercoiled DNA and the complexity of shapes that can result from supercoiling. As shown in Figure 3.1, the rubber band can exist in its normal form as a circular *relaxed* structure. Alternatively, grasping the rubber band on opposite sides between the thumb and forefinger of each hand, pulling the hands slightly apart, and twisting one side of the rubber band results in the introduction of right-handed (clockwise) turns in one part and left-handed (counterclockwise) turns in the other part of the rubber band. Note that grasping the rubber band separates the molecule into two *topological domains* that, when twisted, have the opposite handedness of twisting. When the fingers are brought close together, each twisted region of the rubber band coils around itself forming a *supertwist* or a *supercoil.* Careful inspection of the rubber band shows that the right-handed twisted region supertwists in a left-handed fashion and the left-handed twisted region supertwists in a right-handed fashion. This process of *supertwisting* requires an input of energy since it is not the preferred low energy state of

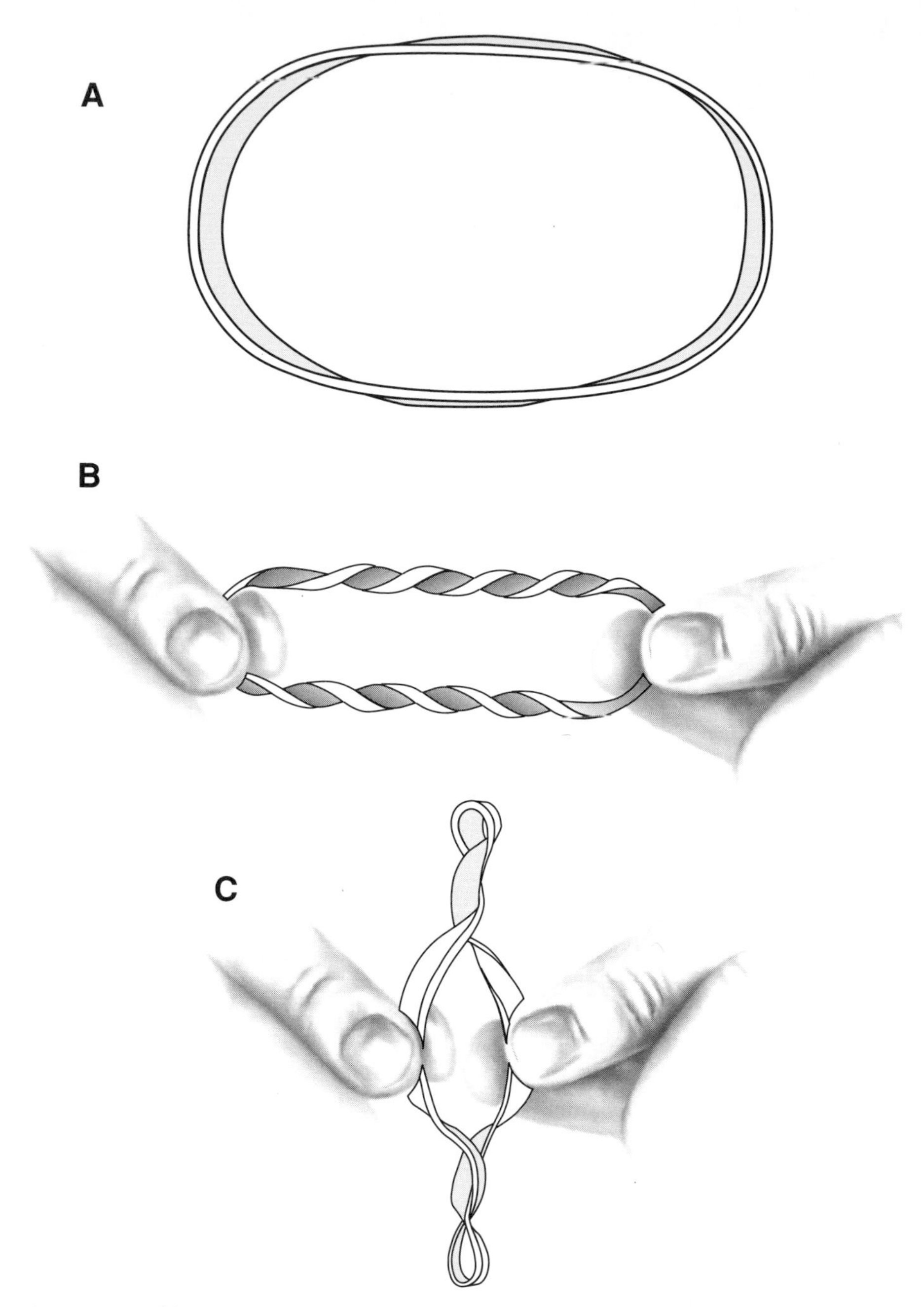

Figure 3.1 Rubber band model for DNA topology. (A) A rubber band in a relaxed form. (B) The band is held firmly between the thumb and forefingers of each hand following twisting by one hand. Left-handed (counterclockwise) turns have been introduced into the top section of the band. Right-handed (clockwise) turns have been introduced into the bottom section of the band. (C) When the hands are brought together, the rubber band forms supertwists. The top section of the band, which contained left-handed turns, forms a right-handed supertwist. The bottom section of the band forms a left-handed supertwist.

existence of the rubber band. When released from the torsional constraints on the rotation of the rubber band (constraints from the thumb and forefinger), the rubber band should return to its original configuration. (Sometimes a little prodding or shaking may be necessary if the rubber band has become organized in a high-energy metastable state.)

In twiddling with the rubber band it should be evident that the basic physical structure of the rubber band has not changed. It consists of a circular band with two flat sides. On the other hand, by twisting the band one introduces the supertwisting or *writhing* of the band. The twisting and writhing can change in relationship to one another, whereas the overall topology of the band (the fact that it is circular with two distinct sides) must remain constant (unless the band breaks).

The structure of DNA becomes more intriguing when one considers that the DNA in most organisms normally exists in a *supercoiled* form. Like the rubber band, supercoiled DNA can exist in a wide variety of conformations with variations in twisting and writhing. This chapter will describe DNA supercoiling, the ramifications of supercoiling on the structure and energetics of the helix, and the biological consequences of supercoiling.

B. Heterogeneity in Forms of a DNA Molecule

A DNA molecule can exist in a number of different forms or topological configurations. A simple plasmid molecule purified from a bacterial cell will exist as a naturally occurring *covalently closed circular* DNA molecule (sometimes called cccDNA) that is negatively supercoiled. Covalently closed means that the two phosphodiester backbones are intact or covalently continuous. The negatively supercoiled plasmid DNA was originally called *form I DNA*. Supercoiled DNA appears as a compacted molecule in which the DNA helix has wound about itself (Figure 3.2). If a covalently closed DNA molecule contains a single nick in one of the strands, the nick provides a swivel and supercoils are lost. The nicked (or open) circular molecule has been called *form II DNA*. Nicked DNA does not contain the supercoils or supertwists of the helix and will appear as a circular ring with the electron microscope (Figure 3.2). Nicked DNA could be called "relaxed" since there are no supercoils. However, the term *relaxed DNA* should be reserved for a covalently closed circular molecule that contains *no* supercoils. As seen with the electron microscope, relaxed DNA will be indistinguishable from nicked DNA. DNA that contains breaks in both phosphate backbones at the same point (or nearly the same point) along the helical axis will form a *linear DNA* molecule.

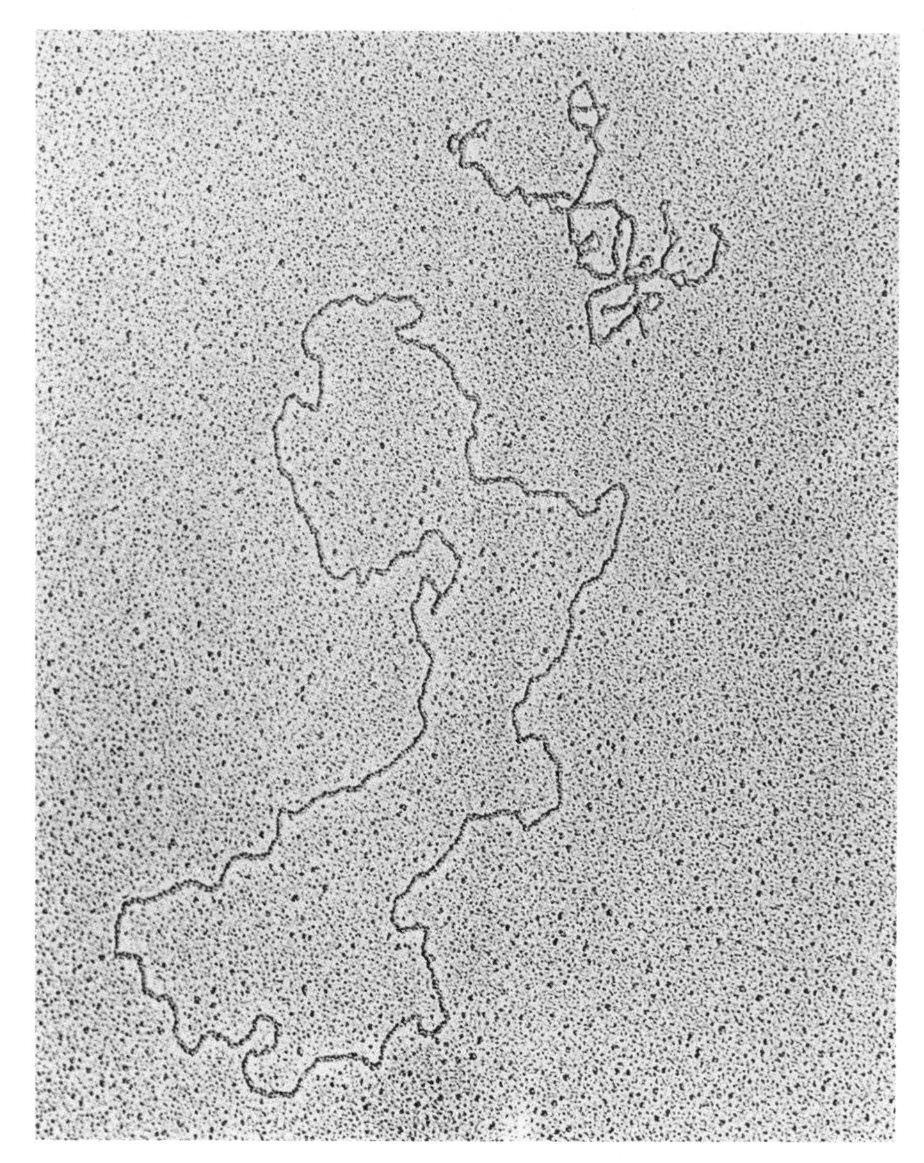

Figure 3.2 Electron micrograph of two forms of DNA. The tangled, twisted molecule is supercoiled DNA, originally called Form I DNA. When circular molecules are relaxed (or nicked) (Form II DNA), they lose the twists. A linear molecule (not shown) is called Form III. The plasmid molecules shown are 9000 bp in length. Courtesy of Jack D. Griffith.

C. Supercoiled Forms of DNA

1. $L = T + W$

The behavior of DNA is much like that of the rubber band just described. One topological property of a DNA molecule is its *linking number, L*. L is defined as the number of times one strand crosses the other when the DNA is made to lie flat on a plane. L must be an integer. The *relaxed* DNA molecule shown in Figure 3.3A has a linking number $L = 20$. The linking number cannot change unless the phosphodiester backbone is broken by chemical or enzymatic cleavage. Although L cannot change, the number of twists or turns of the double helix, as well as the number of supercoils or writhes of the double helix, *can* change. The topology of DNA is described by the simple equation:

$$L = T + W \tag{1}$$

where L, the linking number, is as defined; T is the number of helical turns in the DNA; and W is the writhing number of DNA, which describes the supertwisting or coiling of the helix in space. Cozzarelli *et al.* (1990) provide an in-depth mathematical analysis of the geometry and topology of DNA supercoiling. Scovell (1986) provides a simpler explanation of DNA supercoiling.

The introduction of a *negative supertwist* (a right-handed coil) into the DNA molecule with $L = 20$, shown in Figure 3.3B, changes the value of W by -1 ($W = -1$). Considering Eq. (1), since L cannot change ($L = 20$), T must increase by $+1$ to a value of 21. Conversely, a decrease in the twist number by -1 to a value of $T = 19$ would require a compensating introduction of a left-handed or *positive supertwist* ($W = +1$), as shown in Figure 3.3C. DNA is a dynamic molecule that can exist in myriad states with different values of helical turns and supercoils. This structural heterogeneity results from the flexibility in the winding of the DNA helix and the bendability of the helix to allow the supertwisting or coiling of the helix. Although the linking number is invariant and must be an integer, the number of twists can vary in positive and negative increments with offsetting negative and positive changes in the writhe number.

2. Relaxed DNA

The DNA helix in solution will adopt a preferred helical repeat that is a function of the base composition (as briefly mentioned in Chapter 1). On average, the helical repeat is about 10.5 bp per helical turn of DNA. In linear DNA, in which the ends of the molecule are free to rotate, the DNA will adopt this preferred helical repeat of 10.5 bp per turn. This helical repeat will

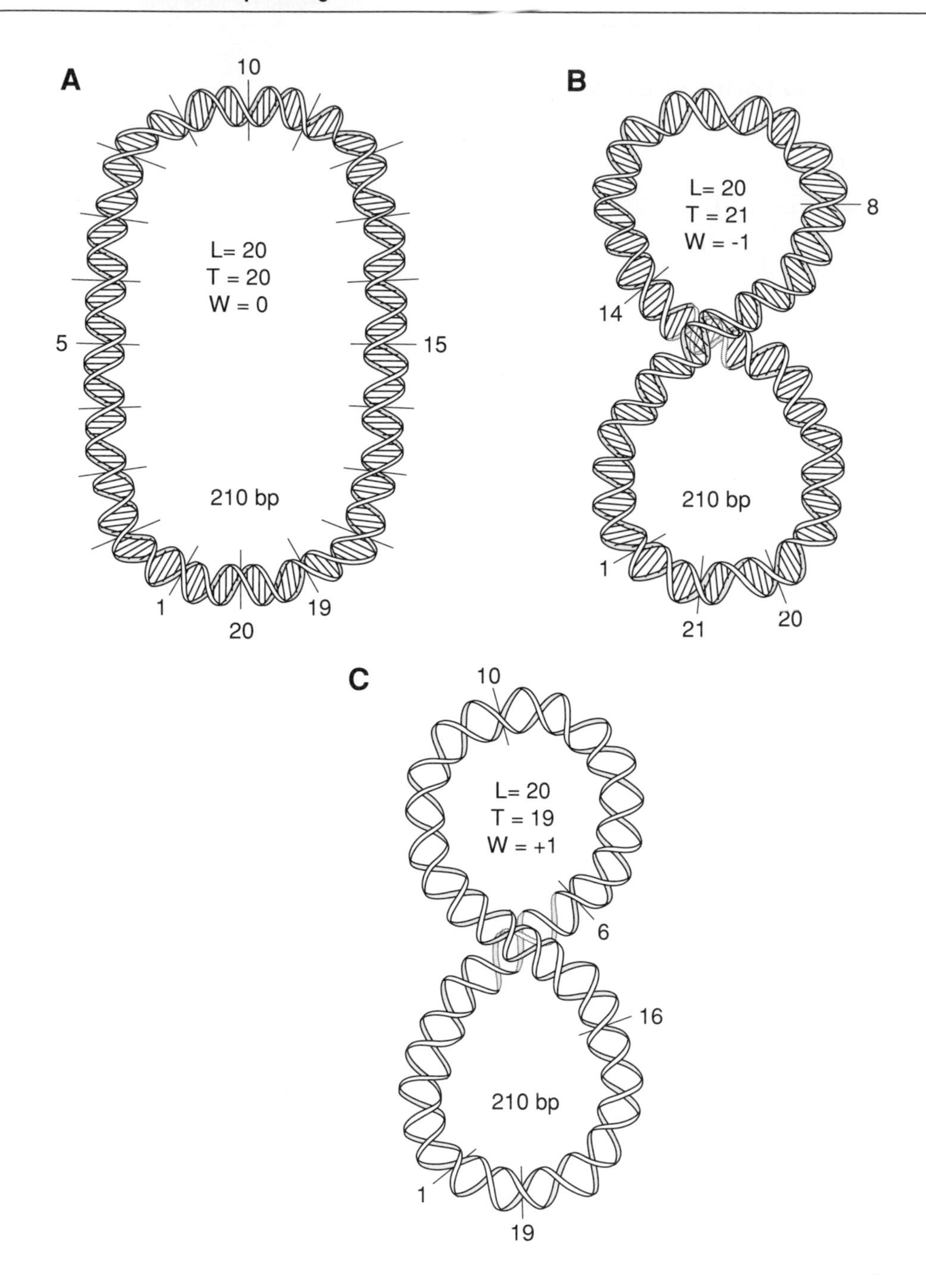
A
10
L= 20
T = 20
W = 0
5
15
210 bp
1
19
20
B
L= 20
T = 21
W = -1
8
14
210 bp
1
21
20
C
10
L= 20
T = 19
W = +1
6
16
210 bp
1
19

also exist in nicked DNA in which the nick provides a swivel around which one strand can rotate about the other.

The preferred helical repeat of a linear DNA molecule represents the lowest energy form of the molecule. When this state of helical twist exists in a covalently closed molecule, the molecule is *relaxed* and contains no supercoils. In the relaxed DNA shown in Figure 3.3A, the *linking number* equals the *twist number* ($L = T = 20$) and $W = 0$. The linking number of relaxed DNA, L_0, is defined as

$$L_0 = N/10.5 \quad (2)$$

where N is the number of base pairs in the DNA molecule and 10.5 refers to the helical repeat. The DNA molecule shown in Figure 3.3 has 210 bp; thus, there are 20 helical turns in the DNA ($L_0 = 20$). One strand crosses the other 20 times in the plane of the page of this book. This linking number will be the same whether the DNA molecule exists in a linear, nicked, or covalently closed relaxed form.

To reiterate, by definition, covalently closed DNA is relaxed if $L = L_0$ (where L refers to the linking number of the DNA molecule). This may seem an obvious and unnecessary point to make but, as you will see, changes in temperature and salt concentration affect L_0 and introduce a new twist into the DNA structure story.

3. Negatively Supercoiled DNA

DNA is said to be negatively supercoiled when it has a deficiency in the linking number compared with relaxed DNA, or $L < L_0$. Negatively supercoiled DNA is *underwound* because it has fewer helical turns than the molecule would contain as a linear or relaxed molecule. This underwinding in the number of helical turns results in more base pairs per helical turn. This results in a decrease in the angle of twist (or the rotation per residue) between adjacent base pairs. Underwinding, therefore, creates *torsional tension* in the winding of the DNA double helix.

The topology of negatively supercoiled DNA is illustrated in Figure 3.4. In the 210-bp DNA molecule shown in Figure 3.4A, there are 20 helical turns

Figure 3.3 Relationship between twist and writhe. (A) A covalently closed, relaxed DNA molecule. The DNA is 210 bp long and contains 20 helical turns (210 ÷ 10.5). The lines drawn perpendicular to the helix axis represent individual base pairs. In this example $L = 20$, $T = 20$, $W = 0$. (B) The introduction of a right-handed supertwist or writhe into the DNA results in a change of writhe ($\Delta W = -1$). A right-handed supertwist is a *negative* supercoil. Since L is invariant (in the absence of topoisomerase action), twist must change by +1 so $T = 21$. (C) A left-handed (or *positive* supercoil) supertwist is introduced into DNA ($\Delta W = +1$). Then the number of helical turns must decrease by −1 so $T = 19$.

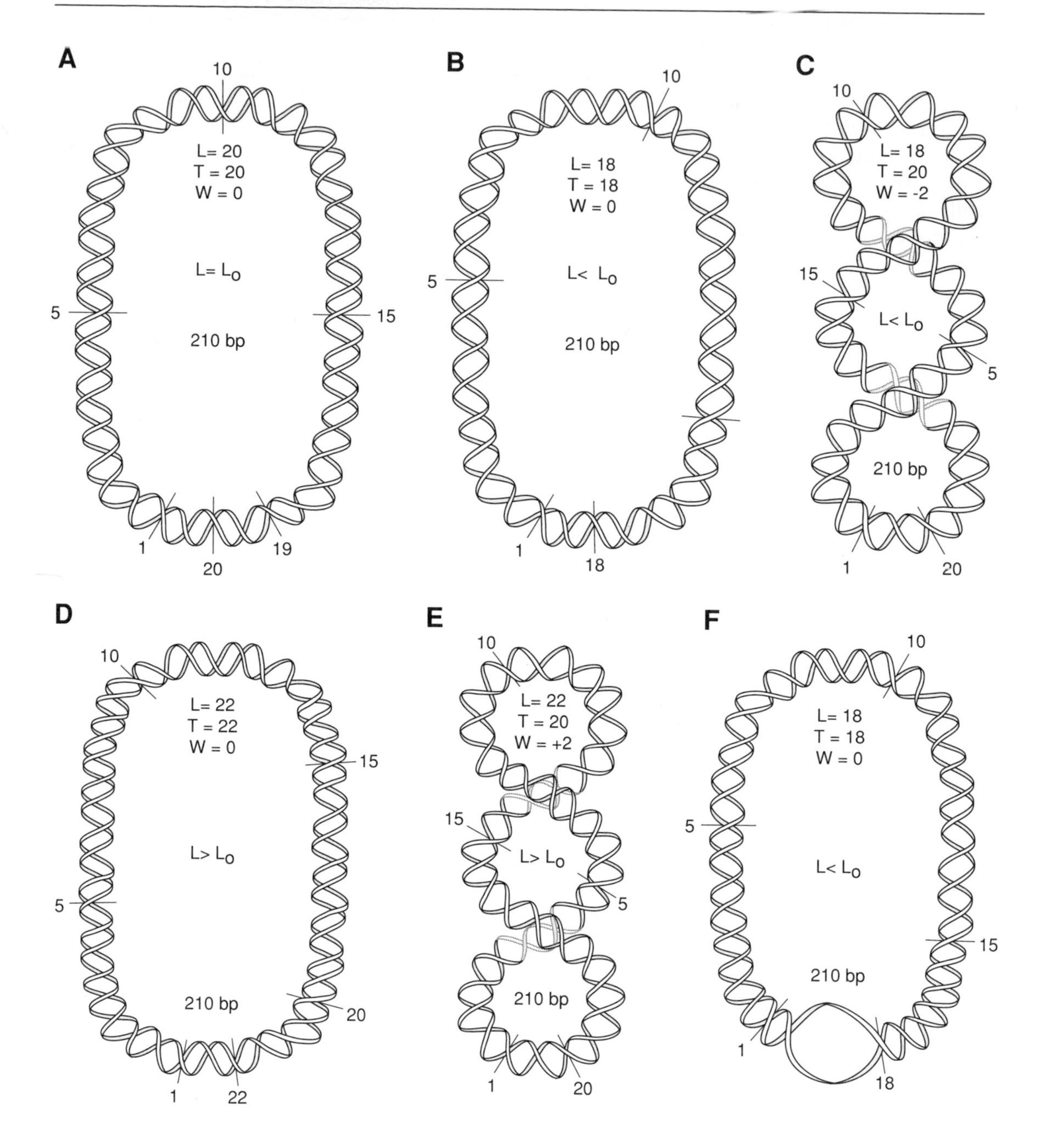
A
10
L= 20
T = 20
W = 0
L= L_o
5
15
210 bp
1
20
19
B
10
L= 18
T = 18
W = 0
5
L< L_o
210 bp
1
18
C
10
L= 18
T = 20
W = -2
15
L< L_o
5
210 bp
1
20
D
10
L= 22
T = 22
W = 0
15
L> L_o
5
210 bp
20
1
22
E
10
L= 22
T = 20
W = +2
15
L> L_o
5
210 bp
1
20
F
10
L= 18
T = 18
W = 0
5
L< L_o
15
210 bp
1
18

in relaxed DNA and $L_0 = L = 20$. In the molecule shown in Figure 3.4B with $L = 18$ there are only 18 helical turns. This will change the average rotation per residue from 34.29°[(20 × 360°)/210] to 30.86° [(18 × 360°)/210]. Energetically, this represents an unfavorable winding of the DNA double helix.

The molecule in Figure 3.4B is negatively supercoiled by definition since $L < L_0$, but it does not contain the familiar supercoil shown in Figure 3.4C. In Figure 3.4B the linking number deficit is manifested as a twist deficit. The torsional strain in the winding of the DNA, which comes from the decreased twist angle of DNA, drives the supertwisting of the DNA into the familiar picture of supercoiled DNA shown in Figure 3.4C. Negatively supercoiled DNA forms a right-handed (or clockwise) supercoil. Since L_0 does not change ($L_0 = 20$), the DNA should be supercoiled by the amount; $\Delta L = (L - L_0) = -2$. In forming two negative supercoils ($\Delta W = -2$), two additional helical turns are wound into the helix ($\Delta T = +2$) with no change in L. One may view the driving force for forming the supercoils as the underwinding in the number of helical turns and the resulting torsional strain inherent in the decreased angle between adjacent bases in the winding of the double helix. Anthropomorphically, the chemical desire of the DNA helix to adopt its "classical B-form" (which, in general, is energetically the most stable form of DNA) drives the writhing or supertwisting of the DNA double helix.

An excellent model system to illustrate this property is a 2- to 3-ft length of rubber tubing. If the two ends of the rubber tube are twisted two full turns and then joined (using a tubing adaptor), the circular tube should form two supercoils as the tube twists back to its original state. Laying the tube flat in its relaxed, circular form and coloring the top blue and the bottom red will illustrate the preferred twist of the tube. When the linking number is changed and no supercoils are allowed to form, the colors will twist. By

Figure 3.4 Introduction of negative and positive supercoils into DNA. (A–C) The introduction of negative supercoils into DNA. (A) A relaxed 210-bp DNA molecule. In this molecule $L = 20$, $T = 20$, $W = 0$. (B) The linking number has decreased by 2 ($L = 18$, two helical turns have been removed by the action of a topoisomerase). This results in an increase in the number of base pairs per turn, a decreased angle of rotation, and $L < L_0$. (C) Two negative supercoils are introduced into this molecule, resulting in $L = 18$, $T = 20$, $W = -2$. (D–E) The introduction of positive supercoils into DNA. (D) Two additional helical turns are wound into the DNA, resulting in an increase in the linking number to $L = 22$. This results in fewer base pairs per turn, an increased angle of rotation, and $L > L_0$. The introduction of two positive supercoils, $W = 2$, allows $T = 20$ (with $L = 22$). (F) The result of unwinding 21 bp in a negatively supercoiled molecule. This results in a relaxation of two negative supercoils. Since negatively supercoiled DNA is underwound (fewer helical turns than preferred by B-form DNA), unwinding 21 bp redistributes the 18 helical turns over fewer base pairs. This creates a more "relaxed" situation (where $T = L_0$). Denaturation is thermodynamically more favorable in negatively supercoiled DNA than in relaxed or positively supercoiled DNA.

introducing two supercoils, the colors will again exist on either side of the tube.

4. Positively Supercoiled DNA

Most DNA isolated from natural sources is negatively supercoiled. However, DNA can exist in a positively supercoiled form when it has a greater linking number than relaxed DNA, or $L > L_0$. Positively supercoiled DNA is *overwound* in terms of the number of helical turns. This overwinding creates a situation in which there are fewer base pairs per helical turn, resulting in an increase in the winding angle between adjacent base pairs. This creates torsional tension in the winding of the DNA double helix (as found in negatively supercoiled DNA).

Overwinding the 210-bp DNA molecule shown in Figure 3.4A by two turns to $L = 22$ creates a situation in which the average rotation per residue changes from 34.29° in relaxed DNA to 37.71° [(22 × 360°)/210] (Figure 3.4D). This, like the change in winding in negatively supercoiled DNA, represents an unfavorable state of the DNA double helix. The tension in the winding of the helix is relieved by the positive supercoiling of the DNA forming a left-handed (counterclockwise) supercoil, as shown in Figure 3.4E. As two positive supercoils form ($\Delta W = +2$) two helical turns are removed, returning the number of helical turns to $T = 20$, which is the preferred conformation for the DNA double helix.

Positively supercoiled DNA has been isolated in what appears to be a naturally occurring form. A bacteriophage-like plasmid molecule from a *Sulfolobus* species, an archebacterium living at high temperature and low pH, has been shown to contain positive supercoils (Nadal *et al.*, 1986). Positively supercoiled DNA would resist unwinding of the helix by heat and acid. Packaging DNA in a positively supercoiled form may be one mechanism for protecting the genetic information from denaturation.

D. Supercoils: Interwound or Toroidal Coils?

Supercoils in DNA need not physically exist as *interwound* supercoils as seen with an electron microscope and shown in Figure 3.5A. Physically, negative supercoils can exist as left-handed toroidal coils, as shown in Figure 3.5B. Initially it may appear that the requirement for the DNA double helix to cross itself, a condition required for supercoiling, is not met. However, toroidal coils topologically satisfy the requirement for W. Although in a toroidal coil the helix does not cross itself in the fashion of an interwound su-

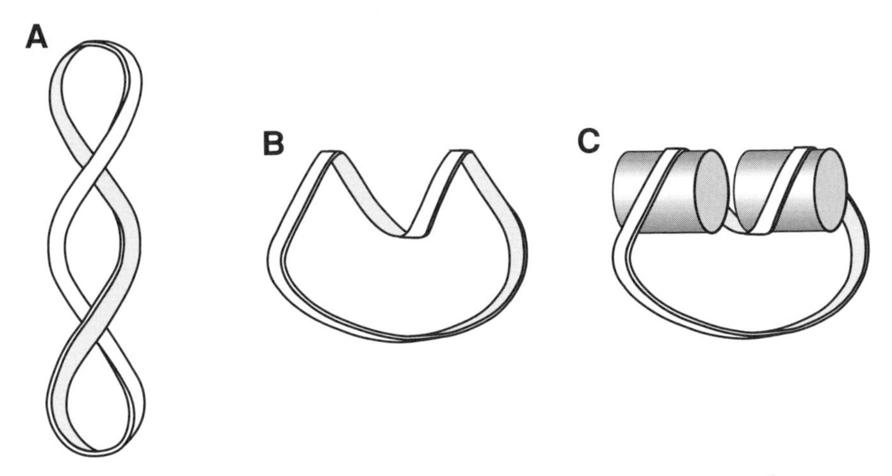

Figure 3.5 Different forms of writhing. (A) A molecule containing two interwound negative supercoils. However, supercoils can also take the form of a toroidal coil. (B) In this example, the two interwound right-handed supercoils (of A) are isomerized into two left-handed toroidal coils. (This can be modeled with a rubber hose.) In biological systems, DNA is frequently wrapped around proteins. (C) The toroidal coils of B are associated in a stable fashion around proteins represented by the cylinders. In this figure, DNA is illustrated as a rubber band rather than as the helix representation shown in Figures 3.3 and 3.4. The edges of the rubber band represent the phosphodiester backbones.

percoil, it does cross itself in the plane of the toroidal coil. In solution, supercoils are likely to be distributed in part by a decreased angle of twist and a mixture of interwound and toroidal coils. For DNA in solution, about 70% of the deficiency in linking number may be distributed as a writhe change and about 30% as a change in twist (Boles *et al.*, 1990).

Toroidal coils of DNA are very important biologically. The organization of DNA in nucleosomes in eukaryotes and in DNA–HU complexes in bacteria involve the toroidal coiling of the DNA around proteins. The DNA is maintained in a stable supercoiled form in which the energy of supercoiling is *restrained* in the stable writhing around protein, as shown in Figure 3.5C. The structure of DNA in nucleosomes and the restraint of supercoils are discussed in detail in Chapter 9.

E. DNA Knots and Catenanes

A circular DNA molecule can fold into a knot. This requires the introduction of a double-stranded break and the wrapping of the DNA around itself before resealing the double-stranded break. Two knotted structures are shown in Figure 3.6A. Two circular DNA molecules can also link together to form catenanes (or interlocked rings), as shown in Figure 3.6B. Knots and catenanes can have a negative or positive sign, depending on the order of the strand crossings or *nodes*. The convention for determining the sign of nodes is described in Figure 3.6. DNA knots and catenanes are found *in vivo* as the products of certain topoisomerases and enzymes involved in site-specific recombination (Spengler *et al.*, 1985; Wasserman and Cozzarelli, 1986; Adams *et al.*, 1992). For example, catenanes can arise at the end of a round of DNA replication in a circular molecule. In addition, the high concentration of DNA in *Escherichia coli* or in a eukaryotic nucleus and the presence of topoisomerases that allow "strand passing" reactions lead to extensive catenation.

F. Superhelical Density and the Specific Linking Difference of DNA

Certain terms and relationships are important for understanding DNA supercoiling. *Superhelical density* is a term that is used very frequently. Superhelical density, σ, is the average number of superhelical turns per helical turn of DNA.

$$\sigma = 10.5\ \tau/N \tag{3}$$

where τ was originally described as the titratable number of superhelical turns,[1] N is the number of base pairs in the molecule, and 10.5 represents the average number of base pairs per turn. τ rather than W is used for historical reasons (see Footnote 1 below). Since there are now many ways to measure the number of supercoils (electron microscopy, two-dimensional agarose gels, as well as the original solution titration methods), τ will be defined as the number of measurable supercoils.

The specific linking number difference, σ_{sp}, is a term that is also used to describe the level of superhelical density:

$$\sigma_{sp} = (L - L_o)/L_o \tag{4}$$

σ_{sp} is very similar to σ. However, mathematically a slightly different value is usually obtained. σ_{sp} refers to the inherent *specific linking difference* and the

[1]Before the use of $L = T + W$, the similar equation $\alpha = \beta + \tau$ was used to describe supercoiling. α is the linking number, β is the duplex winding number, and τ the number of superhelical turns. Essentially τ and W can be used interchangeably in the equation $\alpha = \beta + \tau$ (see Cozzarelli *et al.*, 1990).

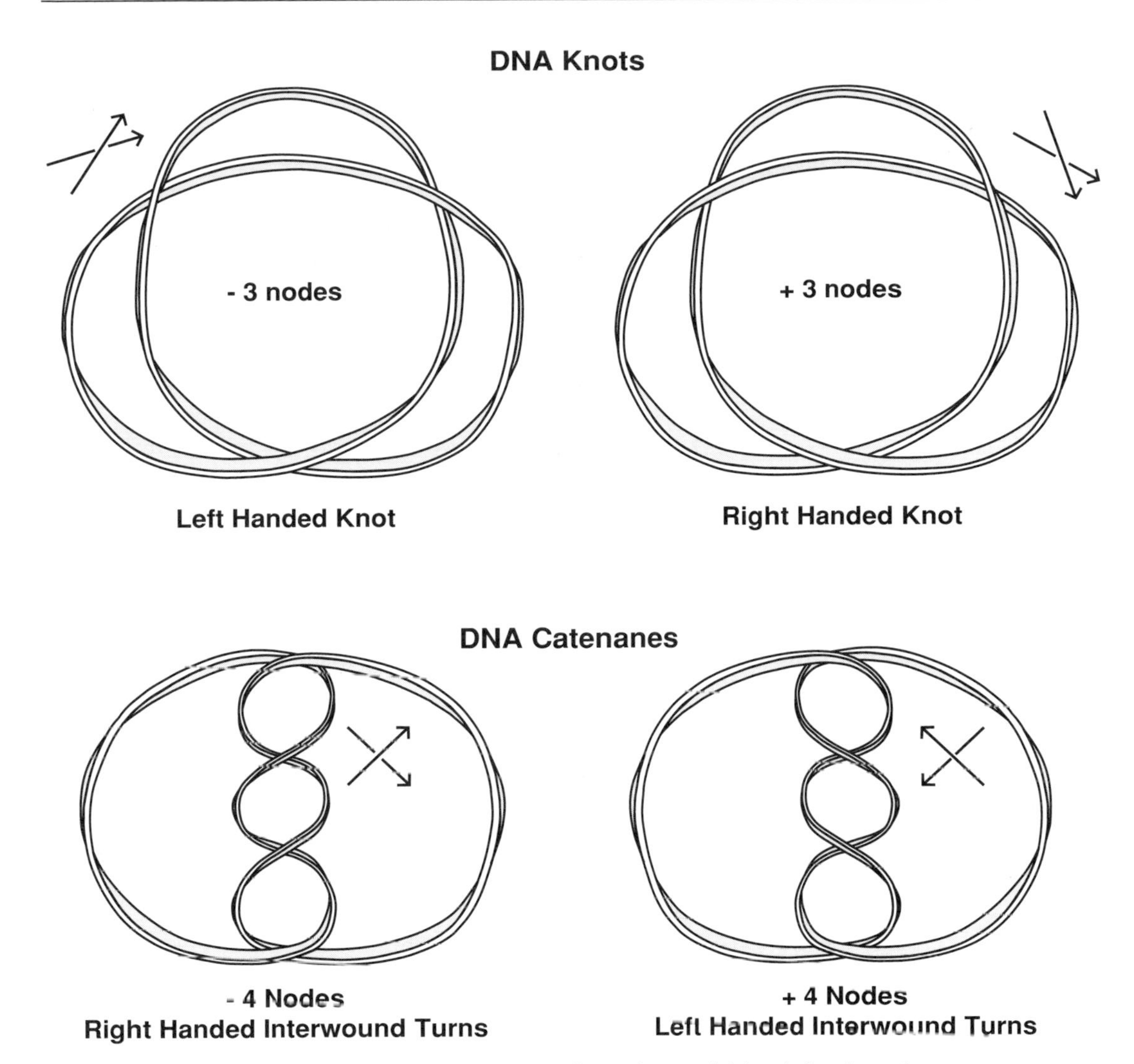

Figure 3.6 DNA knots and catenanes. (Top) Two forms of DNA knots, a left-handed and a right-handed knot. The *nodes* (places where one helix crosses the other) can be either positive or negative. By convention, if an arrow drawn along the *top strand* is rotated $< 180°$ in a clockwise direction to align with an arrow drawn in the same direction along the bottom double helix of the node, then the node is negative. If the top arrow must be rotated counterclockwise $< 180°$ to align with the bottom arrow, the node is positive. (This convention of assigning signs to nodes applies as well to the negative and positive supercoils shown in Figure 3.3B and 3.3C.) (Bottom) Left-handed and right-handed antiparallel catenanes with four nodes. The positive nodes are formed by left-handed (positively supercoiled) interwound turns, whereas the negative nodes are formed by right-handed interwound turns.

value of σ_{sp} can actually be different from that of σ. If stable DNA secondary structures have formed, the superhelical density, σ can be quite different from σ_{sp}. In this special case σ better describes the energetic characteristics of the DNA molecule.

Table 3.1 lists the values of L, T, W, σ, and σ_{sp} for the examples shown in Figure 3.4. (In this example σ and σ_{sp} are identical. Note that if a different helix repeat number, 10.4 for example, were used, or if the molecule were 215 bp in length, σ and σ_{sp} would not be identical.)

G. The Energetics of Supercoiled DNA

The free energy of supercoiling is proportional to the square of the linking number difference in the DNA:

$$\Delta G = (1100\ RT/N)(L - L_o)^2 \tag{5}$$

where R is the gas constant and T is the temperature in degrees Kelvin. N, as before, is the number of base pairs in the DNA molecule. Supercoiled DNA contains a large amount of free energy that can be used to drive biological reactions. Any reaction that lowers the free energy of supercoiled DNA will be thermodynamically favored. The biological processes of transcription and DNA replication require an input of energy to open or unwind the DNA double helix to expose the chemical identity of the bases at the center of the helix. Some of the energy for these process comes from the free energy of supercoiling. The unwinding of the DNA double helix is thermodynamically favored in supercoiled DNA. Unwinding 21 bp of DNA in the supercoiled DNA shown in Figure 3.4C will remove two superhelical turns. Now 18 helical turns are distributed over 189 bp of DNA (an average of 10.5 bp per turn) and the 189 bp of DNA are essentially in a relaxed state. This unwound topoisomer is shown in Figure 3.4F.

Table 3.1
Topological Parameters for DNAs Shown in Figure 3.4

	DNA molecule					
Topological parameter	A	B	C	D	E	F
L	20	18	18	22	22	18
T	20	18	20	22	20	18
W	0	0	− 2	0	+ 2	0
$(L - L_0)$	0	− 2	− 2	+ 2	+ 2	− 2
σ	0	− 0.1	− 0.1	+ 0.1	+ 0.1	− 0.1
σ_{sp}	0	− 0.1	− 0.1	+ 0.1	+ 0.1	− 0.1

H. The Binding to Supercoiled DNA by Intercalation

Certain planar aromatic chemicals bind to DNA by *intercalation.* In this process, the compound inserts between two stacked base pairs of DNA and positions itself within the center of the DNA double helix. The intercalated molecule is stabilized by hydrophobic stacking interactions with adjacent base pairs. The chemical structures of several intercalating compounds (ethidium bromide, 4,5′,8-trimethylpsoralen, and chloroquine) are shown in Figure 3.7. Examples of intercalation are shown in Figure 1.22 for psoralen binding to DNA and in Figure 2.16 for the intercalation of an unpaired nucleotide into a heteroduplex.

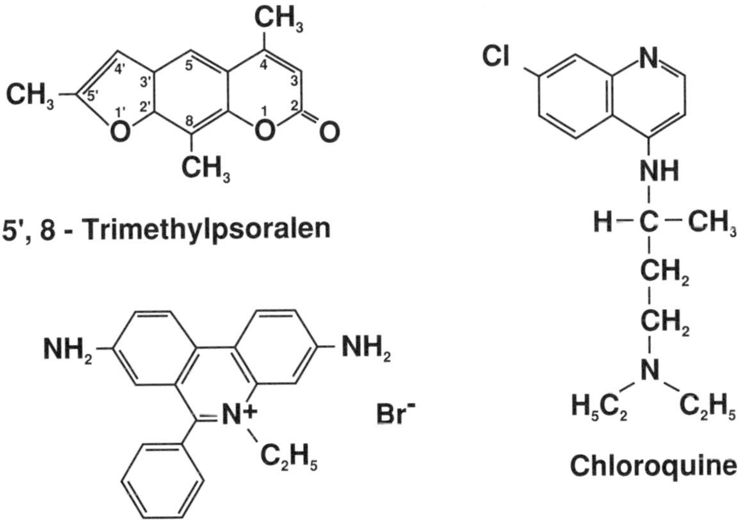

4, 5', 8 - Trimethylpsoralen

Chloroquine

Ethidium Bromide

Figure 3.7 Intercalating molecules. Many planar molecules bind to DNA by intercalation. In this process the flat, planar molecule slides between the stacked base pairs. Three common intercalating drugs are shown here. Ethidium bromide binds tightly to DNA. When irradiated with 360-nm ultraviolet (UV) light, ethidium bromide that is bound to DNA fluoresces bright red. This fluorescence is commonly used to visualize DNA in agarose or acrylamide gels or in cesium chloride (CsCl) density gradients. Chloroquine, an antimalarial drug, is used to unwind DNA and relax supercoils for topological analysis on agarose gels. It binds much more weakly than ethidium bromide. 4,5′,8-Trimethylpsoralen, and other members the psoralen family, bind weakly to DNA. However, on irradiation with 360-nm UV light, psoralens form cyclobutane rings between the 3,4 pyrone and 4′,5′ furan double bonds and the 5,6 double bond of pyrimidine bases. This results in the formation of monoadducts to single strands and, in some cases, interstrand cross-links (Figure 1.22). Psoralens are quite permeable to both prokaryotic and eukaryotic membranes and have been widely used for a number of *in vivo* studies on DNA structure.

Binding by intercalation results in several changes in the shape and flexibility of DNA. The binding of an intercalating drug necessarily results in a lengthening of the DNA helix, since two adjacent base pairs must physically separate to accommodate the intercalated molecule. In addition, intercalation results in an *unwinding* of the DNA double helix. The unwinding is required to allow the physical separation of the adjacent base pairs.

The unwinding from the binding of intercalating drugs relaxes negative supercoils. The unwinding angles for ethidium bromide and psoralen are about 26° and 28° per intercalated molecule, respectively. It takes 12.8 intercalated molecules to unwind the helix by 360° or remove one turn of the DNA double helix with an unwinding angle of 28°. Unwinding from intercalation lowers the value of L_0. Therefore, as an increasing number of molecules intercalates into DNA, the value of L_0 decreases. Since supercoiled DNA contains a deficiency of helical turns ($L < L_0$), a reduction in L_0 will reduce the level of negative supercoiling. At some point ($L = L_o$) the DNA will become relaxed. In addition, for a drug with a high binding constant, as the level of bound drug continues to increase eventually L_o will be lower than L and the DNA will become positively supercoiled.

The binding of intercalating drugs is proportional to the negative superhelical density of the DNA. Supercoiled DNA, as discussed earlier, represents an unfavorable state of winding of the DNA double helix. Any event that relieves this unfavored state will be thermodynamically favored. The free energy inherent in supercoiled DNA therefore favors intercalation and unwinding to reduce the value of $(L - L_0)$. The higher the level of negative supercoiling, the more favored the binding of intercalating drugs (Bauer, 1978).

I. Effects of Temperature and Salt on Helical Winding and DNA Supercoiling

As temperature decreases, the level of negative supercoiling increases because changes in temperature affect the winding of the DNA double helix. As temperature decreases, the twist angle in DNA increases, resulting in fewer base pairs per helical turn and an increase in L_0 . In a covalently closed DNA molecule (either relaxed or supercoiled DNA) in which L cannot change, the temperature-induced change in winding will affect L_0 which will concomitantly affect the superhelical density σ (or $L - L_0$). ΔL_0 varies by 0.012 helical degrees (°) per base pair per °C (Depew and Wang, 1975; Pulleyblank *et al.*, 1975).

The Titration of Superhelical Density by Sedimentation in Ethidium Bromide–Sucrose Gradients

The negative superhelical density of covalently closed DNA molecules can be determined by their sedimentation velocity in a sucrose gradient. Circular plasmid molecules are compact when negatively supercoiled and will sediment rapidly through a sucrose density gradient when subjected to the high gravity forces from centrifugation. The sedimentation coefficient (*s*) will be maximal for a highly supercoiled plasmid molecule. When the plasmid preparation is sedimented through a series of sucrose density gradients containing increasing concentrations of ethidium bromide, the supercoils will be unwound, the plasmid will be less compact, and the sedimentation coefficient will decrease. At some concentration of ethidium bromide (called the equivalence point or the critical dye concentration) all supercoils will be lost, the plasmid will appear relaxed, and the rate of sedimentation will achieve a minimum value. In the presence of higher concentrations of ethidium bromide, positive supercoils will be introduced into the plasmid, and the sedimentation velocity will begin to increase. This relationship is shown in Figure 3.8.

The salt concentration also may influence the winding of the DNA. As the concentration of a monovalent cation (Na^+) or a divalent cation (Mg^{2+}) increases to high levels, the DNA double helix becomes wound less tightly, that is, the twist angle increases. In linear or supercoiled DNA the increase in the twist angle results in an increase in L_0. In a covalently closed DNA molecule L is fixed. Therefore a change in L_0 will result in a change in supercoiling [a measure of $(L - L_0)$]. An increase in the cation concentration results in an increase in the negative supercoiling (a larger negative value) (Bauer, 1978). A recent reinvestigation into the effects of salt on the helical twist of DNA has not observed a significant change in twist for NaCl concentrations of 0–162 m*M* (Taylor and Hagerman, 1990).

J. The Biology of Supercoiled DNA

1. Topological Domains of DNA: A Requirement for Supercoiling

Most DNA in its natural state, including chromosomal and plasmid DNA in bacteria as well as DNA in human cells, is negatively supercoiled.

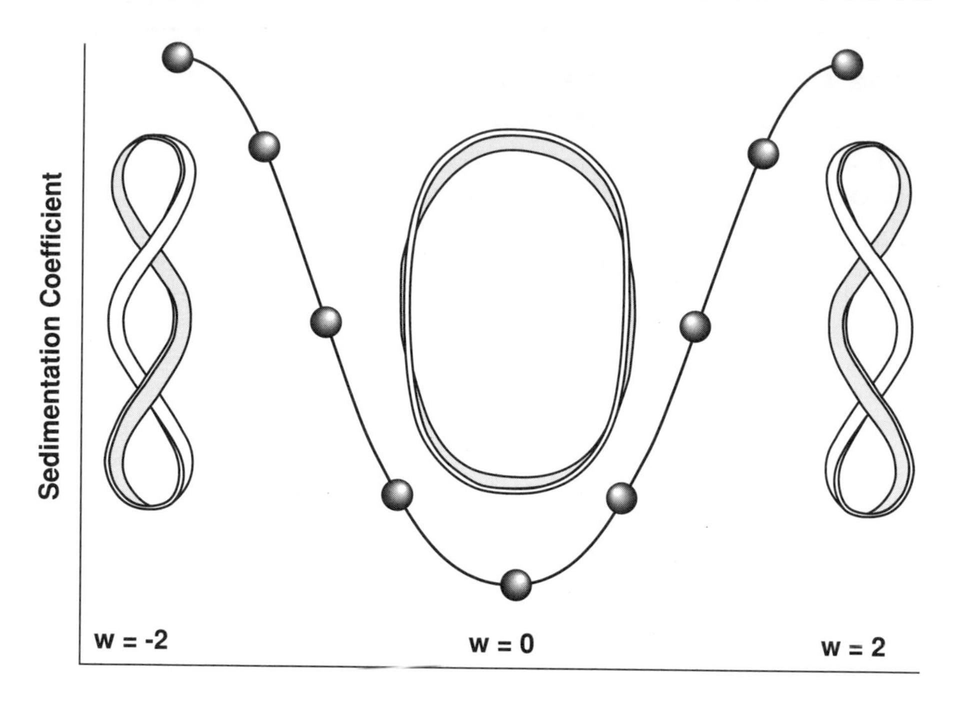

Figure 3.8 Titration of supercoils with intercalating molecules. The sedimentation coefficient (*s*) of DNA is dependent on the level of DNA supercoiling. Relaxed DNA exists in an open, extended form (see Figure 3.2) that sediments under centrifugal force slowly in a sucrose gradient. As DNA becomes negatively or positively supercoiled, the molecule becomes much more compact and sediments more rapidly than relaxed DNA. If negatively supercoiled DNA is sedimented in sucrose gradients containing an increasing concentration of an intercalator, then *s* will decrease. A minimum value is reached at the *equivalence point* where sufficient dye is bound to remove all negative supercoils ($W = 0$). On addition of more intercalator, *s* increases as positive supercoils are introduced into DNA. This method can be used to measure the level of negative supercoiling in large DNA molecules. For example, differences in the level of supercoiling in the *E. coli* chromosome in cells containing mutations in topoisomerase genes were detected using this method (Pruss *et al.*, 1982; Steck *et al.*, 1984).

Plasmid and circular bacterial chromosomes are *topologically closed* and thus have a defined linking number. This property of being a topologically closed system is essential for DNA supercoiling. If one cuts a rubber band and twists one end, the band rotates to remain in an untwisted state. Similarly, if one introduces a twist change at one end of a linear DNA molecule, the other end can simply rotate so there is no net change in the helical winding (or linking number) of the DNA. The *E. coli* chromosome exists as a large (2.9 ×

10^6 bp) closed circle that constitutes one topological domain. (*A topological domain is defined as region of DNA bounded by constraints on the rotation of the DNA double helix.*) The large circle is subdivided into about 45 independent topological domains *in vivo* (Figures 3.9A and 3.9B). The nature of the forces defining the topological domains *in vivo* remains to be established. It is possible that a domain is defined by the attachment of DNA to the bacterial membrane through specialized binding proteins (or DNA gyrase or other topoisomerases).

Eukaryotic chromosomes are believed to exist as long linear molecules. To exist in a supercoiled state, linear DNA must become organized into one or several topological domains. An analysis of the 120-kb bacteriophage T4 genome provided one of the first demonstrations that a linear molecule could be supercoiled *in vivo*. Apparently the ends of the T4 chromosome become fixed to a protein(s) or perhaps the bacterial cell membrane, thus defining a single topological domain. The model for the organization of human (eukaryotic) chromosomes is shown in Figure 3.9D. Independent loops are believed to be formed by the interaction of specific regions of DNA with defined proteins that attach to the nuclear matrix.

2. DNA Topoisomerases

Topoisomerases are enzymes that transiently break and reseal phosphodiester bonds in DNA. The transient break allows one strand to pass through or cross the other strand (effectively providing a swivel or a point of rotation of one strand around the other). These enzymes, therefore, change the linking number, *L*, of DNA in living cells. In bacterial cells, negative DNA supercoils are introduced by the enzyme DNA gyrase. Negative supercoils are removed in cells by DNA topoisomerase I and, under some conditions, by DNA gyrase. A general mechanism of action of these enzymes involves breakage of the phosphodiester bond and formation of a covalent bond between the ends of DNA and the enzyme. The covalent bond is formed between a specific tyrosine in the protein and a phosphate group on the 5′ or 3′ end of the DNA. In some cases the enzymes require ATP as an energy source. Table 3.2 provides a summary of the characteristics and properties of several DNA topoisomerases.

a. Type I and Type II Topoisomerases

Topoisomerases have been categorized into two types. A type I topoisomerase breaks only one strand of the DNA whereas a type II topoisomerase breaks both strands of the DNA. Because of this mechanistic difference, the linking number of DNA, *L*, will change in increments of 1 or 2 for

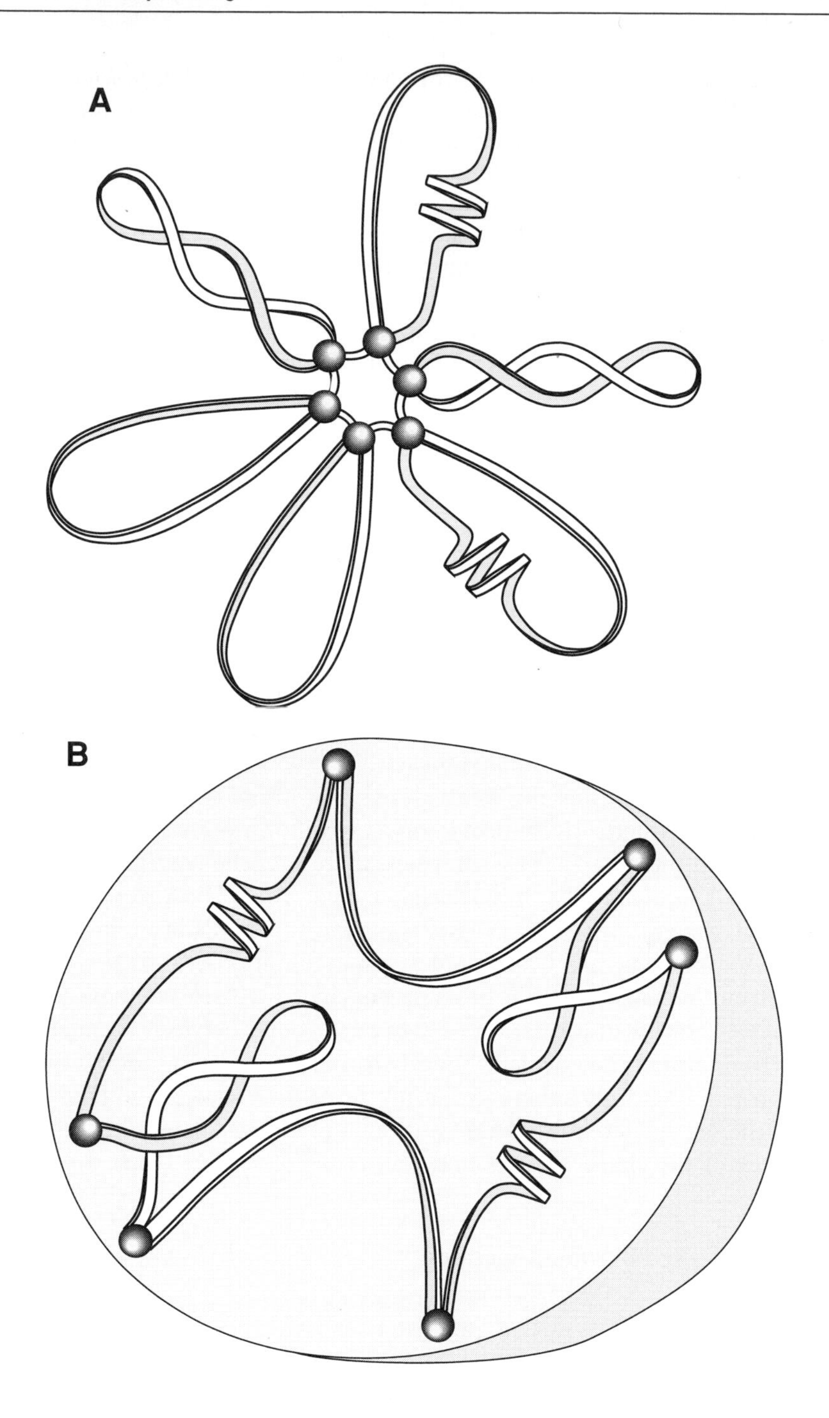
A
B

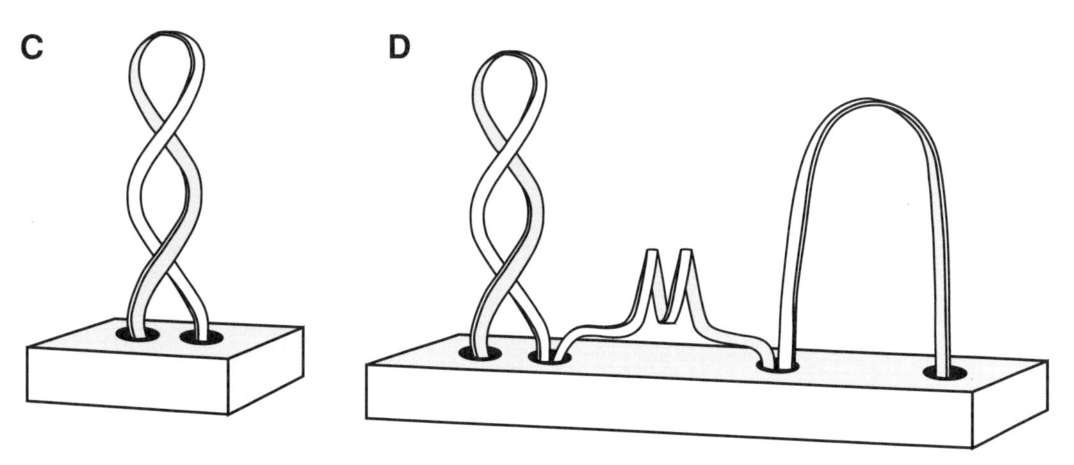

Figure 3.9 Organization of DNA into independent topological domains. (A) A circular DNA molecule organized into six topologically independent domains, which are defined in an unknown fashion inside cells by a mechanism that prevents rotation of the double helix. It is not yet certain what macromolecule in living bacterial cells is responsible for this domain organization. (B) The bacterial chromosome may be organized into topological domains by the interaction of DNA with proteins that are associated with the bacterial membrane. A number of reactions involving DNA, such as replication, may occur at the membrane. (C) A linear DNA molecule can form a topological domain when the ends of the DNA attach to some structure. In the case of bacteriophage T4, this may be the cell membrane. (D), In eukaryotic cells, the long linear chromosomes are attached at various intervals to the nuclear matrix or to a "scaffold" of the metaphase chromosome. This periodic attachment divides the linear DNA into multiple domains. Note that in A, B, and D DNA in some domains is relaxed and DNA in others is supercoiled. This situation is possible when independent domains are defined. Such organization may be important in the control of gene expression which can be dependent on the level of DNA supercoiling.

type I and type II topoisomerases, respectively. Figure 3.10 shows a model for the topological result of the action of *E. coli* topoisomerases I (a type I enzyme). Figure 3.11 shows a model for the topological result of the action of *E. coli* DNA gyrase, a type II topoisomerase.

The properties of type II topoisomerases vary considerably, although they all can change the linking number of DNA. Certain type II topoisomerases called gyrases are the only enzymes that have been shown to catalytically introduce negative supercoils into relaxed DNA. To negatively supercoil DNA, gyrases decrease the linking number, L. As might be expected from Eq. (5), energy from ATP hydrolysis is required for the generation of negative supercoils. In the absence of ATP, gyrase can relax DNA (increase L), thereby

Table 3.2
Properties of DNA Topoisomerases

Enzyme	Type	Gene	Size (kDa)	ΔL	Cofactors	Activities
Prokaryotic *E. coli* Topoisomerase I	I	*topA*	97	increase L; ΔL = 1	Mg^{2+}	Relaxes negatively supercoiled DNA; (will form knots and catenated DNA in nicked molecules); DNA transiently bound by a 5′ phosphotyrosine bond
E. coli DNA gyrase	II	*gyrA* *gyrB*	105 95	decrease L; ΔL = − 2 increase L; ΔL = 2	ATP, Mg^{2+}	ATP-dependent negative supercoiling (to $\sigma <$ -0.1); ATP-independent relaxation of negatively supercoiled DNA; relaxes positively supercoiled DNA; (will form knotted and catenated DNA); responsible for supercoiling the chromosome; DNA transiently bound by a 5′ phosphotyrosine bond
E. coli Topo III	I	*topB*	73.2	increase L; ΔL = + 1	Mg^{2+}	Decatenase activity, also relaxes negatively supercoiled DNA; active in chromosome decatenation following replication *in vitro*
E. coli Topo IV	II	*parC* *parE*	75 70	increase L; ΔL = + 2	ATP, Mg^{2+}	Similar to DNA gyrase; has a DNA relaxing activity but will not negatively supercoil DNA; involved in chromosome decatenation following replication
Bacteriophage T4 topoisomerase	II	gene 39 gene 52 gene 60	64 51 12	increase or decrease L; ΔL = ±2	ATP, Mg^{2+}	Relaxes negatively or positively supercoiled DNA; does not possess DNA supercoiling activity
Eukaryotic (Yeast) Topo I	I	*top1*	90	increase or decrease L; ΔL = ±1	none	Relaxes positively or negatively supercoiled DNA; DNA linked by a 3′ phosphotyrosine bond (human enzyme)
Topo II	II	*top2*	164	increase or decrease L; ΔL = ±2	ATP, Mg^{2+}	Relaxes positively or negatively supercoiled bond DNA; DNA linked by a 5′ phosphotyrosine
Topo III	I	*top3*	74	increase L; ΔL = +1	Mg^{2+}	Weak relaxation activity; only partially relaxes negative supercoiled DNA; strong activity on DNA with single-strand heteroduplex

lowering the free energy in supercoiled DNA. Gyrases have only been purified from bacteria. Other prokaryotic type II topoisomerases do not negatively supercoil DNA in an energy-dependent reaction. These include *E. coli* topo IV and a type II topoisomerase from bacteriophage T4. No type II topoisomerases purified from eukaryotic cells have been shown to negatively supercoil DNA.

Escherichia coli topoisomerase I was the first DNA topoisomerase to be discovered. It was identified by Wang in the early 1970s. The predominant activity of the bacterial enzyme is to relax negative supercoiled DNA (by increasing the linking number). Eukaryotic type I topoisomerases can relax negative or positive supercoils. No energy source is required for the action of type I topoisomerases. A second type I topoisomerase activity, called topoisomerase III, has been purified from *E. coli.* Bacterial topoisomerase III appears to have properties of eukaryotic topoisomerase I enzymes. Its biological role remains to be clearly established but it may be a specialized enzyme involved in resolving products of DNA replication.

Reverse DNA gyrases have been isolated from various species of *archebacteria.* These are organisms that live at high temperatures, low pH, high salt, and, in some cases, anaerobic environments. Reverse gyrases introduce positive supercoils into DNA. The biological role of reverse gyrases and positive supercoiling remains to be fully understood (Kikuchi, 1990).

b. The Biological Importance of DNA Topoisomerases

All biological reactions involving DNA probably require the participation of topoisomerases. For example, following replication in *E. coli,* the two circular chromosomes become entangled by catenation. A type II topoisomerase activity (DNA gyrase or topo IV) is required to transiently break one double strand of DNA and allow the double strand of a second chromosome to pass through the first chromosome. Without this strand passing reaction, the two daughter chromosomes could not physically separate following replication, and the two new bacterial cells could not physically separate. As discussed later in this chapter, the movement of RNA polymerase during transcription generates superhelical tension in the DNA. Topoisomerases are needed to relax tension generated by the movement of proteins through the DNA. Without these enzymes, transcription (and replication) would slow tremendously.

Topoisomerases are required to maintain a precise level of supercoiling in *E. coli* DNA. Mutations in DNA gyrase that reduce the supercoiling reaction result in a decrease in the level of supercoiling in living cells. Similarly, mutations in topoisomerase I result in an increase in the level of supercoiling in cells. The importance of maintaining a precise level of supercoiling is

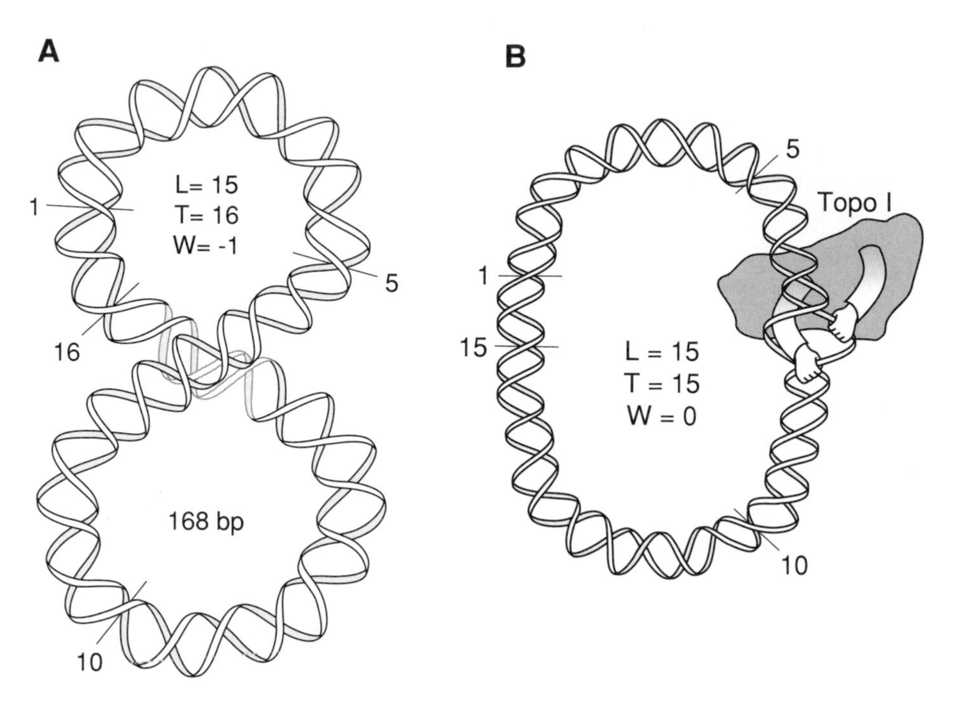

Figure 3.10 Increase in linking number by type I topoisomerase. (A) A 168-bp DNA molecule containing one negative supercoil with $L = 15$, $T = 16$, $W = -1$. (B) One helical turn has been unwound (this is for illustrative purposes, the enzyme does not actually do this). The unwinding results in a relaxation of the supercoil. A topoisomerase I molecule binds to the DNA and associates its active site with one strand of DNA.

demonstrated by the fact that mutations that totally destroy the activity of either DNA gyrase or topoisomerase I can be lethal to *E. coli*. Only partial or "leaky" mutations are viable. There is an exception to this rule, however. A deletion mutation of topoisomerase I is viable if the cell obtains a compensatory mutation in DNA gyrase that reduces the ability of this enzyme to supercoil DNA. A number of reviews on DNA supercoiling and the regulation and genetics of the level of supercoiling in *E. coli* are available (Gellert, 1981; Drlica, 1984,1990,1992; Wang, 1985).

3. Mechanisms of Supercoiling in Cells

Most DNA in its native state is believed to be negatively supercoiled. Since this represents an unfavorable energetic situation, how does this occur? DNA could become supercoiled in living cells in a number of ways: (1) through the action of a DNA gyrase, (2) by wrapping DNA into a nega-

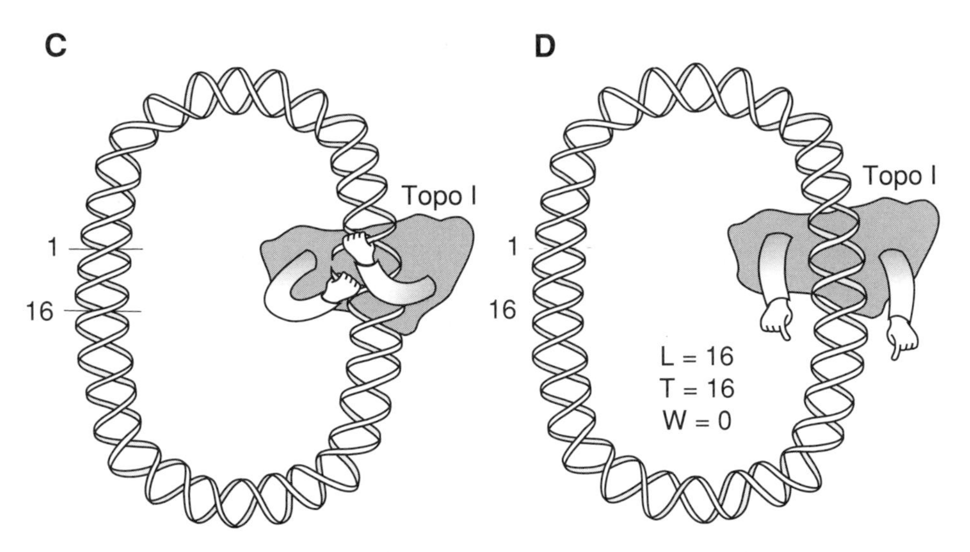

Figure 3.10 *Continued.* (C) The catalytic activity results in the transient breaking of one strand. The DNA becomes linked through its 5′ end via a phosphate to a tyrosine in the topoisomerase. In the active site of the enzyme, the linking number L is increased from $L = 15$ to $L = 16$ by passing one end of the DNA under and the other end over the unbroken strand. (D) Ligation of the DNA re-forms a covalently closed relaxed molecule with $L = 16$, $T = 16$, $W = 0$.

tive supercoil around a protein (in eukaryotic cells these proteins are histones) and subsequently removing a resulting positive supercoil, and (3) by the act of transcription. These mechanisms are illustrated in Figure 3.12.

a. Supercoiling in Bacterial Cells Is Driven by the Activity of DNA Gyrase

The first enzyme that was identified and purified that could introduce negative supercoils into DNA was the *E. coli* enzyme DNA gyrase, discovered by Gellert. In a test tube, purified DNA gyrase will supercoil DNA to a negative superhelical density of $\sigma \approx -0.1$ in a reaction requiring ATP hydrolysis (Figure 3.12A). This level of supercoiling is about twice that observed for DNA purified from cells.

In living cells the level of supercoiling may be regulated by the opposing activities of DNA gyrase and topoisomerase I. Gyrase supercoils DNA and topoisomerase I relaxes DNA. The balance of these two activities keeps the DNA at a precise, finely tuned level of supercoiling in living cells. The ATP : ADP ratio may also be very important in maintaining the level of supercoiling *in vivo*. The concentration of these two nucleotides influences the activity of DNA gyrase.

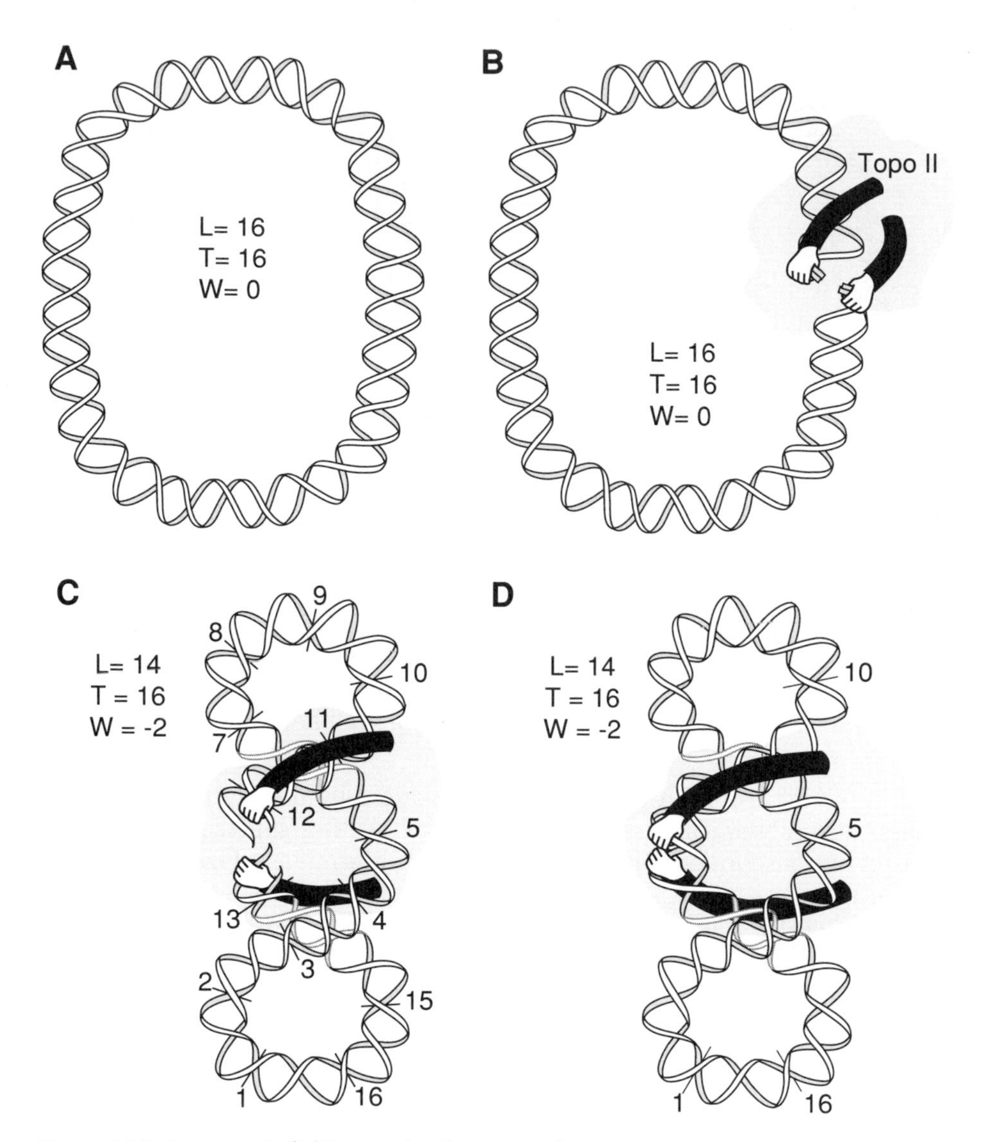

Figure 3.11 Decrease in linking number by action of *E. coli* DNA gyrase, a type II topoisomerase. (A) A relaxed DNA molecule of 168 bp with $L = 16$, $T = 16$, $W = 0$. (B) Gyrase binds DNA, transiently breaks both strands of the DNA, and forms covalent bonds with both 5′ ends of the DNA and a tyrosine on the enzyme. (C) Another part of the plasmid passes through the break in the duplex in a "strand passing" reaction. (D) On ligation, there is a decrease in the linking number of $\Delta L = -2$ to $L = 14$.

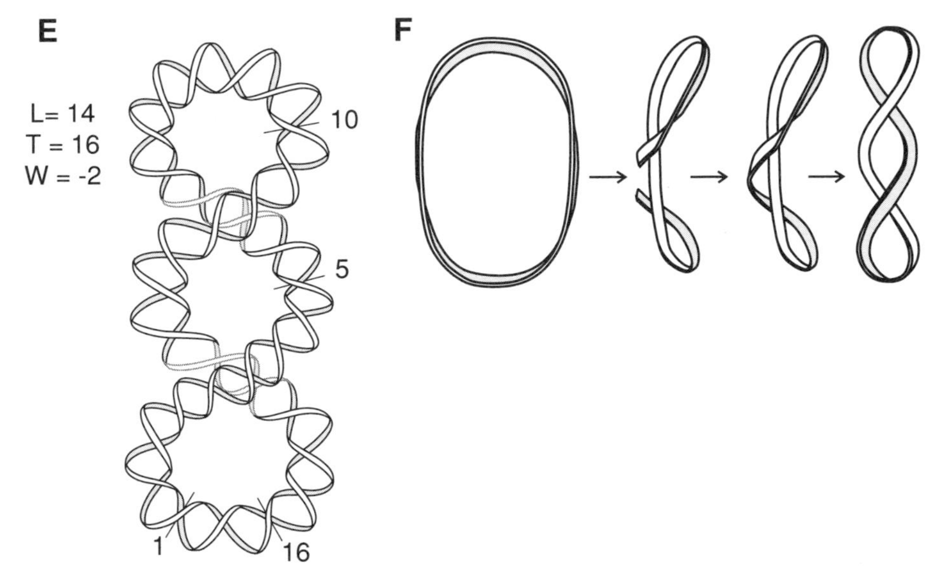

Figure 3.11 *Continued.* (E) The molecule with its two negative supercoils. (F) The change in linking number of $\Delta L = -2$ by a strand passing reaction is illustrated in the absence of the protein using the rubber band model. Although it may not be intuitively obvious that this results in a linking number change of -2, experimenting with appropriate models should convince the reader. This figure is meant to illustrate mechanistically the topology of the reaction of DNA gyrase. Although not illustrated here, the DNA may be wrapped by DNA gyrase into a left-handed toroidal coil, a break is made within that DNA, and another part of the duplex passes through the break (see Kirchhausen *et al.*, 1985).

b. Supercoiling in Eukaryotic Cells Can Be Introduced by Wrapping DNA around Nucleosomes and by the Action of a Topoisomerase

A DNA supercoil can be organized as an interwound supercoil or a toroidal coil as shown in Figure 3.5. In eukaryotic cells, DNA is wrapped in a left-handed fashion around an octomer of two each of four different histones. The DNA–histone complex, called a nucleosome, is described in more detail in Chapter 9. The left-handed toroidal coiling of DNA in space or the stable left-handed wrapping of DNA around a protein results in the introduction of a positive or left-handed supercoil (or writhe) into the the same topological domain of DNA, with no concomitant change in *L*. Either eukaryotic type I or type II topoisomerase can relax the positive supercoil, resulting in the introduction of one negative supercoil. This process is illustrated in Figure 3.12B.

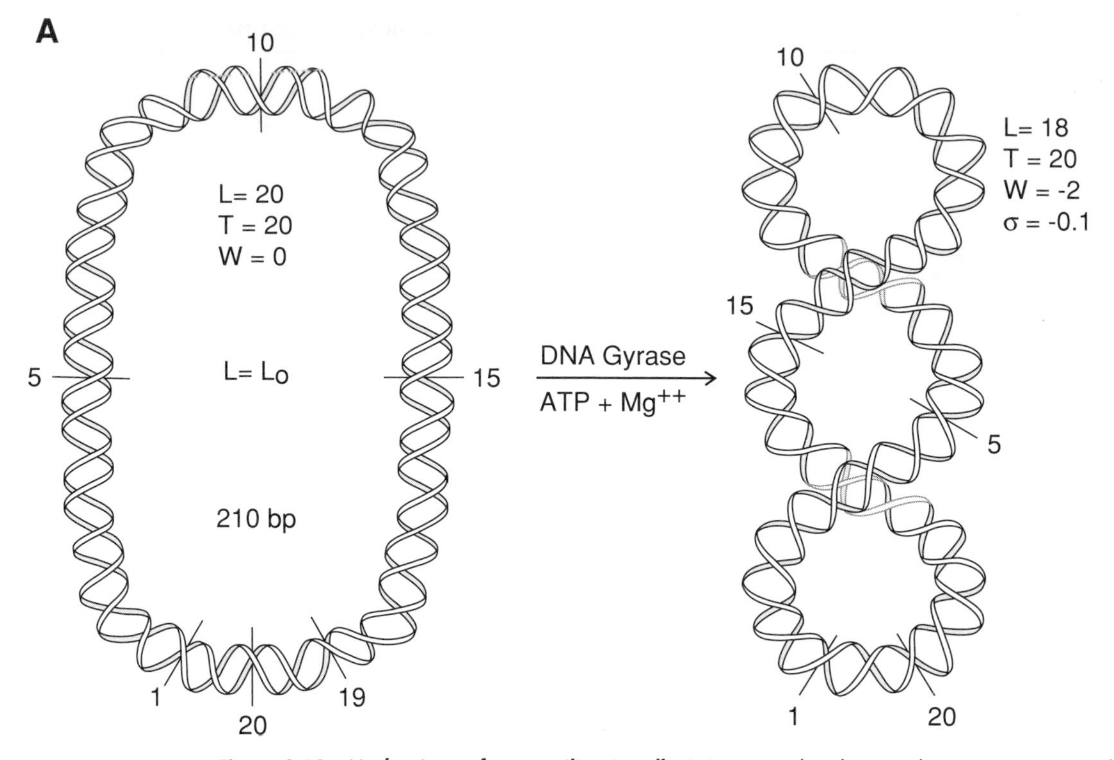

Figure 3.12 Mechanisms of supercoiling in cells. (A) Supercoiling by *E. coli* DNA gyrase. *E. coli* DNA gyrase introduces negative supercoils into DNA in an ATP dependent reaction. *In vitro,* gyrase will introduce supercoils into relaxed DNA (*left*) to a density of $\sigma = -0.1$ (*right*) which is about twice that found in DNA purified form natural sources. (B) The organization of DNA into nucleosomes in eukaryotes may be used to introduce a state of negative supercoiling into DNA. The relaxed molecule with $L = 20$, $T = 20$, $W = 0$ can wrap two counter-clockwise (left-handed) toroidal turns around a nucleosome represented by the cylinder. This results in the introduction of one positive supercoil in the DNA that is not physically associated with the nucleosome (*top right*). [Because there is a change in *T,* the twist of DNA wrapped around the nucleosome, only 1 negative supercoil is restrained ($\Delta L = 1$) by wrapping DNA twice around the nucleosome ($\Delta W = 2$). Thus, only 1 positive supercoil is introduced into the rest of the DNA. This is explained in greater detail in Chapter 9.] Eukaryotic type II topoisomerases can relax this single positive supercoil (*bottom left*). On removal of the nucleosome one negative supercoil has been introduced into the DNA (*bottom right*). (C) The effect of transcription on DNA supercoiling. A topological domain containing 210 bp of DNA is divided into two subdomains by an RNA polymerase molecule (*top left*). Movement of RNA polymerase by 20 bp in the absence of any rotation of the helix at the site of transcription results in a deficiency in helical turns ($L < L_0$) behind the polymerase and an excess of helical turns ($L > L_0$) in front of the polymerase (*top right*). Note that rapid rotation may be sterically prevented by the size of the polymerase, RNA transcript, and associated ribosomes and nascent proteins. The situation results in the formation of negative and positive supercoils behind and in front of RNA polymerase, respectively (*bottom left*). For simplicity, DNA is illustrated using the rubber band model. (*Bottom right*) A eukaryotic type I or prokaryotic or eukaryotic type II enzyme relaxes the positive supercoils, resulting in the net introduction of negative supercoils. Such a mechanism can change the level of supercoiling in cells when certain topoisomerases are inhibited. Normally, the actions of DNA gyrase and topoisomerase I maintain a certain level of negative supercoiling, and only under certain situations is the local level of supercoiling perturbed. Transcription *in vivo* from divergent promoters has been shown to influence the local level of supercoiling.

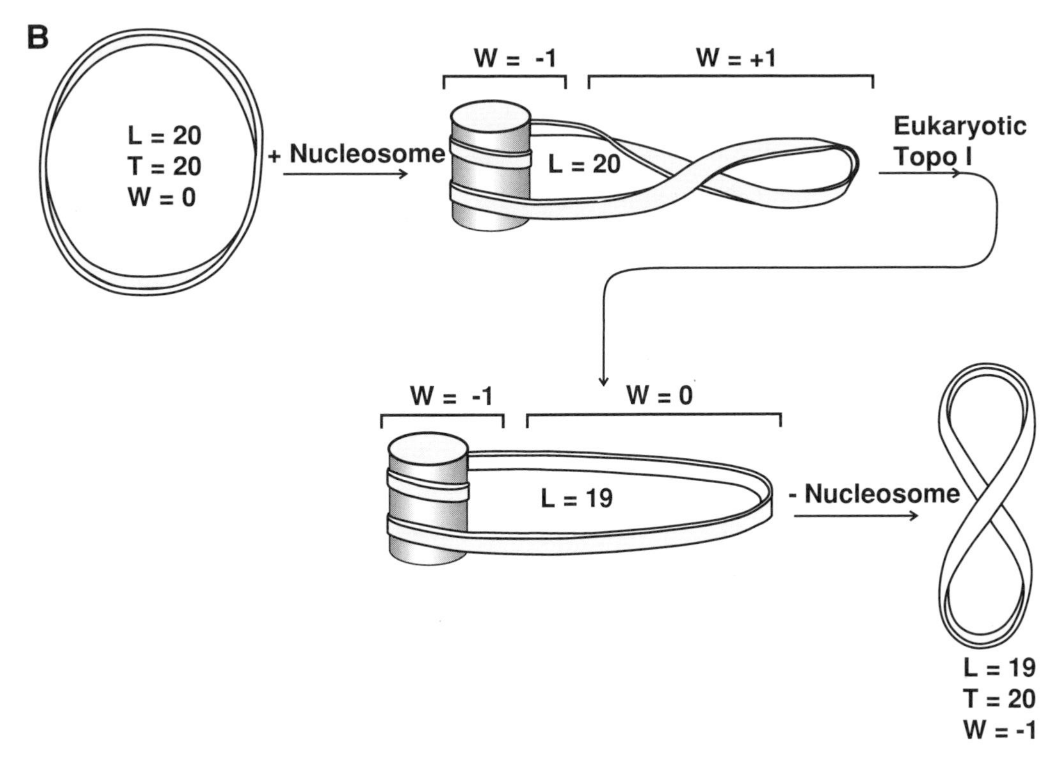

Figure 3.12 *Continued.*

There is an apparent inconsistency in Figure 3.12B. The wrapping of the DNA twice around the protein core ($\Delta W = -2$) results in the introduction of a single positive supercoil which, following the action of a topoisomerase, results in a linking number change of -1. Following the linking number change, removal of the protein with the two left-handed wraps of DNA results in the introduction of one negative supercoil. One toroidal coil in space corresponds to one supercoil, but it takes two turns around the nucleosome to equal one supercoil. This *linking number paradox,* where $\Delta L = -1$ and $\Delta W = -2$, can only be resolved if there is a change in T as the DNA wraps around the nucleosome. The change in T, ΔT, must equal $+1$. Therefore the helix repeat must change from about 10.5 to a value <10.5 in DNA that is involved in coiling around the nucleosome.

c. Transcriptional Effects on Supercoiling

Two remarkable observations led to an appreciation of the significant consequences that transcription can have on the level of supercoiling *in vivo* and to an appreciation of the importance of topoisomerase activities for tran-

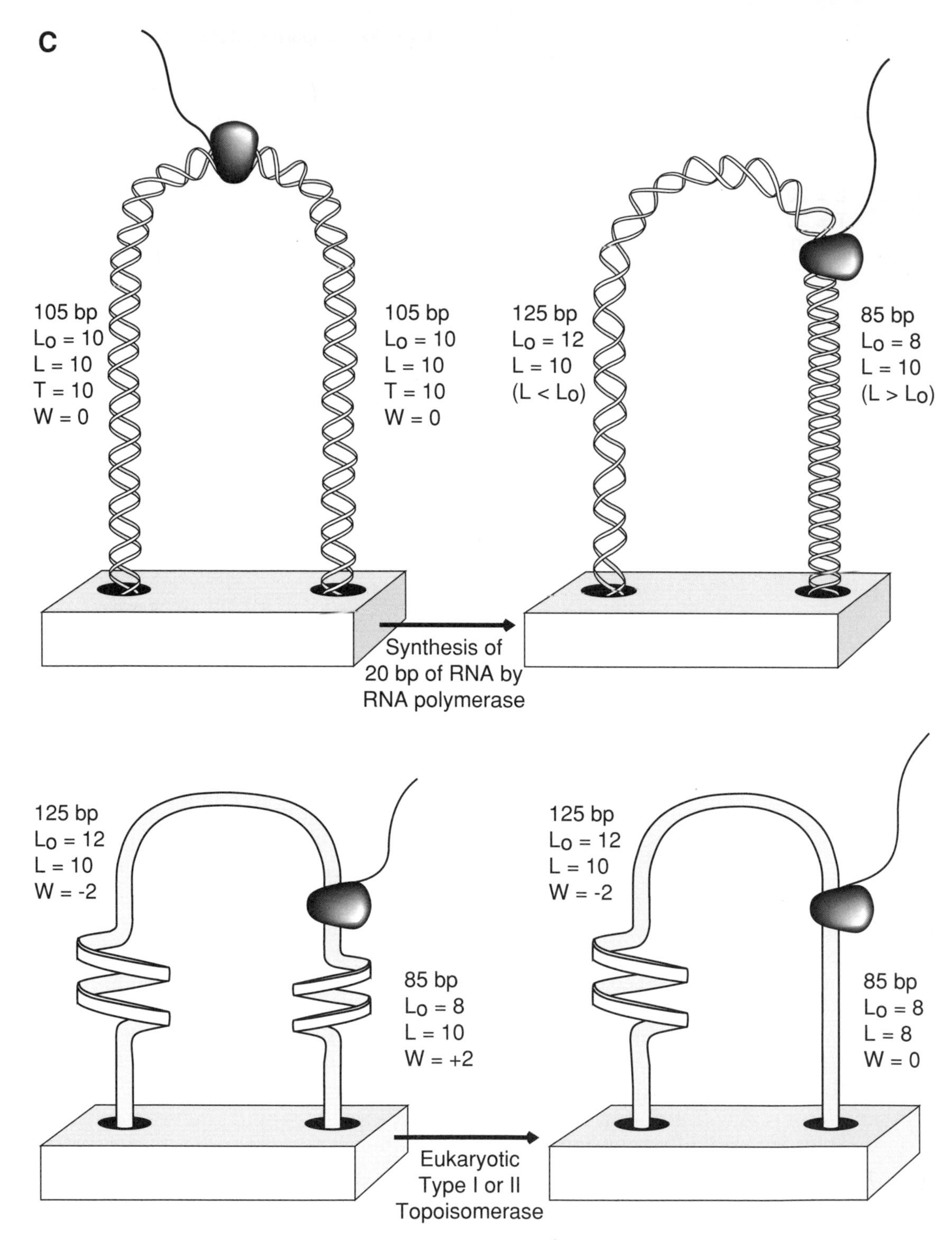

Figure 3.12 *Continued.*

scription. First, Pruss and Drlica (1986) discovered that the negative superhelical density of plasmid pBR322 was twice that normally found in cells when purified from *E. coli* containing a mutation in topoisomerase I. Moreover, the typical Boltzmann Gaussian distribution of topoisomers found in wild-type cells was replaced with a very broad, heterogeneous distribution. The sequences in the DNA responsible for this unusual distribution were those encoding the promoter region for the tetracycline resistance gene. The unusual distribution of topoisomers was dependent on transcription from the tetracycline gene. Second, Lockshon and Morris (1983) described the purification of *positively* supercoiled plasmid DNA from *E. coli.* Positively supercoiled plasmid DNA could be purified from *E. coli* cells treated with novobiocin, an antibiotic that inhibits the activity of DNA gyrase.

Based on these observations, Liu and Wang (1987) pointed out a most obvious and simple consequence of transcription on supercoiling. They argued that movement of an RNA polymerase during transcription within a topologically closed domain would generate negatively supercoiled DNA behind the RNA polymerase and positively supercoiled DNA in front of the RNA polymerase. This twin-domain model of supercoiling is illustrated in Figure 3.12C. Consider a topological domain of DNA in which rotation at the ends of the domain is prevented. Assume that RNA polymerase does not rotate as it moves through the DNA. Under this situation RNA polymerase actually divides the domain into two topologically separate subdomains. Initially, RNA polymerase is positioned in the center of the 210-bp domain creating two 105-bp subdomains with $L_0 = L = 10$, $T = 10$, and $W = 0$. The movement of RNA polymerase by 20 bp now creates a 125-bp domain behind the polymerase in which $L_0 = 12$ (remember that L_0 is the linking number of relaxed DNA or the preferred state of twist of a B-form DNA helix. In front of the RNA polymerase there are 85 bp of DNA with $L_0 = 8$. The actual linking number, L, cannot change if the DNA is not broken and resealed. Moreover, if RNA polymerase does not rotate, L cannot be redistributed between the two subdomains. This creates two subdomains in which $L < L_0$ behind the polymerase and $L > L_0$ in front of the polymerase. These are the conditions of negative supercoiling and positive supercoiling, respectively. To introduce a net negative supercoiling into DNA, a topoisomerase need relax only the positive supercoils ahead of RNA polymerase.

In most biological situations RNA polymerase bound to DNA could rotate, although it is a reasonably large molecule (Cook *et al.*, 1992). However, as the RNA transcript becomes large and, in bacterial cells, becomes bound to ribosomes, rotation around the DNA would be difficult. When the protein made from a plasmid interacts with the membrane, the physical association of membrane, protein, ribosome, mRNA, and RNA polymerase with the DNA can prevent rotation of the DNA. In a plasmid molecule containing one moving RNA polymerase, the DNA behind and in front of RNA polymerase could rotate and the negatively and positively supercoiled regions would cancel. Transiently, however the local level of torsional tension in specific regions is likely to be quite different from the "average superhelical density" or σ of the total plasmid. If two RNA polymerase molecules are transcribing in opposite directions in a topological domain, the differences in levels of supercoiling upstream and downstream of transcription can be significant. In fact, in one particular plasmid (pBR322) with divergent transcription units, a difference in the level of supercoiling upstream and downstream of two transcription units has been detected in living *E. coli* cells (Rahmouni and Wells, 1989; Zheng *et al.,* 1991).

Although transcription can lead to local changes in the level of supercoiling and to global changes if the activities of topo I or DNA gyrase are impaired, this does not necessarily imply that supercoils are normally introduced by transcription. Bacteriophage λ DNA, for example, enters *E. coli,* circularizes, and becomes negatively supercoiled prior to transcription and replication. Moreover, the inhibition of transcription by the addition of the antibiotic rifampicin does not reduce the level of supercoiling in the *E. coli* chromosome. The majority of the negative supercoils in the *E. coli* chromosome are probably introduced by the action of DNA gyrase.

4. Summary of the Effects of Supercoiling on the Structure of DNA

Supercoiled DNA is torsionally strained in the winding of the double helix. The energy in supercoiled DNA can manifest itself as changes in the twist and writhe of DNA. This can be reflected as a state of twisting or shape of the helix that is different from that of relaxed DNA. In addition, the free energy in the supercoiled molecule can drive reactions that result in transient or stable unwinding of the DNA double helix (Sinden, 1987). Unwinding reactions include breathing and the formation of stably unwound regions of DNA. If particular sequence arrangements are present in DNA, cruciform structures, Z-DNA, intramolecular triplex DNA, or slipped mispaired DNA structures may form. Subsequent chapters will address the superhelical-density-dependent formation of these alternative DNA structures and will discuss current ideas regarding their biological significance.

5. The Role of Supercoiling in Gene Expression

Extensive reviews have been written on the topic of the regulation of gene expression by DNA supercoiling (Drlica, 1984; Esposito and Sinden,

1988; Freeman and Garrard, 1992). At this point, a few relevant ideas and examples will be presented to illustrate the significance of supercoiling in gene expression.

The conformation of DNA can have an influence on the access or utilization of information associated with that particular region of DNA. The formation of cruciforms, Z-DNA, or triple strands presents a very different structure to the cell than the usual B-form of DNA. Alternative configurations of DNA might provide switches that could turn genes on or off. The actual switching could be implemented by a number of mechanisms. For example, there could be a protein that must bind to the B-form of a region of DNA to turn on a gene. If the region existed in the alternative configuration, the gene could not function. In fact, such a model system using a cruciform-containing sequence has been constructed and shown to function *in vivo* (Horwitz and Loeb, 1988).

The regulation of gene expression by DNA supercoiling need not be as dramatic as by the formation of an alternative DNA conformation. The energy in the helix may promote the opening of the *promoter* region of the DNA helix to allow RNA polymerase to bind and unwind the DNA and begin transcription. A promoter is the region of DNA upstream from a gene that contains the sequence information signals to bind RNA polymerase, allows it to get into DNA, and, under the right conditions, tells it where to begin synthesizing RNA. A low level of energy may be needed to open some promoter regions whereas a higher level may be needed for others. This simplified model is probably accurate in some cases, but in general the regulation of gene expression is much more subtle. Without addressing the structure of a promoter, in detail, basically two regions contain important signals; these regions are separated by 15–17 bp of DNA. Changes in the sequence of the regions or the spacing of the regions allow enormous variation in the characteristics of the promoter. These sequence and structural elements affect at least five or six kinetically distinct steps in the process of initiation of transcription. DNA supercoiling will affect the rotational relationship between the two important regions and the ability of the regions to breathe. DNA supercoiling can also differentially affect individual kinetic steps in transcription initiation, as discussed by McClure (1985). It should not be surprising, then, that different promoters have been shown to be "tuned" to work best at different levels of DNA supercoiling.

The regulation of the *E. coli* DNA gyrase and topoisomerase genes provides an excellent example of gene regulation by DNA supercoiling. If you were designing an organism, it would seem reasonable to have the gene responsible for the production of the supercoiling activity (DNA gyrase) turned on by a low level of supercoiling in the chromosome and turned off when supercoiling was too high. If the level of supercoiling becomes too high, you would also want to turn on production of an enzyme to relax supercoils (topoisomerase I). By balancing the level of production of these two enzymes, it should be possible to maintain a precise level of DNA supercoiling in the cell. Elegant experiments by Gellert and colleagues have demonstrated that the genes for DNA gyrase are regulated in just such a supercoil-dependent fashion (Menzel and Gellert, 1983). As you might also suspect, the gene for topoisomerase I is turned on at high levels of supercoiling (Tse-Dinh and Beran, 1988). In part by regulating the production of these enzymes, which change linking number in the opposite direction, *E. coli* maintains a finely tuned level of torsional tension in its DNA *in vivo*.

Many other genes may be regulated by the level of supercoiling in DNA. The organization of the chromosome into topological domains allows for the possibility that different regions of the chromosome are differentially supercoiled. On the other hand, sets of genes throughout the chromosome may respond simultaneously to a global change in the level of supercoiling. It is known that changes in growth conditions including temperature, osmotic conditions, or media can lead to changes in the overall level of DNA supercoiling (Balke and Gralla, 1987; Hsieh *et al.*, 1991).

6. The Role of Supercoiling in DNA Replication

Arthur Kornberg has been a leader in the field of the biochemistry of DNA replication since he purified the first DNA polymerase, called DNA polymerase I, in 1962. After identifying many of the proteins that are involved in the process of initiation of DNA replication and chain elongation, Kornberg realized that the state of supercoiling of the DNA was an important factor in triggering the initiation of the replication process. The DNA, through its shape and torsional flexibility, works in concert with the replication proteins to begin the process of replication.

A supercoiled DNA molecule is required for precise initiation of the process of DNA replication at the unique origin of DNA replication in *E. coli,* termed *ori*C. Briefly, the DnaA proteins bind to four 9-bp repeat sequences. The DNA is then wrapped into a toroidal coil around the DnaA proteins. Concomitant with the hydrolysis of ATP, a region of DNA encompassing three 13-bp AT-rich repeated sequences are unwound into a stable denaturation bubble. This exposes the nucleotides to recognition and binding by a specific protein complex composed of the DnaB and DnaC proteins which contain a *helicase* activity (see Figure 7.5). Once the origin region is unwound by the helicase activity, more replication proteins bind to the DNA, and the seemingly monotonous process of elongation begins.

K. Details of Selected Experiments

1. The Goldstein and Drlica Experiment

In a simple but elegant experiment, Goldstein and Drlica (1984) demonstrated that plasmids isolated from *E. coli* grown at different temperatures have different linking numbers. Knowing the coefficient of variation in twist as a function temperature, they argued that *E. coli* maintained a finely tuned, constant level of torsional tension (or unrestrained supercoiling) in its DNA. Goldstein and Drlica found that the change in plasmid supercoiling was that expected from a temperature-dependent change in twist. The bacteria maintain a finely tuned level of supercoiling by changing the linking number of the DNA following a temperature shift. This was a profound observation made using a very simple experiment, a working knowledge of DNA supercoiling, and $L = T + W$.

The results of the temperature shift experiment showed that plasmid purified from cells grown at 17°C was less negatively supercoiled than plasmid purified from cells grown at 37°C (the normal temperature for growth of *E. coli,* a natural inhabitant of the human gut). *This may seem contradictory since a reduction in temperature increases the level of DNA supercoiling.* Why then is plasmid isolated from cells at a lower temperature less negatively supercoiled than DNA from cells grown at a higher temperature? Since this seemingly simple concept can be confusing, the steps of this process are detailed in Figure 3.13.

Consider plasmid pBR322 (the one used by Goldstein and Drlica) which contains 4362 bp and $L_O = 415 = (4362 \div 10.5)$. This plasmid typically contains about 25 negative supercoils [$(L - L_0) = -25$] when isolated from cells grown at 37°C. If the DNA contains 25 negative supercoils then $L = 390$. Following a temperature shift of 20°C from 37°C to 17°C, L_0 *will increase* by $\Delta L_o = (0.012°/\text{bp °C})(4362 \text{ bp})(20°\text{C}) = 1047°$. Dividing 1047° by 360° (the angle of rotation of one complete helical turn of DNA) converts the total twist angle change into DNA turns, or $\Delta L_0 = 2.9$. At 17°C, L_0 will be equal to 417.9. Since the level of supercoiling is reflected by $W = (L - L_0)$ and negatively supercoiled DNA is underwound ($L < L_0$), a larger value of L_0 will result in a *greater* level of negative supercoiling in the living *E. coli* cell. Thus, the initial drop in temperature makes the DNA *more* negatively supercoiled.

Why then is DNA purified from cells grown at a lower temperature *less* negatively supercoiled? The answer to this lies in the biology of *E. coli.* Goldstein and Drlica argued that *E. coli* maintain a precise level of supercoiling, or a defined level of torsional tension, in the winding of the DNA. If the

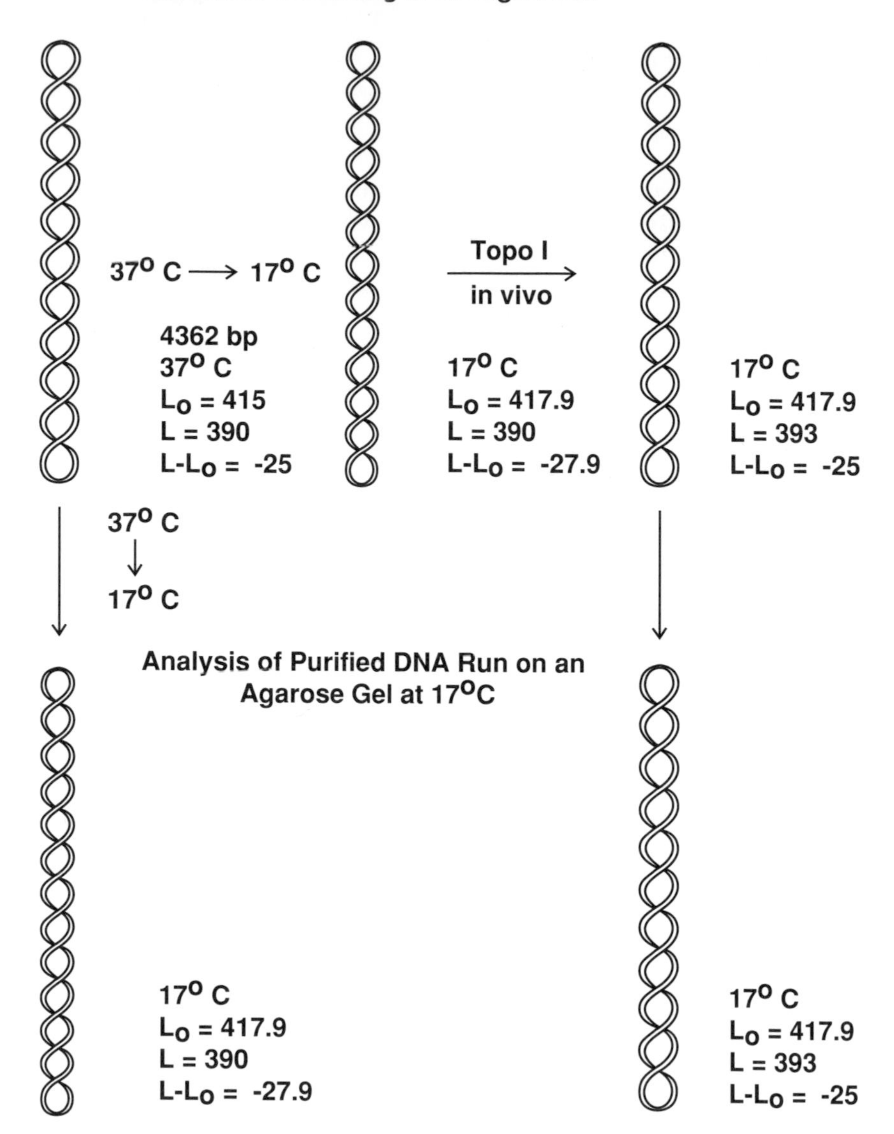

Figure 3.13 The Goldstein–Drlica temperature shift experiment, a demonstration that *E. coli* maintains a precise level of supercoiling in DNA. (*Top left*) the molecule represents a 4362-bp pBR322 molecule at 37°C with $L_0 = 415$, $L = 390$, and $L - L_o = -25$. (*Top center*) The same molecule following a 20°C decrease in temperature, L_o is increased to 417.9 resulting in $(L - L_o) = -27.9$. This molecule is more negatively supercoiled than the original molecule. (*Top right*) *In vivo* the action of topoisomerase I rapidly increases the linking number to $L = 393$ to maintain a value of $(L - L_o) = -25$. (*Bottom left*) Analysis on agarose gels at 17°C of DNA from cells grown at the original temperature of 37°C and (*bottom right*) from the cells shifted and grown at 17°C shows 2.9 fewer supercoils in DNA from cells grown at 17°C. The purification of DNA from cells grown at 37°C and analysis on gels at 17°C results in an increase of L_o to 417.9 which, with the original

level of supercoiling increases *in vivo* following a temperature shift and *E. coli* maintain a precise level of supercoiling ($L - L_0$), then the linking number, L, must increase to re-establish the original value of ($L - L_0$). The temperature shift to 17°C changes the level of supercoiling to $(L - L_0) = (390 - 417.9) = -27.9$. To re-establish $(L - L_0) = -25$, L must increase to $L = 393$. Cells increase L through the activity of topoisomerase I described previously.

One final point must be considered to understand the experimental observation that DNA purified from cells grown at 17°C is less supercoiled than DNA from cells grown at 37°C. The DNA topoisomers from cells grown at two different temperatures are analyzed on agarose gels at the *same* temperature, usually "room temperature," about 20°C. To simplify the discussion we will use 17°C as the temperature of the gel. Therefore, for the plasmid purified from cells at 17°C; $L = 393$, $L_0 = 417.9$, and $(L - L_0) = -25$. The DNA purified from cells grown at 37°C has a linking number of $L = 390$. At 37°C, L_0 was equal to 415. However, by reducing the temperature to 17°C (the condition for the gel), $L_0 = 417.9$. The value of $(L - L_0)$ now equals $(390 - 417.9) = -27.9$. Consequently, on the agarose gel at 17°C, DNA purified from cells at 37°C contains 2.9 *more* negative supercoils than the DNA from cells grown at 17°C. This was the result of Goldstein and Drlica: an incredibly simple experiment which, with a sophisticated understanding of supercoiling, resulted in an understanding of the dynamics of the regulation of supercoiling in *E. coli*!

2. Agarose Gel Electrophoresis

Goldstein and Drlica analyzed the changes in linking number using agarose gel electrophoresis of plasmid DNA. Figure 3.14 shows an agarose gel containing a number of forms of DNA. DNA is loaded into the well of an agarose gel (a gelatinous Jello®-like material) and an electric current is applied from electrodes, one at each end of the electrophoresis chamber. Since DNA is negatively charged, it migrates in the electric field toward the cathode (+ charge). Agarose gels separate DNA on the basis of size and shape. Molecules of the same size (supercoiled, linear, or nicked) will migrate proportionally differently on the basis of shape. For example, smaller supercoiled

value of $L = 393$, results in a level of supercoiling of $(L - L_o) = -27.9$. It is imperative that the linking number of the DNA (L) must not change during the purification protocols or this experiment would not work. The excellent agreement between the experimental data and the calculated expected value suggests that linking number changes are not occurring during DNA purification. (If linking number changes do occur during purification, the difference between linking numbers in cells grown at the two different temperatures must be maintained during purification.)

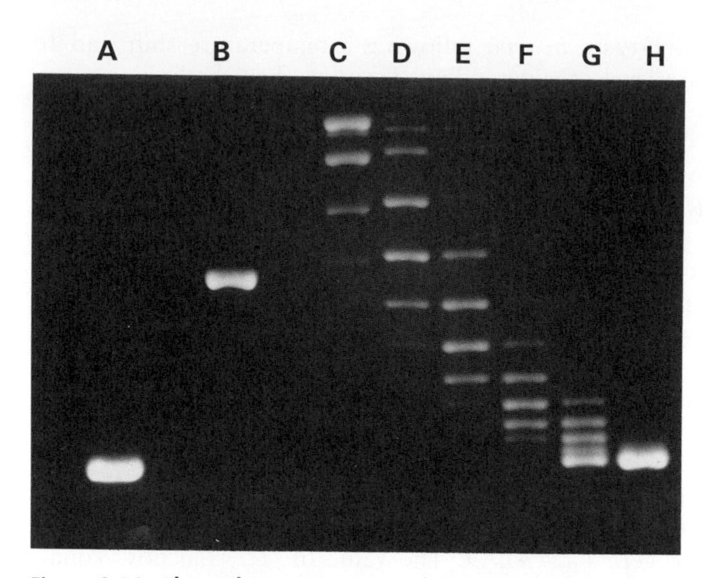

Figure 3.14 Electrophoretic migration of DNA in agarose gels. DNA samples are applied to wells (not shown) at one end of a 1.75% agarose gel in electrophoresis buffer (40 m*M* Tris acetate, 25 m*M* Na acetate pH, 1 m*M* EDTA, pH 8.3). An electric current is applied and the DNA migrates toward the cathode (+ charge). In agarose, DNA migrates as a function of size and shape. All bands in the gel shown are plasmid pUC8 containing 2671 bp. Greater than 95% of the naturally purified DNA sample in lane A is negatively supercoiled and migrates most rapidly in this gel. Although not resolved, there are about 6 individual topoisomers, from $W = -12$ to -18, migrating as a single band. Lane B shows the migration of linear pUC8 (after restriction digestion with *Eco*RI which cuts once within the plasmid). Under these gel conditions, linear DNA migrates midway between negatively supercoiled and nicked DNA. Lanes C–H are individual topoisomer mixtures prepared by adding increasing concentrations of ethidium bromide to DNA in the presence of a topoisomerase. In the absence of an intercalator, all supercoils are relaxed (lane C). In fact, a few positive supercoils are seen in lane C (these migrate slightly more quickly than the corresponding negative supercoils in lane D). In the presence of increasing dye concentration (lanes D–H), supercoils are removed by the change in the total twist angle in DNA, created by the unwinding on dye intercalation. On addition of topoisomerase, only the supercoils not removed by dye binding are relaxed. Lane D contains predominately five negative supercoiled topoisomers ($W + -1$ to -5); lane E contains molecules with $W = -4$ to -7; lane F, molecules with $W = -6$ to -9; lane G, molecules with $W = -8$ to -11; and lane H, molecules with $W \leq -11$. Each individual topoisomer migrates at a unique position within the gel. Relaxed or nicked DNA (no supercoils) migrates most slowly; as the number of supercoils increases, the migration rate increases until the gel no longer resolves the number of supercoils ($W \leq -11$).

molecules will migrate more rapidly than larger supercoiled molecules. In addition, different forms of the same DNA molecule will migrate differently depending on the form of the molecule. Molecules that are more compact will migrate more rapidly. Considering the forms of a plasmid, supercoiled DNA, being most compact, migrates most rapidly in an agarose gel. DNA that is relaxed or nicked will migrate most slowly. Linear DNA will migrate somewhere between the position of supercoiled and nicked DNA. The actual position of migration of linear DNA will depend on a number of factors including agarose concentration, the buffer used, and the rate of migration, which is determined by the amount of the current (V/cm) and the temperature.

Agarose gels are particularly useful for analyzing supercoiling in plasmids since they separate DNA topoisomers on the basis of writhe. A molecule with one supercoil will run a little faster than a relaxed molecule. Specific individual bands are observed for individual DNA topoisomers with 0, 1, 2, 3, etc. negative supercoils. At some point, however, the gel fails to resolve individual DNA topoisomers with increasing numbers of supercoils. In the gel shown in Figure 3.14 the level of resolution is about $(L - L_0) = -11$.

Goldstein and Drlica ran their agarose gels in the presence of the intercalating drug chloroquine, which partially removes negative superhelical turns. This allowed "naturally supercoiled DNA" which migrates as a single "supercoiled" band on an agarose gel to be resolved into the individual 6–8 topoisomers that make up the DNA sample. The difference between -25 and -28 supercoils would not be evident on a standard agarose gel where the level of resolution of individual turns is 0 to 15. Sufficient supercoils must be removed to change the naturally occurring topoisomers into forms that will migrate in the region of the gel that will resolve differences of a single supercoil.

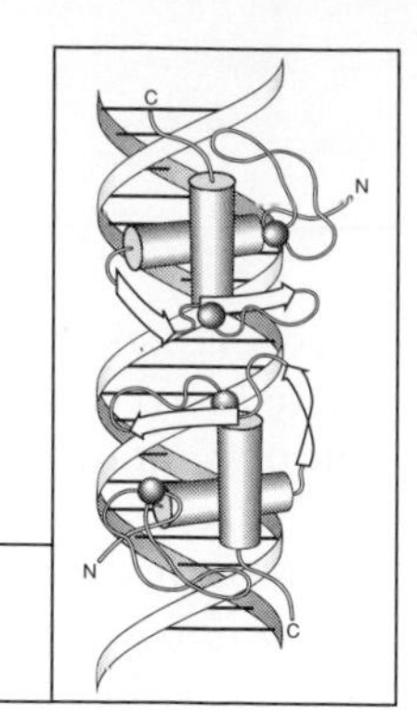

4 CHAPTER

Cruciform Structures in DNA

A. Introduction

The genetic information of all organisms contains unique sequence information necessary to encode proteins and RNA molecules. Within the noncoding regions, and on occasion within coding regions, there are defined ordered DNA sequences (dosDNA) that contain various symmetry elements. The symmetry elements inverted repeats, mirror repeats, and direct repeats are summarized in Figure 4.1. A number of non-B-DNA conformations can form within dosDNA sequences including cruciforms, intramolecular triplex DNA, and slipped mispaired structures.

Before the advent of DNA sequencing, hybridization analysis of DNA revealed snap-back sequences that form a double-stranded region after denaturation and rapid renaturation. On rapid renaturation, double-stranded DNA does not form. However, if an *inverted repeat* is present, a snap-back region or hairpin can form as shown in Figure 4.2. The inverted repeat orientation has also been called a *palindrome*. Linguistically a palindrome is a sentence that reads the same in either direction. For example, "madam, I'm Adam" is a palindrome. Since the DNA has polarity and is "read" by RNA polymerase in the 5′ to 3′ direction, the term palindrome can be used for perfect inverted repeats. Inverted repeat regions that are not completely symmet-

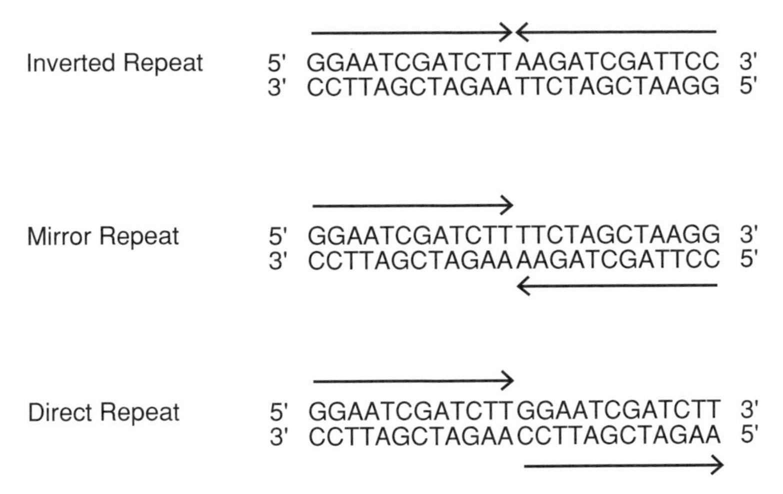

Figure 4.1 Symmetry elements. This figure shows the organization of three different types of symmetry elements in DNA: inverted repeats, mirror repeats, and direct repeats. The arrows above and below the base sequences show the organization of symmetrical complementary sequences in DNA. In the case of an inverted repeat, not only are the two strands of the DNA complementary to each other, but each single DNA strand is self-complementary within the inverted repeated region. In this configuration the sequence of DNA in either the top or bottom strand can form a double-stranded molecule. Inverted repeats have also been called palindromes. A mirror repeat has identical base pairs in one strand equidistant from a center of symmetry. Arrows are drawn above and below the lines to show complementary base pairs. However, these pairs could not form Watson–Crick hydrogen bonds because the complementary base pairs are parallel, not antiparallel, in their orientation. As shown in Chapter 6, certain sequences with mirror repeat symmetry can form triplex structures. Direct repeats are regions of DNA in which a particular sequence is repeated or duplicated. The repeat can be either adjacent to or located at some distance from the first repeat. These sequences may form slip-mispaired structures as described in Chapter 7 (Figure 7.1).

rical or that have a center region that is not an inverted repeat should not be called palindromes (imperfect palindromes or quasi-palindromes are acceptable terms).

DNA sequencing has uncovered many palindromic regions of DNA. Often inverted repeats occur near putative control regions of genes or at origins of DNA replication. This localization has made it tempting to speculate that the ability of DNA to exist in an alternative conformation could provide a molecular switch for controlling transcription or replication.

The idea for cruciform structures (Figure 4.2B) was first proposed by Platt in 1955 and Gierer in 1966. In the early 1970s, Lebowitz proposed that cruciform structures would explain their chemical modification data (Beerman and Lebowitz, 1973; Woodworth-Gutai and Lebowitz, 1976). To

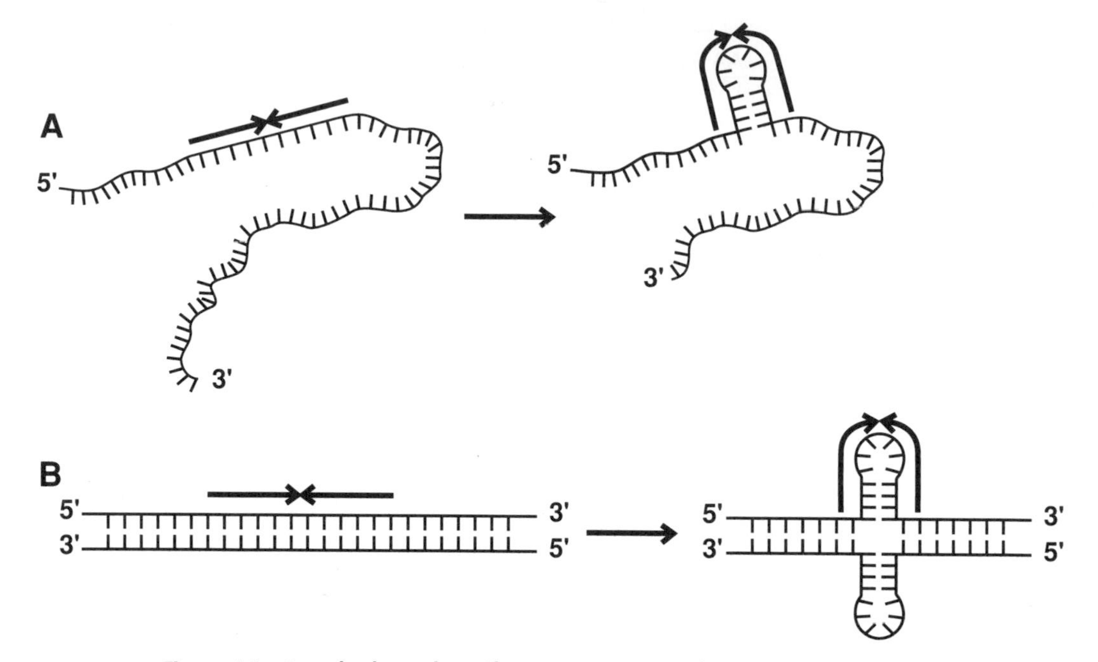

Figure 4.2 Snap backs and cruciforms. (A) An inverted repeat in a single-stranded region of DNA is shown forming a hairpin or snap-back region. The thermodynamic stability inherent in double-stranded relative to single-stranded DNA drives the formation of the hairpin structure. The double-stranded hairpin helix within the region of single-stranded DNA represents the preferred, most stable conformation. (B) When the DNA exists in double-stranded form, the two snap-back regions in single strands form a cruciform structure containing a four-way junction. However, this structure is not as favorable thermodynamically as the linear form of the DNA because hydrogen bonds are lost, since unpaired bases exist at the loop of the cruciform arms. One might expect helix distortions at the base of the cruciform loops. However, the four-way junction can form with very little deformation of the phosphate backbone and no loss of hydrogen bonding. In linear (or relaxed) DNA molecules, an inverted repeat prefers to exist in a linear form and not in the cruciform form. DNA supercoiling is required for cruciform formation.

form a cruciform the interstrand hydrogen bonds in the inverted repeat must melt and intrastrand hydrogen bonds must form between complementary bases in each single strand (Figure 4.3). On cruciform formation, there will necessarily be a loop of 3–4 unpaired bases at the tip of the hairpin, resulting in a loss of hydrogen bonding and base stacking interactions that provide stability to the linear double helix. From thermodynamic and energetic considerations, cruciforms initially seemed an unlikely structure and the suggestion of their existence was met with some degree of skepticism. It was nearly 10 years before the insightful suggestion of Lebowitz was finally accepted. In part, this acceptance paralleled a growing appreciation of the energetics and dynamics of negatively supercoiled DNA.

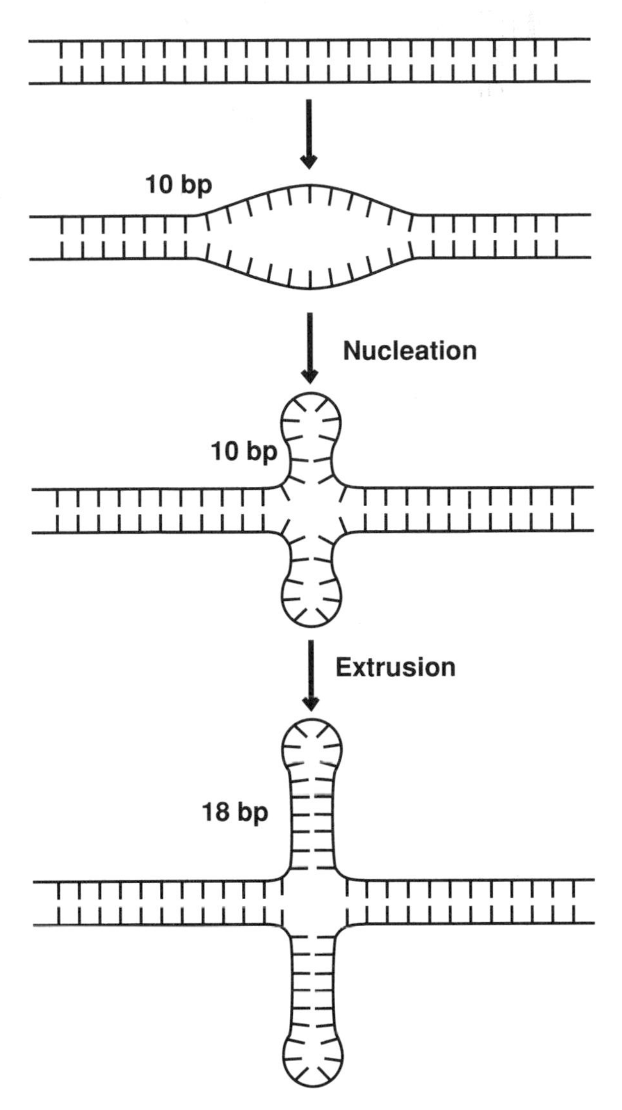

Figure 4.3 Mechanism of cruciform formation. To form a cruciform, 10 bp at the center of symmetry of an inverted repeat must unwind in a prenucleation step. This event allows nucleation to occur by *intrastrand* hydrogen bond formation near the center of symmetry. Once nucleation has occurred, branch migration occurs in the process of extrusion of the cruciform until all base pairs involved in the inverted repeat exist as hairpins or cruciform arms.

B. Formation and Stability of Cruciforms

1. Superhelical Energy Requirement for Cruciform Formation and Cruciform Stability

Energy is required to melt the center of the inverted repeat to allow intrastrand nucleation and to stabilize extrusion into the cruciform form. This energy comes from negative DNA supercoiling. Cruciforms form in supercoiled DNA but not in relaxed or linear DNA (Lilley, 1980; Panayotatos and Wells, 1981; Mizuuchi *et al.*,1982). The energy required to form a cruciform has been shown to be critical. Analysis of cruciform-containing DNA using two-dimensional agarose gel electrophoresis has shown that a very specific minimal level of energy is required for cruciform formation (Gellert *et al.*, 1983). Within a family of topoisomers, the formation of cruciforms is an all-or-none phenomenon. No cruciforms exist at or below one particular level of supercoiling. The addition of one more negative supercoil results in the formation of cruciforms in all molecules of that (and more negative) superhelical density. This level of energy is defined as σ_c, the *critical superhelical density* for cruciform formation. (A two-dimensional gel showing a cruciform transition is presented at the end of this chapter.)

A probable mechanism for cruciform formation is shown in Figure 4.3. About 10 bp must unwind at the center of symmetry to allow cruciform formation (Murchie and Lilley, 1987; Courey and Wang, 1988; Zheng and Sinden, 1988). Figure 4.4 shows the superhelical density dependence for cruciform formation in plasmid populations at different average levels of DNA supercoiling. Since DNA supercoiling promotes breathing in the double helix, it is expected that more cruciforms would form in DNA samples with higher levels of negative DNA supercoiling. This is the case.

Cruciforms are stable and thermodynamically favored in negatively supercoiled DNA because, when cruciforms are formed, negative supercoils are relaxed as shown in Figure 4.5. To form a cruciform, helical turns between the two strands of the DNA (interstrand turns) are unwound, reducing the value of T. Intrastrand helical turns form in the cruciform arms but, at a first approximation, these do not enter the supercoiling equation $L = T + W$.[1] The formation of cruciforms results in the relaxation of one negative supercoil for every 10.5 bp of the inverted repeat that participates in cruciform formation. (This is shown in the two-dimensional gel in Figure 4.19.) The stability of cruciforms results from a loss of the free energy (ΔG) inherent in supercoiled DNA. Thus the stability of cruciforms is proportional to their length (Benham, 1982; Vologodskii and Frank-Kamenetskii, 1982).

[1]Actually the topology of cruciforms can be quite complicated. If interested, see White and Bauer (1987).

Once formed, cruciforms are quite stable even in relaxed DNA because removal of a cruciform in relaxed DNA requires the introduction of one negative supercoil for each 10.5 bp of the cruciform arm converted back into a linear duplex. Thus, removal of a cruciform from relaxed DNA would increase the free energy of the molecule.

2. Effects of Base Composition on the Formation of Cruciforms

Cruciform formation requires an unwinding at the center of symmetry for nucleation of intrastrand base pairing to occur. The rate of unwinding a region of DNA will depend on base composition, temperature, ionic strength, and superhelical density of the DNA. Since 10 bp must unwind at the center of symmetry, the thermal stability of these central 10 bp may be of critical importance in defining the rate of cruciform formation. The rates of cruciform formation in supercoiled DNA vary considerably from half times ($T_{1/2}$) of minutes at 37° to hours at 50–60°C. In addition, in general, there is a correlation between the calculated T_m of the entire inverted repeat and the ease of cruciform formation. The box entitled "Effect of Base Composition on Cruciform Transitions" describes experiments that show a dominant influence of the sequences at the center of symmetry on the formation of cruciforms.

Effect of Base Composition on Cruciform Transitions

The influence of base composition at the center of symmetry on cruciform formation is demonstrated by the S and F series inverted repeats shown in Figure 4.6. The F series consists of two inverted repeats that differ only in having a 10-bp A + T- or 10-bp G + C-rich center. As might be expected, inverted repeat F2 with the 10-bp A + T-rich center has the faster rate of cruciform formation (Figure 4.7). This result is expected since the A + T-rich center should melt more easily than the G + C-rich center (due to lower stacking energies and fewer hydrogen bonds in A + T base pairs). However, this result is not entirely definitive since the overall T_m values of the F1 and F2 inverted repeats are different. To rule out an effect of overall T_m, the S series inverted repeats were constructed. These sequences contain identical base compositions and have very similar *Tm* values. The sequences vary in the relative position of A + T tracts and G_4 and C_4 blocks. As expected, the S3 inverted repeat, with a large A + T-rich center, forms cruciforms more rapidly than the S2 and S1 inverted repeats in which the G_4 and C_4 blocks are progressively moved to the center of the inverted repeat (Zheng and

continues

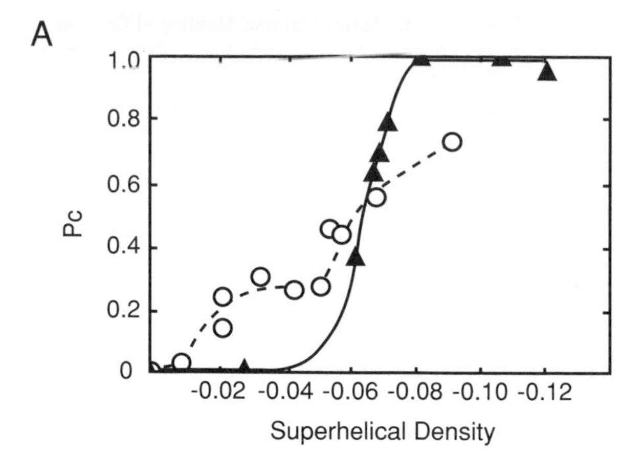

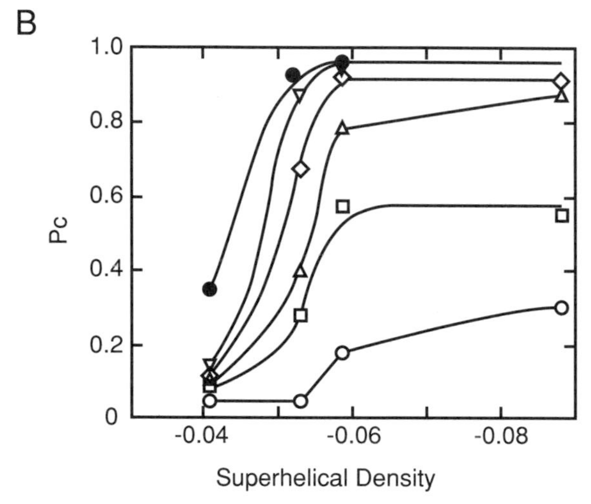

Figure 4.4 The formation of cruciforms is dependent on superhelical density and temperature. It might be expected from analysis of Figure 4.3 that factors that would influence the melting of DNA should influence the formation of cruciform structures. Melting is a function of base composition, ionic environment, temperature, and the level of DNA supercoiling. (A) The formation of cruciforms, designated P_c (the probability of cruciform formation), is plotted as a function of superhelical density. The level of DNA supercoiling (and therefore ΔG) increases with larger negative numbers, thus negative supercoiling increases from left to right. In one experiment, supercoiled DNAs initially containing no cruciforms were incubated 18 hr at 37°C to allow cruciforms to form (▲). Above a particular level of supercoiling ($\sigma = -0.06$) cruciforms were detected. The level of cruciforms was quantitated using the psoralen cross-linking assay. At superhelical density $\sigma = -0.07$, cruciforms existed in all the DNA molecules. In a second experiment, when DNA containing pre-existing cruciforms was relaxed (○), cruciforms remained in DNA of rather low superhelical density (for example, $\sigma = -0.02$). Thus, cruciforms are not in rapid equilibrium between the linear and the cruciform form, since cruciforms cannot form at this level of superhelical energy. Once formed, cruciforms can be quite stable in DNAs with low levels of DNA supercoiling. (B) The effect of temperature on the formation of cruciforms is shown for DNAs of different superhelical densities. Temperature and superhelical density promote the melting of DNA at the center of the inverted repeat. As the temperature increases, a lower level of superhelical density is required to form cruciforms. These results are

continued

Sinden, 1988). Several other experimental systems have also demonstrated that the rate of cruciform formation can be profoundly influenced by changes in base composition in the central 8–10 bp of an inverted repeat (Murchie and Lilley, 1987; Courey and Wang, 1988).

DESIGN OF A "TORSIONALLY TUNED" FAMILY OF INVERTED REPEATS

The formation of a cruciform involves opening the central 8–10 bp at the center of symmetry. This opening is driven by the presence of a critical level of DNA supercoiling, and the level of supercoiling required depends on the base composition at the center of symmetry. A population of plasmid topoisomers containing an inverted repeat is incubated for about 18 hr at 37°C to reach an "apparent equilibrium" that, in large part, may reflect rates of extrusion into a cruciform rather than a true thermodynamic equilibrium. Preserving the base composition at the center of symmetry but destroying the inverted repeat symmetry at the center of the inverted repeat should reduce the rate of cruciform formation. Thus, a higher level of supercoiling is required to form cruciforms at a rate equal to that for the perfect inverted repeat at the lower σ_c. A set of "torsionally tuned" inverted repeats (called the npF series in Figure 4.6) was constructed in which the base composition of a 14-bp A + T-rich center was preserved whereas the symmetry of the central 6, 8, 10, 12, or 14 bp was changed from an inverted repeat to a mirror re-peat (Sinden *et al.*, 1991; Zheng *et al.*,1991) The npF series is "torsionally tuned" because higher levels of DNA supercoiling are required for cruciform formation as the length of the mirror repeat increases. A level of supercoiling of σ = -0.04 is sufficient to drive cruciform formation in the F14C perfect inverted repeat by supporting the opening of 10 bp at the center of symmetry. For the F10S inverted repeat, with the 10-bp mirror repeat symmetrical center, opening the central 10 bp of the inverted repeat will not allow cruciform formation since nucleation cannot occur within this region. Consequently, a higher level of negative supercoiling is required to open a larger denaturation bubble at the center of symmetry to permit intrastrand nucleation and cruciform formation. As shown in Table 4.1, there is a correlation between the size of the nonpalindromic center and σ_c, the level of supercoiling required to drive cruciform formation at similar rates.

consistent with melting at the center of symmetry as a prerequisite for cruciform formation. The temperatures used in this experiment are 0°C(○), 10°C(□), 20°C(△), 30°C(◇), 40°C(▽), and 50°C(●). Figures reproduced with permission from Sinden *et al.*, (1983a) and Sinden and Pettijohn (1984).

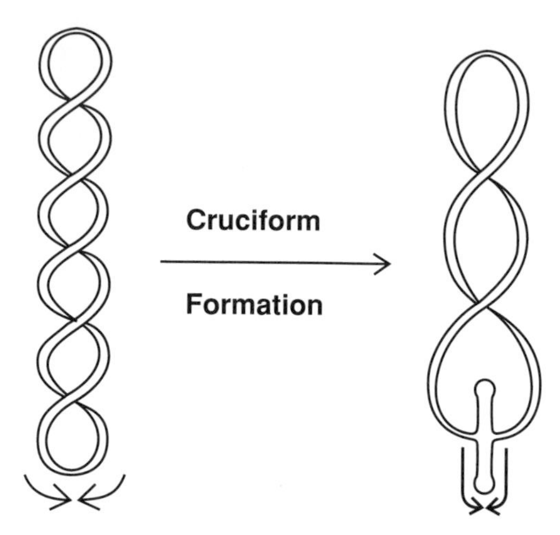

Figure 4.5 Cruciform formation in supercoiled DNA. The formation of cruciforms requires DNA supercoiling to promote melting at the center of symmetry and to stabilize the extruded structure. On the formation of cruciforms, superhelical turns in the DNA are unwound, resulting in a concomitant decrease in the ΔG of the supercoiled state and energetically stabilizing the cruciform. Approximately one superhelical turn is relaxed for every 10.5 bp of DNA that forms a cruciform. There are five negative supercoils in the DNA molecule shown here. The transition of a 30-bp inverted repeat into a cruciform structure with two 15-bp cruciform arms would result in a relaxation of about three superhelical turns.

3. Effects of Temperature on the Formation and Thermal Stability of Cruciforms

A specific energy of activation is required for the formation of a cruciform from an inverted repeat. Presumably this energy of activation reflects unwinding about 10 bp and reorganization of single strands to allow nucleation of base pairing in the hairpin stem (as shown in Figure 4.3). Figure 4.4B shows the effect of temperature on the superhelical density dependence of cruciform formation. As the temperature increases, less energy from DNA supercoiling is required to drive cruciform formation. This is reflected in a shift of the midpoint of the transition ($P_c = 0.5$) to a lower negative superhelical density value. Moreover, at higher temperatures the overall amount of the inverted repeat that forms cruciforms in the DNA sample increases.

Interesting things can happen to the rate and extent of cruciform formation at higher temperatures. For most processes requiring energy, an increase in thermal energy speeds reactions. One unusual characteristic for several different inverted repeats is that the rate of cruciform formation can decrease with increasing temperature. This behavior is shown in Figure 4.7 in which

S Series Inverted Repeats

```
      1          11         21         31         41
S1    GAATTCTATA TATATATAGG GGCCCCTATA TATATATAGA ATTC
S2    GAATTCTATA TATAGGGGTA TATATACCCC TATATATAGA ATTC
S3A   GAATTCGGGG TATATATTTA TATATATATA TATACCCCGA ATTC
S3B   GAATTCGGGG TATATATTAT ATATAATATA TACCCCGAAT TC
```

F Series Inverted Repeats

```
      1          11         21         31         41         51         61         71         81         91
F1    GAATTCCCAA TTGATAGTGG TAAAACTACA TTAGCAGAGG GGCCCGGCCG CGGCCGGGCC CCTCTGCTAA TGTAGTTTTA CCACTATCAA TTGGGAATTC

F2    GAATTCCCAA TTGATAGTGG TAAAACTACA TTAGCAGAGG GGCCCATATA TATATGGGCC CCTCTGCTAA TGTAGTTTTA CCACTATCAA TTGGGAATTC
```

npF Series Inverted Repeats

```
      1          11         21         31         41         51         61         71         81         91         101
F14C  GAATTCCCAA TTGATAGTGG TAAAACTACA TTAGCAGAGG GGCCCGATAT TTATAAATAT CGGGCCCCTC TGCTAATGTA GTTTTACCAC TATCAATTGG GAATTC

F6S   GAATTCCCAA TTGATAGTGG TAAAACTACA TTAGCAGAGG GGCCCGATAT TTAATTATAT CGGGCCCCTC TGCTAATGTA GTTTTACCAC TATCAATTGG GAATTC

F8S   GAATTCCCAA TTGATAGTGG TAAAACTACA TTAGCAGAGG GGCCCGATAT TTTATTATAT CGGGCCCCTC TGCTAATGTA GTTTTACCAC TATCAATTGG GAATTC

F10S  GAATTCCCAA TTGATAGTGG TAAAACTACA TTAGCAGAGG GGCCCGATAT TTAATTTAAT CGGGCCCCTC TGCTAATGTA GTTTTACCAC TATCAATTGG GAATTC

F12S  GAATTCCCAA TTGATAGTGG TAAAACTACA TTAGCAGAGG GGCCCGATAT TTAATTTATT CGGGCCCCTC TGCTAATGTA GTTTTACCAC TATCAATTGG GAATTC

F14S  GAATTCCCAA TTGATAGTGG TAAAACTACA TTAGCAGAGG GGCCCGATAT TTAATTTATA CGGGCCCCTC TGCTAATGTA GTTTTACCAC TATCAATTGG GAATTC
```

Figure 4.6 Inverted repeat DNA sequences. The S series inverted repeats were chemically synthesized as 44-bp sequences with identical base composition. They vary only in the position of G_4 and C_4 blocks (underlined) relative to the $(TA)_n$ blocks. The F series and npF series inverted repeats were constructed from a piece of SV40 DNA that was ligated head to head with an *Apa*I site inserted in the center. The *Apa*I site was then used to insert various centers containing G + C or A + T base pairs. In the npF series, F14C is perfectly palindromic, whereas the central 6, 8, 10, 12, and 14 A + T-rich base pairs in sequences F6S, F8S, F10S, F12S, and F14S, respectively, contain mirror repeat symmetry (doubly underlined).

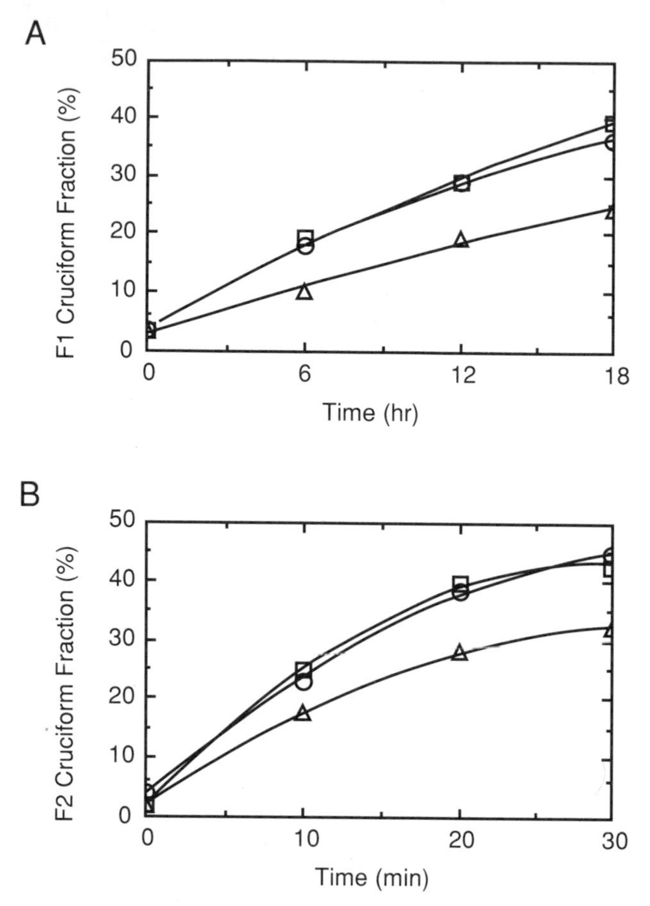

Figure 4.7 The effect of base composition at the center of symmetry on the rate of cruciform formation. The F1 and F2 inverted repeats shown in Figure 4.6 were decruciformed and the rate of cruciform formation measured at various temperatures. (A) The F1 inverted repeat with a 10-bp G + C-rich center formed cruciforms very slowly with about 35% of the inverted repeat forming cruciforms after 18 hr at 37°C (○). The rate of formation at 41°C (□) was very similar to that at 37°C whereas, at 45°C (△), the rate of cruciform formation was reduced. (B) F2, which contains 10 bp of A + T-rich DNA at the center of symmetry, forms cruciforms much more rapidly than F1. By 30 min at 37°C, about 40% of the inverted repeat formed cruciforms. The increased rate of cruciform formation likely results from the comparative ease in melting the center of F2 relative to the center of F1. In the F2 inverted repeat (like F1), the rate of cruciform formation decreased at 45°C. The decrease in the rate of cruciform formation at an elevated temperature may be explained by the formation of a second alternative conformation somewhere in the plasmid DNA. This would reduce the level of supercoiling in the DNA, effectively lowering the level of energy available to drive the cruciform transition at this higher temperature. Alternatively, a decreased rate of formation at a higher temperature may represent a lower probability of occurrence of the nucleation event. Figure reproduced with permission from Zheng and Sinden (1988).

Table 4.1
Effect of Center Sequences on the Kinetics and Superhelical Density Dependence of the Cruciform Transition

Inverted repeat	Sequence at the center of symmetry	Superhelical density required for formation (σ_c)	Rate of formation (k_c)[a]
bke[b]	GAATTC		1.7×10^{-4}
16T[b]	GATATC		1.3×10^{-3}
14T16T[b]	ATATATAT		very fast
15C16C[b]	GCCGGC		very slow
F1[c]	GGCCGCGGCC	−0.046	6.5×10^{-4}
F2[c]	ATATATATAT	−0.043	2.3×10^{-2}
S1[c]	TATATATAGGGGCCCCTATATATA	−0.046	6.5×10^{-4}
S2[c]	TATAGGGGTATATATACCCTATA	−0.043	1.7×10^{-3}
S3B[c]	TATATATTATATATAATATATA	−0.039	2.6×10^{-3}
F14C[d]	ATATTTATAAATAT	−0.038	
F6S[d]	ATATTTAATTATAT	−0.043	
F8S[d]	ATAATTAATTATAT	−0.047	
F10S[d]	ATATTTAATTTAAT	−0.049	
F14S[d]	ATATTTAATTTATA	−0.065	

[a]$k_c = 0.6931/t_{1/2}$ where $T_{1/2}$ is the apparent half-life of the linear-to-cruciform transition.
[b]Murchie and Lilley (1987).
[c]Zheng and Sinden (1988).
[d]Zheng *et al.* (1991); Sinden *et al.* (1991).

the rate of cruciform formation is slower at 45° than at 37 or 41°C. One explanation for this decrease in rate is that other regions of the DNA double helix within the plasmid molecule may unwind at elevated temperatures (going from 37 to 45°C, for example). Unwinding within the plasmid would result in a relaxation of negative supercoils and a corresponding reduction in the free energy available from supercoiling to drive the cruciform transition. Thus, sequences at one site within the topological domain may affect the rate of the structural transition at the inverted repeat at a second, perhaps distant, site. This is an example "long-range communication" or long-range effects of one region of DNA on another. This type of long-range structural communication may occur over thousands of base pairs.

Another possibility for a decrease in the rate or extent of cruciform formation at high temperature is that the nucleation process could be melted or destabilized. This is discussed in the box entitled "Why Do Different Cruciforms Have Different Temperature Optima?"

Why Do Different Cruciforms Have Different Temperature Optima?

At a certain superhelical density, an increase in temperature generally promotes the formation of a higher fraction of cruciforms within a population of topoisomers (as shown in Figure 4.4B). However, the level of cruciform formation may decrease with increasing temperature if melting of the nucleation event occurs. Analysis of the extent of cruciform formation as a function of temperature for the four S series inverted repeats is shown in Figure 4.8. S2 behaves as might be expected with a higher level of cruciform formation at higher temperatures. The behavior of S3A and S3B is unusual, however, because these molecules have relatively low temperature optima for cruciform formation at 25°C. Presumably, above 25°C the rate of cruciform formation decreases in these "apparent equilibrium" studies. One explanation for the decrease in cruciform formation at temperatures above 25°C is that above this temperature the nucleation event involving A·T base pair formation within the central 10 bp of the inverted repeat begins to be melted. Thus, although the A + T-rich center can melt easily, the nucleation event within a hairpin arm may also be easily melted. A similar melting of the nucleation event may occur in the S1 inverted repeat above 32°C.

As mentioned earlier, cruciforms are very stable in relaxed covalently closed circular DNA because the conversion back to a linear form requires the introduction of negative supercoils into DNA, which is energetically unfavorable. Cruciforms are also reasonably stable in nicked DNA (Mizuuchi *et al.*, 1982). This results from a kinetic trapping in the cruciform state, since cruciforms are thermodynamically unstable in nicked DNA. At 0°C cruciforms are extremely stable in nicked DNA. However, as the temperature increases they are rapidly converted back into the linear form of the inverted repeat with a half-life of less than 1 min at 37°C. Agents that stabilize the DNA double helix, such as divalent cations and spermidine, will stabilize cruciforms in nicked DNA (Sinden and Pettijohn, 1984). Stabilization may also result from an interaction between the two cruciform arms or the loops of the arms.

4. Effects of Salt on the Cruciform Transition: C-Type Cruciform Formation

The preceding discussion of the characteristics of cruciform formation has described experiments in buffer that approximates physiological ionic strength, typically 50 m*M* NaCl. Lilley has termed this type of transition an

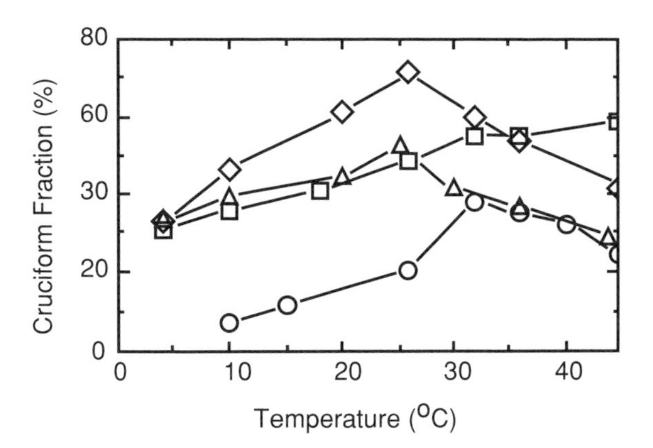

Figure 4.8 The temperature dependence of cruciform formation for the S series inverted repeats. The S1, S2, S3A, and S3B inverted repeats are shown in Figure 4.6. In this experiment, plasmids containing no cruciforms were incubated 18 hr at different temperatures before the fraction of the inverted repeat existing as cruciforms was measured. Sequence S1 (○) with the G + C-rich center of symmetry forms cruciforms less well than the S2 and S3 sequences; it has a temperature optimum of about 32°C, above which the rate of cruciform formation decreases. Sequence S2 (□) has a near linear response in terms of the fraction of cruciforms that exist as a function of temperature. This is the behavior expected if a higher temperature promotes a higher level of melting at the center of symmetry. Sequences S3A (△) and S3B (◇) with the A + T-rich centers of symmetry form cruciforms most easily with an optimum temperature of 25°C. The decrease in the extent of cruciform formation at the higher temperatures with S3A and S3B may reflect melting of intrastrand A·T base pairs that would be required to form the nucleation event prior to extrusion into cruciform arms. Figure reproduced with permission from Zheng and Sinden (1988).

S-type mechanism (where S refers to *s*alt dependent). The S-type transition is dependent on supercoiling, temperature, ionic conditions, and the divalent cation (Singleton, 1983; Sinden and Pettijohn, 1984; Sullivan and Lilley, 1987). Lilley has characterized a second pathway by which cruciforms can form in solutions lacking salt. He has termed this a C-type mechanism. "C" refers to ColE1, the plasmid containing the sequence in which this behavior was first observed (Figure 4.9; Lilley, 1985; Sullivan and Lilley, 1986).

C-Type cruciform formation is very different from the transition occurring in the presence of counter ions. Under physiological conditions, as discussed earlier, the rate of cruciform formation is dependent on the base composition at the center of symmetry. Under C-type conditions the rate of cruciform formation is *independent* of base composition. A variety of inverted repeats placed into a site in a ColE1 plasmid showed nearly the same characteristics of cruciform formation. These characteristics follow. (1) The C-type cruciform transition occurs at low temperatures <30°C (in contrast to

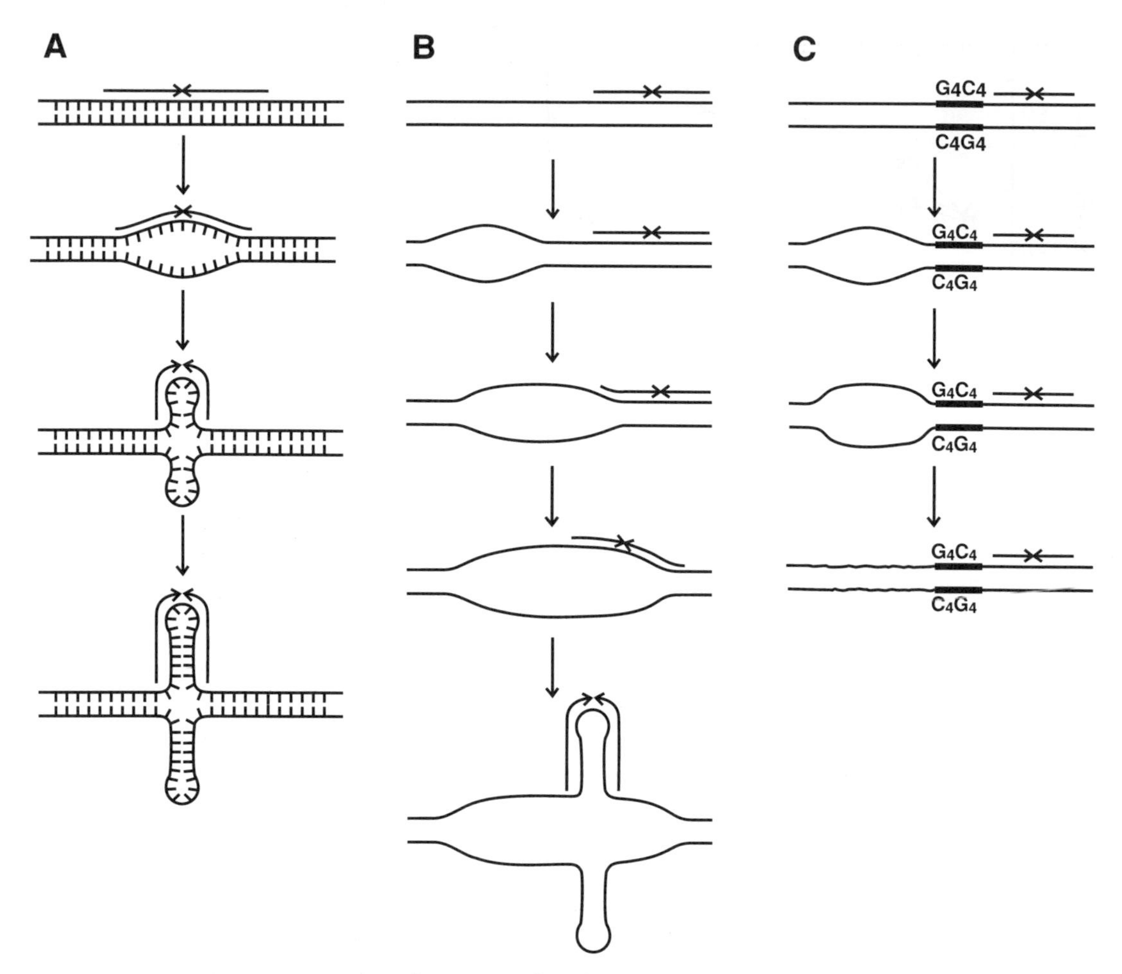

Figure 4.9 S-Type and C-type cruciform kinetics. (A) This mechanism leading to cruciform formation (see also Figure 4.3) has been called S-type by David Lilley. S-Type refers to salt type, the type of reaction that occurs at physiological ionic strengths. In this case, 10 bp at the center of symmetry must melt before nucleation and cruciform extrusion can occur. (B) The C-type kinetic pathway for cruciform formation. For C-type kinetics, which occurs in solutions with no or very low salt, an A + T-rich region of DNA flanking the inverted repeat breathes and forms a denaturation bubble, as shown in the second structure. The initial denaturation bubble can occur hundreds of base pairs away from the inverted repeat. The bubble is enlarged, encompassing the inverted repeat, and forms a stable open region. Within this open region the hairpins form, resulting in formation of a cruciform structure. (C) If a G_4C_4 tract is introduced between the inverted repeat and an A + T-rich (or C-type-inducing) sequence, expansion of the stable denaturation bubble toward the inverted repeat is blocked by the thermal stability of base pairing of the G_4C_4 tract. Since the bubble cannot transmit through this block, cruciforms do not form. Thus, a requirement for C-type kinetics is a C-type-inducing sequence (the A + T-rich region) and a C-type transmitting sequence (DNA through which the open stable region can expand).

S-type formation which generally requires higher temperatures). (2) C-Type cruciform formation requires very low ionic strengths. Counter ions stabilize the DNA double helix, and in a solution of very low ionic strength the DNA double helix is much more readily unwound. (3) C-Type cruciform formation has a large energy of activation of about 180 kCal/mol, several times that required for S-type formation (about 40 kCal/mol). These properties are summarized in Figure 4.10.

Initially, C-type behavior was only observed in the ColE1 plasmid. Thus, *C-type kinetics was a property of the plasmid, not the inverted repeat.* On further study Lilley and co-workers discovered that C-type behavior was due to A + T-rich DNA sequences found within several hundred base pairs of the inverted repeat (Sullivan and Lilley, 1986). Many different A + T-rich regions have now been identified that will confer a C-type mechanism to an inverted repeat (Sullivan *et al.,* 1988). Thus, there is no specific sequence involved; rather, the A + T richness of nearby DNA is important. The A + T-rich sequences are called *C-type-inducing sequences.* In very low salt concentration, a large region of DNA at the A + T-rich C-type-inducing sequence is believed to unwind and form a large denaturation bubble that likely includes the inverted repeat (Bowater *et al.,* 1991). Within this denaturation bubble, the hairpin can form (Figure 4.9). A *C type-transmitting sequence,* which can unwind, must be present for C-type cruciform formation (Sullivan *et al.,* 1988). The insertion of a G_4C_4 sequence between the A + T-rich C-type-inducing sequence and the inverted repeat will block C-type cruciform formation. Presumably, the high T_m of G + C-rich DNA will resist melting and will stop unwinding.

The C-type cruciform transition story is a fascinating one. However, *in vivo,* where the concentration of counter ions is relatively high, cruciform formation occurs by an S-type mechanism (Zheng *et al.,* 1991). Nevertheless, the C-type cruciform transition provided one of the first examples of long-range effects of neighboring DNA sequence on the structure, properties, and characteristics of a particular site in DNA. This long-range communication occurs over hundreds of base pairs.

5. The Removal of Cruciforms from Supercoiled DNA

There are three methods by which cruciforms can be reconverted into the linear form: (1) introduction of positive DNA supercoils, (2) heating above the T_m of the inverted repeat, and (3) removal by transcription or DNA replication. These are illustrated in Figure 4.11.

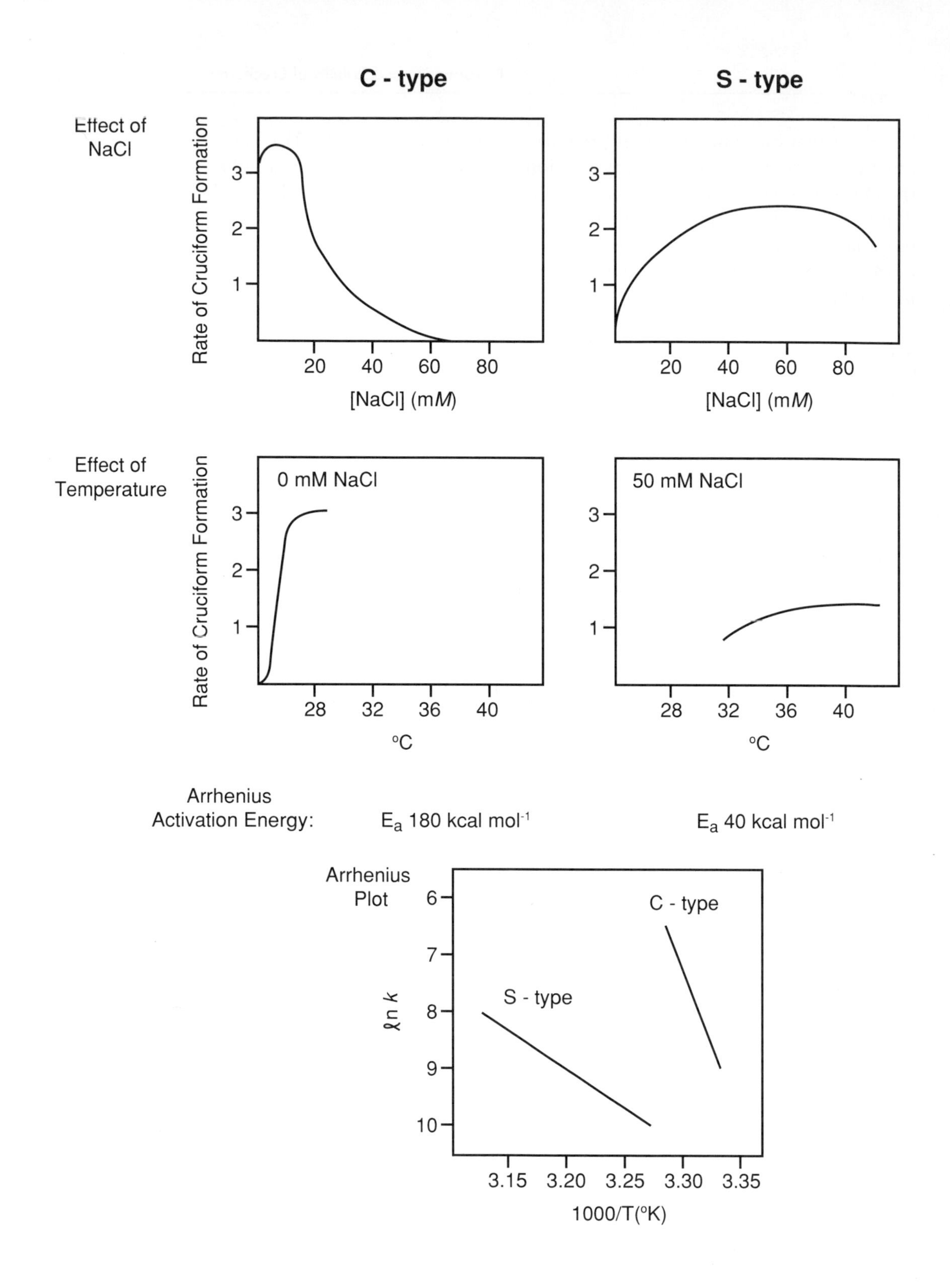

C - type
S - type
Effect of NaCl
Rate of Cruciform Formation
3
2
1
20
40
60
80
[NaCl] (mM)
Effect of Temperature
0 mM NaCl
50 mM NaCl
28
32
36
40
°C
Arrhenius Activation Energy:
Ea 180 kcal mol-1
Ea 40 kcal mol-1
Arrhenius Plot
ℓn k
6
7
8
9
10
C - type
S - type
3.15
3.20
3.25
3.30
3.35
1000/T(°K)

a. Introduction of Positive Supercoils

Since cruciforms are stable in negatively supercoiled DNA, the introduction of positive supercoils into DNA should drive cruciforms back into the linear form. Positive supercoils can be introduced by the addition of the intercalating drug ethidium bromide. (The binding of intercalating drugs and their effects on DNA supercoiling were discussed in Chapter 3.) The removal of ethidium bromide results in the reintroduction of negative supercoils. Removing the ethidium bromide at 0–4°C generally prevents the formation of cruciforms [except for $(AT)_n$ inverted repeats which form cruciforms extremely rapidly at 4°C].

b. Heating above the T_m

Cruciforms are stable in supercoiled DNA at temperatures up to the melting temperature (T_m) of the hairpin arm. Once the hairpin arm is melted, on rapid cooling the inverted repeat returns to the linear form. This behavior is expected considering the process of hybridization. Nucleation is the critical rate-limiting step for DNA hybridization. For cruciforms to form, intrastrand nucleation must occur within the hairpin arm. Although the complementary sequences in a single strand of an inverted repeat are in very close proximity, two nucleation events are provided by the plasmid DNA flanking the inverted repeat. "Zippering" or double helix formation from the ends of the inverted repeat results in rapid formation of linear DNA, preventing formation of the cruciform. Although thermodynamically cruciforms are favored in supercoiled DNA, the reannealing into duplex DNA occurs much more rapidly than nucleation within an unwound hairpin arm.

Figure 4.10 Characteristics of C-type and S-type cruciform kinetics. (*Top*) Comparison of the effects of sodium chloride on the rate of cruciform formation for C-type and S-type cruciform kinetics. C-Type kinetics have a salt optimum of <10 m*M* NaCl and the rate of cruciform formation is reduced significantly at >50 m*M*. S-Type kinetics, however, are most rapid at 50–80 m*M* NaCl. On the *y* axis, the rate of cruciform formation is in arbitrary units. (*Center*) The effect of temperature on the rate of cruciform formation under C-type kinetic conditions (0 m*M* sodium chloride) and S-type kinetic conditions (50 m*M* sodium chloride). C-Type kinetics occur at a low temperature and the dependence on temperature is very sharp. S-Type kinetic profiles have a less steep dependence on temperature. (*Bottom*) The Arrhenius activation energies and Arrhenius energy plots. The Arrhenius plot shows rates (ln*k*) as a function of temperature 1000/T(°K). The activation energy can be derived from the slope, according to $k = Ae^{-E_a/RT}$ where *k* is the rate constant, A is a constant, E_a the activation energy, R is the gas constant, and *T* the temperature (in Kelvin). Modified, with permission, from Lilley (1989).

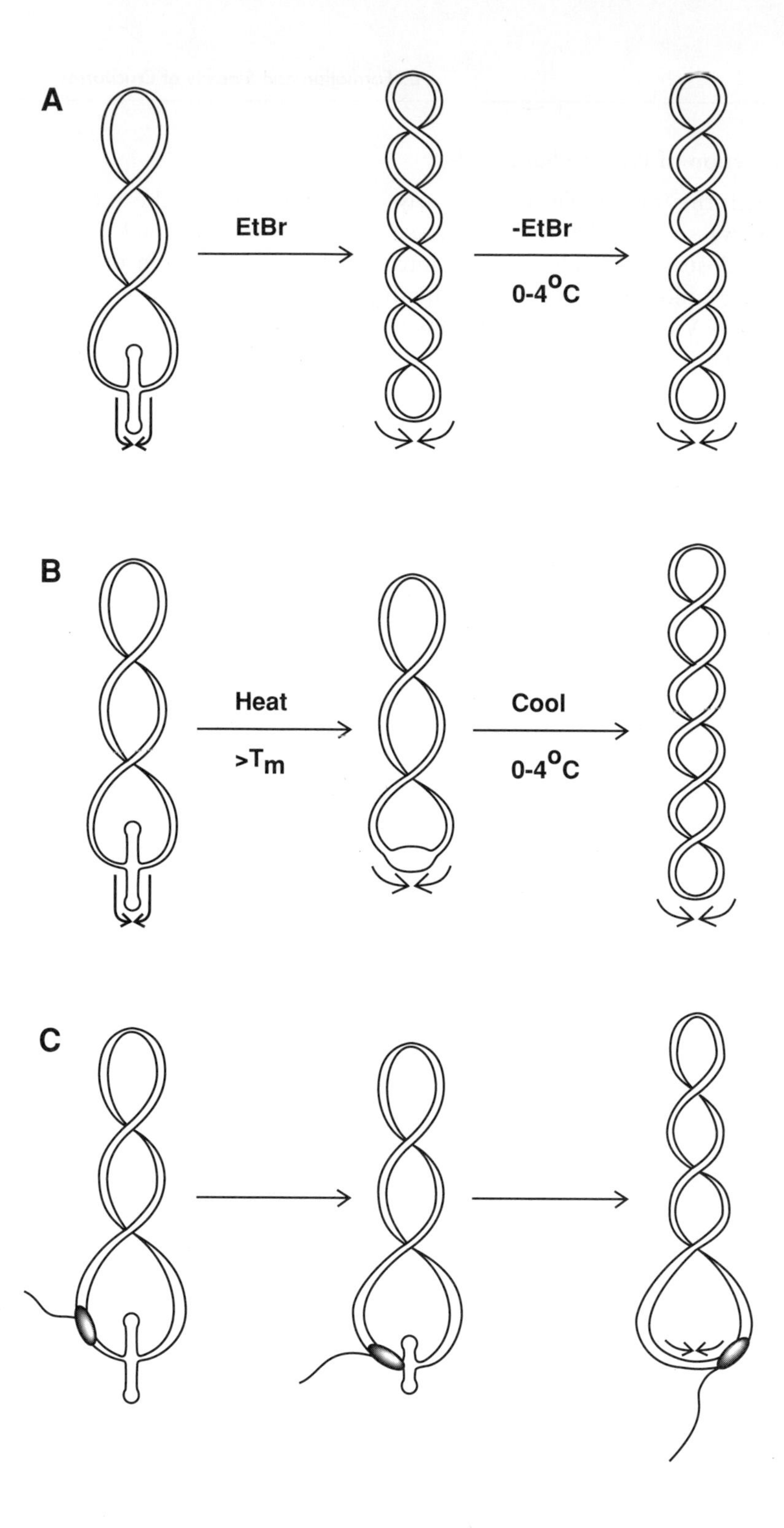
A
EtBr
-EtBr
0-4°C
B
Heat
>Tm
Cool
0-4°C
C

c. The Removal of Cruciforms by Transcription and DNA Replication

The movement of RNA polymerase through a cruciform will result in the formation of the linear form of the inverted repeat. This situation is analogous to removal of cruciforms by heating above the T_m. As RNA polymerase transverses the cruciform it will necessarily unwind the base pairing in the hairpin arms. Double-stranded DNA behind the RNA polymerase provides a nucleation site for hybridization of the complementary strands and the DNA rapidly reforms a double helix in the wake of a departing RNA polymerase. Morales *et al.* (1990) demonstrated that transcription *in vitro* removed cruciforms. Transcription and replication are probably major factors responsible for removing cruciforms from DNA *in vivo* (Zheng *et al.*, 1991).

C. Assays for Cruciform Structures in DNA

Many assays for cruciforms have been developed. This section will discuss several assays to illustrate some of the principles and approaches used to detect cruciforms. Many of these approaches have also been applied to the analysis of Z-DNA, intramolecular triplex structures, and other non-B-DNA structures.

1. Nuclease Cutting at the Loop of Hairpin Stems

Some of the first evidence for the existence of cruciforms came from studies examining the sensitivity of DNA to single-strand-specific nucleases

Figure 4.11 Removal of cruciform structures from DNA. (A) One method of removing cruciform structures from DNA is incubating DNA in the presence of a high concentration of ethidium bromide. Ethidium bromide binds to DNA by intercalation and removes negative supercoils. At high concentrations when positive supercoils are introduced into the DNA, cruciforms are driven back into the linear form. Removal of ethidium bromide (by extraction with phenol or butanol) reintroduces negative supercoils into the DNA. Maintaining DNA between 0 and 4°C kinetically prevents the formation of cruciforms in DNA. (B) Cruciforms can also be removed by heating DNA above the T_m of the inverted repeat. When heated above the T_m, the cruciform structure melts forming a denaturation bubble. On rapidly cooling DNA to 0–4°C, the inverted repeat forms a linear structure rather than the cruciform conformation, probably due to rapid annealing that begins from nucleation points at either end of the inverted repeat. This occurs more rapidly than renaturation within individual single strands to form the cruciform arms. (C) Cruciforms are removed by transcription. Movement of RNA polymerase through a cruciform drives the cruciform back into the linear form. The cruciform is unwound as polymerase transverses the structure. Interstrand renaturation, initiated from one end of the inverted repeat, promotes the formation of the linear form of the inverted repeat more rapidly than intrastrand base pairing within a hairpin arm, which would lead to cruciform formation.

(nucleases that cut only single strands of DNA). Nuclease S1, purified from *Aspergillus oryzae,* is a single-stranded endonuclease that requires Zn^{2+} for activity and has a pH optimum of ~ 4.5. S1 nuclease fails to cut relaxed plasmid DNA containing the linear form of an inverted repeat. However, digestion of supercoiled plasmid DNA containing a cruciform with S1 nuclease results in the introduction of a double-stranded break at the center of the inverted repeat (Figure 4.12). The location of this S1 nuclease cut can be mapped by digestion with a restriction endonuclease that cuts the plasmid DNA once at a known site. From analysis of the sizes of the two fragments, the S1 nuclease-sensitive site can be determined (Lilley, 1980; Panayotatos and Wells, 1981).

2. Chemical Modification at the Loop of Hairpin Stems

The unpaired bases that exist at the tips of hairpin arms have a different chemical reactivity than bases in double-stranded DNA. This differential reactivity provides an assay for cruciforms. The reactions with DNA of many chemicals used to detect cruciforms were described in Chapter 1. These chemicals include chloroacetaldehyde, dimethylsulfate, osmium tetroxide, and diethylpyrocarbonate. Following chemical modification of single-stranded DNA at the loop of the hairpin arm (the center of the inverted repeat), it is necessary to identify the site of modification at base-pair resolution.

A number of schemes have been developed to map the site of chemical modification. Following chemical modification of most bases, the glycosidic bond is more labile than the bond of an unmodified base. The modified nucleotide can be readily displaced, forming an apurinic or apyrimidinic (AP) site lacking a base on the deoxyribose sugar. Base displacement is followed by elimination of the phosphate from the deoxyribose. These reactions are catalyzed by heating DNA in a solution containing piperidine (see Figure 1.20). To map the sites of chemical modification (Figure 4.13), DNA can be cut with a restriction enzyme (*Bam*HI) and the 5′ ends labeled with ^{32}P. A cut is then made with a second restriction enzyme and the fragment containing the inverted repeat is purified. The purified fragment is subjected to piperidine treatment, which cleaves the DNA, and the reaction products are analyzed by electrophoresis on a DNA sequencing gel. Alternatively, a *primer extension* method can be used to map the sites of modification (Figure 4.13). In this procedure, a ^{32}P-labeled DNA synthesis primer is added 3′ to the site of modification. DNA polymerization will occur until the modified base is encountered. The pause or stop sites reflect the site of chemical modification.

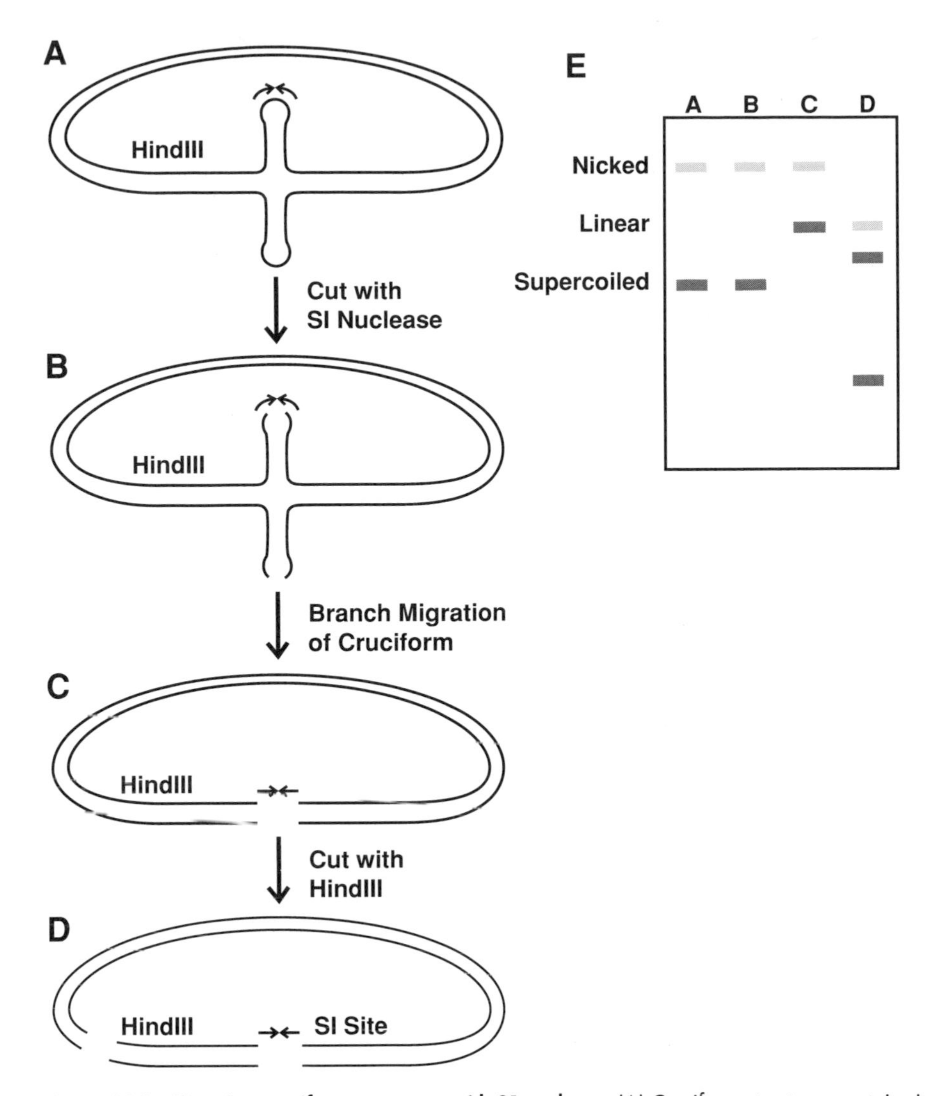

Figure 4.12 Mapping cruciform structures with S1 nuclease. (A) Cruciform structures contain single-stranded loops at the tips of the cruciform arms that can be cut with S1 nuclease, which is specific for single-stranded DNA. (B) In a supercoiled molecule, a cruciform will be stable if it has nicks at the ends of the cruciform arms. (C) If DNA is relaxed, or if DNA is heated above the T_m of the inverted repeat, branch migration will drive the cruciform back into the linear form of the inverted repeat. Thus, S1 nuclease treatment can convert a supercoiled DNA molecule into a linear form. (D) The site of the double-stranded cut can be mapped by cutting DNA with a second restriction enzyme whose recognition site location is known. (E) The analysis of the products of these reactions are shown in an idealized gel. Lanes A–D refer to the products of the reactions present in A–D. Nicked DNA is typically present in supercoiled DNA preparations and is not linearized by treatment with S1 nuclease (Lane C). It is linearized, however, by treatment with *Hin*dIII (Lane D). From the sizes of the *Hin*dIII–S1 bands, the exact location of the center of the inverted repeat (the site of the cruciform) can be identified.

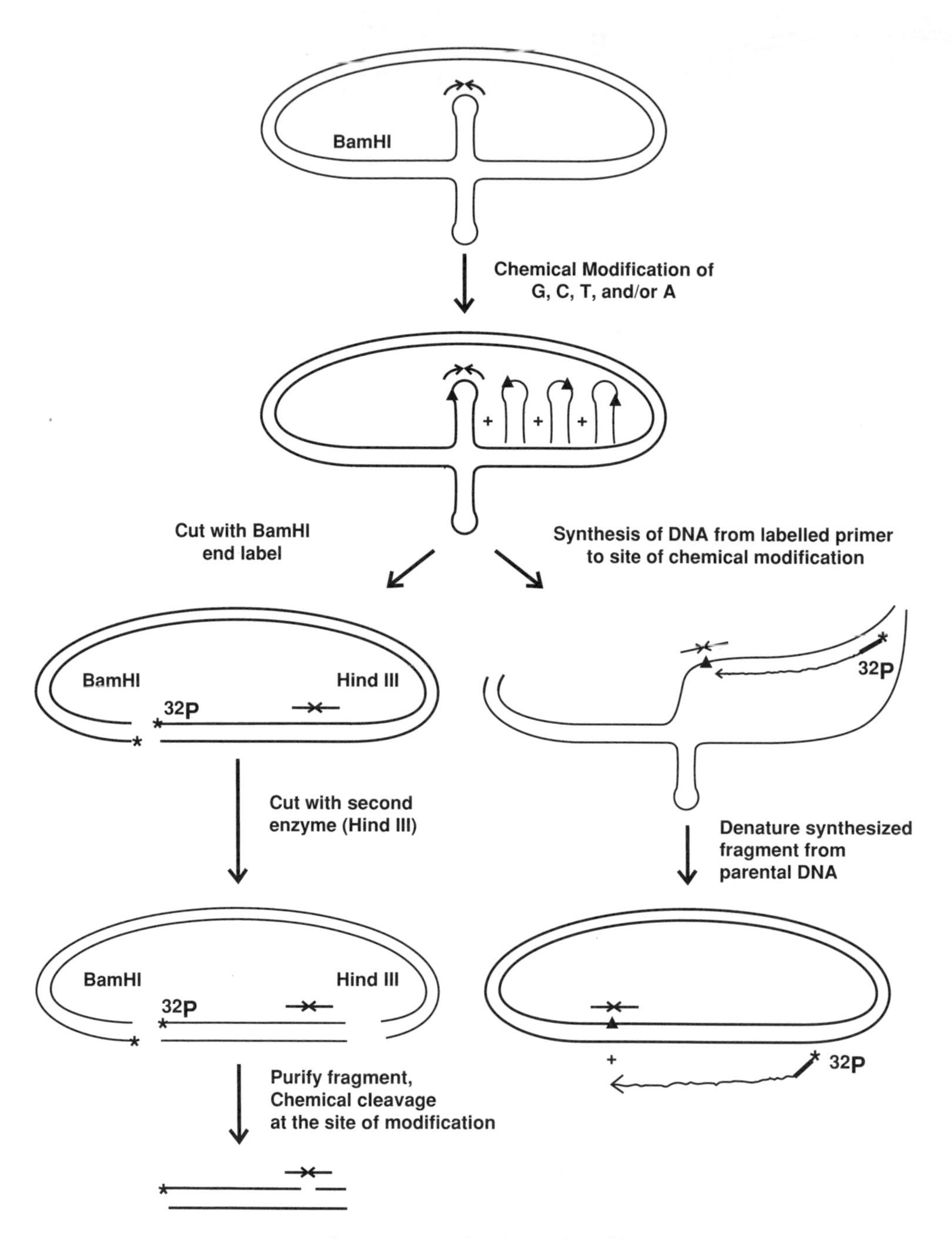

Figure 4.13 Mapping cruciform structures by chemical modification. Single-strand loops at the tips of cruciform arms are reactive with a variety of chemical probes. (*Left*) One method for the identification of the site of chemical modification involves chemical cleavage at the site of modification. DNA is cut with a restriction enzyme and end-labeled with radioactive ^{32}P at the site of restriction enzyme cleavage. Typically, DNA is then cut with a second restriction enzyme on the other side of the inverted repeat. The linear fragment is purified and subjected to a chemical cleavage reaction (piperidine treatment). The products of the cleavage reactions are denatured and analyzed by elec-

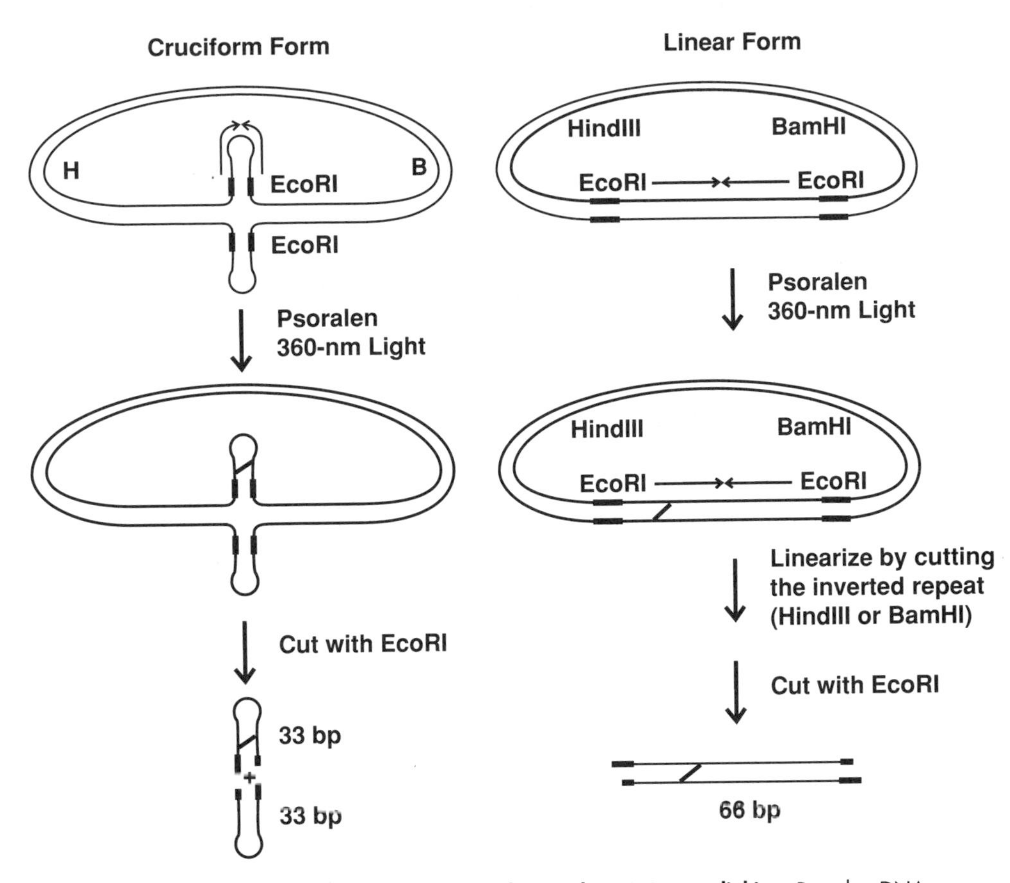

Figure 4.14 Detection of a cruciform structure with psoralen DNA cross-linking. Psoralen DNA cross-linking can be used to identify the conformation of an inverted repeat that can exist as a cruciform. (*Left*) If cruciforms are present, treatment with psoralen and light introduces intrastrand cross-links into the hairpin arm. This "locks" the DNA into the cruciform conformation. If cloned inverted repeats are cut with restriction enzymes that cut at the base of the cruciform arms, two half-size fragments are produced. In the example shown, a 66-bp inverted repeat is cut into two 33-bp hairpin arms. (*Right*) Cross-linking the linear form of the inverted repeat, followed by cutting with restriction enzymes that flank the inverted repeat, generates a full length (66-bp) restriction fragment. The 66- and 33-bp DNA fragments are resolved by electrophoresis on nondenaturing DNA polyacrylamide gels. The cross-linking assay need not be restricted to inverted repeats that can be cut out as restriction fragments. Cruciform structures in DNA induce a bend that will cause a restriction fragment containing a small cross-linked cruciform to migrate anomalously slowly during electrophoresis in a polyacrylamide gel.

---

trophoresis on a denaturing DNA sequencing gel. If the cruciform is chemically modified, a band of length equal to the distance from the first restriction cleavage site to the center of the inverted repeat will be observed. (*Right*) The *primer extension* approach to map the site of chemical modification. In this procedure, a radioactively labeled primer is extended with DNA polymerase to the site of the chemically modified base, producing a length of DNA from the primer to the site of chemical modification. DNA polymerases typically stop or pause at sites of chemical modification. The size of this single-stranded DNA is analyzed by electrophoresis on a DNA sequencing gel.

3. Analysis Using Cruciform Resolvases

Resolvases, which process recombinational intermediates, can be used to identify cruciforms in DNA. These enzymes, described in Section D,3, cut cruciforms at the base of the arms.

4. The Psoralen Interstrand Cross-Linking Assay for Cruciforms

The intercalating drug psoralen can form interstrand cross-links between the two strands of a DNA duplex on exposure to 360-nm light (see Figure 1.22). Thus, the DNA can be covalently locked into either the linear form or the cruciform form of the inverted repeat. As shown in Figure 4.14, the identification of a cross-linked hairpin arm is diagnostic for the existence of the cruciform structure. The restriction fragment corresponding to a *cloned* inverted repeat will be half its original length when cross-linked as a cruciform arm. Detection of this novel, half-sized band provides a positive indication for the presence of cruciforms. An advantage of the psoralen cross-linking assay for cruciforms is that it is quantitative (see box). Another significant advantage is that the assay can be performed in living cells.

Quantitation of the Level of Cruciforms Using the Psoralen Cross-Linking Assay

The fraction of an inverted repeat in a plasmid population that exists as cruciforms can be quantitated using the psoralen cross-linking assay. As illustrated in Figure 4.14, cross-links are introduced into a sample of DNA containing cruciforms. Following cross-linking, the plasmid DNA is cut with a restriction enzyme to relax superhelical tension. This converts all non-cross-linked cruciforms into the linear form. Next, the DNA is cut with a restriction enzyme that will cut out the inverted repeat. This generates two types of molecules: full-length linear inverted repeats and half-sized cruciform arms. The assay was developed using a 66-bp inverted repeat cloned as an *Eco*RI fragment into a plasmid (Sinden *et al.*, 1983a,b). Digestion of the linear form with *Eco*RI produced a 66-bp linear molecule whereas digestion of a cruciform produced two 33-bp hairpin arms. The percentage of the inverted repeat migrating as the 33-bp band represents F_c, the fraction of the inverted repeated cross-linked as cruciforms. F_l, the fraction of the inverted repeat cross-linked in the linear form, is determined by denaturing and quickly reannealing a sample of DNA following *Eco*RI digestion. (Following denaturation, non-cross-linked linear 66-bp molecules will snap back into 33-bp hairpin arms.

Only linear 66-bp molecules containing at least one cross-link will rapidly renature into the 66-bp linear form of the inverted repeat.) From the values of F_c and F_l, the fraction of the inverted repeat existing as cruciforms, P_c, can be determined, assuming a Poisson distribution of cross-links:

$$P_c = [F_c/(F_c + F_l)]. \tag{1}$$

P_c is independent of the total level of cross-links introduced into DNA.

Cruciforms introduce a bend into DNA, and cross-linking cruciforms locks this bend into DNA. This provides an additional psoralen-based assay for cruciforms (Zheng and Sinden, 1988). As described in Section E,2, the four-way junction at the base of the cruciform adopts a preferred orientation in solution. The relationship of the two long helices on either side of the cruciform arms is bent relative to DNA that does not contain a cruciform. Thus, a long piece of DNA containing a stable cruciform is bent and migrates anomalously during acrylamide gel electrophoresis (as discussed in Chapter 2). This property provides an assay for cruciforms in DNA (Gough and Lilley, 1985). Although cruciforms are not stable in linear DNA, a cross-linked cruciform cannot reform linear DNA and is therefore locked into the bent conformation. This procedure is applicable to inverted repeats that are not immediately flanked by restriction enzyme cutting sites.

5. Removal of Negative Superhelical Turns by Cruciform Formation

The formation of a cruciform results in the relaxation of one negative supercoil for about every 10.5 bp extruded into a cruciform. The relaxation of negative supercoils provides an assay for cruciforms. The precise measurement of the number of superhelical turns removed provides an indication of the length of the inverted repeat that has formed cruciforms. Mizuuchi *et al.* (1982) constructed very large inverted repeats, supercoiled these with DNA gyrase *in vitro*, and visually analyzed cruciform formation by electron microscopy. Not only could the four-stranded cruciform structure be visualized, but the circular plasmid DNA was relaxed.

The change in electrophoretic mobility on an agarose gel provides a very sensitive assay for cruciform formation. These gels distinguish individual topoisomers on the basis of the number of superhelical turns in DNA. The relaxation of negative supercoils on cruciform formation can be monitored by analysis on agarose gels by measuring a reduction in the migration of supercoiled DNA following cruciform formation. Identification of the linking number at which a cruciform forms provides an indication of σ_c, the critical superhelical density required for cruciform formation. The analysis of cruciforms using two-dimensional gels is shown at the end of this chapter (Figure 4.19).

6. Loss of Restriction Digestion at the Center of an Inverted Repeat

A clever assay for cruciforms involves the construction of a restriction enzyme site at the center of the inverted repeat (Mizuuchi *et al.*, 1982). If this restriction site is unique in the plasmid, then the restriction enzyme will linearize the plasmid when the inverted repeat exists in the linear form. Upon formation of a cruciform, the double-stranded restriction site is destroyed since the recognition sequences now exist as single-stranded loops at the ends of the hairpin arms. In this configuration the plasmid is not cut by the restriction enzyme.

7. Antibodies That Bind to Cruciform Structures

Antibodies have been raised against cruciform structures. The antibody binding to cruciforms can be detected by a variety of methods. The method utilized by Frappier *et al.* (1987) involves analyzing the electrophoretic mobility of a ^{32}P-labeled four-way junction molecule on an acrylamide gel. On specific binding by a cruciform antibody, the mobility of the cruciform DNA is greatly reduced.

D. Evidence for the Existence of Cruciforms *in Vivo*

1. Considerations of an Ideal *in Vivo* Assay for DNA Structure

It is important to understand the structure of DNA and how it functions inside living cells. Is DNA supercoiled? Does DNA contain unrestrained torsional tension in the winding of the DNA double helix? Do cruciforms, Z-DNA, and intramolecular triplexes exist? How are proteins organized on the DNA? These questions are being answered through analysis of DNA *in vivo* using a wide variety of chemical and enzymatic probes of DNA structure. There are three important concerns regarding the relevance of *in vivo* assays for DNA structure: (1) the condition of the cells during the assay, (2) the viability of the cells after the assay, and (3) the effect of the probe on DNA structure.

First, it is important to consider the effect of the probe or assay conditions on the biological system. The least disruptive conditions are most likely to provide data that reflect conditions in a *living cell.* Consider two extremes: the irradiation with ultraviolet light to form pyrimidine–pyrimidine dimers is completely noninvasive whereas many reactive chemical probes (dimethylsulfate, osmium tetroxide, potassium permanganate, and chloroacetaldehyde) react with cell membranes and many other cellular components, in addition to DNA.

The second consideration for an *in vivo* assay is the viability of the cells after the assay. It would be ideal to make measurements under conditions that allow cells to survive the assay procedure. In the case of many chemical probes (such as osmium tetroxide or chloroacetaldehyde), cells are no longer viable. Procedures that require purification of nuclei from eukaryotic cells clearly do not preserve viability. On the other end of the spectrum, procedures involving low level exposure to UV light or psoralen and UV light do not significantly reduce viability. *In vivo* studies should refer to treatments in which cells are alive during and preferably (but not necessarily) after the assay. *In situ* studies should refer to analyses in cells that are not capable of forming colonies following the assay procedure. Although many chemical probes are lethal, one hopes that they provide a "snapshot" of the DNA in its natural state.

A third consideration is the effect of the assay procedure on the DNA structure under analysis. Certain alternative DNA conformations may be in dynamic equilibrium with the linear B-form of the DNA. This is true for Z-DNA and some intramolecular triplexes. Long cruciforms, however, are not in rapid equilibrium with the linear form. Certain assays will drive the structural equilibrium in one direction or the other. For example, chemical and antibody assays for Z-DNA drive the equilibrium toward Z-DNA and can result in an overestimation of Z-DNA *in vivo.* The psoralen photobinding assay for Z-DNA drives the equilibrium toward B-DNA.

Many independent and alternative approaches to the analysis of DNA structure in cells have been developed. Each individual assay has its advantages and its particular disadvantages. It is important to assay DNA structure *in vivo* with several different assays since conclusions are strengthened when multiple approaches provide similar answers.

2. Application of the Psoralen Cross-Linking Assay for Quantitating Cruciforms

Psoralen is an excellent probe for *in vivo* studies because it is freely permeable into eukaryotic cells and reasonably permeable into bacterial cells (Cimino *et al.,* 1985; Sinden and Ussery, 1992; Ussery *et al.,* 1992). Because of the low solubility of many psoralen derivatives and the relatively low binding constant to DNA, little psoralen binds to DNA by intercalation in its "equilibrium dark-binding" mode, even when a solution is saturated with psoralen. On absorption of 360-nm light, covalent cyclobutane rings are formed between one or both ends of the psoralen molecule and pyrimidine bases (Figure 1.22). The extent of DNA cross-linking can be precisely controlled by varying the light dose, the psoralen concentration, or both. Psoralen reacts predominantly with DNA, although there is some reaction with other cellular components including RNA, protein, and lipid. In terms of cell viability, the reaction with DNA appears to have the most significant biological consequences.

To cross-link cruciforms in living *Escherichia coli*, psoralen is added to the cells and the cells are irradiated with 360-nm light to introduce monoadducts and interstrand cross-links. In the absence of 360-nm light, there is no reduction in the viability of the cells. On irradiation, as damages are introduced, there is a corresponding decrease in the viability. A wild-type *E. coli* cell (with a full complement of DNA repair enzymes) can correctly repair and survive, on average, about 60 cross-links per chromosome (Sinden and Cole, 1978). To assay cruciform structures in cells it is necessary to introduce cross-links into about 20% of the inverted repeats to provide a reasonable signal. (Cells do not readily survive this level of treatment.) Following the cross-linking, plasmid DNA containing the cloned inverted repeat is purified from cells, and the assay described in Section C,4 is performed.

Using the psoralen cross-linking assay, Zheng *et al.* (1991) demonstrated the superhelical-density-dependent existence of the npF series of cruciforms (shown in Figure 4.6) in *E. coli.* Cruciform formation *in vivo* was dependent on supercoiling since more cruciforms were found in cells containing a mutation in topoisomerase I that leads to an increase in the level of DNA supercoiling *in vivo.* In addition, within the "torsionally tuned" npF series of inverted repeats, cruciforms existed at higher levels for those inverted repeats that formed cruciforms at the lowest superhelical density.

The fraction of the inverted repeat existing as a cruciform *in vivo* depended not only on the sequence of the inverted repeat but on the location within the plasmid DNA. As discussed earlier, transcription removes cruciforms. When cloned within a transcription unit, the fraction of the F14C inverted repeat existing as cruciforms was about 0.5% in the *topA10* strain compared with 20–50% outside a transcription unit. For these experiments, cells were incubated 12–24 hr in growth medium after the bacteria had reached stationary phase. Presumably the process of replication had been arrested and transcription reduced, allowing cruciforms to accumulate to these high levels.

Estimation of Unrestrained Supercoiling in *E. coli* from Analysis of Cruciforms

Since the formation of cruciforms is superhelical density dependent, the level of cruciforms detected in living cells provides an indication of the *in vivo* level of unrestrained supercoiling. From analysis of the fraction of the inverted repeats existing as cruciforms, *in vivo* superhelical densities of $\sigma = -0.034$ and $\sigma = -0.04$ were estimated for wild-type and *topA10 E. coli.* The *in vivo* level of cruciforms was quite low (0.01–0.6%) when cloned within a transcription unit. When the F14C inverted repeat was cloned upstream or downstream from the tetracycline gene in plasmid pBR322, about 5 times more cruciforms were found upstream than downstream of the gene (Zheng *et al.*, 1991). This result is consistent with the twin-domain supercoiling model of Liu and Wang (Chapter 3, Section J,3,c), which suggests that negative supercoils will be introduced behind a moving RNA polymerase whereas positive supercoils will be introduced in front of the polymerase. From analysis of the minimum length that would form cruciforms, an *in vivo* level of supercoiling of about $\sigma = -0.041$ was suggested for osmotically shocked cells (McClellan *et al.*, 1990). The level of cruciforms in this study varied under growth conditions, demonstrating that intracellular tension supercoiling can change in response to the environment.

3. Nuclease Assays for Cruciforms in Cells

If a nuclease existed in cells that specifically recognized and cut cruciforms, then identification of specific cutting at inverted repeats should provide an indication of the existence of cruciforms. Panayotatos and Fontaine (1987) described experiments in which they cloned the T7 endonuclease VII gene onto a plasmid under transcriptional control of a regulated promoter. When the gene was turned on and the nuclease was expressed in cells, cuts were made at the inverted repeats. In addition, cuts were made throughout the entire chromosome, resulting in a complete destruction of the intracellular DNA. This experiment is certainly consistent with the interpretation that cruciforms exist in cells. It is difficult to rule out, however, that the nuclease was also recognizing other structural features in chromosomal DNA.

4. Detection of Cruciforms in Cells Using Chemical Probes

Certain chemicals that modify the unpaired bases at the loop of a hairpin arm can enter cells and preferentially modify these regions of DNA. Following modification, the DNA can be purified and analyzed to identify the site of modification. A disadvantage of this approach is that the cells are killed following treatment with many chemical probes. Osmium tetroxide, for example, severely damages living cells. This approach has been used by McClellan *et al.* (1990) to modify the loops of cruciform arms in a series of $(TA)_n$ inverted repeats. Sequences greater than 30 bp were chemically modified but 24-bp sequences were not reactive. Moreover, the level of chemical modification was higher with increasing lengths of the inverted repeats. Since the level of negative supercoiling required for cruciform formation increases with shorter $(TA)_n$ inverted repeats, the longest inverted repeats would be expected to exist as cruciforms at the highest levels in cells.

5. Analysis of DNA Supercoiling as an Indication of Cruciform Formation in Cells

The formation of a cruciform in a plasmid results in the relaxation of one negative supercoil for about every 10.5 bp of DNA extruded into a cruciform. *Escherichia coli* maintains a precise level of negative supercoiling in cells. Therefore, the relaxation of negative supercoils in cells by the formation of a cruciform should theoretically induce a decrease in the linking number in plasmids containing extruded cruciforms and should lead to the restoration of the original level of supercoiling by the action of DNA gyrase. Following plasmid purification from cells, providing that no change in the linking number occurred during purification, a population of topoisomers should be present that has more negative supercoils than plasmids in which the cruciform did not form. The increase in the number of supercoils on cruciform formation *in vivo* (ΔL) should be approximately equal to $N/10.5$, where N is the number of base pairs in the inverted repeat that have extruded into the cruciform. Typically, the formation of cruciforms *in vivo* does not occur in all topoisomers of a plasmid population. In the case in which cruciforms formed in 50% of the topoisomers, rather than a single Gaussian distribution of topoisomers (as shown in Figure 3.14), there would be two overlapping Gaussian distributions. The higher negatively supercoiled population, resulting from cruciform formation, would be shifted by ΔL. Haniford and Pulleyblank (1985) used this system to present one of the first indications for formation of cruciforms in *E. coli*.

6. Application of Cruciform Antibodies to the Analysis of Cruciforms in Eukaryotic Cells

Antibodies have been raised and isolated that recognize structural features of cruciforms but not the DNA sequence of specific inverted repeats. *In vitro,* antibody binding to DNA can be detected by the precipitation of DNA or the retardation of cruciform-containing DNA during gel electrophoresis (Frappier *et al.,* 1987). When antibodies are applied to isolated nuclei, they can be detected by binding a second fluorsecent antibody to the cruciform antibody. The localization of the fluorescence by microscopy reveals the sites and relative abundance of cruciforms.

In eukaryotic cells, cruciforms are detected most strongly at the G_1/S boundary of the cell cycle, just before the beginning of the period of DNA synthesis (S phase). Ward *et al.* (1990,1991) estimated that there might be as many as 3×10^5 cruciforms per eukaryotic nucleus and suggested that inverted repeats may form cruciforms as a prerequisite for the initiation of DNA replication. As discussed subsequently, inverted repeats are quite common at origins of DNA replication. Perhaps the formation of a cruciform triggers initiation of DNA replication at a specific origin. Alternatively, cruciforms may simply accumulate at the G_1/S boundary when no replication is occurring. As soon as synthesis begins, cruciforms may be converted by replication through the inverted repeat back into the linear form.

Caution should be exercised in interpreting the cruciform antibody (and Z-DNA antibody; see Chapter 5) studies of alternative conformations in eukaryotic cells. These analyses require the isolation and purification of nuclei to allow the antibody to enter the nucleus. The cells must be gently lysed, removing the nuclei from their natural environment. It has been shown by Jackson *et al.* (1990) that even the gentlest isolation procedures result in significant changes in the chromatin organization of DNA. In eukaryotic cells, negative supercoiling in the bulk of chromosomal DNA is restrained by the organization of DNA into nucleosomes. Relaxed DNA should not support the formation of cruciforms (or Z-DNA or triplex structures). However, if the chromatin structure is disrupted and nucleosomes are lost, unrestrained supercoils may be introduced into DNA as artifacts and may not reflect the natural *in vivo* situation of the DNA. However, as discussed in Chapter 9, gene regions that are active in transcription are torsionally stressed in eukaryotic cells.

E. The Biology of Cruciforms: Do Cruciforms Have a Biological Function?

As yet there are few definitive biological roles for cruciform structures, although the prevalence of inverted repeats in bacterial, eukaryotic, and viral DNA cries out for a biological role for cruciforms. On the other hand, in the linear form, inverted repeats play important biological roles as binding sites for dimeric proteins. This section will discuss how cruciforms could play biological roles in transcription or DNA replication. These descriptions will range from speculative to those based on existing preliminary data.

1. The Biology of Inverted Repeated DNA Sequences

a. Inverted Repeats as Protein Binding Sites

Many short inverted repeats, ranging in size from 4 to 20 bp, represent the binding sites for specific proteins. One class of binding sites are restriction endonuclease cleavage sites, which are also modification sites for corresponding methylases. Many 4-, 6-, 8-, or 10-bp recognition sites for restriction enzymes are palindromic. Recognition and cleavage (or methylation) of these sites occur in the linear form. Moreover, most restriction sites are too short to form a cruciform (which requires a 3- to 4-bp loop at the end of the hairpin stem).

Another class of binding sites are quasi-inverted repeated operator sequences. Most dimeric repressor proteins bind these sequences, with each monomer repressor molecule recognizing a 6- to 10-bp DNA sequence.

b. Inverted Repeats Occur at Replication Origins

Inverted repeats are frequently associated with origins of DNA replication. The sequence of several inverted repeats associated with origins are listed in Table 4.2. The 254-bp *E. coli* origin of replication contains many inverted repeat elements. It has been tempting to speculate that DNA at the origin could adopt a very elaborate secondary structure with many cruciforms or quasi-palindromic secondary structures, since an unusual structure at the origin might facilitate the assembly of the proteins required for the initiation of DNA replication. The origin region does adopt an organized three-dimensional shape but, to the disappointment of DNA structure enthusiasts, the origin is organized as a loop of the linear duplex form by the association of specific proteins involved in the initiation of replication. However, many origins do have an alternative secondary structure called a DNA unwinding element (discussed in Chapter 7).

Hairpins in Single-Stranded Bacteriophage Genomes. A region containing three inverted repeats is required for the initiation of DNA replication in the single-stranded bacteriophage G4. The phage genome contains a region with 44-, 21-, and 20-bp inverted repeats which in single-stranded DNA will exist as snap-back or hairpin regions. The double-stranded structures of the 44- and 21-bp inverted repeats (the 21-bp sequence has an 8-bp nonpalindromic center) are required for proper origin function (Table 4.2). Following

Table 4.2
Palindromic Replication Origins

Origin	Sequence
ϕX-174[a]	GGTTATAACGCCGAAGCGGTAAAAATTTTAATTTTTGCCGCTGAGGGGTTGACCAAGCGAAGCGCCCTACGTTTTCT (I, II)
G4[a]	AACAAAGGCCGCCCCTCACTGGTCAGATACCTGCCCAATGTGGGGCGGCCGTGC (III) CTACGGAGATACTCGAGTCTCCGATACATGGACGGCGAAAGCCGCCGTCCCTACT (II, I)
pT181[a]	TAGACAATTTTTCTAAAACCGGCTACTCTAATAGCCGGTTGGACGCACATACTGTGTGCATATCTGATCCAAA
HSV-*ori*L[b]	−70 −60 −50 −40 −30 −20 −10 AGCCGGTGGGCGTGGCCGCTATTATAAAAAAAGTGAGAACGCGAAGCGTTCGCACTTTGTCCTAATAATATA 10 20 30 40 50 60 70 TATATTATTAGGACAAAGTGCGAACGCTTCGCGTTCTCACTTTTTTTATAATAGCGGCCACGCCCACCGGCT
HSV-*ori*S[b]	AAGCGTTCGCACTTCGTCCCAATATATATATATTATTAGGGCGAAGTGCGAGCA
SV40[c]	10 0 5230 5220 5210 CAGAGGCCGAGGCCGCCTCGGCCTCTGAGCTATTCCAGAAGTAGTGAGGAGGCTT 27 bp palindrome early palindrome (15 bp)
EBV[d]	9021 9031 9041 9051 9061 9071 9081 ACTAACCCTAATTCGATAGCATATGCTTCCCGTTGGGTAACATATGCTATTGAATTAGGGTTAG

[a]Kornberg and Baker (1992).
[b]Weller *et al.* (1985); Lockshon and Galloway (1986); Elias and Lehman (1988).
[c]Toose (1982).
[d]Hudson *et al.* (1985).

infection of the phage, the DnaG primase protein binds to the double-stranded sequence of the 20-bp inverted (repeat region I) as a prerequisite for initiation of synthesis of an RNA primer. Although the secondary structures (hairpins) of this region are required for replication within a single-stranded DNA molecule, it is not known if the G4 cruciforms exist in supercoiled DNA or if they would be important biologically.

Eukaryotic Viruses. Inverted repeats are associated with origins of replication of many eukaryotic viruses including SV40, herpes simplex virus (HSV), and Esptein–Barr virus (EBV). The SV40 origin of replication contains a perfectly symmetrical 27-bp inverted repeat and a 15-bp "early palindrome" which is an imperfect inverted repeat (Table 4.2). The inverted repeats are required for replication but in the linear form. Several regions have been identified as origins of replication in HSV. Two regions, ori_{L1} and ori_{L2}, both contain large palindromes that are very A + T rich. ori_{L1} is a 144-bp inverted repeat with a central 20-bp region of perfectly symmetrical, pure A + T-rich sequence (Weller *et al.,* 1985). This sequence forms a cruciform extremely rapidly and at a low superhelical density. ori_{L2}, at 136 bp, is very similar to ori_{L1} but not quite as long. It, too, forms cruciforms very easily. The ori_{L2} repeat was cloned into *E. coli* (Lockshon and Galloway, 1986) but was very unstable and underwent deletion readily. The ori_{L1} inverted repeat could not be cloned into *E. coli* but was cloned into yeast, where long inverted repeats appear to be more stable than in *E. coli*. ori_{S} contains a shorter inverted repeat which is very similar to ori_{L1}.

Plasmid pT181—Is a Cruciform Cut to Begin Replication? Plasmid pT181 represents perhaps the most exciting possibility yet for a biological regulatory role for a cruciform. Novick and co-workers found that a small inverted repeat (Table 4.2) that represents the origin of replication in plasmid pT181 can form a cruciform *in vivo*. This cruciform may be involved in the initiation of plasmid replication (Noirot *et al.,* 1990). In this replication system the initiation protein RepC binds to the origin region of either single-stranded or double-stranded DNA. This protein introduces a specific nick at the center of the inverted repeat. In supercoiled DNA (but not single-stranded DNA) replication begins at the nick.

Noirot et al. (1990) demonstrated that the binding of the RepC protein greatly enhanced the formation of an unwound structure, possibly the cruciform, at the origin. In pT181 there are sequences that unwind, forming S1 nuclease-sensitive sites, at lower superhelical densities than that required for cruciform formation at the origin. When these competing sequences were deleted, a higher level of cruciforms was observed *in vivo* in this plasmid than in the wild-type plasmid. Moreover, replication of the modified plasmid proceeded more efficiently *in vivo* as indicated by a higher plasmid copy number (or plasmid : RepC protein ratio).

What is the biological function of the RepC protein? The RepC protein may recognize and bind to the cruciform, thereby stabilizing it. A nick would then be introduced into the loop of the cruciform arm, which acts as a replication primer. Alternatively, the RepC protein may participate in the formation and stabilization of the cruciform. A model for the role of the cruciform in the initiation of plasmid replication is presented in Figure 4.15.

c. Involvement of an Inverted Repeat in N4 Virion RNA Polymerase Promoter Recognition.

The bacteriophage N4 requires a phage-encoded RNA polymerase packaged in its virion for the transcription of early messages from the 72-kb linear N4 genome. Transcription of early messages requires a supercoiled DNA template on which the *E. coli* SSB protein acts as a transcriptional activator (Markiewicz *et al.,* 1992). The promoters of these early genes contain inverted repeats that are required for transcription. Remarkably the inverted repeat symmetry and not a defined base sequence is required for transcription (Glucksmann *et al.,* 1992). A model has been proposed in which DNA is supercoiled by DNA gyrase, producing a single-stranded region at the inverted repeat and allowing the formation of a short cruciform structure (or stem–loop structures in unwound DNA). The SSB protein may stabilize the alternative secondary structure at the promoter and facilitate the binding of the N4 RNA polymerase (Glucksmann *et al.,* 1992). This system represents probably the most convincing evidence for a system in which there is the involvement of a cruciform structure in the regulation of gene expression.

2. Cruciforms as Models of Holliday Recombination Intermediates

Genetic recombination is a process that results in the novel linkage of duplex DNA of homologous chromosomes. The process of recombination is shown in Figure 4.16. Two homologous chromosomes (or two nearly identical regions of DNA) align (Figure 4.16A), and strands are exchanged between the two DNA duplexes (Figure 4.16B). In Figure 4.16C, the exchanged region has moved to the right following *branch migration.* This four-stranded structure is called a Holliday recombination intermediate (after Robin Holliday who first described it). This structure can be isomerized to the four-way junction structure (Figure 4.16D). Although it appears that the exchanging DNA

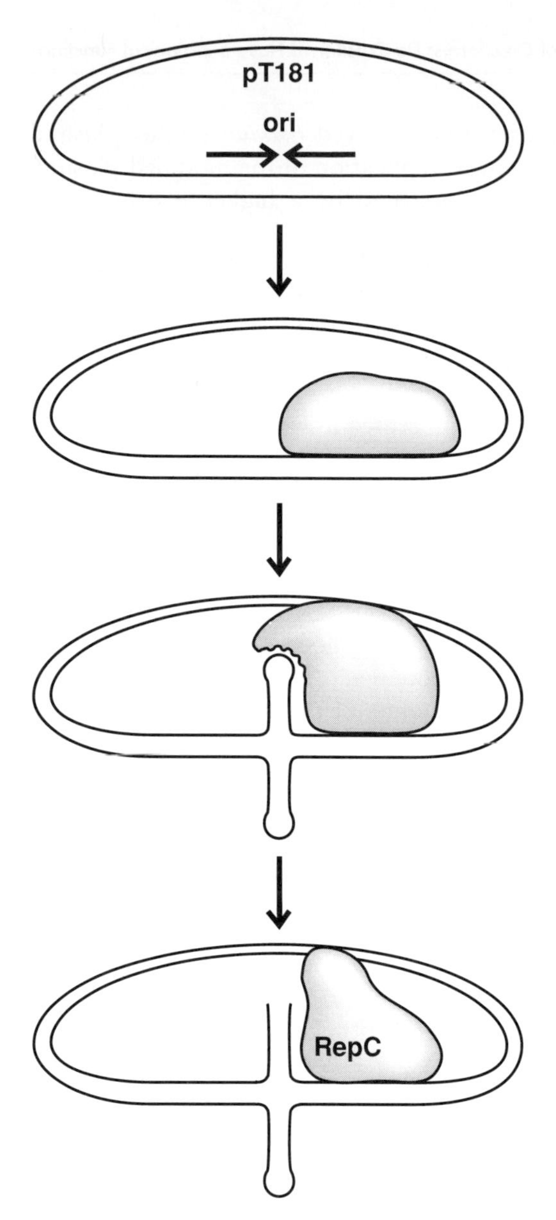

Figure 4.15 Potential role for a cruciform structure in replication at the origin of plasmid pT181. The origin of plasmid pT181 contains an inverted repeat (Table 4.1). The RepC protein may bind adjacent to the replication origin when the inverted repeat exists in the linear form. Superhelical energy may promote formation of a cruciform structure within the inverted repeated. The loop of the cruciform structure may be susceptible to endonucleolytic digestion dependent on RepC. Once the center of the inverted repeat is cut, it provides a primer (3′-OH end) for extention by DNA polymerase.

strands are "stretched" between the original helices, the construction of molecular models suggests that a four-stranded recombinational intermediate could form easily.

Analysis using chemical and nuclease probes has shown no unpaired bases at the four-way junction (Duckett *et al.,* 1988). The laboratories of Hagerman and Lilley have constructed a number of different four-way junctions and used electrophoretic mobility and various biophysical techniques to elucidate the three-dimensional structure of the four-way junction (Cooper and Hagerman, 1987,1989; Duckett *et al.,* 1988; Murchie, 1989). The DNA adopts a preferred spatial orientation that is not tetrahedral or square planar, but in which pairs of helices at the four-way junction stack into an X structure (Figure 4.17).

Changes in the nucleotides at the base of the four-way junction can influence which strands of the Holliday structure stack. In nature, once a four-way junction has been formed between two homologous chromosomes, it is mobile and the four-way junction can branch migrate. Changes in DNA sequence at the base of the junction may influence the shape of the junction, which then may direct the resolution of the Holliday structure into one of the two recombinants shown in Figure 4.16 (molecules E and F). Recombinant structure F consists of two molecules containing a heteroduplex at the B/b′ (and B′/b) region. Recombinant E, which has been cut differently than F, has a new covalent linkage of flanking outside markers. In addition to the heteroduplex B/b′ (and B′/b) regions in recombinant E, A is linked to C and C is linked to A.

The enzymes that cut four-way junction or Holliday structures are called *resolvases.* These have been purified from a number of sources ranging from bacteriophage T4 to mammals (Mizuuchi *et al.,* 1982; de Massey *et al.,* 1984,1987; Symington and Kolodner, 1985; West and Körner, 1985; West *et al.,* 1987; Elborough and West, 1990; Taylor and Smith, 1990; Connolly *et al.,* 1991). Plasmids containing cruciforms as well as synthetic four-way junctions have been used as substrates for resolvases. These enzymes typically cut 4–5 bp up the stem on either the 3′ or the 5′ end of a cruciform arm. Cuts are always made on opposite sides of the four-way junction to resolve the recombinant into two symmetrical molecules. Figure 4.17 shows the cutting pattern for the T4, calf, and yeast resolvases (Bhattacharyya *et al.,* 1992).

Cruciforms can exist at high levels in plasmid DNA in *E. coli.* Therefore RuvC resolvase, which cuts products of genetic recombination, apparently must not recognize and cut cruciforms in DNA *in vivo.* This certainly allows the possibility that cruciforms could form and play important roles in the control of gene expression or in the initiation of DNA replication.

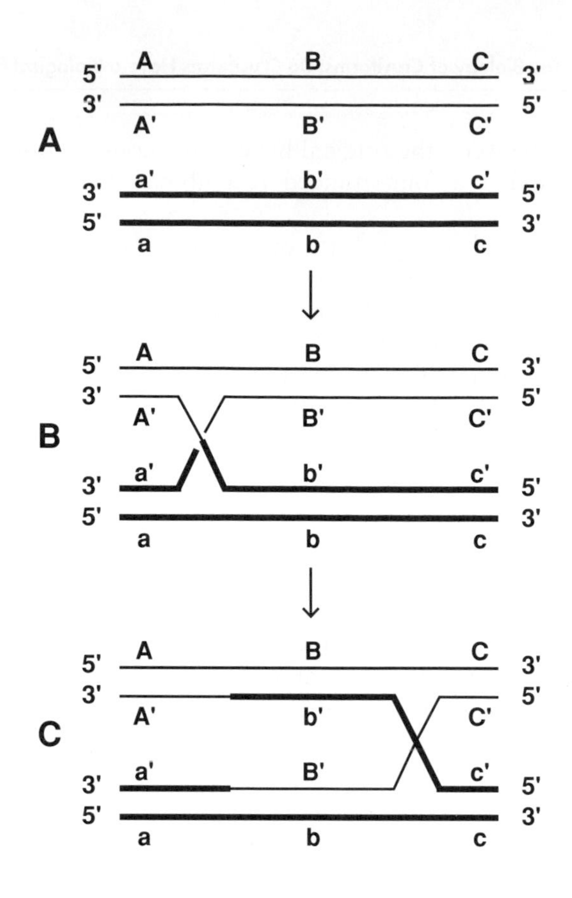
A
5' A B C 3'
3' A' B' C' 5'
3' a' b' c' 5'
5' a b c 3'
B
5' A B C 3'
3' A' B' C' 5'
3' a' b' c' 5'
5' a b c 3'
C
5' A B C 3'
3' A' b' C' 5'
3' a' B' c' 5'
5' a b c 3'

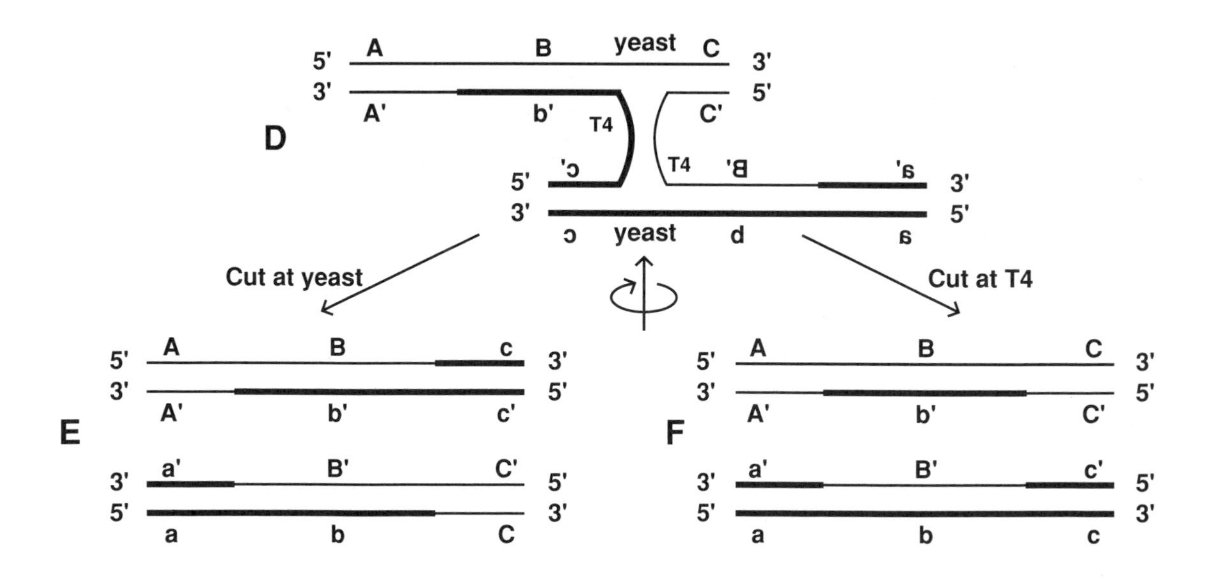
D
5' A B yeast C 3'
3' A' b' C' 5'
T4
T4
5' c' B' a' 3'
3' c yeast d a 5'
Cut at yeast
Cut at T4
E
5' A B c 3'
3' A' b' c' 5'
3' a' B' C' 5'
5' a b C 3'
F
5' A B C 3'
3' A' b' C' 5'
3' a' B' c' 5'
5' a b c 3'

3. The Genetic Instability of Long Inverted Repeats in *Escherichia coli*

It is difficult to clone and maintain long inverted repeats in plasmids in *E. coli* (Lilley, 1981; Mizuuchi *et al.*, 1982; Leach and Stahl, 1983). In fact, it is virtually impossible to clone perfect inverted repeats longer than 150 bp in *E. coli*. This is not due to homologous recombination, since cloning is done in *rec*A$^-$ cells to prevent RecA-mediated recombination. There are probably two mechanisms involved in the genetic instability of long inverted repeats First, long inverted repeats may be deleted by intramolecular recombination that does not involve RecA (Warren and Green, 1985). Since intramolecular recombination would require the formation of a DNA loop, this may require a length of DNA in the inverted repeat longer than the persistence length of DNA, about 100–200 bp. Interactions with proteins, for example, the wrapping around the histone-like HU protein, could shorten this length. The second mechanism for the instability of *cloned* inverted repeats involves slipped misalignment during DNA replication in which the misalignment event is stabilized by the formation of a hairpin arm (Figure 4.18). Slipped mispairing between direct repeats during DNA replication leads to deletion of the DNA between the direct repeats and one copy of the direct repeat. Deletion can occur between direct repeats that are separated by a few to over several hundred base pairs of DNA (Albertini *et al.*, 1982). Cloned inverted repeats represent a special situation in which the direct repeats constitute the duplicated restriction sites used for cloning. The DNA between the direct repeats is palindromic and can form a hairpin stem that stabilizes and facilitates the misalignment.

The deletion of inverted repeats has been systematically studied in several laboratories (Collins, 1980; DasGupta *et al.*, 1987; Williams and Müller, 1987; Balbinder *et al.*, 1989; Weston-Hafer and Berg, 1989,1991; Sinden *et*

Figure 4.16 Resolution of Holliday recombination intermediates. (A) Two homologous chromosomes align. One chromosome is denoted by thin lines with uppercase markers (letters) and one by thick lines with lowercase markers. (B) A nick has been introduced into complementary strands of homologous chromosomes; the strands have been exchanged, and the nick sealed. (C) Branch migration to the right has occurred, creating two heteroduplex regions: Bb′ and B′b in the upper and lower chromosomes, respectively. (D) This Holliday structure can be isomerized into the structure shown by rotating the lower chromosome 180° through the plane of the paper. (The end containing a′a which was on the left in C is now on the right.) This four-way junction, which is structurally analogous to a cruciform structure, can be resolved (or cut) two different ways by resolvases. (E) The yeast resolvase cuts on the top and bottom of the structure in D, resolving the molecule into two recombinants in which AA′ is now linked to cc′ and a′a is linked to C′C. This represents a recombination event in which the regions flanking the heteroduplex are derived from different original chromosomes. (F) The T4 enzyme cuts on the left and right side of the structure shown in D, and the DNA sequences flanking the heteroduplex are preserved. For example, AA′ is still linked to CC′. Therefore cutting at the T4 site does not result in the new genetic linkage of flanking markers.

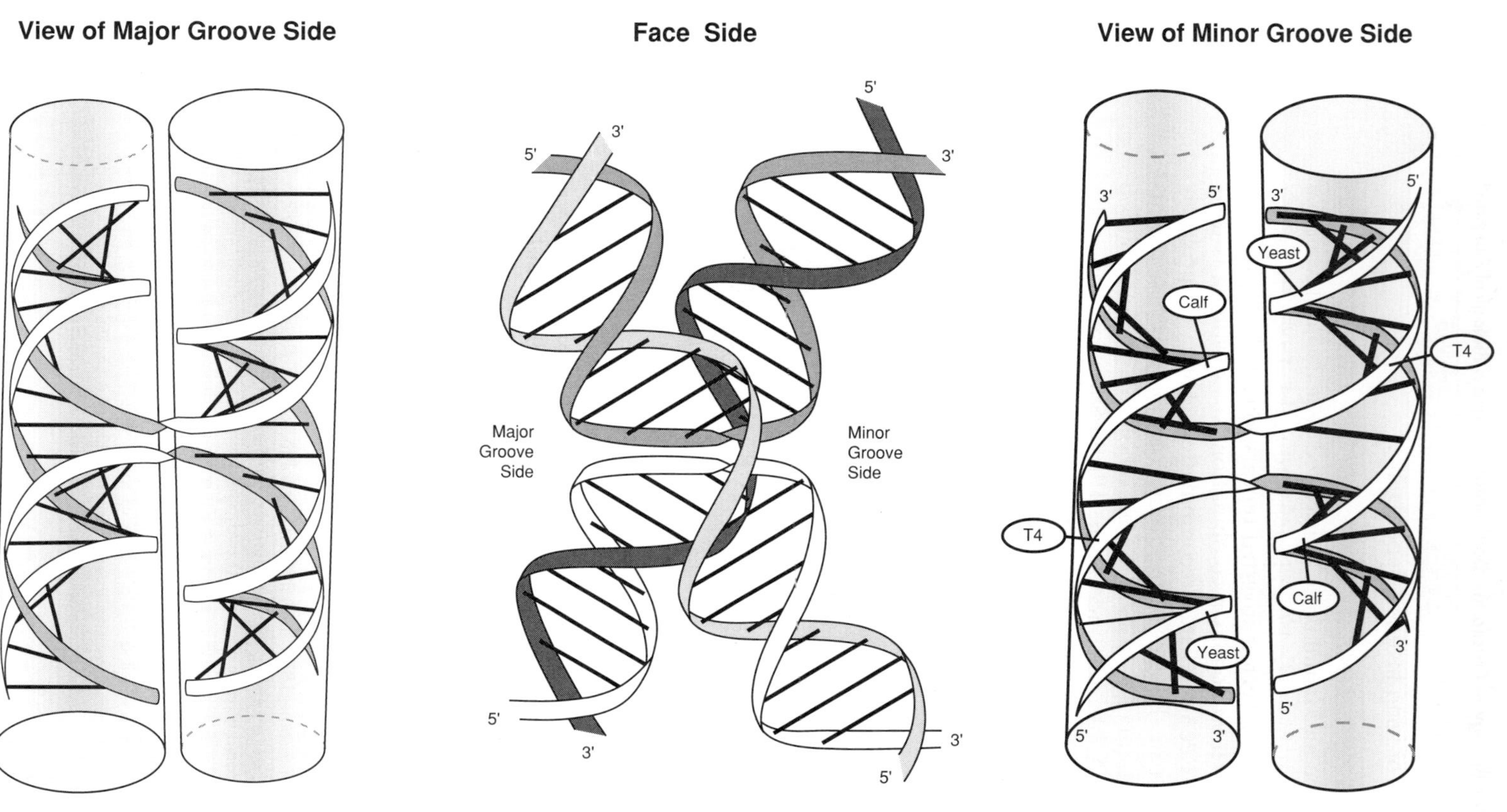

Figure 4.17 Helical representations of the Holliday structure. (*Center*) A helical representation of the "face side" of a Holliday recombination intermediate, with the four individual strands indicated by different degrees of shading. There are two stacked helices organized in a crosswise X structure. Each stacked helix contains one continuous strand and two others that curve 90° as they move into an adjacent helical arm of the Holliday structure. (*Left*) The major groove side of these two DNA helices. The bottom end of the helix on the left and the top end of the right helix are pointing out from the plane of the paper. As the strands cross between helices, there is only a slight bend. All Watson–Crick hydrogen bonds are maintained at the 4-way junction. (*Right*) The minor groove side of a Holliday junction. The sites of cleavage of the yeast, calf, and T4 Holliday structure resolvases are indicated. Note that although all enzymes have different cutting locations, the sites of cutting are symmetrically placed with respect to the center of the Holliday junction. The yeast and T4 enzymes cut opposite strands of the cruciform structure and resolve the two molecules differently (as shown in Figure 4.16). Modified, with permission, from the *Annual Review of Biophysics and Biomolecular Structure,* Volume 22, ©1993, by Annual Reviews Inc.

al., 1991; Trinh and Sinden, 1991,1993). The frequency with which deletion of an inverted repeat occurs varies as a function of the inverted repeat. The deletion frequency depends on length and thermal stability, that is, the more stable the hairpin stem, the higher the frequency of deletion. As mentioned earlier, cruciforms are stable in plasmid DNA *in vivo* and can exist at a high frequency (up to 20–50%). The formation of stable cruciforms in cells can increase the frequency of deletion by presenting a preformed hairpin arm to the replication apparatus (Sinden *et al.*, 1991). DNA polymerases are known to stop or pause at hairpin arms (Weaver and DePamphilis, 1984). Synthesis up the arm copies the first copy of the direct repeat, misalignment can occur, and replication continues resulting in deletion as shown in Figure 4.18.

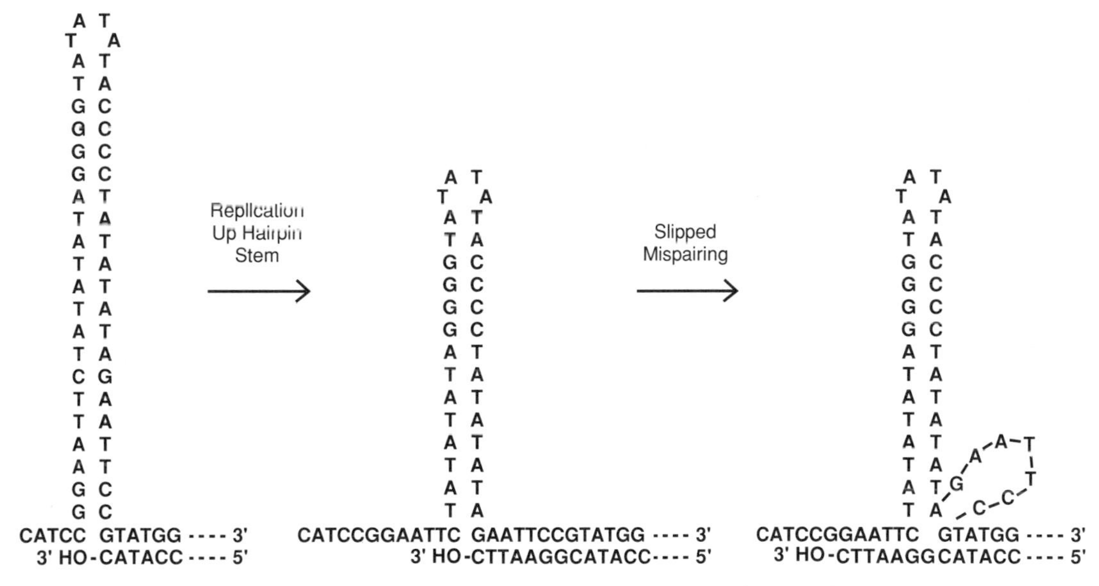

Figure 4.18 Mechanism of slipped mispairing responsible for deletion of inverted repeats flanked by direct repeats. The structure of the S2 inverted repeat is shown, under which is shown the replication intermediate in which DNA polymerization has occurred up to a hairpin arm formed from the inverted repeat in the template strand (*left*). (*Center*) Replication up the stem results in the synthesis of the first direct repeat, the *Eco*RI site GAATTC. (*Right*) Slipped mispairing, in which the first *Eco*RI site in the progeny strand pairs with the second *Eco*RI site in the template strand, results in the formation of the structure shown. This structure contains a hairpin stem with a short bubble of DNA consisting of an *Eco*RI site. Continued replication from this structure results in deletion of the inverted repeat and one copy of the direct repeat in the progeny strand. Figure modified with permission from Sinden *et. al.* (1991).

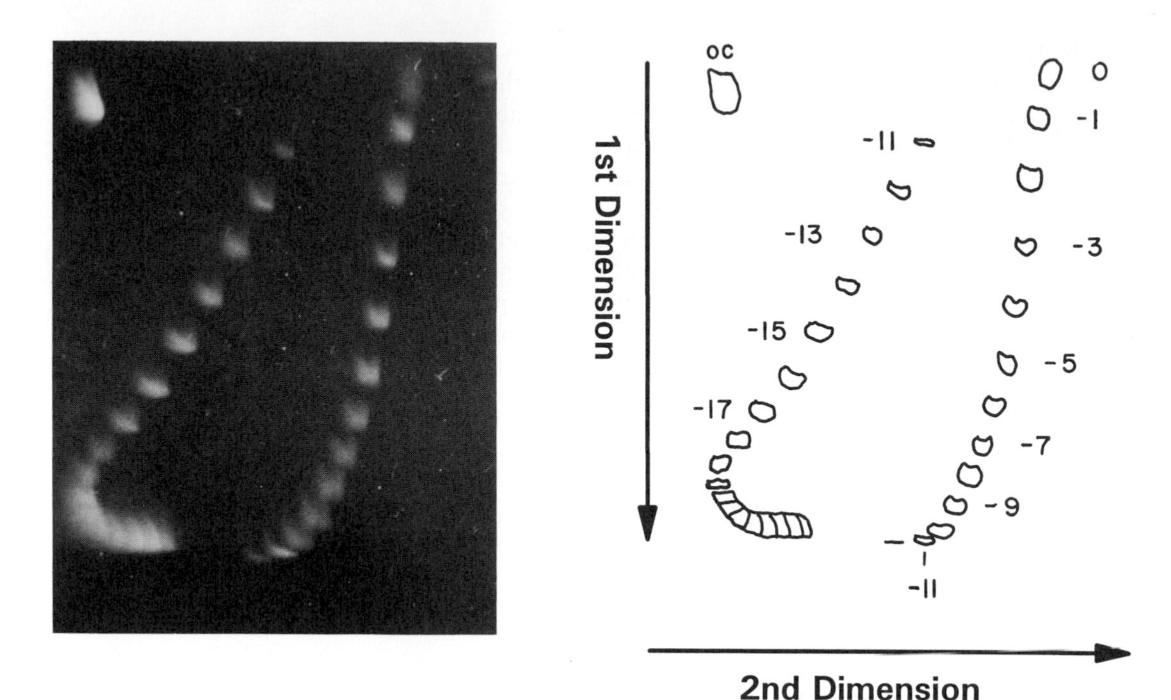

Figure 4.19 Detection of a cruciform transition using two-dimensional agarose gel electrophoresis. A set of topoisomers from $L - L_0 = 0$ to $L - L_0 = -27$ was prepared by treating different samples of plasmid pUC8F14C with topoisomerase I in the presence of various concentrations of the intercalator ethidium bromide. Following purification and removal of the ethidium bromide, the samples were mixed together to give the "complete topoisomer population" shown. (*Left*) A photograph of the two-dimensional gel. (*Right*) A representation in which the individual spots are identified. The direction of electrophoretic migration was from top to bottom in the first dimension and from left to right in the second dimension. The intense spot in the upper left hand corner is nicked or open circular (oc) DNA which migrates most slowly in both the first and the second dimension. This gel ($22 \times 22 \times 0.4$ cm, 1.75% agarose, 40 m*M* Tris, 25 m*M* acetate, 1 m*M* EDTA, pH 8.3) was subjected to electrophoresis for about 24 hr at 100 V in the first dimension, during which 11–12 isomers were resolved. The expected migration for topoisomers with $L - L_0 < -12$ is at a position equal to or slightly faster than that for $L - L_0 = -12$. Rather than migrating faster than topoisomer -12, topoisomer -13 migrates with a very reduced mobility in the first dimension. In fact, topoisomers -11 and -12 were split, with some molecules migrating rapidly in the gel and some migrating more slowly. The retardation of migration is an indication that negative superhelical turns were removed from the plasmid prior to electrophoresis in the first dimension. Topoisomer -13 migrates at a position equal to that of topoisomer -3, with a difference of 10 supercoils. Part of topoisomer -12 migrates at a position close to topoisomer -2 in the first dimension. Thus, at $L - L_0 < -13$, some or all topoisomers migrate with a mobility equivalent to a topoisomer with 10 fewer negative superhelical turns. pUC8F14C contains a 106-bp perfect inverted repeat. The formation of a cruciform relaxes about 1 negative supercoil for 10.5 bp. Therefore, this pattern is consistent with the formation of a 106-bp cruciform beginning at a negative superhelical density of $\sigma = -0.04$ (topoisomer -11). Following the formation of the cruciform, the addition of more negative supercoils makes the plasmid migrate more quickly, heading to an increase in migration at higher negative su-

F. Specific Experiments

1. Two-Dimensional Agarose Gels

Two-dimensional agarose gels are used to determine the superhelical density of plasmid DNA and to identify the individual topoisomer at which a transition to an alternative DNA conformation occurs. The formation of cruciforms, Z-DNA, intramolecular triple-stranded DNA, or a stably unwound region associated with a replication origin results in the relaxation of negative superhelical turns. Two-dimensional agarose gels have been used to identify all these structural transitions.

Figure 4.19 shows a two-dimensional agarose gel of plasmid pUC8 containing the 106-bp F14C inverted repeat. The sample is composed of a "complete mixture" of topoisomers from $(L - L_0) = +3$ to $(L - L_0) = -20$. Gel electrophoresis is conducted in two dimensions (directions). Samples are loaded in a well in the top left region of the gel and electrophoresis is from top to bottom in the first dimension. Following electrophoresis in the first dimension, individual topoisomers migrate as discrete spots from the relaxed $(L = L_o)$ to the supercoiled position. In this gel, topoisomers with $(L - L_o) \leq -12$ migrate at a single position called the "fully supercoiled band." Electrophoresis in the second dimension allows resolution of the multiple topoisomers running at the fully supercoiled position. For the second dimension, the gel is turned 90° prior to electrophoresis. To allow resolution of topoisomers with $(L - L_o) < -12$, an intercalating dye is added to unwind about 21 negative superhelical turns from the DNA. The number of turns unwound can be controlled by varying the concentration of the intercalator. Typically ethidium bromide or chloroquine is used as an intercalator. Chloroquine is preferred since it is not as toxic as ethidium bromide and the

---

perhelical densities. At about $L - L_0 = -23$, the plasmid does not migrate more rapidly on addition of more supercoils.

To resolve the bands that would overlap when separated in a single lane in the first dimension, the gel is run for 24 hr at 100 V in a second dimension 90° to the first. Moreover, chloroquine is added to unwind about 20 negative supercoils to facilitate the resolution of individual topoisomers. The relaxation of 20 negative supercoils means positive supercoils will be introduced into topoisomers with $L - L_0 > -20$. The introduction of positive supercoils removes cruciform structures from plasmid DNA so that the migration in the second dimension is proportional to linking number. For example, topoisomer 0, which migrated most slowly in the first dimension, migrates most rapidly in the second dimension. Topoisomer −20, which migrated rapidly in the first dimension, migrates slowly in the second dimension.

Two-dimensional gels are frequently used to detect alternative secondary structural transitions in plasmid DNA. Cruciforms, Z-DNA, intramolecular triplex structures, and regions of unwound DNA (see Figure 7.4) can be detected using two-dimensional gels.

level of unwinding as a function of the relationship between drug concentration and extent of unwinding is easier to control.

In the first dimension, topoisomer $(L - L_o) = -12$ does not migrate faster than isomer -11, as expected. Rather topoisomer -12 migrates slowly, at a rate equal to topoisomer $(L - L_o) = -2$. This difference reflects the loss of 10 negative supercoils on formation of the 106-bp cruciform in the F14C inverted repeat (Figure 4.6). Topoisomers $(L - L_o) = -11$ to -21 all migrate as molecules with 10 fewer supercoils. In the second dimension, with 21 negative supercoils unwound, the originally relaxed DNA (which migrated most slowly in the first dimension) migrates more rapidly since it now contains 21 positive supercoils. DNA with $(L - L_o) = -21$ initially migrates at the position of relaxed DNA in the second dimension, since all supercoils have been relaxed by binding of the intercalator. DNA topoisomers with $(L - L_o) < = -21$ will still contain negative supercoils and, as the linking number decreases (i.e., $L - L_o = -22, -23, -24 \ldots$), the individual topoisomers will migrate more rapidly with increasing numbers of negative supercoils.

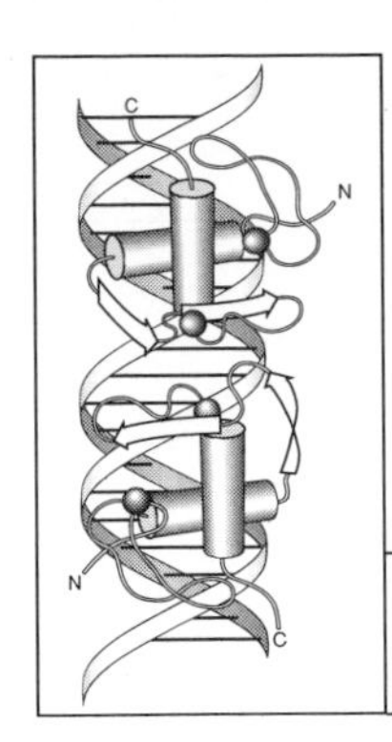

CHAPTER

5

Left-Handed Z-DNA

A. Introduction

The idea of a left-handed DNA helix has been around for a long time. In 1970 Mitsui *et al.* suggested that the polymer polyd(I–C)·polyd(I–C) could adopt a left-handed conformation under specific conditions based on the X-ray diffraction pattern and circular dichroism (CD) spectra of polyd(I–C)·polyd(I–C). However, much later Sutherland and Griffin (1983) found that the polymer analyzed by Mitsui *et al.* adopted an unusual non-B-DNA right-handed configuration called D-DNA, not a left-handed configuration. In 1972, Pohl and Jovin presented the CD spectrum of a left-handed alternating copolymer poly(dG–dC)·poly(dG–dC). This CD spectrum was very different from that for classical B-form DNA. Idealized CD spectra for B-form DNA and left-handed Z-DNA are shown in Figure 5.1. The classical characteristic B-form DNA CD spectrum shows a positive molar ellipticity (θ) peak at 270–280 nm, a negative ellipticity at about 250 nm, and a crossover point (where the ellipticity equals 0) at about 260 nm. The Z-DNA spectrum shows an inversion of the peaks relative to the spectrum of B-DNA. In 1979, Rich and co-workers solved the X-ray crystal structure of d(CpGpCpGpCpG) (Wang *et al.*, 1979). Quite unexpectedly, this 6-bp

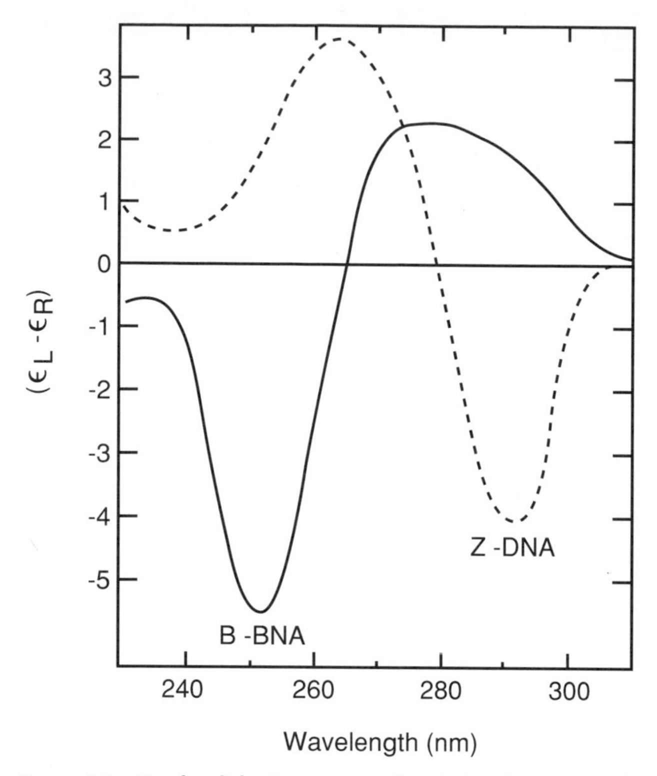

Figure 5.1 Circular dichroism spectra of B-DNA and Z-DNA. At low salt concentrations polymers of $(dGdC)_n$ exist as B-DNA, giving a typical B-DNA circular dichroism spectrum shown by the solid line with a peak of positive ellipticity at 270–280 nm and a negative peak at about 250 nm. When this DNA is incubated in 3–4 *M* salt solutions, there is a "reversal" of the CD spectrum, giving the spectrum shown by the dashed line which is characteristic of left-handed or Z-DNA. On Z-DNA formation, there is a positive ellipticity peak at about 265 nm and a negative peak at about 290 nm; the crossover point shifts to a higher wavelength.

oligonucleotide (the first crystal structure of a DNA molecule to be solved) existed as a left-handed helix.

Following the realization that certain alternating purine–pyrimidine sequences could exist as left-handed helices, an enormous scientific effort ensued to characterize the DNA sequences and environmental conditions required for Z-DNA formation. The discovery of Z-DNA is important because the ability of DNA to adopt non-Watson–Crick structures could have profound implications for the processes of replication, recombination, and transcription. Z-DNA can exist readily in bacterial cells and, although sequences that can form Z-DNA are widely found in human DNA, nature has yet to divulge the biological role for Z-DNA in eukaryotic cells.

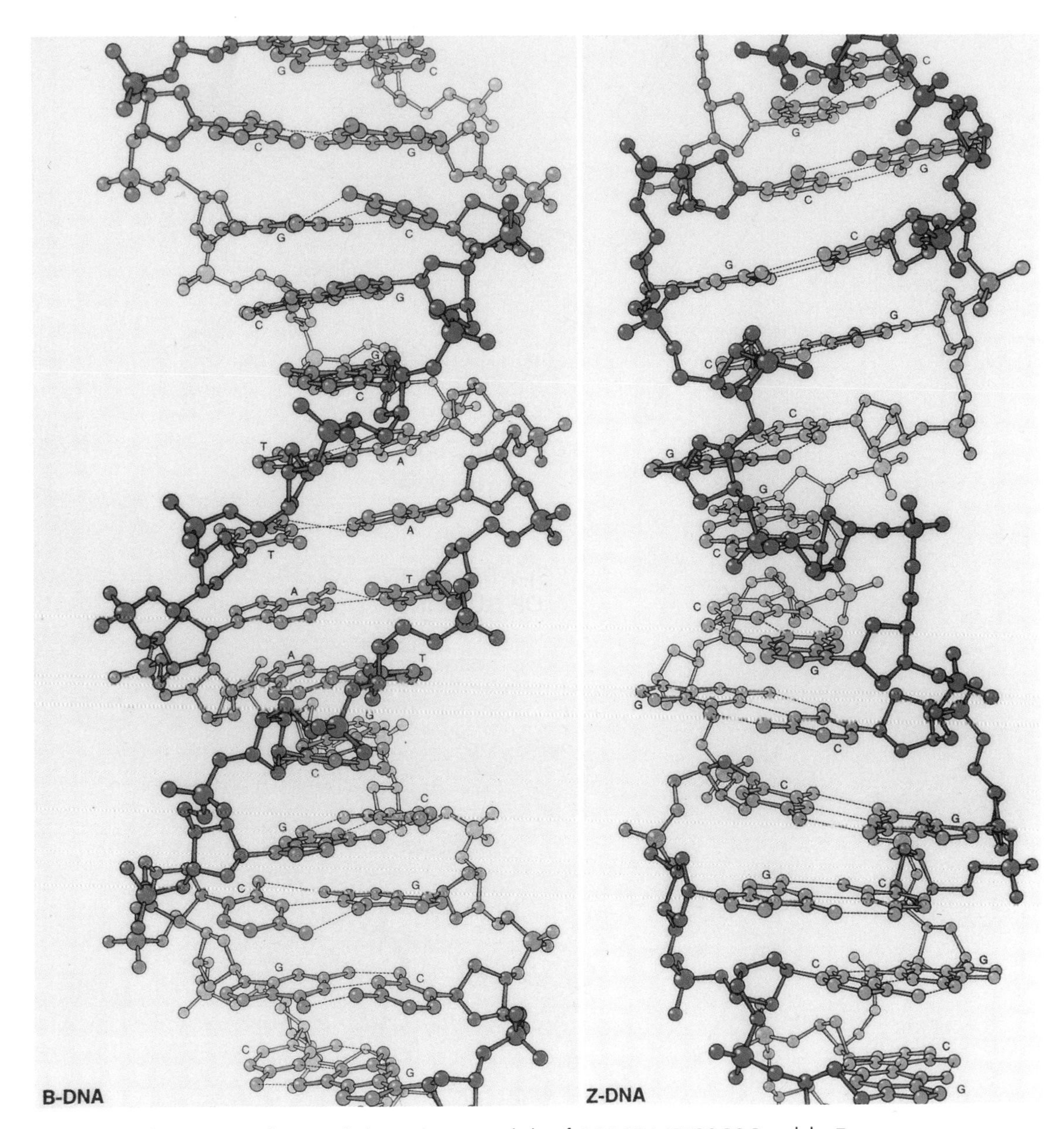

Figure 5.2 The B-DNA and Z-DNA helices. The B-DNA helix of GCGCGAATTCGCGC and the Z-DNA helix of $(GC)_6$ are shown. Copyright by Irving Geis. Used with permission.

Anti **POSITION OF GUANINE**

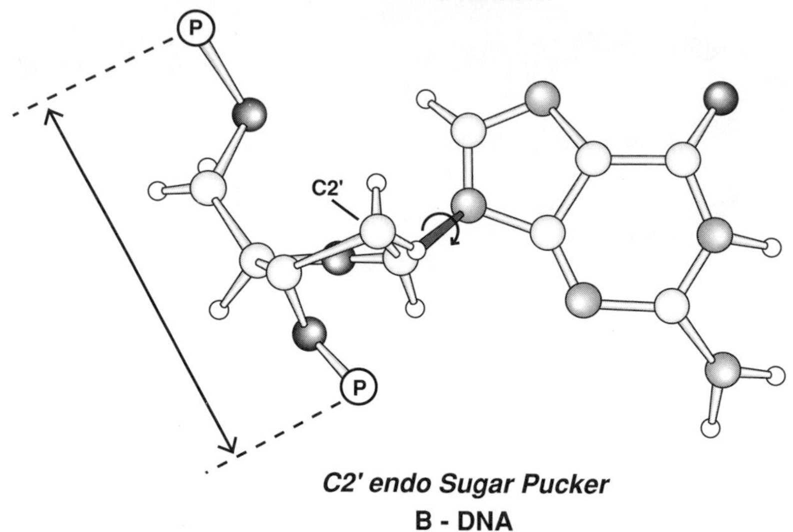

Syn **POSITION OF GUANINE**

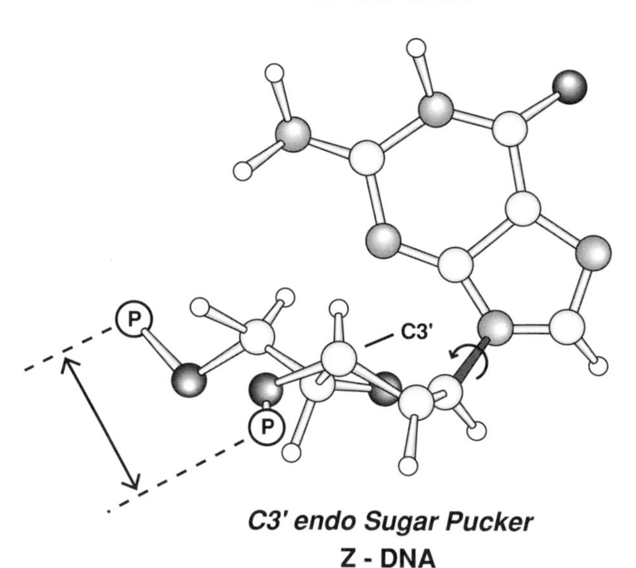

Figure 5.3 The structure of guanosine in B-DNA and Z-DNA. (*Top*) In B-DNA, the glycosidic bond (black) between the N9 position of the purine ring and C1 carbon of the ribose sugar is in the *anti* position and the deoxyribose contains a C2′ endo pucker. (*Bottom*) When Z-DNA forms, the glycosidic bond rotates guanine 180° into the *syn* position and the deoxyribose contains a C3′ endo pucker. The sugar pucker change results in a very different distance between the phosphates

B. The Structure of Z-DNA

1. The Left-Handed Z-DNA Helix

The structure of Z-DNA is shown in Figure 5.2 (B-DNA is shown for comparison). There are only a few similarities between B-DNA and Z-DNA, including the double-stranded, antiparallel nature of the two strands and Watson–Crick hydrogen bonding. Apart from these similarities, there are a number of differences between B-DNA and Z-DNA. First, there is a zig-zag phosphate backbone in Z-DNA compared with a smooth backbone in B-DNA. Second, B-DNA has a distinct major and minor groove. In Z-DNA, the major groove has all but disappeared into a nearly flat surface. The one visible groove is deep and narrow. This groove is structurally analogous to the minor groove in B-DNA. Third, the Z-DNA helix, at 18 Å in diameter, is narrower than B-DNA (20 Å in diameter). Fourth, the helix repeat in Z-DNA is 12 bp per turn compared with 10.5 bp per turn for right-handed B-DNA. Finally, in B-form DNA the helix pitch is, on average, 34 Å with an average rise of 3.24 Å. The helix pitch of Z-DNA is 44.6 Å with a rise of 3.72 Å. (The helix parameters for B-DNA and Z-DNA are listed in Table 1.3.)

2. Sugar Pucker and Glycosidic Bond Rotation in Z-DNA

Two major structural differences between B-DNA and Z-DNA are the sugar pucker and the configuration of the glycosidic bond in dG. In B-form DNA all sugar residues exist in the C2′ endo configuration (discussed in Chapter 1). In Z-DNA the sugar pucker for dC remains C2′ endo but the pucker changes to C3′ endo in dG residues. The structures of C2′ endo, *anti* dG and C3′ endo, *syn* dG are shown in Figure 5.3. The zig-zag phosphate backbone is the result of the alternating C2′ endo conformation in pyrimidines and C3′ endo conformation in purines. One major consequence of the change in sugar pucker is the effect this has on the distance between the phosphate groups attached to the C5′ and C3′ ribose positions (Figure 5.3). Changing the sugar pucker from C2′ endo to C3′ endo reduces this distance

---

on the C3′ and C5′ deoxyribose carbons. In B-form DNA, this longer distance acts to extend the paired bases into the center of the DNA helix. When the sugar undergoes a C3′ endo conformational change, the distance between the phosphate groups shortens substantially. This acts to pull the base away from the center of the helix resulting in the positioning of the base pairs on the outside of the helical axis in Z-DNA (see Figure 5.6). Reproduced, with permission, from the *Annual Review of Biochemistry,* Volume 53, ©1984, by Annual Reviews Inc.

dramatically. This shortening has the result of pulling the base away from the center of a B-DNA helix and closer to the phosphate backbone, which is on the outside of the DNA helix. The other significant change occurring in dG in Z-DNA is a 180° rotation around the glycosidic bond. The configuration goes from the *anti* configuration found in B-form (and A-form) DNA to the *syn* configuration in Z-DNA. Some of these changes can be seen in the representation of Z-DNA shown in Figure 5.4.

3. Bases Do a Backflip to Form Z-DNA

A major structural rearrangement required for the B-DNA-to-Z-DNA transition is a flipping or inversion of the bases relative to the helix axis. Within a region of Z-DNA the bases flip over 180° (Figure 5.4). This flipping is accompanied by the rotation of the glycosidic bond of the purines from *anti* to *syn* and the sugar pucker change. For the pyrimidine nucleotides in Z-DNA, the sugar accompanies the base in its 180° rotation.[1]

4. The Dinucleotide Repeat in Z-DNA and Base Stacking

In B-form DNA each base pair is oriented in the helix similarly with respect to its adjacent base pairs, so all dinucleotide base pairs are structurally similar. The dinucleotides CpG and GpC have twist angles of 29.8° and 40.0°, respectively (see Table 1.4) which are not similar to the average helical repeat of 34.3°. Another prominent feature of B-DNA is that the base pairs are stacked regularly on top of one another. As shown in Figure 5.5, there is reasonable overlap between the bases in the GpC dinucleotide. There is somewhat less physical overlap between the bases in the CpG stack. In B-form DNA the bases are hydrogen bonded in a position almost perfectly centered within the double helix (the dot in Figure 5.5 denotes the center of the DNA helix). As illustrated in Figure 5.6, the bases in B-form DNA stack into a cylinder within the center of the double helix, with the sugars and phosphates on the outside of the helix. Because all base pairs are positioned similarly with respect to other base pairs, B-form DNA has a base pair repeating unit of one.

In Z-DNA the repeating base pair unit is two. The CpG dinucleotide is very different from the GpC dinucleotide. There are 12 bp per turn of Z-DNA which should produce an average helix twist of –30° (the value is negative to denote the left-handed rotation in Z-DNA compared with a positive rotation found in right-handed helices). However, the twist angle for CpG is

[1]Models for left-handed DNA other than the one described by Rich and colleagues have been proposed. One model called Z[WC]-DNA, described by Ansevin and Wang (1990), does not require 180° flipping of the bases.

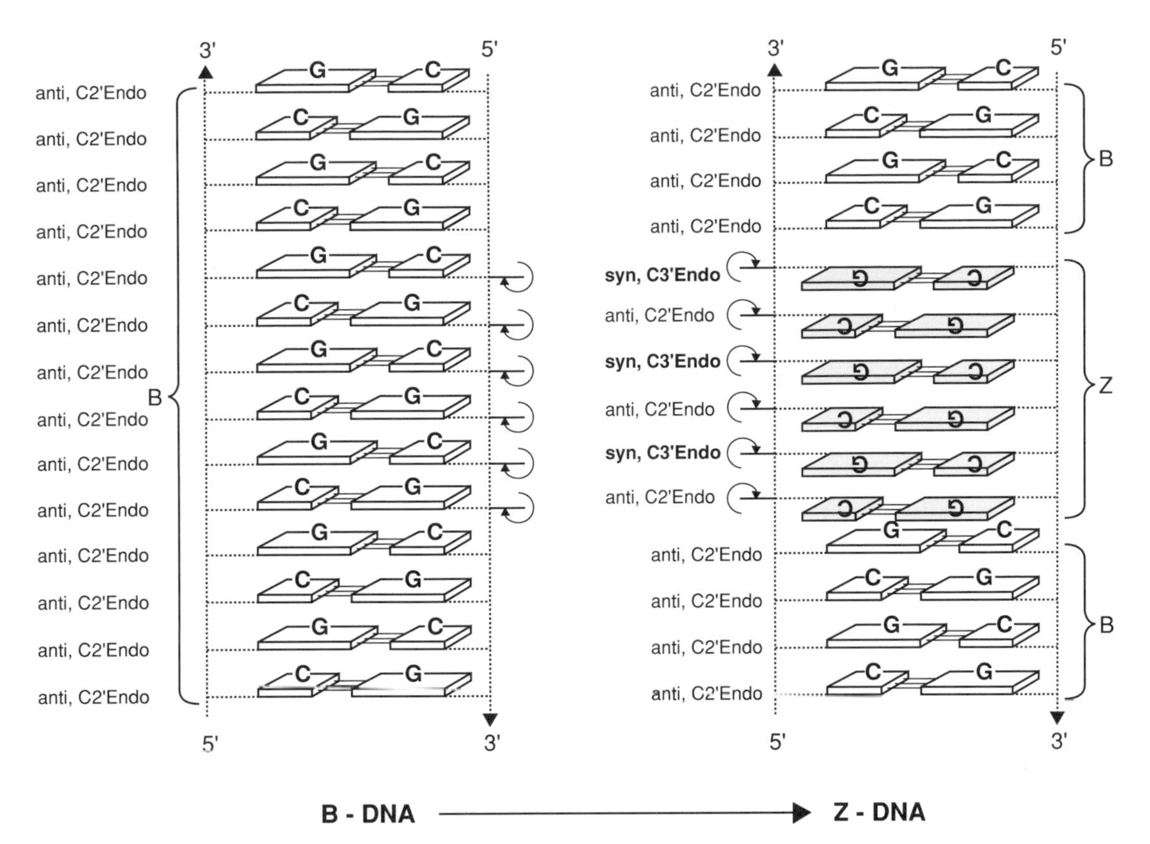

Figure 5.4 Topological changes in the positions of the base pairs on Z-DNA formation. On the formation of Z-DNA within a region of B-DNA, there is an inversion or flipping of the base pairs 180° with respect to the helical axis. This is indicated by the upside-down base pairs in the Z-DNA region on the right side of the figure where the central $(CG)_3 \cdot (CG)_3$ has formed Z-DNA. Also on Z-DNA formation, as indicated in Figure 5.3, the glycosidic bond of quanosine changes to the *syn* conformation and the sugar pucker changes to C3′ endo. This produces alternating *anti–syn* and C2′ endo–C3′ endo conformations in each single strand in Z-DNA. Modified, with permission, from the *Annual Review of Biochemistry,* Volume 53, ©1984, by Annual Reviews Inc.

only −9°, whereas the twist angle for GpC is −51° (Figure 5.5). Consequently, in contrast to B-DNA, each base pair in Z-DNA is not oriented in the helix in a similar fashion with respect to the adjacent base pairs. However, every two base pairs are oriented within the helix in a fashion identical to the next two adjacent base pairs. The twist angle for two base pairs relative to the next two base pairs is −60°, producing a helix repeating unit of two.

Another major difference between B-DNA and Z-DNA is that in Z-DNA the base pairs are sheared with respect to their stacking positions. The

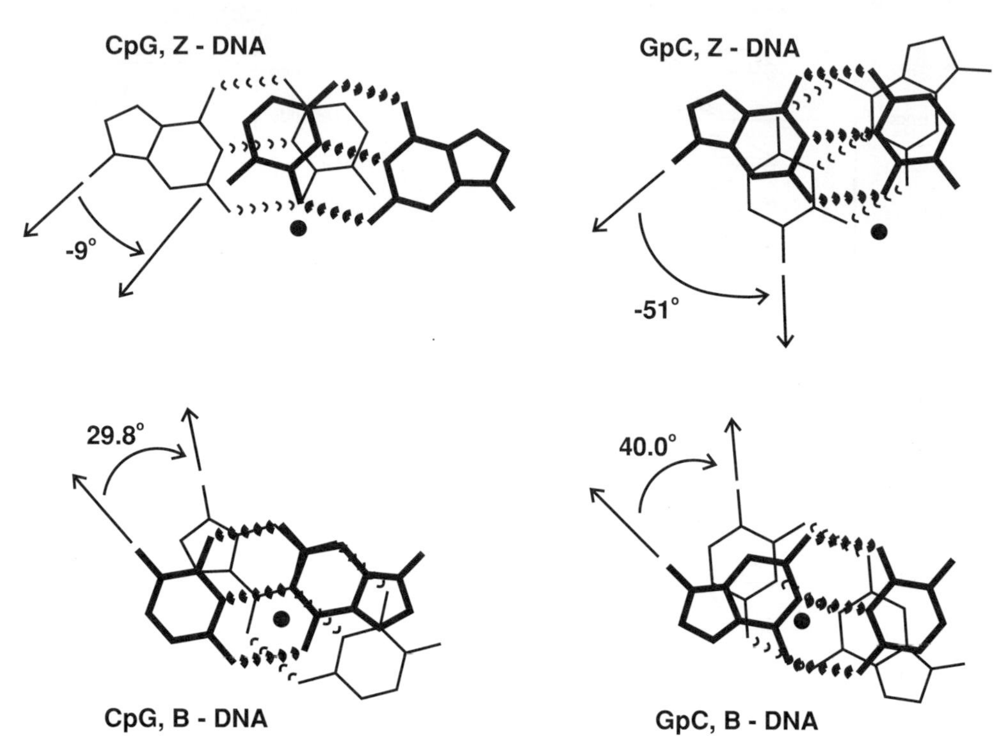

Figure 5.5 Base stacking in Z-DNA and B-DNA. The C·G base pairs drawn with thick lines are stacked above the C·G base pairs drawn with thin lines. (*Left*) CpG dinucleotide base pairs. (*Right*) GpC dinucleotide base pairs. (*Top*) The stacking interactions present in Z-DNA. (*Bottom*) The stacking interactions in B-DNA. Note the good overlap between base pairs in B-DNA, especially in GpC. There is extremely poor overlap in the CpG dinucleotide in Z-DNA. The thick dot represents the center of the B-DNA or Z-DNA helix. The twist angles between the adjacent base pairs are indicated for the CpG and GpC dinucleotides. The twist angles are positive for the right-handed B-DNA helix and negative for the left-handed Z-DNA helix. Modified, with permission, from the *Annual Review of Biochemistry,* Volume 53, ©1984, by Annual Reviews Inc.

orderly stacking in the center of the helix found in B-DNA is nonexistent in Z-DNA. Analysis of the CpG dinucleotide in Figure 5.5 shows that the two cytosines are stacked but the guanines are positioned under and over the ribose sugars. Thus, in the CpG dinucleotide the bases are sheared by being effectively pulled away from each other with respect to the center of the helix. In the GpC dinucleotide in Z-DNA, the bases are relatively well stacked.

The position of bases within the helix also distinguishes DNA from B-DNA. In B-DNA the bases form a cylinder within the double helix, with the hydrogen bonds at the center of the helix. However, as shown in Figures 5.5 and 5.6, the bases in Z-DNA are positioned toward the outside of the Z-

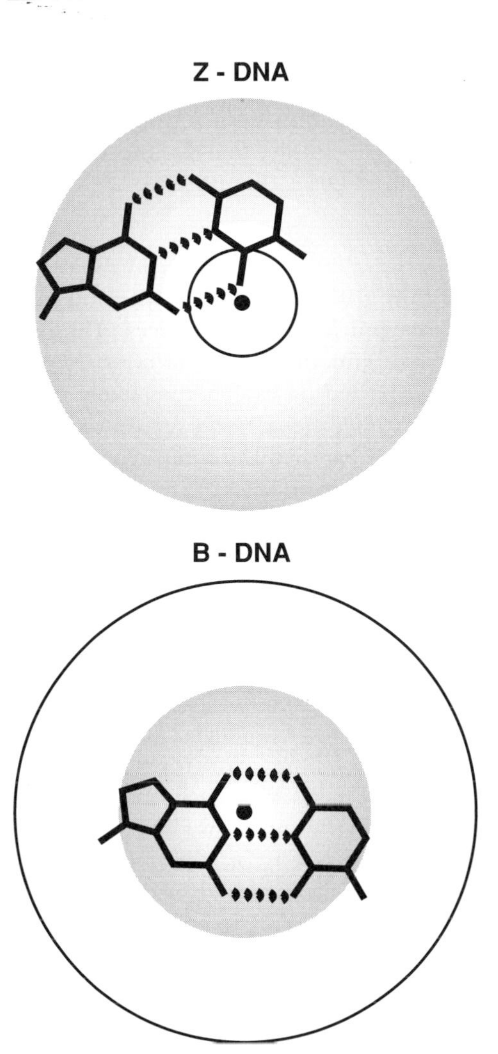

Figure 5.6 Position of the base pairs within the helix. End-on views show the location of base pairs within the helix. The shaded regions are areas that the base pairs occupy as they rotate through the helix axis. In B-DNA, the bases are stacked and centered in the helix with the phosphate backbone organized around the periphery of the helix. (The unshaded area contains the deoxyribose groups and phosphate groups.) In Z-DNA, the base pairs are displaced from the center of the helix, reducing stacking interactions and placing the N7 position of purines on the outside of the helix.

DNA helix so that neither the bases nor the hydrogen bonds overlap the center of the helix. In B-DNA, the bases are protected from the solvent by their location at the center of the helix. In Z-DNA, certain ring positions are much more chemically reactive than in B-DNA. For example, the N7 and C8 positions of guanine are exposed to the solvent in Z-DNA (Figure 5.6).

5. The B–Z Junction

When a region of Z-DNA exists within a larger B-form DNA molecule, there must be a junction between the right- and left-handed helices. The best estimation of the physical structure of a B–Z junction is that it consists of a region of probably 3–4 unpaired or weakly paired bases. There is likely not a sharp transition from the left- to the right-handed helix but a region of a few base pairs that is partially unwound. The selective or differential reactivity of many chemicals at B–Z junctions has provided sensitive assays for Z-DNA.

C. Formation and Stability of Z-DNA

1. The Sequence Requirement for Z-DNA

Z-DNA can form in regions of alternating purine–pyrimidine sequence; $(GC)_n$ sequences form Z-DNA most easily. $(GT)_n$ sequences also form Z-DNA but they require a greater stabilization energy for formation than $(GC)_n$. $(AT)_n$ generally does not form Z-DNA since $(AT)_n$ easily forms cruciforms. $(AT)_n$ can form Z-DNA under two special conditions. First, up to 10 alternating $A \cdot T$ base pairs embedded within a $(GC)_n$ or $(GT)_n$ region will form Z-DNA (Klysik *et al.*, 1988). Second, conditions of high negative supercoiling, high salt, and the presence of $NiCl_2$ cause $(AT)_n$ to form left-handed DNA (Nejedly *et al.*, 1989).

2. Chemical Modifications Stabilize Z-DNA

Certain chemical modifications of DNA will drive the B-DNA $\rightleftarrows$ Z-DNA equilibrium in favor of Z-DNA. Bromination at the C5 position of cytosine or the C8 position of guanine allows the stabilization of Z-DNA at lower salt concentrations than required to stabilize a nonbrominated polymer. The addition of a bulky group at the C8 position of guanine favors the *syn* conformation, which places this group on the outside of the Z-DNA helix (see Figure 5.6). In the *anti* conformation, the bulky group would be inside the cylindrical axis of the DNA near the phosphate backbone. Methylation of the N7 position of guanine also favors the formation of Z-DNA. Bromination and especially methylation (or ethylation) of DNA from environmental mutagens will affect the stability of the B or Z conformation of the DNA double helix.

In bacterial and eukaryotic DNA many bases are methylated. Methylation is used in bacteria for restriction modification systems and the methyl-directed mismatch repair system. In eukaryotic cells, the C5 position of cytosine in the CpG dinucleotide is frequently methylated, as a general mechanism preventing transcriptional activity from large regions of eukaryotic chromosomes.

3. The Effect of Cations on the B-to-Z Equilibrium

The formation of Z-DNA was first observed in very high concentrations of salt (3 *M* NaCl). As shown in Figure 5.2, the distance between the negatively charged phosphates is much shorter in Z-DNA than B-DNA. Therefore, Z-DNA will be destabilized by the charge–charge repulsion of the negative charges on the phosphates. High concentrations of monovalent cations (Na^+, K^+, Rb^+, Cs^+, Li^+) can shield the negative charges and stabilize the Z conformation. Divalent cations (for example, Mg^{2+}, Mn^{2+}, and Co^{2+}) are much more effective at shielding negative charges than monovalent cations, and will stabilize Z-DNA at much lower concentrations. For instance, cobalt hexamine chloride will stabilize Z-DNA at millimolar concentrations whereas 2–3 *M* concentrations of NaCl are required to stabilize Z-DNA. In addition spermine, spermidine, and certain polyamines will stabilize Z-DNA.

4. DNA Supercoiling Stabilizes Z-DNA

In the early 1980s, experiments from the laboratories of Wells, Rich, and Wang demonstrated the role of supercoiling in Z-DNA formation (Peck *et al.*, 1982; Singleton *et al.*, 1982). Klysik *et al.* (1981) showed that, when a plasmid containing the sequence $(dCdG)_n \cdot (dCdG)_n$ was analyzed by agarose gel electrophoresis in the presence of 4 *M* salt, relaxation indicative of the formation of Z-DNA was observed. Subsequently, Stirdivant *et al.* (1982) and Peck *et al.* (1982) showed that, as the level of negative supercoiling increased, the concentration of salt required to drive the Z-DNA transition decreased. In fact, the level of negative superhelical energy necessary to drive the B-to-Z transition at physiological ionic strengths was well within the level of supercoiling found in DNA purified from cells. $(dCdA)_n \cdot (dGdT)_n$ also forms Z-DNA in supercoiled plasmids (Haniford and Pulleyblank, 1983a; Nordheim and Rich, 1983). However, $(GT)_n$ Z-DNA forming sequences require a higher level of negative supercoiling to form Z-DNA than $(GC)_n$ tracts of equal size. The level of negative supercoiling required to form Z-DNA is a function of length of the PuPy tract. As a general rule, the longer the Z-DNA forming sequence, the less negative supercoiling required to drive the B-to-Z transition. Table 5.1 shows the level of negative supercoiling required to form Z-DNA for a number of Z-DNA forming sequences.

DNA supercoiling is one physiological condition that can drive the formation of Z-DNA. Z-DNA is quite stable in supercoiled DNA since the formation of Z-DNA effectively unwinds DNA, resulting in the relaxation of supercoils. As discussed in Chapter 3, conditions that unwind DNA are thermodynamically favored. This situation is analogous to that described in the preceding chapter for cruciforms and the formation of intramolecular triplex stranded DNA discussed in the following chapter. As shown in Figure 5.7, when a 12-bp region of B-DNA undergoes a transition from a right-handed helix to one complete left-handed helical turn, the number of twists in the DNA will decrease by 2.14 [1.14 from the removal of B-DNA (12 ÷ 10.5) plus 1 from the formation of Z-DNA]. In terms of 10-bp units, the formation of Z-DNA removes 1.78 turns for every 10 bp that form Z-DNA [(10 ÷ 10.5) + (10 ÷ 12)].

5. Protein-Induced Formation of Z-DNA

The widespread occurrence of alternating purine–pyrimidine sequences in eukaryotic DNA, especially alternating GT sequences, has led to speculation that Z-DNA and Z-DNA binding proteins must exist. The identification of proteins that bind Z-DNA might imply a role for the left-handed conformation of DNA in biology. Proteins that bind to Z-DNA have been identified from many organisms including *Escherichia coli, Drosophila,* and higher

Table 5.1
Supercoiled Density Dependence for Z-DNA Formation

Z-DNA sequence	σ_c
$(CG)_{16}$[a]	−0.031
$(CG)_8GG(CG)_7$[b]	−0.034
GCGCGCGAGCGCGCGCGCTCGCGCGC[b]	−0.042
$CG(TG)_{20}AATT(CA)_{20}CG$[c]	−0.032
$(CG)_{13}AATT(CG)_{13}$[c]	−0.025
$(CG)_6TA(CG)_6$[d]	−0.042
$(TG)_6TA(TG)_6$[e]	−0.057
$(TG)_{26}$[f]	−0.047

[a]Peck *et al.* (1982).
[b]Ellison *et al.* (1985).
[c]Blaho *et al.* (1988).
[d]Sinden and Kochel (1987).
[e]Kochel and Sinden (1988).
[f]Haniford and Pulleyblank (1983a).

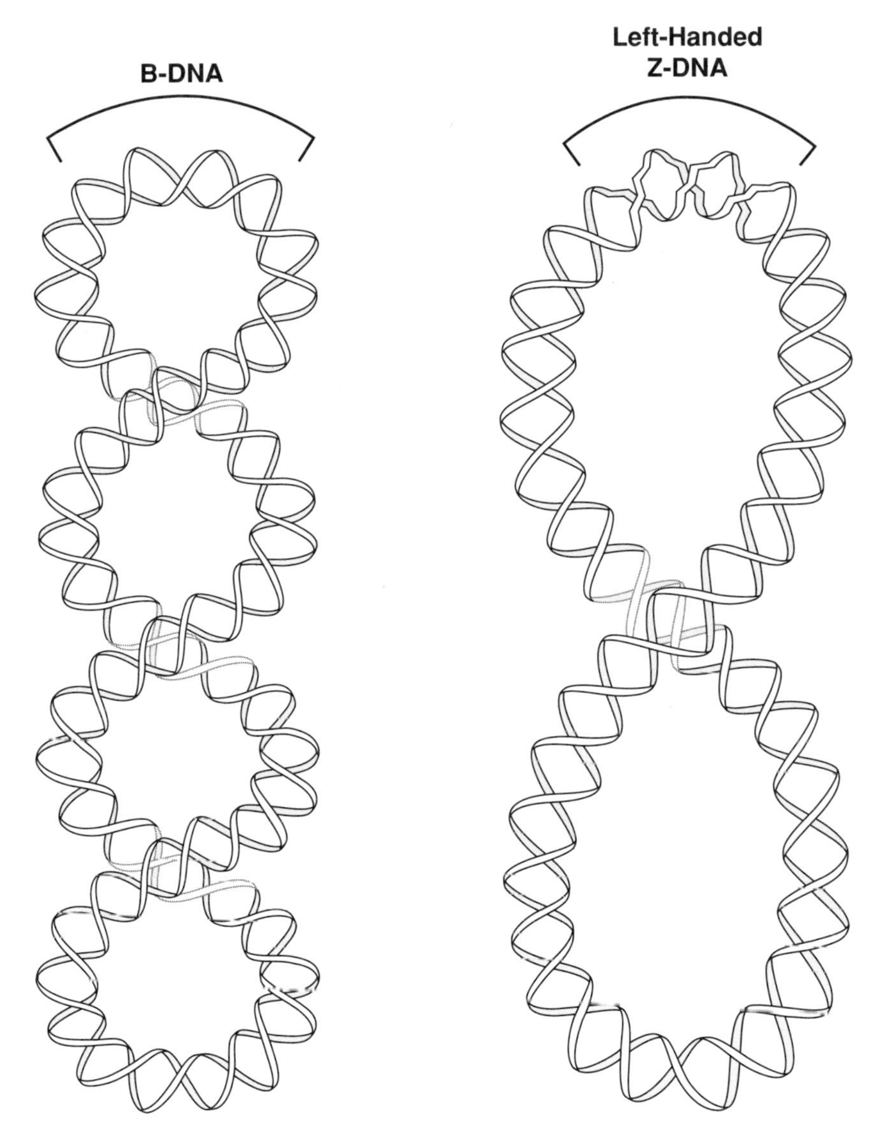

Figure 5.7 Superhelical-density-dependent formation of Z-DNA. Z-DNA can be formed in negatively supercoiled DNA. The energy of negative supercoiling drives the conformational transition. The flipping of DNA from a right-handed to a left-handed helix removes torsional stress in the DNA molecule and is therefore thermodynamically favored. When a region of 10 bp of B-form DNA undergoes a Z-DNA conformational transition, approximately 2 negative supercoils will be relaxed. This is indicated in the molecule shown here and is discussed in greater detail in the text.

eukaryotes. It is not yet known with certainty, however, if any Z-DNA binding protein has a biological role that actually involves binding to left-handed DNA.

The first example of proteins that could recognize and specifically bind to Z-DNA were Z-DNA antibodies. These antibodies recognize left-handed Z-DNA but not right-handed DNA. Lafer *et al.* (1985a) showed that antibody binding could induce the formation of Z-DNA and stabilize Z-DNA by binding the left-handed helix in relaxed (linearized) plasmid DNA. This was a significant finding since it provided an example of a protein that could affect an equilibrium between two alternative helical forms of DNA.

D. Assays for Z-DNA

There are a large number of physical and chemical differences between Z-DNA and B-DNA that have been determined using physical–chemical approaches such as X-ray crystallography, circular dichroism, ultraviolet and Raman spectroscopy, nuclear magnetic resonance, and electron spin resonance. These physical–chemical approaches, especially X-ray crystallography, have provided a picture of Z-DNA at the atomic level. In other assays, changes in the supercoiling of plasmids due to Z-DNA formation can be detected by agarose gel electrophoresis. Differential reactivity of Z-DNA with DNA methylases, restriction endonucleases, and a variety of chemicals has provided very powerful assays for Z-DNA that have been applied *in vivo.* Antibodies against Z-DNA have also been extremely useful as probes for Z-DNA both *in vitro* and *in situ.* An excellent review on Z-DNA by Rich *et al.* (1984) lists 15 types of assays for Z-DNA, and provides an excellent discussion of Z-DNA antibodies and many of the physical chemical assays.

1. Z-DNA Antibodies

A number of different polyclonal and monoclonal antibodies have been raised against Z-DNA. Brominated poly(dG–dC)·poly(dG–dC) is typically used to elicit the antibody response. Different antibodies against Z-DNA recognize different structural features of the Z-DNA helix. For example, some recognize the base pairs that are on the outside of the DNA helix whereas others recognize the sugar residues (Moller *et al.*, 1982).

The binding of antibodies to Z-DNA can be detected by a variety of methods. One involves the binding of DNA to nitrocellulose. Double-stranded DNA does not normally bind to a nitrocellulose filter. However, when a protein is tightly bound to DNA, the protein will stick to nitrocellulose and trap the DNA with it. By using radioactively labeled DNA, the extent of Z-DNA formation can be quantitated by measuring the amount of radioactivity retained on a nitrocellulose filter in the presence of Z-DNA antibody. Antibody binding to DNA is very tight and relatively insensitive to changes in the ionic environment. Thus, antibodies are useful over the wide range of salt concentrations and in the presence of the various mono-, di-, and trivalent cations that stabilize Z-DNA. See the box entitled "Utilizing Antibodies to Localize a Z-DNA Site within a Plasmid" for an application of Z-DNA antibodies.

Utilizing Antibodies to Localize a Z-DNA Site Within a Plasmid

Nordheim *et al.* (1982a) showed that Z-DNA antibodies bound to a $(GC)_n$ Z-DNA region within a supercoiled plasmid. In addition, they used this method to identify a naturally occurring Z-DNA-forming sequence within the plasmid. Proteins can be "fixed" onto DNA by incubation in the presence of glutaraldehyde which forms covalent bonds, or cross-links, between the DNA and the protein. Following cross-linking with glutaraldehyde, the plasmid was cut with a restriction enzyme (*Hae*III) that produces 22 fragments of various sizes. The DNA sample was then passed through a nitrocellulose filter. DNA fragments without covalently bound antibody passed through the filter whereas DNA fragments with bound antibody were retained. The DNA samples that passed through the filter were analyzed by electrophoresis on an acrylamide gel and compared with a *Hae*III digestion pattern of "control" plasmid (not exposed to the antibody–glutaraldehyde treatment).

The antibody-treated fragments that passed through the nitrocellulose revealed "missing bands" or bands reduced in intensity, called "reduced bands." The "missing bands" and "reduced bands" that were retained on the nitrocellulose filter contained Z-DNA. This type of experiment can be used to localize Z-DNA to a small region of a DNA molecule.

Why did the experiment of Nordheim *et al.* (1982a) require cross-linking Z-DNA antibody to the DNA with glutaraldehyde? In the presence of 3–4 *M* NaCl or DNA supercoiling, Z-DNA is stabilized relative to the B conformation. However, Z-DNA-forming regions are in an equilibrium between the B and Z forms. In supercoiled plasmid, the equilibrium is shifted well toward the Z form. Relaxation of DNA supercoils by digestion with *Hae*III causes regions of Z-DNA to be converted back to the B form. As mentioned earlier, some antibodies can bind to DNA inducing or stabilizing regions of Z-DNA in relaxed DNA. For the most part, however, antibodies would not induce Z-DNA and in time, without cross-linking, antibodies would fall off the DNA as the Z-DNA region reformed into B-DNA.

2. Relaxation of Negative Supercoils as an Assay for Z-DNA

In plasmids containing regions of alternating $(GC)_n$ or $(GT)_n$, DNA supercoiling will drive the B-to-Z transition and stabilize Z-DNA within these purine–pyrimidine runs. As discussed in Section B,4, the formation of Z-DNA within a plasmid results in the relaxation of about 1.78 negative supercoils for every 10 bp of B-DNA that form Z-DNA. Two-dimensional agarose gels, discussed in Figure 4.19, can be used to visualize the Z-DNA transition. The critical superhelical density at which Z-DNA forms is determined from the topoisomer number for which migration in the first dimension is slower than for the preceding topoisomer. As with cruciforms, the number of superhelical turns removed provides an indication of the length of the Z-DNA region.

3. Chemical Probes of Z-DNA

Numerous chemical probes have been applied as assays for Z-DNA. The basis for these assays lies in the differential chemical reactivity of the Z and B forms of the double helix. In Z-DNA, the N7 and C8 positions of guanine are on the outside of the helix. Consequently, these positions are more reactive to some chemicals in the Z form than in the B form. Chemicals that specifically react with unpaired bases will frequently react at the B–Z junctions. The assays used to identify the sites of chemical reaction, which were described in Chapter 4, can also be applied to the analysis of Z-DNA. The sites of modification can be mapped by cleavage of the DNA with S1 nuclease (see Figure 4.12). In addition, sites of chemical modification can be mapped by chemical cleavage at the site of the modified base or by a primer extension assay (see Figure 4.13)

Chapter 1 Section H provides more detailed information on the following chemical probes that can be used to assay Z-DNA. Many of these chemical probes are not specific for Z-DNA, but can be used to analyze cruciforms, intramolecular triplexes, or other non-B-DNA structures.

a. Bromoacetaldehyde, Chloroacetaldehyde, Osmium Tetroxide, and Hydroxylamine

Bromoacetaldehyde (BAA), chloroacetaldehyde (CAA), osmium tetroxide (OsO_4), and hydroxylamine react specifically with unpaired bases. Their reaction products with DNA are shown in Figure 1.21. These reagents will chemically modify the B–Z junctions.

b. Diethylpyrocarbonate

Diethylpyrocarbonate (DEPC) reacts with purines at the N7 position forming a carbethoxy group (see Figure 1.21). DEPC will react with purines in B-form DNA; however, reactivity is enhanced in Z-DNA. In $(GC)_n$ Z-DNA regions, guanines are hyperreactive. In $(GT)_n$ Z-DNA regions, guanines (in the GT strand) are not hyperreactive; however, adenines (in the CA strand) are hyperreactive (Johnston and Rich, 1985; McLean *et al.*, 1988). The specificity of these reactions has been used as a chemical probe for DNA conformation.

c. Aminofluorenes

The aminofluorene derivatives constitute a group of chemical carcinogens that covalently modify DNA (Grunberger and Santella, 1981). *N*-2-Acetylaminofluorene (AAF) covalently binds to the C8 position on guanosine. Although this position is relatively protected in B-form DNA, AAF reacts with B-form DNA. When a region of $(GC)_n$ or $(GT)_n$ exists in the Z-DNA conformation, AAF does not covalently bind to the C8 position of guanosine despite the fact that this atom is relatively exposed to the solvent. In a typical B-form DNA helix, a bulky modification on the C8 position of guanosine is difficult to incorporate into a B-form helix. In fact, the AAF-modified guanosine is probably bulged out of the helix (Santella *et al.*, 1981). Within an alternating purine–pyrimidine region, the AAF-modified G can easily exist in the *syn* configuration in a Z-DNA helix. Modification of B-form poly(dG dC)·poly(dG-dC) with AAF facilitates its transition to Z-DNA.

Aminofluorene derivatives react in a hypersensitive fashion with guanosines at B–Z junctions (Rio and Leng, 1986). The structure at the junction must facilitate the reactivity with aminofluorene derivatives. The characteristic reactivity of guanosine in B-DNA compared with the selective hyperactivity of B–Z junction regions in Z-DNA provides an assay for Z-DNA.

d. Diepoxybutane

Diepoxybutane is a relatively simple compound that covalently binds to the N7 position of guanosine. It forms only monoadducts to B-form DNA. However, in the Z-DNA conformation, interstrand cross-links are formed between guanosines of adjacent base pairs in opposite strands (Castleman *et al.*, 1983; Kang *et al.*, 1985). This would seem to be an ideal assay for Z-DNA if it could be applied *in vivo*. Unfortunately, diepoxybutane is relatively reactive with the N7 of guanosine, making the glycosidic bond labile. On formation of an apurinic site, the phosphodiester bond is susceptible to hydrolysis which leads to the introduction of nicks in the DNA. If Z-DNA is stabilized by supercoiling in cells, chemicals that nick DNA will likely lead to the loss of Z-DNA.

e. Phenanthroline Metal Complexes

Chiral (tris)-phenanthroline metal complexes have been developed as probes of DNA structure (Barton and Raphael, 1985). These compounds contain four modified phenanthroline rings coordinated to a transition metal atom. The stereochemistry of the metal complexes and the ability of the phenanthroline rings to intercalate into DNA allow compounds to be designed that bind to specific shapes of DNA. For example, structures with a right-handed propeller twist of the phenanthroline moiety can bind to right-handed B-DNA. One reagent, Λ-tris(4,7-diphenyl-1,10-phenanthroline) cobalt(III), which has a left-handed propeller-like twist, binds specifically to left-handed Z-DNA. (Other compounds have been designed that bind selectively to A-DNA and to cruciform structures.) An additional advantage of these reagents is that the use of cobalt as the transition metal makes these stereospecific DNA binding chemicals photoactive. Irradiation of DNA in the presence of Λ-tris(4,7-diphenyl-1,10-phenanthroline) cobalt(III) results in the introduction of nicks in the DNA at regions of alternating purine–pyrimidine runs that have the potential to form Z-DNA. This cleavage provides a potential assay for Z-DNA.

f. Psoralen Photobinding

The psoralen derivative 4,5′,8-trimethylpsoralen has provided an extremely useful assay for DNA that is applicable in living cells. As discussed in Chapter 1, psoralens bind to B-form DNA by intercalation and, on absorption of 360-nm light, form monoadducts and interstrand cross-links (see Figure 1.22). The structure of Z-DNA is not conducive to cross-link formation, since the base-stacking interactions are sheared and the twist angles are irregular. Thus, the 5,6 double bonds of pyrimidines in Z-DNA are not positioned correctly for cross-linking to occur. Indeed, psoralen and other intercalating drugs bind poorly to Z-DNA (binding of psoralen and ethidium bromide have been shown to actually drive the B-to-Z equilibrium toward B-DNA).

The sequence GAATT$(CG)_6$TA$(CG)_6$AATTC was designed as a Z-DNA-forming sequence that was assayable using psoralen. Psoralen preferentially photobinds to the 5′ TA at the center of the 26-bp Z-DNA-forming region when in the B-form but not when it exists as Z-DNA. The level of psoralen photobinding can be measured with base-pair resolution by an exonuclease III mapping procedure (Figure 5.8) or by a primer extension assay (see Figure 4.13). A "standard relationship" has been determined between the superhelical density of the DNA, which defines the fraction of the Z-DNA-forming sequence stabilized as Z-DNA, and the relative rate of photobinding to the 5′ TA in the Z-DNA-forming sequence. The extent of photobinding to the Z-DNA 5′ TA is measured relative to a nearby 5′ TA that can only exist as B-DNA. Analysis of this relative ratio in living cells will provide an estimate of the effective level of supercoiling and level of existence of Z-DNA (Sinden and Kochel, 1987; Kochel and Sinden, 1989).

The reactivity of the *Eco*RI sites flanking the Z-DNA-forming region also provides an indication of the conformation of the DNA. The central

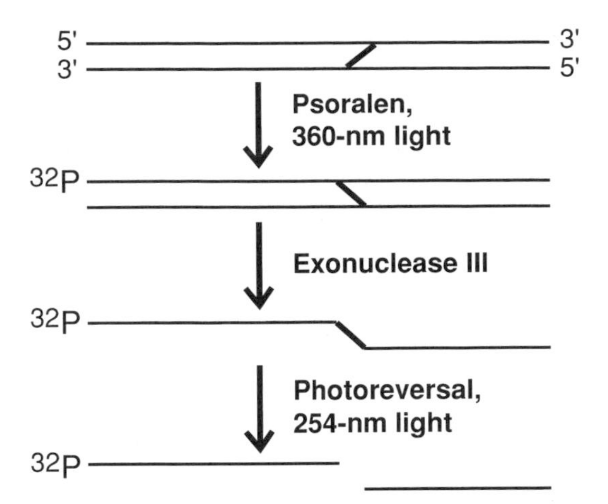

Figure 5.8 Psoralen exonuclease III assay for Z-DNA. Psoralen photobinding can be mapped with base-pair resolution by treating DNA containing a psoralen cross-link with exonuclease III, which digests DNA from a 3′OH end. Digestion stops quantitatively at an intrastrand cross-link or at a monoadduct in a single strand. After psoralen cross-links are photoreversed by treating with 254-nm light, the length of DNA from the end label to the site of the psoralen cross-link can be measured by electrophoresis on a DNA sequencing gel. Psoralen will photobind DNA in the B conformation but not in the Z conformation. In addition, psoralen photobinding to B–Z junctions is hypersensitive, providing an excellent assay for the existence of Z-DNA.

AATT sequences are not reactive with psoralen when they exist as B-DNA. However, when Z-DNA forms, the *Eco*RI sites exist as B–Z junctions and the TT dinucleotides are hyperactive with psoralen. In the Z-DNA-forming sequence GAATT$(CG)_6$TA$(CG)_2$$(TG)_8$AATTCC, psoralen photobinding occurs to only one strand of each B–Z junction (Hoepfner and Sinden, 1993). The reactivity of the junctions is a function of supercoiling (and the extent of Z-DNA formation), providing additional "standard relationships" that can be used to estimate the extent of Z-DNA formation in living cells.

The psoralen photobinding assay has several advantages as an *in vivo* probe for Z-DNA compared with other Z-DNA assays. First, this assay can be done at levels of photoaddition of about one psoralen adduct per 8000 bp of DNA. Although psoralen is a potent mutagen, >95% of an *E. coli* culture survives this level of DNA damage. Thus, the assay is truly done in "living cells." Second, the photobinding of psoralen to Z-DNA actually drives the $B \rightleftharpoons Z$ equilibrium toward B-DNA. Consequently, the assay itself cannot introduce the Z-DNA conformation. This is not true for some assays involving chemical probes or Z-DNA antibodies. Third, the assay can be done under a variety of growth conditions (in different media and at various temperatures). Fourth, the photobinding protocol is very rapid. Fifth, this approach can be applied to DNA in living eukaryotic cells, since psoralens are freely permeable to eukaryotic cells and nuclei.

4. Assays Involving the Restriction/Modification System

Restriction endonucleases and their corresponding methylases usually recognize 4, 6, or 8 bp of DNA. The endonucleases introduce a nick into each of the two DNA strands and the methylases methylate the DNA at the same sequences. Methylation then prevents subsequent digestion by the corresponding endonuclease. The endonucleases and methylases recognize the B-form of DNA and, in general, do not recognize the site in the Z-DNA form or in a B–Z junction.

The validity of this approach was first demonstrated by Singleton *et al.* (1983) using the restriction endonuclease *Bam*HI. The *Bam*HI recognition site (GGATCC) was placed adjacent to a run of alternating GC base pairs in plasmid DNA. This site was cut by the enzyme in relaxed DNA. When the DNA was supercoiled, Z-DNA formed in the GC sequence and the *Bam*HI site existed as a B–Z junction; *Bam*Hl cutting was reduced by more than 80%. The *Hha*I restriction enzyme fails to cleave its recognition site (GCGC) when the site exists as Z-DNA (Azorin *et al.*, 1984; Vardimon and Rich, 1984; Zacharias *et al.*, 1984). In addition, the *Hha*I methylase, designated M*Hha*I, fails to methylate cytosine when the recognition site exists in the Z-DNA conformation. As described in the next section, restriction enzymes and methylases have proved to be useful tools in the analysis of Z-DNA *in vivo*.

E. Z-DNA *in Vivo*

1. Psoralen Photobinding to DNA in Living Cells

One of the first indications that Z-DNA could exist in living cells came from the analysis of psoralen cross-linking and photobinding to the Z-DNA-forming sequence $(CG)_6TA(CG)_6$. The rate of photobinding to this sequence in living cells was equivalent to that observed in purified DNA with an *in vitro* superhelical density of $\sigma = -0.035$ (Sinden and Kochel, 1987; Zheng *et al.*, 1991). At this negative superhelical density, about 40% of the topoisomers contained the Z-DNA-forming sequence in the Z conformation in living wild-type *E. coli* cells. In cells with a defective topoisomerase I (*topA*10), the Z-DNA-forming sequence existed as Z-DNA in about 80% of the plasmids, suggesting a superhelical density of $\sigma = -0.048$. These experiments demonstrated that Z-DNA can exist at high levels in living *E. coli* cells and provided one measurement of supercoiling levels in living cells.

2. Application of Restriction/Modification Assays in Living Cells

Jaworski *et al.* (1987) provided evidence for the existence of Z-DNA *in vivo* by analyzing the activity of the *Eco*RI methylase (M*Eco*RI) at the *Eco*RI site (GAATTC) in B–Z junctions in living *E. coli.* These investigators cloned a gene encoding temperature-sensitive *Eco*RI methylase into a plasmid. The methylase was active when cells were grown at the permissive temperature (5–32°C) but not at the restrictive temperature (42°C). Following a temperature shift from 42°C to between 5 and 32°C, the M*Eco*RI methylated all GAATTC sites in a plasmid lacking Z-DNA sequences within about 30 min. However, if the methylation recognition sequence existed in a Z-DNA region, the enzyme could not methylate the GAATTC sequence (Figure 5.9).

In this initial study, two plasmids showed an inhibition of methylation. These contained two blocks of $(CG)_{13}$, one with an *Eco*RI site between the $(CG)_{13}$ blocks: $(CG)_{13}$AATT$(CG)_{13}$. The other sequence contained a central *Bam*HI site but had *Eco*RI sites flanking both $(CG)_{13}$ blocks: GAATT$(CG)_{13}$GATC$(CG)_{13}$AATTC. In the latter case, the *Eco*RI sites were in B–Z junctions. In plasmids containing either of these Z-DNA-forming sequences, 30–55% inhibition of methylation was observed. Methylation occurred to >90% levels at *Eco*RI sites in control plasmids lacking Z-DNA-forming sequences or in a plasmid containing an *Eco*RI site distant from the Z-DNA-forming sequence.

3. Application of a Linking Number Assay for Z-DNA in Living Cells

DNA in living cells is maintained at a set level of negative supercoiling, as discussed in Chapter 3. Haniford and Pulleyblank (1983b) reasoned that, if the transition to left-handed Z-DNA occurred in plasmid DNA living cells, the number of supercoils in the plasmid would decrease. Goldstein and Drlica (1984) demonstrated that a change in supercoiling *in vivo* (as a consequence of a temperature-induced change in the twist angle) results in a re-equilibration of the linking number of DNA to restore the original set level of supercoiling. The topoisomerases responsible for maintaining a precise level of supercoiling *in vivo* should decrease the linking number by about two for every 10 bp of B-DNA that formed Z-DNA. Consequently, topoisomers in which Z-DNA formed should then have a higher level of negative supercoiling. If 50% of a Gaussian distribution of topoisomers formed Z-DNA and if the linking number of these topoisomers was re-equilibrated to a lower level, then a bimodal topoisomer distribution should be observed. The distance between the centers of the two distributions should approximate the number of supercoils unwound by the B-to-Z transition. Haniford and

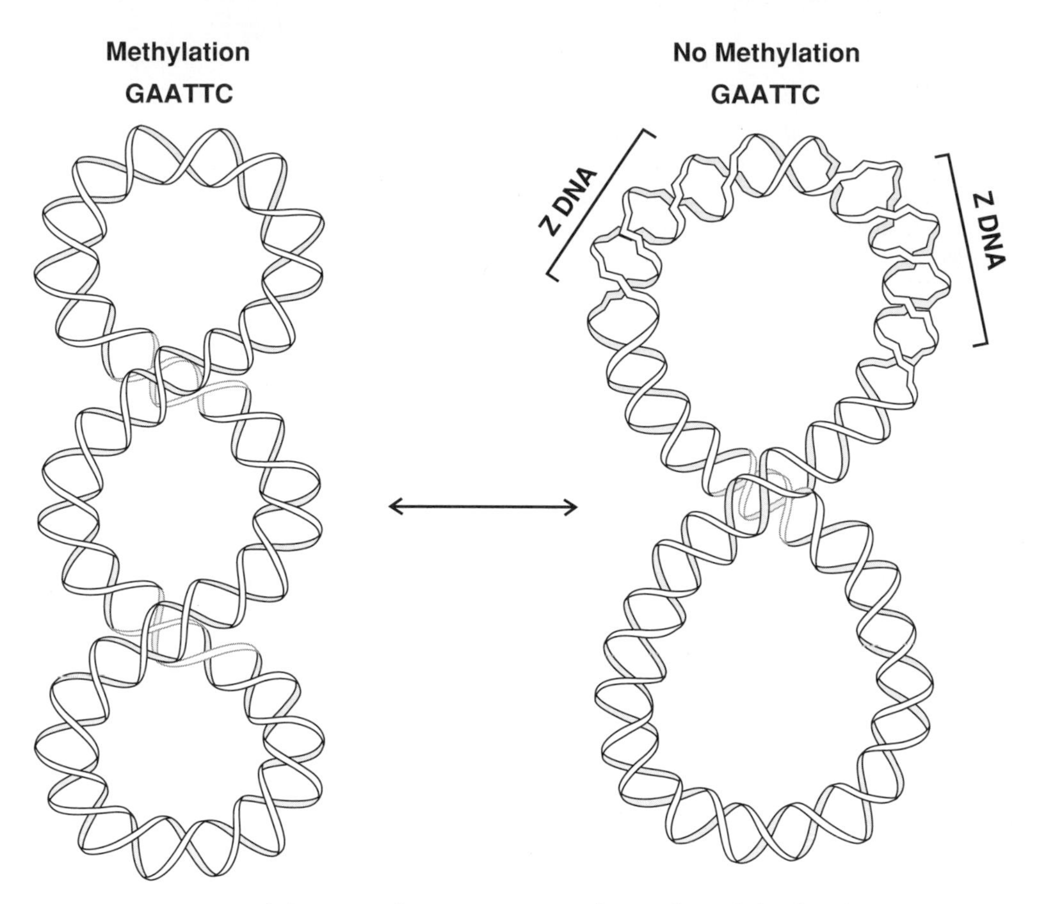

Figure 5.9 Methylase assay for DNA. Sequences that can be methylated, such as the *Eco*RI site GAATTC, require that the DNA exist in the B conformation for the methylase to act. If a methylation site is positioned within or adjacent to a Z-DNA sequence, the transition to Z-DNA disrupts the helical conformation of the methylation site and renders it unrecognizable by the methylase. In the situation shown here, the restriction site was placed between regions of Z-DNA. The appropriate cloning and expression of methylases *in vivo* has been used to detect the existence of Z-DNA in living cells.

Pulleyblank (1983b) were the first to observe this bimodal distribution in plasmid containing a $(GC)_n$ Z-DNA-forming insert. The bimodal distribution was not observed in cells under normal growth conditions but only in cells treated with chloramphenicol to "amplify" the plasmid. (See the box entitled "Chloramphenicol Amplification" for details.) Zacharias *et al.* (1988) used the linking number assay to examine a large number of Z-DNA-forming plasmids. The assay revealed a bimodal distribution in both chloramphenicol-treated cells and log phase cells (not treated with the antibiotic). The results from this approach agreed quite well with the results from the M*Eco*RI assay. The minimum length required to inhibit methylation or to produce the bimodal distribution was about 44 bp of alternating GC base pairs, that is, $(CG)_{22}$. $(CG)_{13}$ did not produce the bimodal distribution.

Chloramphenicol Amplification

Chloramphenicol is an antibiotic that binds to a protein in the ribosome and stops protein synthesis. Chloramphenicol is used to "amplify" plasmid DNA in cells. The initiation of DNA replication in *E. coli* occurs in a coordinately regulated fashion from a unique replication origin called *ori*C. An initiation-specific protein called DnaA must be synthesized *de novo* prior to the initiation of replication at *ori*C. In the presence of chloramphenicol, DnaA is not synthesized and chromosome replication and cell division cease. The replication of most plasmids requires neither DnaA nor *de novo* protein synthesis. Therefore the replication of plasmids in chloramphenicol-treated *E. coli* continues, resulting in an amplification of plasmid in *E. coli*. This method is very commonly used to increase the amount of plasmid for purification purposes or for experiments in which plasmid is analyzed *in vivo*. Within the first 4–10 hours of the addition of chloramphenicol, the level of negative supercoiling in the plasmid in wild-type cells increases. By 24 hours the level of supercoiling usually returns to the original level present in cells before amplification.

4. Application of Chemical Probes for Z-DNA *in Vivo*: Application of OsO_4

Rahmouni and Wells (1989) applied osmium tetroxide (OsO_4), a chemical probe for unpaired pyrimidines, to detect Z-DNA *in situ*. OsO_4 was added to *E. coli* cells (in the presence of 2,2′-bipyridine) to allow chemical modification of unpaired thymines and cytosines (see Figure 1.21). Plasmid was then purified from cells and analyzed by a primer extension assay to map the sites of chemical modification. Hyperactive sites of chemical modification were detected at the thymines within the *Eco*RI sites (GAATTC) between two $(CG)_n$ blocks or flanking a $(CG)_n$ block. In these experiments, some reactivity *in situ* with $(CG)_6$ was observed. Reactivity was not observed with $(CG)_5$ or $(CG)_4$ Z-DNA-forming sequences. Strong reactivity was observed for the $(CG)_{13}$ Z-DNA-forming sequence.

5. Z-DNA Antibodies as *in Vivo* Assays for Z-DNA in Eukaryotic Cells

Nordheim *et al.* (1981) were the first to use Z-DNA antibodies to detect the presence of Z-DNA in eukaryotic cells. This was done using the polytene chromosomes of the *Drosophila* salivary gland. These chromosomes replicate repeatedly, producing more than 1000 copies that, when lined up side-by-side, make the chromosome very easy to see. Cells were attached to a microscope slide and treated with a variety of "fixing" agents including formaldehyde and acetic acid. These procedures lock the intracellular protein–protein

Application of Z-DNA Assays for Measuring the Level of Unrestrained Torsional Tension in *E. coli*

The analyses of the level of Z-DNA in *E. coli* provide information about the level of unrestrained DNA supercoiling in bacterial cells because the formation of Z-DNA under physiological conditions (in the absence of Z-DNA binding proteins) requires negative DNA supercoiling. Since the level of Z-DNA formation is dependent on the level of supercoiling, determination of the level of Z-DNA *in vivo* provides an indication of the level of unrestrained torsional tension maintained in DNA inside cells. Most of our understanding of the energy within the helix in cells comes from quantitative measurement or simple detection of alternative DNA conformations *in vivo*.

The various Z-DNA assays and probe sequences used have provided different estimates of the level of supercoiling in DNA *in vivo*. Sinden and Kochel (1987) and Zheng *et al.* (1991), using psoralen photobinding, estimated a σ *in vivo* of -0.035 for wild-type cells and -0.048 for cells defective in topoisomerase I (Table 5.2). The experiments of Jaworski *et al.* (1987) were consistent with DNA at a level of supercoiling *in vivo* of $\sigma = -0.025$.

Utilizing the linking number assay, the shortest Z-DNA-forming sequence that produced a bimodal distribution *in vivo* formed Z-DNA at $\sigma = -0.025$ *in vitro* (Zacharias *et al.*, 1988). Thus, a superhelical density *in vivo* of $\sigma = -0.025$ was suggested from these experiments. The Z-DNA-forming sequence $(CG)_6TA(CG)_6$ studied by Zheng *et al.* (1991) did not produce a bimodal distribution although >40% of the sequence existed as Z-DNA in *E. coli* as determined by the psoralen assay. How can this apparent discrepancy with the results of Zacharias *et al.* (1988) be reconciled? Keep in mind that the assays employed or subtle procedural differences used may have an influence on the data obtained and the conclusions reached. Moreover, the linking number assay is *indirect* since it requires purification of plasmid from cells and analysis of the

continues

continued

plasmid on agarose gels. One assumes that changes in linking number do not occur on lysis of the cells or during subsequent purification steps. However, there is really no independent way to confirm that linking number changes do not occur. The OsO_4 experiments of Rahmouni and Wells (1989) suggested an *in vivo* negative superhelical density in the range of $\sigma = -0.038$. This is substantially lower (more negative) than previous estimates from the Wells laboratory based on the M*Eco*RI assay or linking number assays, but agrees with the results using the trimethylpsoralen assay.

Using the OsO_4 assay, Rahmouni and Wells (1989,1992) examined the level of Z-DNA in different locations of a plasmid *in situ*. According to the twin-domain model for supercoiling discussed in Chapter 3, the local level of supercoiling can be influenced by transcription. Specifically, DNA upstream of divergently transcribed genes can be more negatively supercoiled when transcription occurs. DNA downstream of divergently transcribed genes can be less negatively (perhaps even positively) supercoiled during transcription. Various $(CG)_n$ Z-DNA-forming sequences were cloned into plasmid pBR322 upstream or downstream from two divergently transcribed genes (the *Tet* gene and the *Amp* gene). Upstream from the genes, a $(CG)_7$ sequence readily formed Z-DNA. Downstream from the genes, a 40-bp Z-DNA-forming sequence could not be detected to exist as Z-DNA. These experiments provided an elegant demonstration of transcription influencing levels of local supercoiling. The existence of Z-DNA *in vivo* can depend on the length (and sequence) of the potential Z-DNA-forming region as well as its location within a chromosome. Moreover, these experiments demonstrate that Z-DNA formation, in some cases, can be coupled to transcription.

Tablo 5.2
In Vivo* Superhelical Densities Predicted from Measurement of Z-DNA *in Vivo

		σ *in vivo*	
Assay	Sequence	Wild type	TopA10
Psoralen photobinding	$(CG)_6TA(CG)_6$	−0.035	−0.048
M*Eco*RI inhibition	$(CG)_{13}AATT(CG)_{13}$	−0.025	
	$GAATT(CG)_{13}GATC(CG)_{13}AATTC$		
Linking number assay	$(CG)_{22}$	−0.025	
OsO_4	$(CG)_{13}$	−0.038	

and protein–nucleic acid interactions into place and preserve the original three-dimensional organization present in cells. These treatments are also necessary to permeabilize the cell and nucleus to allow the entry of antibodies. Following fixation, antibodies are added and allowed to bind to their recognition sites. Next, a second fluorescently tagged antibody that will bind to the first antibody is added. (For example, if the first antibody against Z-DNA was purified from goats, an anti-goat antibody made by and purified from rabbits could be used.) The intracellular location of the antibodies can be visualized using fluorescence. The experiments of Nordheim *et al.* (1981) identified many regions of the *Drosophila* chromosome that reacted with the Z-DNA antibody. Many of these regions were in *interbands,* regions between dark-staining *bands.* Bands are believed to be associated with regions of the DNA that encode proteins whereas the interbands are regions of chromosomes believed not to contain genes. During periods of massive transcriptional activity, many bands can become enlarged or *puff.* Nordheim *et al.* also identified a number of transcriptionally active puffs that reacted with Z-DNA antibodies. Similar results have been obtained for a variety of organisms in several different laboratories.

In certain cases, Z-DNA antibodies from different laboratories gave different staining patterns on *Drosophila* chromosomes. For example, some investigators observed antibody binding to bands, some to interbands, and some to both. These results could have been due to different recognition properties of antibodies but in some cases they were attributed to differences in the fixation procedures. Many components of buffers and fixatives are known to favor the Z-form of DNA. For example, acetic acid is commonly used in the preparation of cells for antibody binding and immunofluorescence. In 1983, Hill and Stollar published an important paper dealing with the specificity of Z-DNA antibody binding to *Drosophila* cells. Microsurgically isolated salivary chromosomes carefully isolated in buffers at physiological ionic strength and pH showed essentially no staining with Z-DNA antibody. If these chromosomes were treated with acetic acid for a short period of time, Z-DNA antibodies would bind to interbands and puffs. Following longer treatment with acetic acid, antibody binding to bands was predominant. What did this result mean? Was Z-DNA induced by the acetic acid treatment? Was Z-DNA present in native chromosomes but just not accessible to the antibody until the acetic acid treatment uncovered it? These are questions that workers in the field were certainly aware of and had considered carefully (Pardue *et al.,* 1983). Hill and Stollar's interpretation was that acetic acid was removing the core histones that organize DNA into nucleosomes in eukaryotic cells.

The removal of nucleosomes from the bulk of eukaryotic DNA converts the DNA from a relaxed state (with unrestrained supercoils) which cannot support Z-DNA formation to a negatively supercoiled state which can drive the B-to-Z transition. If negative supercoils are introduced by the fixation procedure, then relaxation of the supercoils should prevent the formation of Z-DNA (and/or destabilize pre-existing regions of Z-DNA). Hill and Stollar showed that treatment with either topoisomerase I to relax negative supercoils or DNase I to introduce nicks into DNA prevented Z-DNA antibody binding. An important control experiment showed that B-DNA antibodies could bind to *Drosophila* chromosomes at all stages of manipulation, demonstrating that the DNA was accessible to antibodies.

The results of Hill and Stollar, although pointing out potential problems with isolating and fixing polytene chromosomes, do not necessarily mean that Z-DNA does not exist in living cells. Their microdissection procedure and the manipulation of purified chromosomes represents a significant change from an intracellular environment. In fact, Jackson and Cook (1985) have demonstrated that even gentle purification of nuclei from cells results in

Probing Z-DNA in Eukaryotic Cells with Z-DNA Antibodies: Additional Studies

Wittig *et al.* (1989,1991) reinvestigated the Z-DNA question using very different procedures for binding Z-DNA antibodies inside eukaryotic cells. These investigators used a procedure developed for studying chromosomes in living cells in as native or natural a state as possible at that time (Jackson and Cook, 1985). In this procedure eukaryotic cells are first encased in agarose beads. The cells can be gently lysed and the cytoplasm and parts of the membrane can be washed out of the beads. The nucleus remains in the beads and is permeable to macromolecules such as antibodies and various enzymes. This procedure, in 1985, seemed to be the best to date for preserving the "most native" state of the nucleus.

When Z-DNA antibody was added to nuclei in agarose beads, it bound readily after a short lag time, which probably represented the time required for the antibody to diffuse through the agarose. When increasing concentrations of the antibody were added, the amount bound increased until a "binding plateau" was reached. Remarkably, this binding plateau was constant over a 200-fold range of antibody concentration. The binding level was assumed to represent pre-existing Z-DNA in eukaryotic nuclei. The level of antibody binding was thought to represent approximately one Z-DNA sequence every 100 kb of DNA. When the antibody concentration was increased even further, the amount of antibody bound increased dramatically. As discussed earlier, certain Z-DNA antibodies can dramatically influence the B $\rightleftharpoons$ Z equilibrium. In fact, certain Z-DNA antibodies can induce and

continues

continued

stabilize Z-DNA in relaxed DNA. The dramatic increase in antibody binding at the high concentrations was believed to repre-sent the binding to regions of potential Z-DNA that originally existed in the B-DNA conformation. These regions may-have formed Z-DNA as a result of antibody binding.

Was the Z-DNA observed in eukaryotic cells a function of unrestrained DNA supercoils? To address this question, several control experiments were done. Nuclei in microagarose beads were treated with DNase I to introduce nicks into DNA. This treatment resulted in the loss of much of the antibody binding in the plateau region, but at high inducing levels of antibody, binding still occurred. When the topoisomerase I inhibitor camptothecin was added, a higher level of Z-DNA antibody binding was observed. This result is consistent with the interpretation that, by inhibiting an enzyme that can relax negative supercoils, a higher level of supercoiling will exist and will support the formation of Z-DNA.

The experiments described certainly indicate the existence of Z-DNA in eukaryotic nuclei and suggest that Z-DNA was stabilized by negative supercoiling. The experiments were done using, at the time, the "best up-to-date method" for preserving the natural state of the DNA. However, as gentle as these conditions might be, the torsional strain and pre-existing Z-DNA could have been the result of the manipulation of cells to get to the stage of nuclei encased in the agarose beads. In fact, in 1991 Wittig *et al.* used a "new" procedure developed by Jackson *et al.* (1988) that *"more naturally"* preserved the organization of chromosomes in cells than their previous method. In these later experiments, a plateau region of Z-DNA antibody binding was still observed; however, there was little change in antibody binding following the inhibition of topoisomerase I with camptothecin. The level of Z-DNA in nuclei was examined when transcription and/or replication was allowed to occur in the agarose beads. There was a four-fold increase in the levelof Z-DNA as a result of transcription and about a two-fold increase following replication. When inhibitors of transcription, either α-amanitin or actinomycin, were added to this system, the level of Z-DNA decreased significantly. An effect of transcription inhibitors was seen in nuclei in the "new buffer" or in nuclei in which transcription was stimulated by the addition of ribonucleotides. These results suggest the existence of Z-DNA in eukaryotic nuclei. Moreover, they demonstrate that transcriptional activity (and to some extent replication) can introduce Z-DNA into eukaryotic cells.

significant changes in the organization of DNA in chromosomes. Therefore a more *in vivo*-like approach to the analysis of Z-DNA in living cells is required. Further attempts at using antibodies are described in the box entitled "Probing Z-DNA in Eukaryotic Cells with Z-DNA Antibodies: Additional Studies."

F. Possible Biological Functions for Z-DNA

The sequences that form Z-DNA are widely found in eukaryotic genomes. $(GC)_n$ sequences are occasionally found, but $(GT)_n$ sequences are much more frequent. Long alternating purine–pyrimidine tracts have not yet been found naturally in bacterial DNA. Bacterial DNA, at least *E. coli,* is wound with sufficient unrestrained supercoiling to drive the formation of Z-DNA. When cloned into plasmids, Z-DNA can exist in bacterial cells and does not seem to be obligatorily deleted or attacked by nucleases. The best indication that Z-DNA might exist in eukaryotic cells comes from the analysis of Z-DNA antibodies discussed in the preceding section. Although these results are encouraging, they do not yet establish unequivocally that Z-DNA readily exists as a normal component of the eukaryotic genome.

There are many opportunities for the involvement of Z-DNA in biological processes. This section will present a speculative discussion of possible biological roles for Z-DNA. A demonstration of the actual involvement of Z-DNA, as well as other alternative forms of DNA, in biology will await further experimentation.

1. Possible Roles of Z-DNA in Gene Expression

a. Z-DNA May Control the Level of Supercoiling

In bacterial cells, the expression of certain genes can be regulated by the level of unrestrained supercoiling in the DNA. This is not the only means of gene regulation, and not all genes are regulated by supercoiling. However, the regulatory controls for some genes do respond to supercoiling in ways that are logical for the organism (the topoisomerase I and DNA gyrase genes of *E. coli* are examples).

The formation of Z-DNA could act in eukaryotic cells to modulate the level of supercoiling and potentially control the level of expression. In eukaryotic cells, the bulk of superhelical energy is restrained by the organization of DNA into nucleosomes. However, as discussed in Chapter 9, unrestrained torsional stress occurs at active gene regions. In some cases, supercoiling may be needed to facilitate the opening of a promoter region to allow RNA polymerase to bind. Relaxed topological domains in eukaryotic cells may be transcriptionally silent. For transcription to occur, the DNA within the domain, at least within the active gene region, may need to become negatively supercoiled.

How could Z-DNA influence supercoiling-dependent gene expression? The formation of Z-DNA relaxes negative superhelical turns. Conversely, the formation of B-DNA from a region of Z-DNA will introduce negative supercoils. Consider a region of Z-DNA stabilized by a Z-DNA binding protein existing within a relaxed topological domain. Removal of the Z-DNA binding protein would allow the conversion of this region of DNA into the B form. This would topologically be accompanied by the introduction of negative supercoils into the domain, which may then turn on selected genes. This completely hypothetical situation is shown in Figure 5.10. The situation could also work equally well in reverse. DNA could be wound with a level of negative supercoiling higher than optimal for expression from a specific gene. The formation of Z-DNA would lower the level of supercoiling and turn on expression of selected genes.

DNA supercoiling will also change the three-dimensional organization of DNA (facilitating compaction), which could greatly increase the effective concentration of two sites on the DNA. This situation is illustrated in Figure 5.11 where, on introduction of supercoils, an enhancer binding protein and a promoter binding protein physically interact to turn on gene expression. As discussed in Chapter 2, looping can control gene expression and, in many cases, looping requires a supercoiled DNA template. At least in theory, there are many ways in which changes in the level of supercoiling could act to influence gene expression.

b. Z-DNA May Provide a Molecular Switch for Gene Expression

The helix structures of B-DNA and Z-DNA are very different and distinct proteins might be required to bind to each helix. A situation could exist in which the formation of either B-DNA or Z-DNA would recruit the binding of a specific protein required for gene expression (in either a positive or a negative regulatory sense). Thus, by flipping between B-DNA and Z-DNA, a region of DNA could act as a molecular switch. These possibilities are shown in Figure 5.12.

c. Z-DNA May Influence Transcription

Z-DNA could influence transcription at several levels. Z-DNA could act to physically prevent the movement or the binding of RNA polymerase. Alternatively, Z-DNA might facilitate the movement of RNA polymerase. Z-DNA might also act as a gene regulatory unit called an *enhancer*. Unfortunately, gene regulation and the potential involvement of Z-DNA are complicated processes and there is no consensus on the roles of Z-DNA in transcription.

Purified RNA polymerase has been shown to stop or pause at regions of Z-DNA (van de Sande *et al.,* 1982; Durand *et al.,* 1983; Peck and Wang, 1985; Job *et al.,* 1988). For example, Peck and Wang (1985) showed that purified *E. coli* RNA polymerase would transcribe through a $(GT)_{21}$ region that

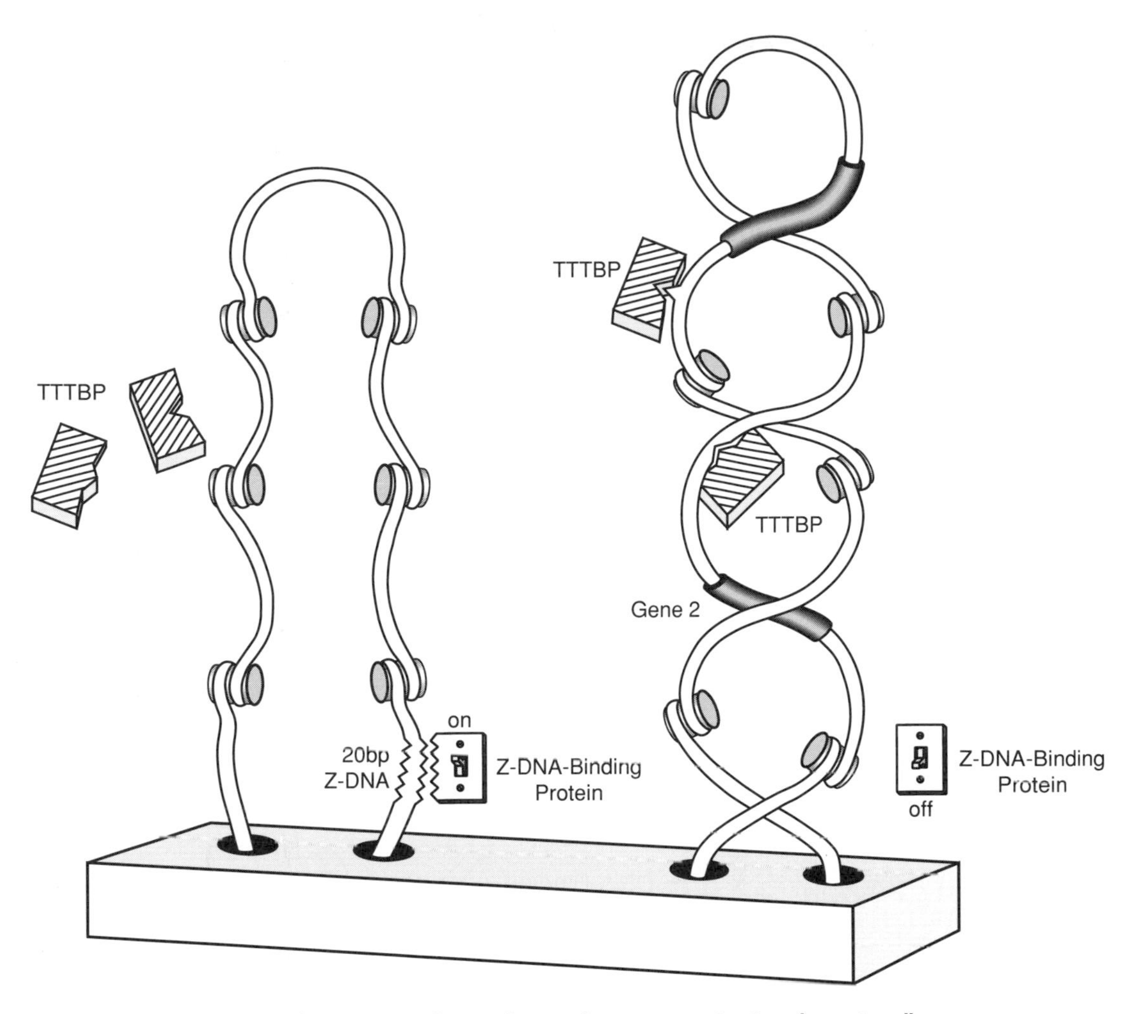

Figure 5.10 Potential role for Z-DNA in the regulation of gene expression in eukaryotic cells: Model 1. Unrestrained negative superhelical tension may be required for gene expression in eukaryotic cells. On average, most negative supercoils are restrained by the organization of DNA into nucleosomes. However, negative superhelical tension exists around genes that are active in transcription (or perhaps primed for transcriptional activity as discussed in Chapter 9). The transition between B- and Z-form DNA could potentially influence the level of supercoiling in the eukaryotic cell. The example shown involves torsionally tuned topological binding proteins (TTTBPs) binding to DNA when it becomes negatively supercoiled. In a completely hypothetical model, a region of Z-DNA stabilized by a Z-DNA binding protein exists in the relaxed topological domain on the left. In response to some inducing signal, the Z-DNA binding protein would come off the DNA and the region of Z-DNA would flip back into a B conformation. This would introduce the negative supercoils restrained in the Z-DNA conformation back into the rest of the topological domain. The subtle change in the helical twist of the DNA resulting from supercoiling may be sufficient to allow the TTTBPs to bind to the DNA and stimulate gene expression. Figure modified with permission from Esposito and Sinden (1987).

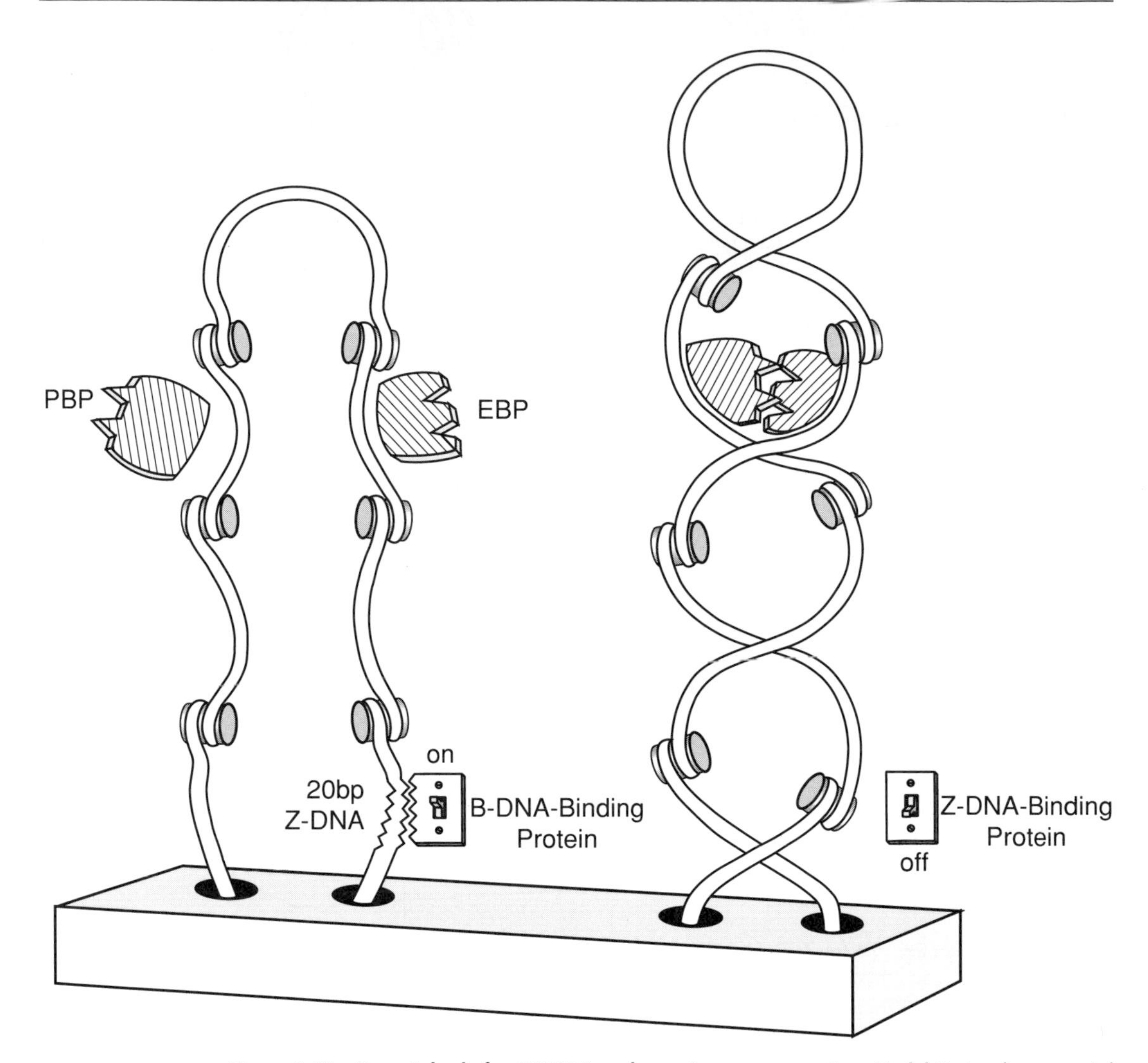

Figure 5.11 Potential role for Z-DNA in eukaryotic gene expression: Model 2. Another potential role for Z-DNA in gene expression may involve a change in the three-dimensional architecture of the topological domain. On supercoiling the DNA, through the removal of Z-DNA binding protein, the structural organization of the topological domain can change. Supercoiled DNA is more compact and less open than relaxed DNA. This supercoiling may bring a promoter binding protein (PBP) in contact with an enhancer binding protein (EBP), both of which are bound at distant locations of the DNA. These proteins may only come together on DNA supercoiling and compaction of the topological domain. Once these proteins contact each other, they may stimulate gene expression from an adjacent gene. In some cases, DNA looping facilitating protein–protein contact requires supercoiled DNA. Figure modified with permission from Esposito and Sinden (1987).

existed as Z-DNA in supercoiled plasmid, but that it would not go through a $(GC)_{16}$ region existing as Z-DNA. On the other hand, using chemically modified templates others have demonstrated that eukaryotic polymerases can proceed through left-handed helices although at a slower rate than through B-DNA. The behavior of RNA polymerase *in vivo* may be different from that in solution. Purified polymerase may lack important subunits or cofactors that modify or control its activity. For example, although *E. coli* RNA polymerase is blocked by a $(GC)_{16}$ region of Z-DNA in supercoiled plasmid *in vitro,* this Z-DNA region does not block transcription *in vivo* (Peck and Wang, 1985). In eukaryotic cells, there are three different RNA polymerases, each of which may act differently when encountering regions of Z-DNA.

Enhancers are elements of regulatory regions that stimulate the level of expression from a gene. They can function over large distances (thousands of base pairs), function in either orientation, and function either upstream or downstream from the gene. DNA sequences that can form alternative helical structures including Z-DNA and triplex regions are frequently found upstream or downstream from genes. These sequences may act as enhancers or have some other regulatory effect on gene expression. They could act by changing supercoiling levels within a domain, which may influence protein–DNA interactions. Alternatively, Z-DNA formation may influence the three-dimensional architecture of a domain, thereby altering protein–protein interactions.

Z-DNA-forming sequences have been cloned in regulatory regions upstream from various genes. There are three possible effects of inserting a Z-DNA region into a regulatory region. (1) The potential Z-DNA region may have no effect on gene expression. (2) Transcription could be inhibited. (3) Transcription could be enhanced. There is no consensus for the role of Z-DNA-forming sequences in the regulation of gene expression, since all three possible results have been obtained (Santoro *et al.,* 1984; Banerjee *et al.,* 1985; Banerjee and Grunberger, 1986; Naylor and Clark, 1990). This is probably not surprising when one realizes that the regulation of eukaryotic gene expression involves multiple elements and interactions. There may be a dozen or more regulatory elements upstream from a gene. Each of these may bind a transcription factor. There may be factors binding to the B-form or the Z-form of a potential Z-DNA-forming sequence. The formation of Z-DNA or B-DNA may influence the level of DNA supercoiling, which could affect transcription factor binding to numerous independent binding sites within a single topological domain. Alternatively, the nucleosome organization throughout a regulatory region could vary if a region of DNA existed in a B- or Z-DNA helix. Perhaps potential Z-DNA regions should be considered one small unit of a large complex circuitry controlling gene expression.

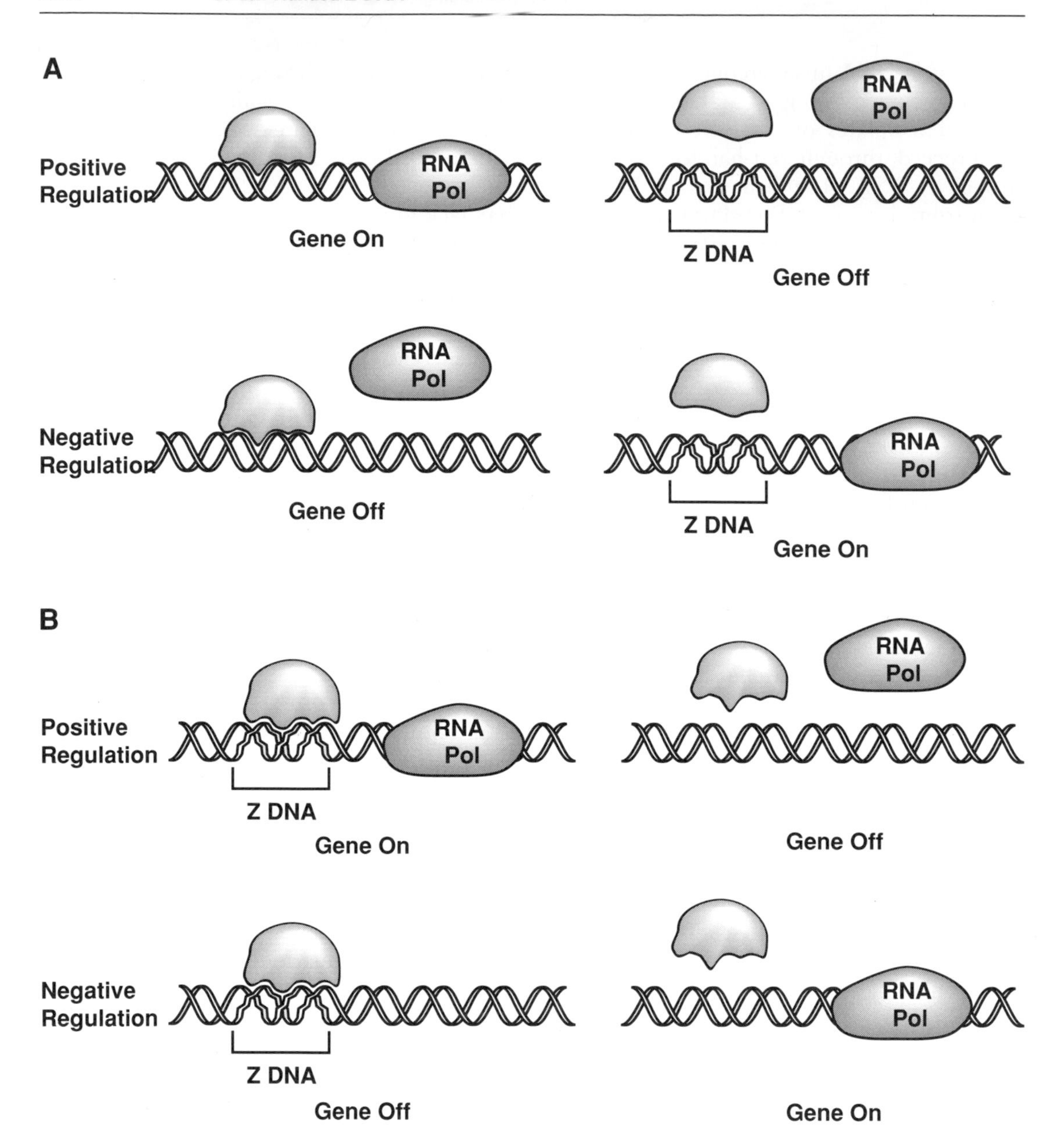

Figure 5.12 Potential regulation of gene expression by Z-DNA. (A) A potential role for Z-DNA in gene regulation by B-DNA binding proteins. For positive regulation by B-DNA binding proteins, a protein could bind to B-form DNA, interact with RNA polymerase, and promote transcription (*left,* "Gene On"). On formation of Z-DNA, the DNA binding protein could no longer bind, precluding interaction with and stimulation of transcription by RNA polymerase (*right,* "Gene Off"). For negative regulation by B-DNA binding proteins, the bound protein may prevent binding and transcrip-

d. Z-DNA May Affect Nucleosome Positioning on Promoters

The ability of DNA to wrap into nucleosomes depends on the sequence of DNA. Certain regions are more flexible than others. Regions of Z-DNA are not easily wrapped into nucleosomes (Garner and Felsenfeld, 1987). The strong preferential binding of a nucleosome to a particular region of DNA will act to phase nucleosomes from that site. In addition, if nucleosomes are excluded from a particular site due to the binding of a protein transcription factor or by the existence of Z-DNA, then nucleosomes may be phased from the region of Z-DNA. This situation is illustrated in Figure 5.13.

2. Possible Roles of Z-DNA in Genetic Recombination

There are strong implications that Z-DNA may be involved in genetic recombination (Blaho and Wells, 1989). Genetic recombination may involve the transient formation of a left-handed helix. Consider two single-stranded circles of DNA that are complementary. These cannot form a complete B-DNA double helix unless one of the circles is broken. When one circle is nicked, that strand can wrap around the circular strand forming double-stranded DNA. If both double-stranded circles are intact, for every region of the complementary strands that pairs into a right-handed helix, a similarly sized region will be forced into a left-handed helix. Such a DNA molecule has been called form V DNA (Brahmachari *et al.*, 1987). When a single strand initiates recombination by base pairing with a homologous duplex, before one of the ends of the single strand can physically wrap around the duplex, a helical turn of left-handed DNA may form for every helical turn of right-handed DNA. Not all regions of DNA will form left-handed Z-DNA. Regions of alternating purine–pyrimidine symmetry will most easily adopt the left-handed helix. In fact, in form V DNA the alternating PuPy regions preferentially form the left-handed regions. In a DNA molecule containing a region that can potentially form left-handed Z-DNA, the DNA adjacent to a

---

tion by RNA polymerase (*left*, "Gene Off"). On Z-DNA formation, the negative regulatory protein could no longer bind, allowing the recognition of the promoter by RNA polymerase and subsequent transcription (*right*, "Gene On"). (B) The hypothetical regulation of transcription by Z-DNA binding proteins. For positive regulation by a Z-DNA binding protein, the protein may bind to a region of Z-DNA, interact with RNA polymerase, and stimulate transcription from a gene (*left*, "Gene On"). In the absence of Z-DNA, the Z-DNA binding protein could no longer bind, and RNA polymerase could not bind to its cognate promoter and promote transcription (*right*, "Gene Off"). For negative regulation by Z-DNA binding proteins, the binding of a Z-DNA binding protein to Z-DNA would prevent transcription (*left*, "Gene Off"). On formation of B-DNA, the negative regulatory protein could not bind, allowing recognition, binding, and transcription by RNA polymerase (*right*, "Gene On"). Current and future experiments will determine if any of these schemes actually operate in cells.

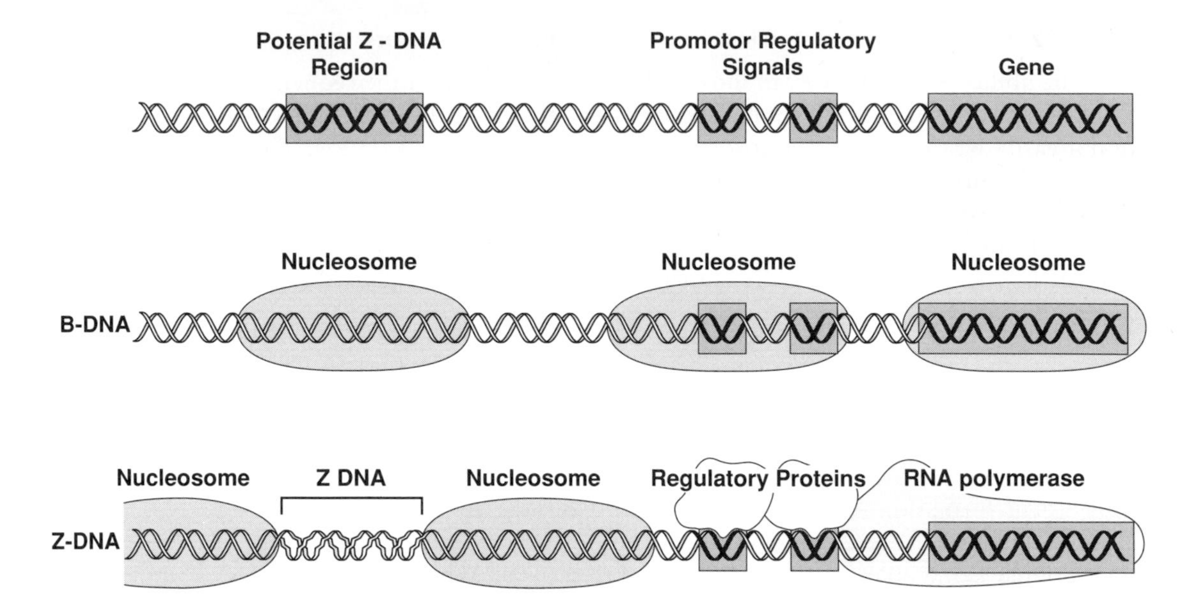

Figure 5.13 Potential positioning of nucleosomes by Z-DNA formation. The formation of a region of Z-DNA may position nucleosomes. The phasing of nucleosomes, positioning them over or away from a promoter region, is one way in which Z-DNA could contribute to the regulation of gene expression. (*Top*) The region of DNA shown contains a potential Z-DNA region, promoter regulatory signals, and the body of the gene. (*Center*) In the B-DNA conformation, nucleosomes may be positioned over the potential Z-DNA region, over the promoter regulatory signals, and over downstream coding regions of the gene. (*Bottom*) On DNA supercoiling, a region of Z-DNA may form and prevent organization of this region of DNA into a nucleosome. This will position nucleosomes adjacent to the region of Z-DNA so the promoter regulatory signals are now accessible to transcription factors that may recruit RNA polymerase and initiate transcription from the gene.

potential Z-DNA forming region may readily participate in genetic recombination. Such a region could be more recombinagenic than sequences not flanked by potential Z-DNA sequences.

There are numerous examples from the literature showing that hot spots for genetic recombination in eukaryotic cells correlate with $(GT)_n$ or $(CA)_n$ tracts. Stringer did a very simple experiment to determine if potential regions of Z-DNA stimulated genetic recombination in *E. coli.* He cloned a $(GT)_n$ sequence into a plasmid and grew the plasmids in *E. coli.* He found that the plasmids underwent genetic recombination, as evidenced by the formation of dimers, trimers, tetramers, and so on at a much higher frequency than the plasmid without the Z-DNA-forming region (Murphy and Stringer, 1986).

Proteins that are involved in recombination bind to Z-DNA. Recombination proteins RecA from *E. coli* and RecI from *Ustilago* mediate the alignment of homologous DNA duplexes and mediate the actual strand exchange reaction. These proteins show a preferential binding to left-handed Z-DNA (Kmiec *et al.*, 1985; Blaho and Wells, 1987,1989; Wahls *et al.*, 1990). In fact, by purifying proteins from human cells based on their ability to bind to Z-DNA, Fishel *et al.* (1988) isolated a protein that was active in the homologous pairing and strand exchange reactions. Numerous studies showing a relationship between the location of a potential Z-DNA-forming region and a site of genetic recombination in eukaryotic cells have implicated Z-DNA in this process (Molineaux *et al.*, 1986; Weinreb *et al.*, 1988,1990; Boehm *et al.*, 1989; Choo *et al.*, 1990; Wahls *et al.*, 1990).

3. Z-DNA Binding Proteins

Proteins that bind to Z-DNA have been found in many different organisms including *E. coli, Drosophila,* yeast, nematodes, chickens, frogs, bull testis, and human cells (Nordheim *et al.*, 1982b; Azorin and Rich, 1985; Lafer *et al.*, 1985b,1988; Gut *et al.*, 1987; Krishna *et al.*, 1988; Zhang *et al.*, 1992; Herbert *et al.*, 1993). Since sequences that form Z-DNA are widely found in eukaryotic genomes, Z-DNA binding proteins might be expected in eukaryotic cells. The *E. coli* genome, however, does not contain long stretches of $(GT)_n$ or $(GC)_n$ that could form Z-DNA at the superhelical densities found naturally in living cells. Nonetheless, in *E. coli* a number of Z-DNA binding proteins have been identified, including the RecA protein that is involved in genetic recombination (Blaho and Wells, 1987; Lafer *et al.*, 1988).

Recently Jovin and colleagues purified *Drosophila* topoisomerase II as a Z-DNA binding protein (Arndt-Jovin *et al.*, 1993). They showed that topoisomerase II binds preferentially to Z-DNA and that binding of GTP to the enzyme weakens its affinity for B-DNA while increasing its affinity for Z-DNA. These authors suggested that in actively transcribed regions, which are wound with torsional tension from unrestrained supercoils, the formation of Z-DNA may recruit the binding of topoisomerase II. Thus, Z-DNA binding by topoisomerase II may be a way to target the enzyme to regions of the chromosome that may require the activity of enzyme.

Are all "Z-DNA binding proteins" actually specific for Z-DNA? The binding of certain Z-DNA binding proteins may be a biological artifact. Krishna *et al.* (1988,1990) described proteins purified from the egg yolk of nematodes, chickens, and frogs that showed tight binding to Z-DNA. However, these are cytoplasmic proteins that probably never enter the nucleus nor interact with DNA. Moreover, the proteins identified in egg yolk bind Z-DNA with a much higher affinity than most Z-DNA binding proteins purified from other sources. The binding of egg yolk proteins to Z-DNA can be competed with by phospholipids, which make up a major part of egg yolks. Thus, these apparent Z-DNA binding proteins may actually be phospholipid-binding proteins. Proteins identified by Gut *et al.* (1987) from bull testis on the basis of preferential binding to brominated poly(dG–dC)·poly(dG–dC) were later shown not to be specific for Z-DNA (Rohner *et al.*, 1990).

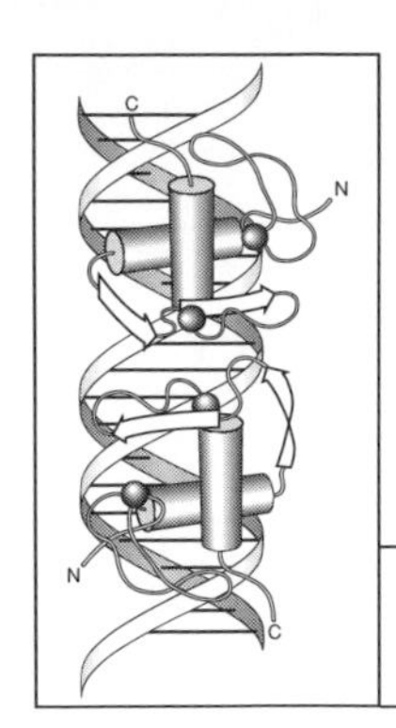

CHAPTER 6

Triplex DNA

A. Introduction

Felsenfeld, Davis, and Rich first reported the formation of triple helices in nucleic acids in 1957. Subsequently, several groups showed that triple-stranded polymers could form (Felsenfeld *et al.*, 1957; Chamberlin, 1965; Chamberlin and Patterson, 1965; Felsenfeld and Miles, 1967; Morgan and Wells, 1968; Arnott and Selsing, 1974). For example, a normal Watson–Crick-paired helix of poly(dA)·poly(dT) would hydrogen bond with an additional strand of poly(dT) forming a $(dA)(dT)_2$ triple helix. Triple helices also formed in RNA strands, for example, $(rA)(rU)_2$. The triple helix will be written as (*dT*)·(dA)·(dT) with the third strand in italics.

1. Intermolecular Triplex DNA

Triple-stranded DNA is formed by laying a third strand into the major groove of DNA (Figure 6.1) which, in the triplex form, may adopt an unwound B-DNA-like conformation. Complementary hydrogen bonding interactions are responsible for the specificity of the third strand interaction. Since the Watson–Crick base-pairing surfaces are already involved in hydrogen bonding within the duplex, the third strand must hydrogen bond to another surface of the duplex. The third strand pairs in a Hoogsteen base-pairing

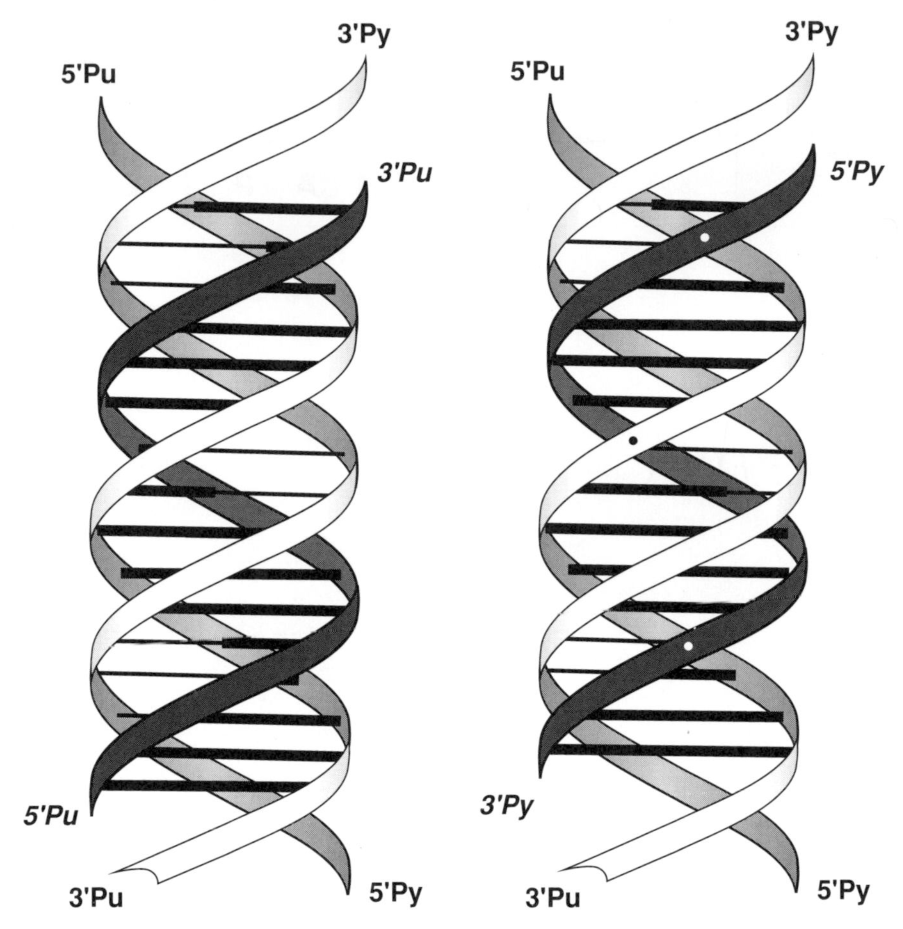

Figure 6.1 An intermolecular triple helix. In the intermolcular *Pu*·Pu·Py triple helix (*left*), the polypurine third strand is organized antiparallel with respect to the purine strand of the original Watson–Crick duplex. The third strand phosphate backbone is believed to be centered in the major groove of a DNA helix of intermediate structure between A-DNA and B-DNA. (For simplicity, the triple helices are shown based on the B-form DNA helix.) In the intermolecular *Py*·Pu·Py triplex (*right*), the polypyrimidine third strand is organized parallel with respect to the purine strand and the phosphate backbone is positioned asymmetrically in the major groove, closer to the pyrimidine strand.

scheme (Figure 6.2). The base-pairing schemes involved in triple-stranded DNA (*T*AT, CGC, *G*GC, and *A*AT) are shown in Figure 6.3. The central strand of the triplex must be purine rich since a pyrimidine does not have two hydrogen bonding surfaces with more than one hydrogen bond. Thus, triple-stranded DNA requires a homopurine·homopyrimidine region of DNA. If the

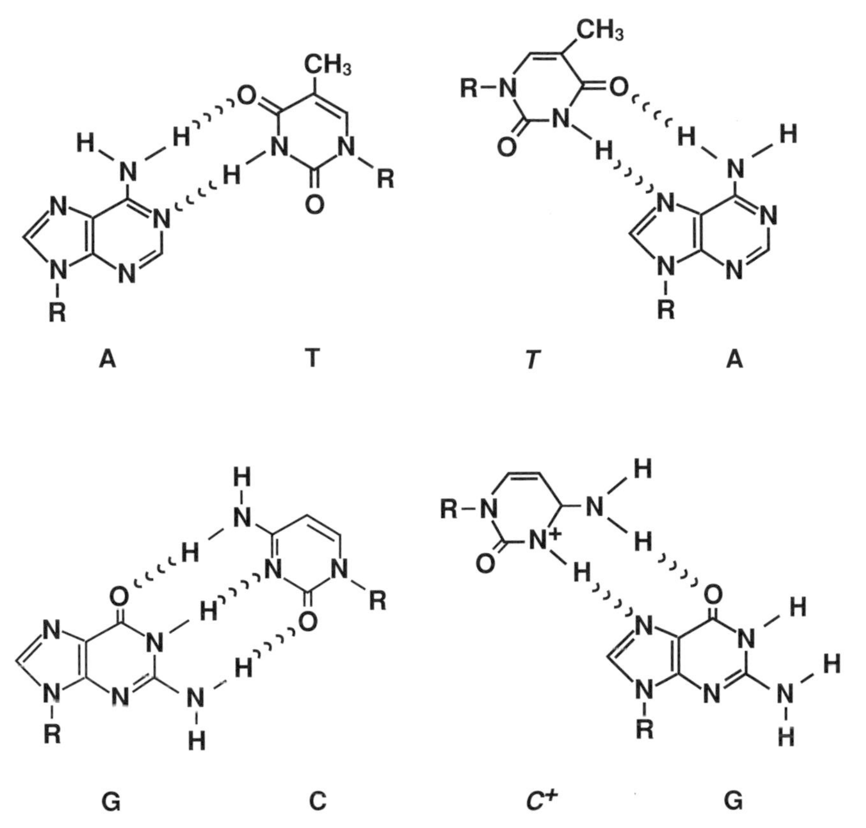

Figure 6.2 Watson–Crick and Hoogsteen base pairs. (*Left*) Traditional Watson–Crick A·T and G·C base pairs. (*Right*) Hoogsteen T · A and C · G base pairs. For cytidine to form a Hoogsteen base pair with guanine, the N3 position of cytosine must be protonated. The Hoogsteen base-pairing surface in the purine molecule is different from the surface involved in Watson–Crick base pairing (Hoogsteen base pairing involves the N7 position in the imidazole ring). The surface involved in Hoogsteen pairing of the pyrimidines is the same as that involved in Watson–Crick base pairing.

third strand is purine rich, it forms reverse Hoogsteen hydrogen bonds in an antiparallel orientation with the purine strand of the Watson–Crick helix. If the third strand is pyrimidine rich, it forms Hoogsteen bonds in a parallel orientation with the Watson–Crick-paired purine strand (Figures 6.1 and 6.3).

2. Intramolecular Triplex DNA

Mirkin *et al.* (1987) showed that an intramolecular triplex could form within a single homopurine·homopyrimidine duplex DNA region in supercoiled DNA. This observation reawakened a general interest in triplex DNA because many sequences in the human genome have the potential to form intramolecular triplex structures. These sequences are commonly associated with regulatory regions of genes.

Frank-Kamenetskii and co-workers showed that, in addition to requiring a continuous strand of purine bases, the homopurine·homopyrimidine region must contain *mirror repeat symmetry,* (see Figure 4.1; Lyamichev *et al.,* 1985,1986; Mirkin *et al.,* 1987). The mirror repeat is a region of DNA that has the same base sequence reading in both the 3′ and the 5′ direction (from a central point) in one strand of DNA (the sequence GAA AAG represents a mirror repeat). The requirement for mirror repeat symmetry arises from the third strand base-pairing rule discussed subsequently.

There are four different ways in which a homopurine·homopyrimidine mirror repeat sequence can fold into an intramolecular triplex (Figure 6.4). The most common structure is the *Py*·Pu·Py configuration in which half of the pyrimidine strand pairs as the third strand and the complementary strand of this region remains unpaired. The two different *Py*·Pu·Py structural isomers that can form, the Hy5 or Hy3 isomers, correspond to structures in which the 5′ or 3′ half of the pyrimidine strand pairs as the third strand (Htun and Dahlberg, 1989). Similar structures Hu5 and Hu3 can form in which the third strand is the p*u*rine strand. Examples of all four possible isomers have been reported in the literature. A helical representation of an intramolecular triplex structure is shown in Figure 6.4B.

As in intermolecular triplexes, when the third strand is the pyrimidine strand, it forms Hoogsteen pairs in a parallel fashion with the central purine strand. In this case, the ribose groups of the pyrimidine strand and the central purine strand are oriented in a cis orientation. When the third strand is the purine strand, it forms reverse Hoogsteen pairs in an antiparallel fashion with the central purine strand and the ribose groups of the third strand are oriented in a trans configuration.

B. Formation and Stability of Intramolecular Triple-Stranded DNA

1. General Sequence Requirements: The Third Strand Pairing Rule and Mirror Repeat Symmetry

There are two general requirements for intramolecular triple stranded DNA formation. First, a homopurine·homopyrimidine region is preferred to provide a continuous purine central strand. Second, the region should contain mirror repeat symmetry. In 1988, Fresco and co-workers defined third strand pairing rules (Table 6.1) (see Letai *et al.,* 1988). The central strand must be a purine, either G or A. A central guanosine can Hoogsteen base pair with C, G, or I.[1] Adenosine can pair with T, A, or I.

The third strand pairing rule prefers that, in general, a homopurine·homopyrimidine region have mirror repeat symmetry to allow formation of an intramolecular triple strand. The upper sequence of Figure 6.5 has perfect mirror repeat symmetry, and all bases in the triplex region are paired (with the exception of the fourth unpaired strand). In the case of a quasi-mirror repeat, shown at the bottom of Figure 6.5, there are three A·G base pairs that cannot form Hoogsteen base pairs. Nonpaired bases within the triplex stem lead to formation of an unstable structure.

The vast majority of intramolecular triplexes studied to date form in regions of polypurine·polypyrimidine sequence that have mirror repeat symmetry. However, Dayn *et al.* (1992) showed that this is not an absolute requirement. A sequence containing G residues with mirror repeat symmetry interspersed with As in one half and Ts in the other half of the sequence (for example, GGAGGGAGGGGA↔TGGGGTGGGTGG) will form intramolecular triplexes with *G*·G·C and *T*·A·T triplex base-pairing schemes. Typically the polarity of the third strand is reversed for a *Pu*·Pu·Py triplex and a *Py*·Pu·Py triple helix, which contain reverse Hoogsteen and Hoogsteen base pairs, respectively. The pairing of a third strand T in a reverse Hoogsteen fashion is possible because of the symmetrical nature of hydrogen bonding contacts in thymine with respect to the C3–C6 axis.

2. Low pH Intramolecular Triplex Structures

For many years, homopurine·homopyrimidine sequences were suspected of forming unusual non-B-DNA structures, based on the observation that S1 nuclease, a single-strand-specific enzyme, would cut at homopurine tracts in many different eukaryotic DNAs. S1 nuclease has a pH optimum between 4 and 5. When Mirkin *et al.* (1987) presented the first convincing evidence for intramolecular triplex DNA in a plasmid, they showed that low pH was required. Cytosine must be protonated to form a Hoogsteen hydrogen bond to a guanosine. This protonation will occur readily at pH 5 but occurs

[1]Inosine is the nucleoside of hypoxanthine, a purine ring with a carbonyl at the C6 position. It is a common biochemical precursor of A and G. Inosine is used frequently when synthesizing hybridization probes because it will base pair with T or C in Watson–Crick hydrogen bonding. In synthetic triple-stranded probes, it will pair with both G and A in a Hoogsteen base pair.

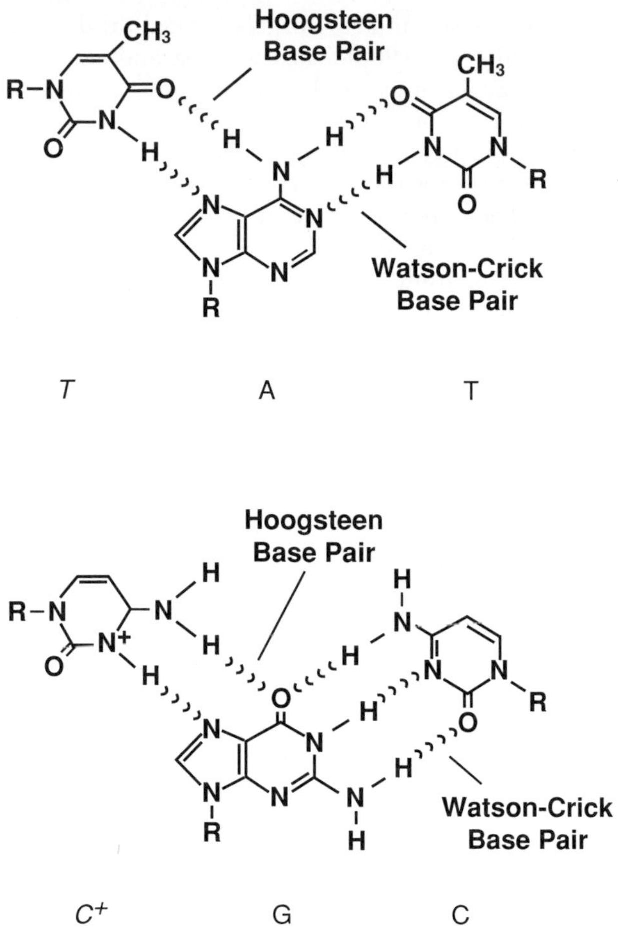

Hoogsteen Base Pair

Watson-Crick Base Pair

T A T

Hoogsteen Base Pair

Watson-Crick Base Pair

C+ G C

Figure 6.3 Base pairing schemes involved in DNA triplexes. (*Top*) Base pairing within a *T*·A·T triplex. The purine (A) is involved in both Watson–Crick base pairing to a T and Hoogsteen base pairing to a second T. (*Bottom*) A *C*+·G·C triplex base pairing scheme. The purine (G) is paired to cytosine in a Watson–Crick base pair and to a protonated cytosine in a Hoogsteen base pair. (*Top, facing page*) The G of a Watson–Crick G·C base pair can also form a Hoogsteen base pair with an additional guanine in a *G*·G·C triplet. (*Bottom, facing page*) An *A*·A·T triplet in which an A forms a Hoogsteen base pair with the adenine of an A·T Watson–Crick base pair. Note that the orientation of the base for the *Py*·Pu·Py triple helices is a Hoogsteen configuration whereas that for the *Pu*·Pu·Py triple helices is a reverse Hoogsteen configuration, in which the ribose groups on the two purines are in a trans orientation. This difference is reflected in the different location and polarity of the third strand, as shown in Figure 6.1.

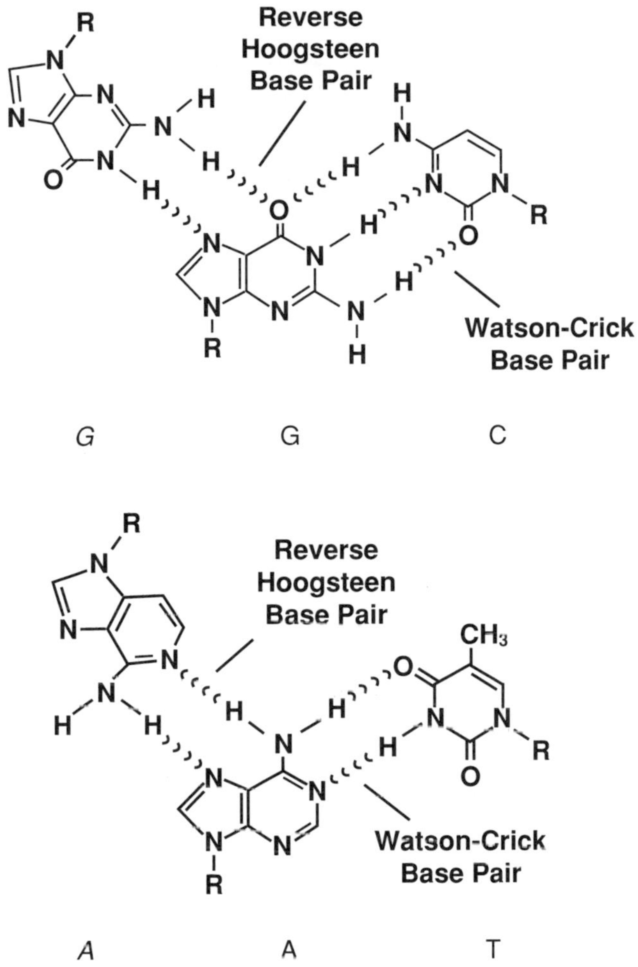

Figure 6.3 *Continued.*

rarely at pH 7 or above. Protonated cytosines, as shown in Figure 6.3, are involved in only two of the possible four isomers (Hy5 and Hy3). These were the first isomers identified, since these conformations form more readily than the *Pu*·Pu·Py isomers which do not require low pH. Mirkin *et al.* (1987) named intramolecular triplex DNA *H-DNA*, in part because of protonation (H^+) of the Hy isomers and in part because Hoogsteen pairs were involved.

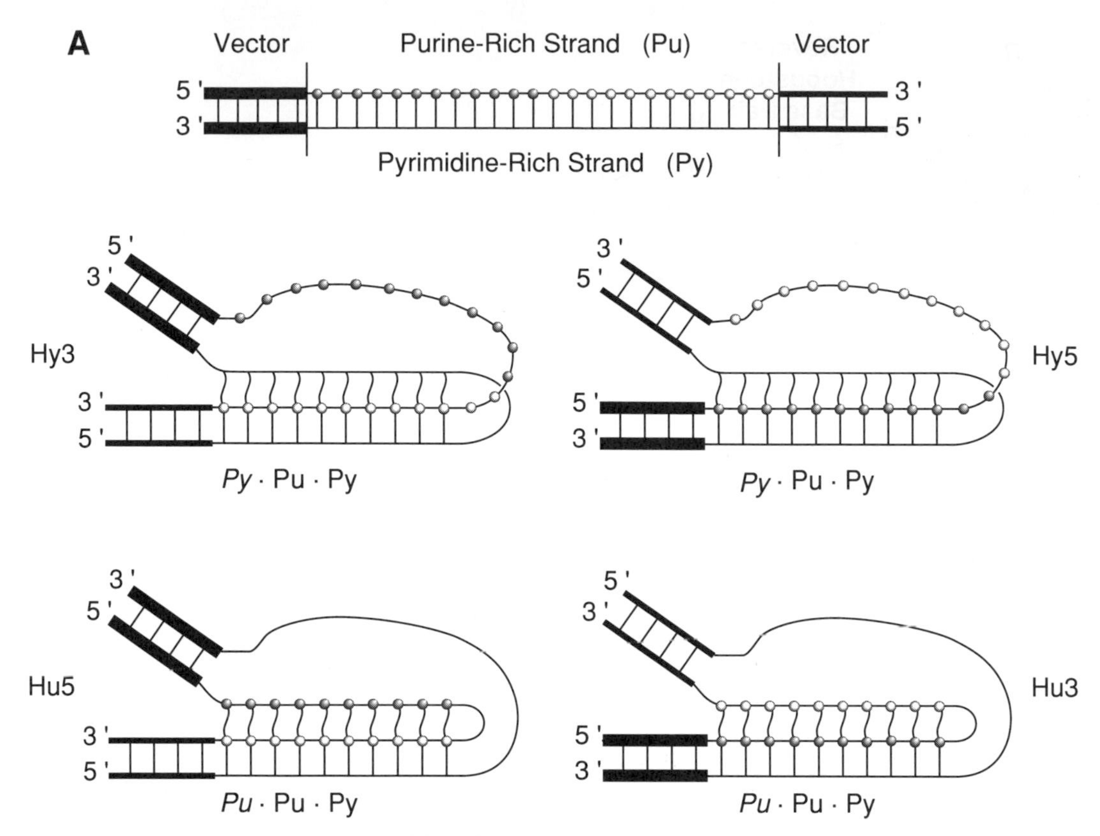

Figure 6.4 Four possibilities for intramolecular triplex formation. (A) Intramolecular triplex structures can form within regions of DNA with predominantly purines in one strand in which the Pu·Py region has mirror repeat symmetry. The four different intramolecular triplex isomers that can form are called Hy3, Hy5, Hu5, and Hu3. H refers to H-DNA, a term used for intramolecular triplex structures; y refers to the pyrimidine-rich strand being involved as the third or triplex strand; u refers to the involvement of the purine-rich strand as the third or triplex strand. The numbers 5 and 3 refer to the 5′ or 3′ end of the purine- or pyrimidine-rich strand that pairs as the third strand. For example, the Hy3 triplex involves a *Py*·Pu·Py triplex in which the 3′ end of the pyrimidine-rich strand folds back and forms the third strand of the triplex. The 5′ half of the pyrimidine-rich strand can also fold back and form the isomeric Hy5 triplex structure. The 3′ or 5′ halves of the purine-rich strand form intramolecular triplex structures Hu3 and Hu5, respectively. (B) A helical representation of intramolecular triplex DNA. The third strand lays in the major groove, whereas its complementary strand exists as a single stranded region. Analysis of intramolecular triplex structures on polyacrylamide gels suggests that DNA containing the intramolecular triplex is significantly bent. Figure modified with permission from Wells *et al.* (1988).

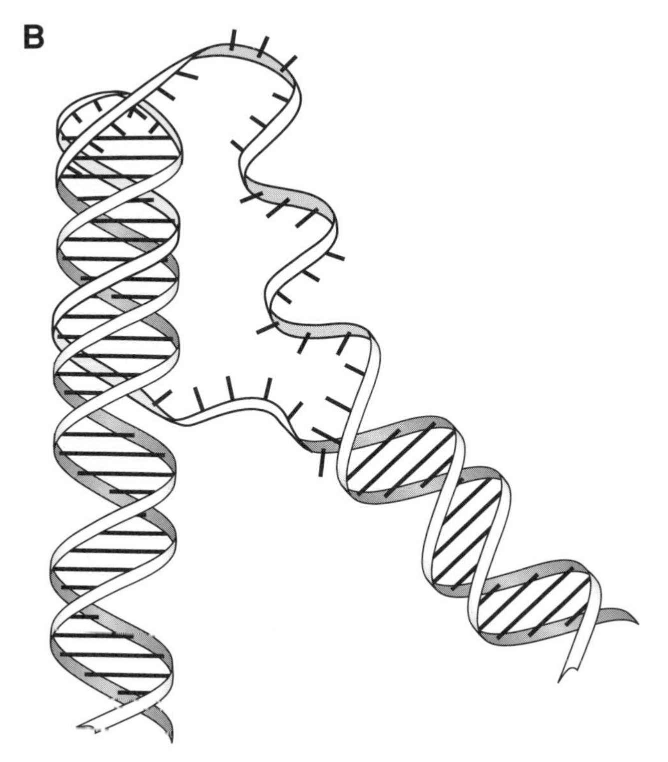

Figure 6.4 *Continued.*

3. DNA Supercoiling Is Required for Intramolecular Triplex Structures

The triple helix does not represent the most thermodynamically stable structure that can be adopted by two complementary poly(Pu)·poly(Py) strands. There is a loss of Watson–Crick hydrogen bonding and base stacking

Table 6.1
Third Strand Pairing Rules[a]

	Third-strand residues				
Central W-C purine strand	A	T/U	I	G	C
A	+	+	+	–	–
G	–	–	+	+	+

[a]From Letai *et al.* (1988).

Mirror Repeat

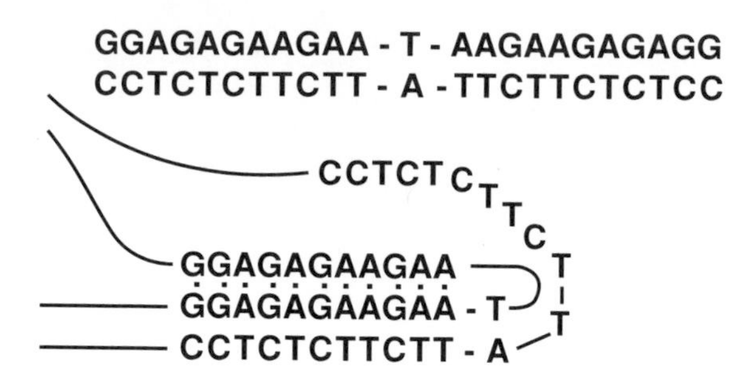

quasi-Mirror Repeat

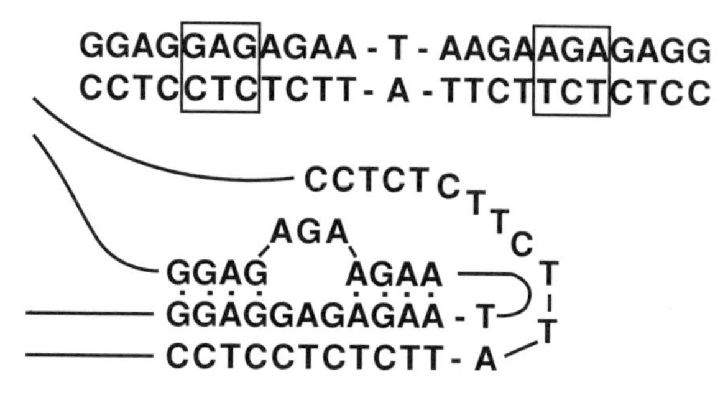

Figure 6.5 Intramolecular triplex structures prefer mirror repeat symmetry. (*Top*) A sequence of DNA containing mirror repeat symmetry. The T·A base pair represents the center of the mirror repeat region, from which the bases going out in either direction are identical in each strand. (*Bottom*) A quasi-mirror repeat in which the two blocks of three base pairs no longer contain mirror repeat symmetry. Were an intramolecular triplex to form in this structure, the 3-bp non-mirror-repeat region would not form Hoogsteen base pairs. This mismatch would significantly reduce the stability of this potential intramolecular triplex structure.

interactions in the unpaired region forming the triplex (half of one strand folds back and pairs as the third strand while the other half of that strand remains unpaired). There must also be a region of DNA adjacent to the triplex region, a duplex–triplex junction, that will be unpaired. The structure of a triplex region of DNA in a duplex molecule is shown in Figure 6.4B.

DNA supercoiling provides the energy to drive the unwinding of the DNA to allow triplex formation. A precise level of supercoiling is required to form various intramolecular triplex regions (Table 6.2). The formation of intramolecular triplex DNA results in the relaxation of about one supercoil for every 10.5 bp of the homopurine·homopyrimidine region (Figure 6.6). The

Table 6.2
Superhelical Density Dependence for Intramolecular Triplex Formation

Sequence	σ_c
$(AG)_{20}$[a]	−0.025
$(AG)_{37}$[a]	−0.018
$(AGGAG)_6$[a]	−0.038
$(AGGAG)_9$[a]	−0.033
$(AGGAG)_{12}$[a]	−0.028
$(GAA)_4TTC(GAA)_4$[b]	−0.028
$(GAA)_4TTCGC(GAA)_4$[b]	−0.036
$(GAA)_4TTAATTCGC(GAA)_4$[b]	−0.043

[a]Collier and Wells (1990).
[b]Shimizu *et al.* (1989).

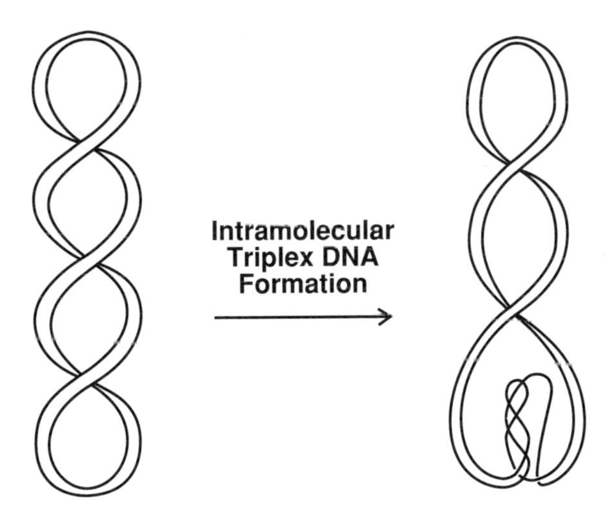

Figure 6.6 DNA supercoiling promotes intramolecular triplex DNA formation. The energy from DNA supercoiling is used to drive intramolecular triplex formation. DNA supercoiling is needed to drive the melting of the DNA at the center of the Pu·Py region. DNA involved in intramolecular triplex formation is unwound resulting in a relaxation of negative superhelical turns. The number of superhelical turns relaxed by the formation of intramolecular triplex DNA structures is complex and can vary depending on topological linking within the loop, as described in Figures 6.10 and 6.11. Basically about one negative supercoil is relaxed for every 10 bp involved in intramolecular triplex formation.

loss of base pairing and base stacking interactions is stabilized by the decrease in ΔG resulting from the relaxation of negative supercoils.

The relaxation of supercoiling by triplex formation is actually rather complicated. One half of the poly(Pu)·poly(Py) region is unwound, relaxing supercoils. In addition, the winding of the third strand back into the major groove results in its crossing the complementary strand in the direction opposite that normally found in B-form DNA. A more detailed discussion of the extent of relaxation is presented later when the mechanisms for triplex formation are discussed.

Although DNA supercoiling is generally required to form and stabilize alternative secondary DNA structures, there are exceptions. Early S1 nuclease studies on purified eukaryotic DNA (Hentschel, 1982) suggested that an alternative helical form may exist *in linear DNA*. Homopurine·homopyrimidine regions were cut by S1 nuclease at low pH (and in the presence of divalent metal ions). As a possible explanation for this characteristic, certain long poly(Pu)·poly(Py) sequences have now been shown to form intramolecular triplex structures in linear DNA at low pH (Michel *et al.*, 1992).

4. Neutral pH Intramolecular Triplex Structures

Certain triplex DNA structures can also form in supercoiled DNA at neutral pH, especially when protonation is not required. This was first shown by Kohwi and Kohwi-Shigematsu (1988) for the intramolecular-triplex-forming sequence $(dG)_{30}\cdot(dC)_{30}$. This sequence formed an Hy3 $C^{+}\cdot G\cdot C$ triple helix in supercoiled DNA at low pH in the absence of Mg^{2+}. A $G\cdot G\cdot C$ triple helix formed at neutral pH and at low pH in supercoiled DNA in the presence of Mg^{2+} (Figure 6.7). The different structures formed at neutral pH with Mg^{2+} and low pH without Mg^{2+} were determined from analysis of the chemical reactivity of intramolecular triplex DNA to chloroacetaldehyde. Intramolecular triplex structures have been reported to form at neutral pH from a variety of poly(Pu)·poly(Py) sequences. In general, the formation of intramolecular triplexes at neutral pH requires DNA supercoiling and divalent cations.

5. $(CT)_n\cdot(GA)_n$ Triple Helices: The Mechanism and Topology of Intramolecular Triplex Formation

a. $(CT)_n\cdot(GA)_n$ Triple Helices

The $(CT)_n$ repeat may be quite significant biologically since it can adopt a large number of non-B-DNA structures and is widely found in the human genome. Evidence that the sequence $(CT)_n \cdot (GA)_n$ could form a non-B-DNA helix was obtained by a number of investigators (Hentschel, 1982; Cantor and Efstratiadis, 1984; Htun *et al.*, 1984). Pulleyblank *et al.* (1985) characterized the S1 nuclease sensitivity of this sequence at low pH in negatively supercoiled DNA. Although the structure they analyzed was probably triplex DNA, their results were interpreted to suggest a non-B-DNA helix that had Watson–Crick A·T base pairs alternating with a Hoogsteen *syn* $G \cdot C^+$ base pair. Cantor and Efstratiadis (1984) proposed that this sequence formed left-handed DNA. In 1988 and 1989, Htun and Dahlberg demonstrated the formation of the Hy5 and Hy3 isomers of this sequence. Subsequently, Pulleyblank's group demonstrated that a number of different triplex structures, as well as non-B-DNA structures that are not triplexes, can form in a $(GA)_n \cdot (TC)_n$ region (Glover and Pulleyblank, 1990; Glover *et al.*, 1990). The structures that can form vary as a function of superhelical density and length of the homopurine·homopyrimidine region.

b. Hy5 and Hy3: Two Structural Isomers Reflect the Mechanism of Supercoil-Dependent Formation of Intramolecular Triplex DNA

Two elegant papers by Htun and Dahlberg in 1988 and 1989 demonstrated differential formation of Hy5 and Hy3 triplex isomers within the $(GA)_n \cdot (CT)_n$ sequence (Figure 6.8), dependent on the negative superhelical density of the DNA. The differential formation of Hy5 and Hy3 provided insight into the mechanism of formation of intramolecular triplex DNA. The models for intramolecular triplex DNA formation for the Hy5 and Hy3 isomers proposed by Htun and Dahlberg (1989) are shown in Figure 6.9. Unwinding occurs at the center of the polypurine · polypyrimidine mirror repeat. "Nucleation" then occurs in which either the 3′ or the 5′ half of the pyrimidine strand begins hydrogen bonding to the purine strand through interaction in the major groove. Hoogsteen base pairs are formed with the purine strand. The process need not involve a complete unwinding of half of the mirror repeat. The twisting or rotation of one helix winds the pyrimidine strand into the major groove. Rotation of the helix at the other end of the mirror repeat in the opposite direction unwinds DNA for the transfer. The concomitant rotation to unwind one half of the mirror repeat, with the opposite winding to spool the DNA up on the other half, can account for the relaxation of about one negative supercoil for every 10.5 bp of the mirror repeat involved in triplex formation. The complex topology of intramolecular triplex DNA and the reasons for the differential formation of Hy3 or Hy5 are described in the box entitled "DNA Supercoiling and the Topology of Hy5 and Hy3."

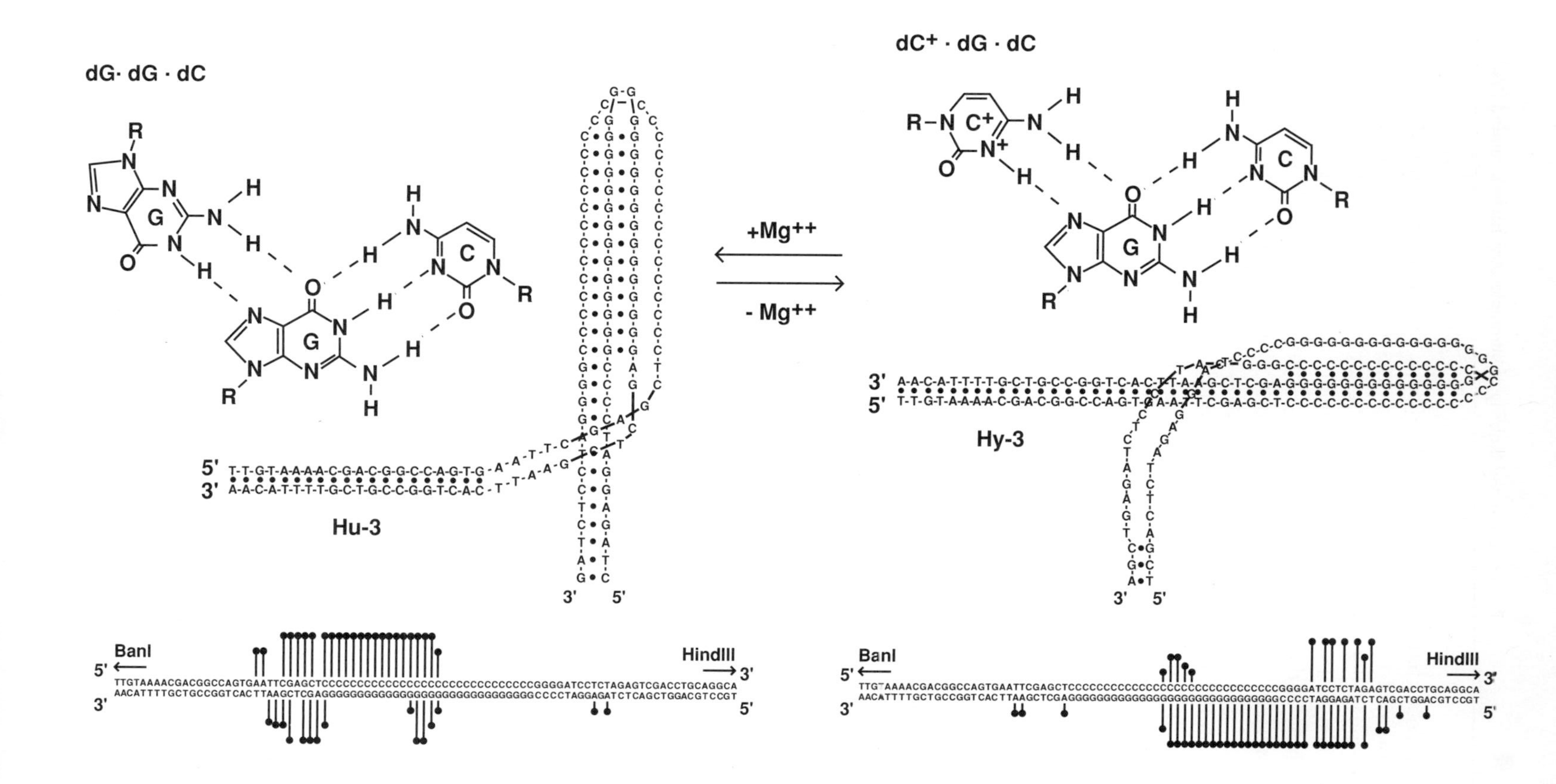

dG· dG · dC
dC+ · dG · dC
+Mg++
- Mg++
Hu-3
Hy-3
5' T-T-G-T-A-A-A-A-C-G-A-C-G-G-C-C-A-G-T-G
3' A-A-C-A-T-T-T-T-G-C-T-G-C-C-G-G-T-C-A-C
3' A-A-C-A-T-T-T-T-G-C-T-G-C-C-G-G-T-C-A-C
5' T-T-G-T-A-A-A-A-C-G-A-C-G-G-C-C-A-G-T-G
BanI
HindIII
5' TTGTAAAACGACGGCCAGTGAATTCGAGCTCCCCCCCCCCCCCCCCCCCCCCCCCCCCCCGGGGATCCTCTAGAGTCGACCTGCAGGCA 3'
3' AACATTTTGCTGCCGGTCACTTAAGCTCGAGGGGGGGGGGGGGGGGGGGGGGGGGGGGGGGCCCCTAGGAGATCTCAGCTGGACGTCCGT 5'

DNA Supercoiling and the Topology of Hy5 and Hy3

The topology of triplex DNA is actually a little more complicated than relaxing about one negative supercoil for every 10.5 bp of the mirror repeat. Htun and Dahlberg argued that Hy3 and Hy5 were topologically distinct. Formation of Hy3 relaxes one more negative supercoil than formation of Hy5. Glover *et al.* (1990) showed that the Hy5 isomer of a $(TC)_{17}\cdot(GA)_{17}$ region relaxed 2.6 negative superhelical turns whereas the Hy3 isomer relaxed 3.6 turns. There are two ways to describe the basis for this difference. The first argument presented by Htun and Dahlberg (1989) is shown in Figure 6.10. The two-cylinder model in Figure 6.10 shows the possible events leading to nucleation during the initiation of intramolecular triplex formation. The first step (B) is the unwinding or denaturation by a negative rotation of half of a helical turn at the center of the mirror repeat. This untwisting can be driven by negative supercoiling. In Step C, two things can happen: the additional negative rotation by half a turn (shown on the left), or a positive rotation offsetting the initial unwinding (shown on the right). In Step D, the cylinders are folded to permit nucleation of Hy3 (far left) or Hy5 (far right). Thus, as transfer of the third strand begins, Hy3 and Hy5 differ topologically by one negative supercoil. Nucleation in the two middle isomeric forms requires passing the third donor strand over the unpaired purine strand, resulting in topological linking in the tip (linked tip). Htun and Dahlberg argued that the existence of

continues

Figure 6.7 Intramolecular triplex forms within poly(dG)·poly(dC). Runs of polyG·polyC can exist as two different intramolecular triplex isomers: Hu3 and Hy3. Hy3 requires low pH conditions to protonate cytidine which forms Hoogsteen base pairs to guanine (the *dC*$^+$·dG·dC triplex). In the presence of magnesium at low or high pH, the 3′ end of the guanine strand folds back and forms Hoogsteen base pairs to the 5′ half of this strand forming the Hu3 isomer (the *dG*·dG·dC triplex). Analysis of DNA structure with chloroacetaldehyde (CAA) provided some of the first clear evidence that different triplex isomer structures could form within this (dG)·(dC) sequence. The sites of chemical modification with CAA are shown below the model structures for Hu3 and Hy3. A vertical line above or below a base represents a site of chemical modification. The height of the line is in proportion to the intensity with which chemical modification occurs. Modification is detected by chemical cleavage at these sites and analysis by electrophoresis on DNA sequencing gels. In the Hu3 structure, the purine strand is involved in triplex strand formation and the pyrimidine strand exists as an unpaired single strand. This single strand 5′ half of the pyrimidine-rich strand is sensitive to covalent modification by CAA. In addition, at the regions of the duplex–triplex junction and at the center of the triplex, there is chemical reactivity on both strands. A different pattern of reactivity is observed for the Hy3 conformation. The 5′ half of the purine-rich strand exists as a single strand and is chemically reactive with CAA. Chemical reactivity is observed at the center and 3′ end of the purine region, reflecting bases at the loop of the intramolecular triplex and at the triplex–duplex junction. These patterns of covalent modification by CAA are diagnostic for the particular conformation. Figure used with permission from Kohwi and Kohwi-Shigematsu (1988).

continued

these two "linked top" forms is unlikely since their formation is probably energetically very expensive and sterically hindered.

Glover *et al.* (1990) presented an alternative topological argument to describe the difference in the relaxation of negative supercoils by the formation of Hy5 and Hy3. These investigators purified the different distinct topoisomers from an agarose gel and, using various chemical assays, identified the rapidly migrating isomer as Hy5 and the slowly migrating isomer (which was less negatively supercoiled) as Hy3. Rather than showing four possible isomeric forms by cylinder rotation and folding, Glover *et al.* (1990) used the closed circular DNA models shown in Figure 6.11. The structures of Hy5 and Hy3 that are believed to exist are shown as structures A and C. In Figure 6.11A, the Hy5 triplex structure is shown on the left. On the right, the triplex has been unwound to reveal a linking number of one. In Figure 6.11C, unwinding Hy3 shows that the two strands of DNA are topologically unlinked. The positive linking number inherent in Hy5 results in the relaxation of one *less* negative supercoil than in Hy3. Structures B and D have a linked tip in which the pyrimidine strand donated as the third strand would wrap around the purine strand. The linked tip structures are analogous to the two central cylinder structures of Figure 6.10D. Glover *et al.* (1990) suggested that these structures are not likely to exist for steric reasons.

The preferential formation of Hy3 at higher negative superhelical densities may be due to the energy of supercoiling favoring the continued negative rotation of Figure 6.10. At less negative superhelical densities, Hy5 may form easily since the positive rotation can occur. Alternatively, because more negative supercoils are lost when Hy3 forms, this structure may be found because it is thermodynamically more stable than Hy5. The differential formation of Hy3 at higher negative superhelical densities may be a combination of thermodynamic stability and the kinetics of formation of the two structures.

There is an additional interesting feature of topology of intramolecular triplex structures. On two-dimensional agarose gels, multiple relaxed species will frequently be observed. For a single linking number there will be a series of species that contain different levels of relaxation of negative supercoils, in contrast with cruciforms or Z-DNA for which a single discrete relaxed molecule is seen. This behavior can be explained if the initial pairing of the third strand does not begin at the center of the polypurine region. If triplex formation does not begin at the center of the polypurine region, the intramolecular triplex will be shorter and will result in the relaxation of fewer than the maximum number of supercoils.

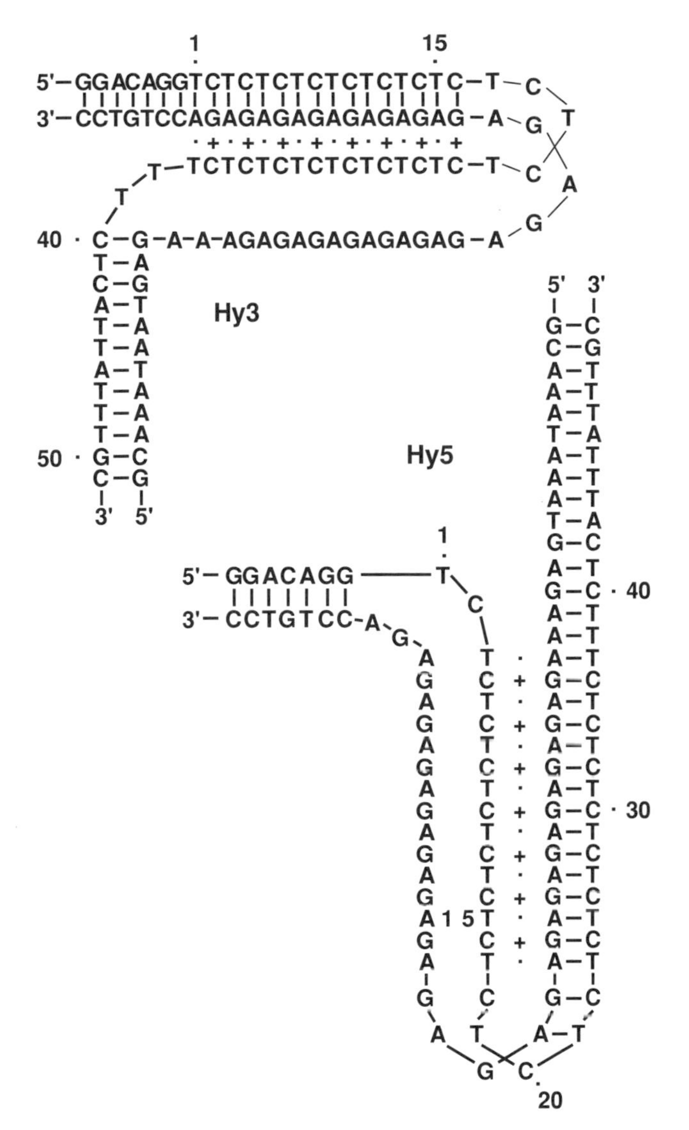

Figure 6.8 Intramolecular triplex isomers of d(CT)$_n$·d(GA)$_n$. Htun and Dahlberg (1988;1989) studied intramolecular triplex structure formation in d(CT)·d(GA) in which both Hy3 and Hy5 isomers form. Both structures require protonation and form in supercoiled DNA at low pH. Analysis of the chemical reactivity of these structures, much like that shown in Figure 6.7, suggested the structure of the intramolecular triplexes shown here. The bases shown at the tips, at the intramolecular triplex loop, and within the single-strand regions of the intramolecular triplex structures were reactive to single-strand-specific chemicals. The Hoogsteen base pairs between T and A are denoted with a dot (·) whereas those between protonated C and G are denoted with a plus (+). Adapted with permission from Htun and Dahlberg (1989), *Science* **243.** Copyright 1989 by the AAAS.

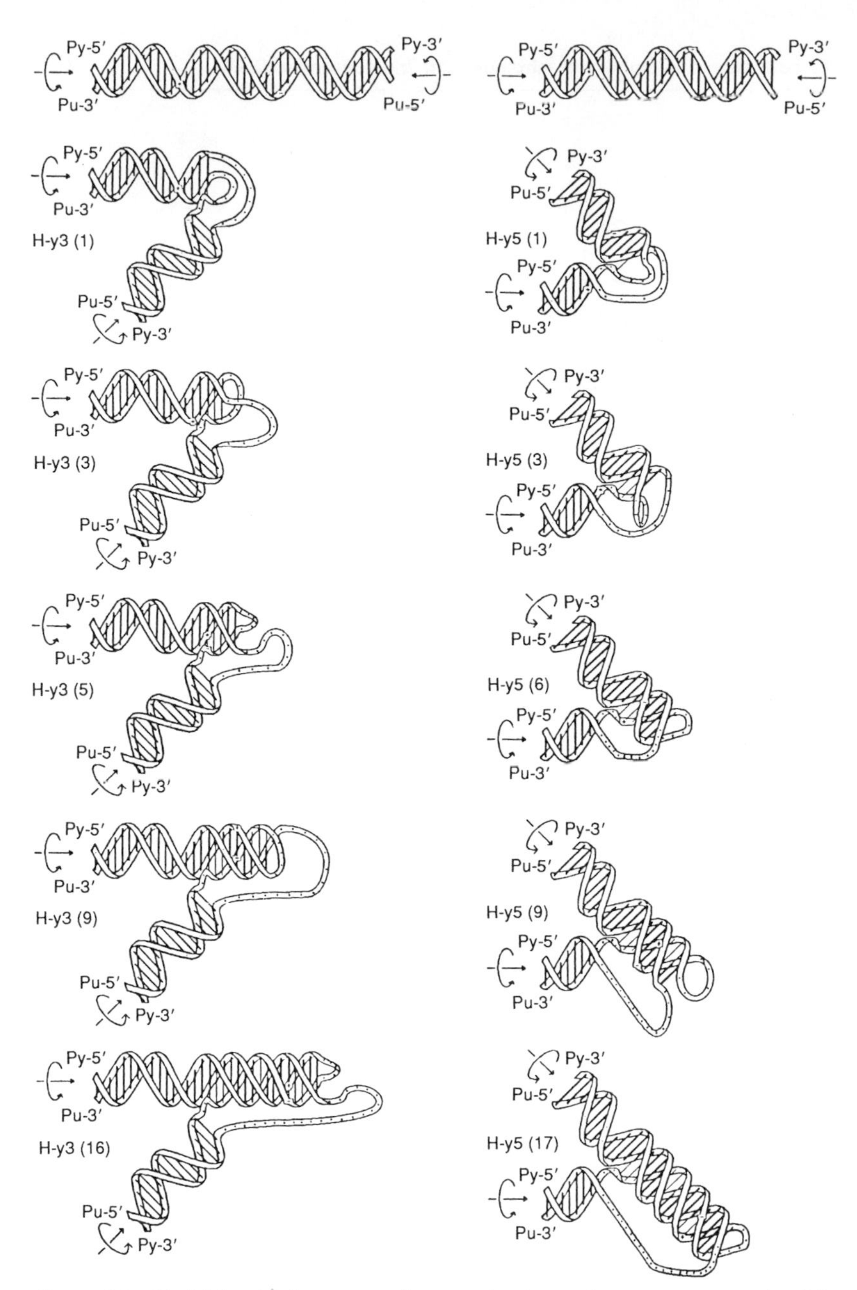

Figure 6.9 Model for the formation of intramolecular triplex DNA. Htun and Dahlberg (1989) suggested that, to form an intramolecular triplex DNA structure, an initial melting of the center of the polypurine mirror repeat occurs and the pyrimidine-rich strand folds back into the major groove, pairing with the purine-rich strand. Depending on which half of the strand inititates pairing, structural isomers Hy3 and Hy5 can be formed. The structure is extended into a longer intramolecular triplex structure by rotation of the Watson–Crick helices, unwinding the Watson–Crick helix in one half of the polyPu·polyPy region while winding the pyrimidine strand into the major groove of the other half of the polyPu·polyPy region. Reprinted with permission from Htun and Dahlberg (1989), *Science* **243.** Copyright 1989 by the AAAS.

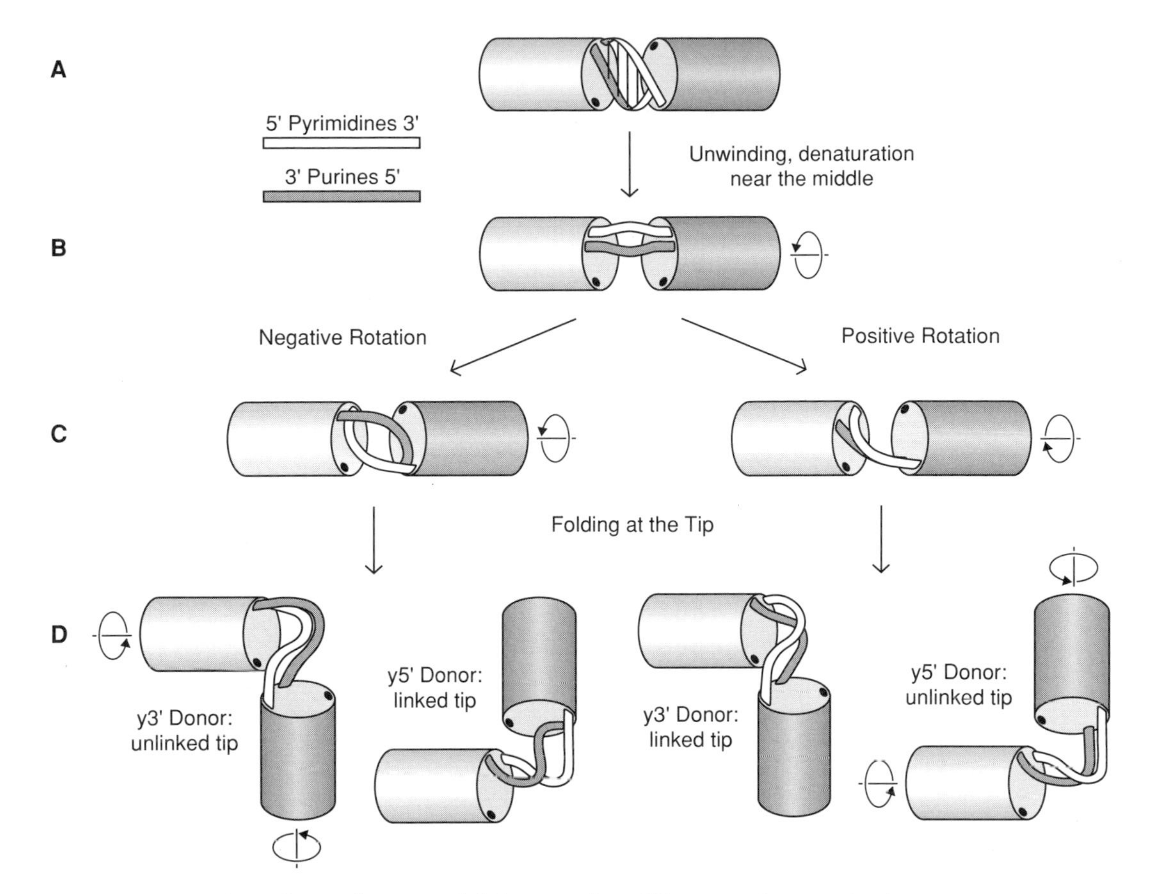

Figure 6.10 Topological considerations of linking number differences of intramolecular triplex structures. DNA supercoiling provides the energy to unwind or denature bases at the center of the purine·pyrimidine-rich strands, as shown in B. Once denaturation has occurred, continued unwinding by DNA supercoiling can result in a negative rotation of two helices with respect to each other (C, *left*). (The dots on the cylinders represent the location of the major groove.) A positive rotation would return the two helices to their original starting positions although base pairs at the center may not reform as shown (C, *right*). (D) Two ways to fold the two helices with respect to each other to initiate intramolecular triplex base pairing and to create the tip of the intramolecular triplex structure. If the darker helix shown for the negative rotation in C is bent down (*far left*), one initiates pairing of the Hy3 isomer in which the tip is topologically unlinked. If one initiates pairing of the Hy5 isomer by folding the darker end of the helix up, there is topological linking of the two strands in the tip. For the positive rotation scheme, a downward bend of the darker helix would initiate pairing of the Hy3 isomer with linkage in the tip. Note that the white strand crosses over the dark strand. If the darker cylinder shown for the positive rotation is bent up (*far right*), the Hy5 isomer would form and the two strands in the tip of the intramolecular triplex would be unlinked. The difference in linking and unlinking results in a linking number difference of one with respect to the number of negative supercoils that are relaxed by intramolecular triplex structure formation. Adapted from Htun and Dahlberg (1989), *Science* **243.** Copyright 1989 by the AAAS.

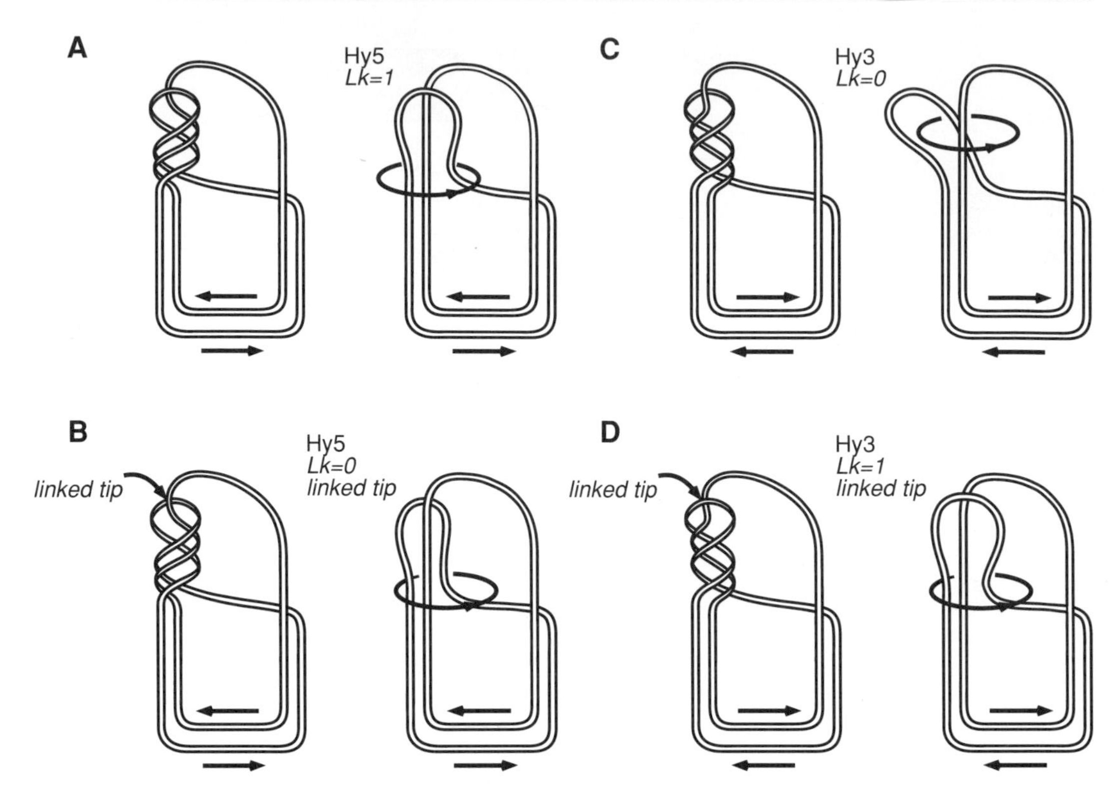

Figure 6.11 Glover, Farah, and Pulleyblank representations of topological linking differences in the tips of intramolecular triplex structures. This figure shows topological linking and unlinking in Hy5 or Hy3 intramolecular triplex structures. The left half of each panel shows the intramolecular triplex structure whereas the right half of each panel shows how unwinding the third strand from the major groove demonstrates the catenation of the two single strands, which would result in a linking difference of 1. (A) Hy5 structure in which the DNA strands in the triplex are topologically linked ($L = 1$). The strands at the top of the triplex structure, however, are unlinked. (B) The Hy5 isomer is shown in which the two strands of DNA are unlinked. By unwinding DNA involved in the intramolecular triplex structure, it is evident that the two single strands of DNA are not catenated and would be free to separate. Although the two strands are unlinked, the strands are linked at the tip of the triplex. One strand at the tip passes through the loop made by the pyrimidine strand as it folds back to Hoogsteen pair with the purine strand. (C) The Hy3 structure which is topologically unlinked and has an unlinked tip. (D) The linked Hy3 structure, in which linking in the tip occurs. The number of supercoils relaxed by formation of Hy3 or Hy5 will depend on whether the two strands are topologically linked ($L = 0$ or $L = 1$). Structures in which the tip is topologically linked are believed to be unfavored for steric and energetic reasons. The unlinked tip structures (A, C) are believed to form *in vitro* in which the formation of Hy3 relaxes one more negative supercoil than formation of Hy5. Adapted with permission from Glover *et al.* (1990). Copyright © 1990 by the American Chemical Society.

6. Effect of Base Composition and Sequence Organization on the Formation and Stability of Triplex DNA in Mirror Repeats

The center of the mirror repeat and the symmetrical halves of the repeat both influence the formation and stability of the triplexes. The formation of triplex DNA requires melting of DNA at the center of the mirror repeat followed by third-strand nucleation using Hoogsteen base pairing. If progressively more bases are inserted between the halves of mirror repeat sequence, increasing levels of negative DNA supercoiling will be required to melt the bases between the halves of the mirror repeat to allow nucleation (Hanvey *et al.*, 1988,1989; Shimizu *et al.*, 1989). Moreover, base composition of the center of the mirror repeat will influence the formation of a triplex region. Shimizu *et al.* (1993) showed that the superhelical density required for intramolecular triplex formation is proportional to the A + T content at the center of the Pu·Py tract. As the A + T content at the center increases, the negative superhelical density required for intramolecular triplex formation decreases. This result suggests that a rate-limiting step in intramolecular triplex formation involves melting the center of the Pu·Py region.

Shimizu *et al.* (1994) showed that a critical determinant governing the formation of either the Hy5 or Hy3 isomer is the size of the melted region at the center of the polypurine·polypyrimidine region. A small melted region, found when high G + C content exists at the center or in the presence of Mg^{2+}, which stabilizes the helix leads to formation of the Hy5 isomer. If a large region at the center can open—in an A + T-rich center, in the absence of divalent cations, or in highly negatively supercoiled DNA—the Hy3 isomer will predominate. Although the Hy3 form is thermodynamically more stable, relaxing one additional negative supercoil, a limited denaturation bubble may only support formation of the less stable Hy5 isomer. The size of the denaturation bubble is critical for determining which isomer will form, because a larger opening is required to wrap the pyrimidine strand into the Hy3 form than into the Hy5 form.

Many different polypurine sequence motifs will form triplex DNA including $(G)_n$, $(AGGA)_n$, $(GA)_n$, $(CGGAA)_n$, $(GAA)_n$, and $(GAAA)_n$ (Hanvey *et al.*, 1988). $(A)_n$ will form an intramolecular triplex, but it requires very long regions of $(A)_n$ (Fox, 1990). In addition, once formed, the intramolecular triplexes with the highest G + C content are more stable thermally than ones with lower G + C content. The limit of the thermal stability of intramolecular triplexes in supercoiled DNA ranges from 50 to 60°C.

Mirror repeat homopurine·homopyrimidine regions can contain mismatches and still form intramolecular triplexes, although the stability of the triplex regions is reduced (Mirkin *et al.*, 1987; Hanvey *et al.*, 1988; Belotserkovskii *et al.*, 1990). Belotserkovskii *et al.* (1990) examined the effect of all possible 16 base-pair combinations at a single site within a mirror repeat sequence on intramolecular triplex formation. Intramolecular triplexes form within mirror repeats in which all possible mismatched base pairs occur, although in all cases more negative superhelical energy was required for formation of a triplex containing a mismatch. Moreover, the stability of the intramolecular triplexes was differentially affected by different mismatches.

7. The Kinetics of Intramolecular Triplex Formation

The kinetics of intramolecular triplex DNA formation can be rapid. Hanvey *et al.* (1989) have shown that, on reducing the pH of a solution of supercoiled DNA from neutral to about 5.0, triplex DNA formed with a $t_{1/2}$ of less than 2 minutes. The actual rate of intramolecular triplex formation will be dependent on pH, supercoiling, and length of the polypurine region.

The kinetics of duplex formation from a triplex region is also fast. When negatively supercoiled DNA containing an intramolecular triplex was relaxed by calf thymus topoisomerase at pH 5.2 at 25°C, the loss of the triplex paralleled the relaxation of negative supercoils. However, intramolecular triplex regions can form in relaxed DNA [for example, in $(TC)_{18}$ at pH 4.5 (Htun and Dahlberg, 1989)]. Therefore, the rate of the triplex-to-duplex transition *in vitro* depends on a number of factors including pH, supercoiling, temperature, and the concentration of divalent cations.

C. Assays for Intramolecular Triplex DNA

Most of the assays developed for triplex DNA are conceptually similar to those described in the chapters on Z-DNA and cruciforms. The formation of intramolecular triplex DNA leads to the relaxation of negative supercoils, which can be detected by two-dimensional agarose gel electrophoresis. Moreover, intramolecular triplex structures contain an unpaired strand that is sensitive to many chemicals and single-strand-specific nucleases. The triplex stem and triplex–duplex junction also show altered sensitivities to a number of chemical or enzymatic probes of DNA structure. The rationale and details behind these assays will only be mentioned briefly here.

1. Two-Dimensional Agarose Gels

The formation of an intramolecular triplex in supercoiled DNA results in relaxation of about one negative superhelical turn for every 11 bp of the mirror repeat involved in triplex formation (Htun and Dahlberg, 1989). Depending on whether the Hy5 or Hy3 isomer forms and whether the winding of the third strand is linked or unlinked at the tip (see Figure 6.11), the number of supercoils relaxed can vary by $\Delta W = +1$. The relaxation of negative supercoils is seen by two-dimensional agarose gel electrophoresis. It is necessary to run the first dimension under conditions that favor triplex DNA formation. Thus, the first dimension may be run at about pH 5.0 in the presence of Mg^{2+}. In the second dimension, the gel is run under more typical conditions of pH 8 in the presence of chloroquine to unwind negative supercoils.

2. Chemical Probes for Intramolecular Triplex Structures

a. Chloroacetaldehyde (CAA) and Bromoacetaldehyde (BAA)

CAA and BAA (see Figure 1.21) react with unpaired cytosines and adenines at the Watson–Crick base-pairing surface. Therefore, the unpaired regions will be sensitive to modification as first shown by Kohwi-Shigematsu and Kohwi (1985; Kohwi and Kohwi-Shigematsu, 1988). Incubation of CAA-modified DNA in the presence of either hydrazine or formic acid, followed by treatment with piperidine, leads to cleavage at the modified base (see Figure 6.7). Cleavage can also be accomplished by treating CAA-modified DNA with S1 nuclease. The sites of chemical cleavage are then identified using gel electrophoresis.

b. Diethylpyrocarbonate (DEPC)

DEPC alkylates the N7 position of unpaired purines or purines in the *syn* conformation. A and G residues that exist in the unpaired strand will be sensitive to chemical modification. DEPC-modified bases are identified by chemical cleavage by treatment with piperidine or by primer extension analysis (Johnston, 1988).

c. Dimethylsulfate (DMS)

DMS modifies guanine residues by reacting predominantly with the N7 position of purines. It modifies guanine readily in the B-DNA conformation. When DNA is involved in the formation of an intramolecular triplex, the N7 position of guanine participates in Hoogsteen base pairing (see Figure 6.2). Therefore guanines involved in Watson–Crick and Hoogsteen base pairing will be protected from DMS modification (Hanvey *et al.*, 1988; Johnston, 1988).

d. Osmium Tetroxide (OsO_4)

OsO_4 reacts strongly with thymine and to a lesser extent with cytosine in single-stranded DNA (Vojtiskova and Paleček, 1987; Paleček, 1992). When the pyrimidine strand in an Hu isomer is single stranded, it will react preferentially with OsO_4. The single-stranded region of DNA in an intramolecular Hy-form triplex is usually not very reactive with OsO_4 since the purine-rich strand is single stranded. (Hy forms are the most common triplex isomers.) In both Hu and Hy isomers, there is reactivity of T residues (and some Cs) present in the loop of the third strand where it folds back to initiate Hoogsteen base pairing. In addition, there is reactivity with OsO_4 at the duplex–triplex junction between normal B-form DNA and the 5′ end of the unpaired purine strand and the 3′ end of the pyrimidine strand in an Hy isomer.

e. Psoralen Photobinding

The photobinding of psoralen to DNA provides an assay for intramolecular triplex structures in DNA. Psoralen will photobind to a 5′ TA preferential psoralen photobinding site at the center of the mirror repeat when the dinucleotide occurs in B-form DNA. When an intramolecular triplex forms, the 5′ TA is single stranded and will not react with psoralen. In addition, changes in photobinding at the duplex–intramolecular triplex junction, which have single-stranded characteristics, also provide an indication of the conformation of the mirror repeat region (Ussery and Sinden, 1993).

3. Nucleases as Probes for Intramolecular Triplexes

Single-strand-specific nucleases would certainly be expected to cut a polypurine·polypyrimidine mirror repeat in the intramolecular triplex form, since unpaired regions of intramolecular triplex DNA provide substrates for the enzyme. Since S1 nuclease has a pH optimum of 4.5–5.0, the reaction conditions are ideal for intramolecular triplex formation. In fact, an initial indication that alternative helical structures of DNA formed at homopurine·homopyrimidine regions came from analysis of the sensitivity of these regions to S1 nuclease (Hentschel, 1982).

4. Photofootprinting

The efficiency of pyrimidine dimer formation within a polypyrimidine region has been used as an indication of intramolecular triplex DNA formation. On absorption of 254-nm light, cyclobutane bonds can form between the 5,6 double bonds of adjacent pyrimidine bases (see Figure 2.13). The predominant photoproducts are thymine-thymine dimers (T^T), thymine–cytosine (T^C), and cytosine–cytosine (C^C) dimers. In addition to these structures, a number of other photoproducts can occur including 6–4 photoproducts (in which a bond is formed between carbon 6 on one pyrimidine and carbon 4 of an adjacent pyrimidine) as well as various hydroxylated bases.

Becker and Wang developed a procedure called photofootprinting in which the pattern of ultraviolet (UV) light damage introduced into DNA reflects DNA–protein interactions as well as the helical conformation of DNA (Becker and Wang, 1984; Becker and Grossmann, 1992). There are several ways to detect UV photoproducts in DNA. 6–4 Photoproducts can be identified by heating DNA in piperidine, which breaks the phosphodiester backbone at the sites of 6–4 photoproducts. Alternatively, a few nanograms of DNA can be analyzed by a primer extension assay (as described in Figure 4.13) in which DNA polymerase stops at the UV photoproducts. Since DNA polymerases will stop at many of these sites of DNA damage, primer extension can identify 6–4 photoproducts as well as T^T, T^C, and C^C dimers. Photofootprinting is applicable *in vitro* over wide ranges of temperature and pH, and is a noninvasive procedure *in vivo*.

Lyamichev *et al.* (1990) and Tang *et al.* (1991) characterized the formation of pyrimidine dimers in intermolecular triplex-forming DNA. UV photoproducts were introduced readily into duplex polypyrimidine regions. However, when involved in triplex formation, pyrimidine dimers could not form between the adjacent pyrimidines in the pyrimidine strand that was Watson–Crick base paired or the pyrimidine strand that was Hoogsteen base paired as the third strand of a triplex. UV dimer formation requires an unwinding of the DNA to a twist angle of 0°, when two bases are fused by a cyclobutane bond. Moreover, DNA containing pyrimidine dimers is kinked or bent (see Figure 2.13). Presumably the flexibility in twisting and bending DNA that is required to facilitate the photodimerization is absent from intramolecular triplexes.

D. Evidence for the Existence of Triplex DNA *in Vivo*

1. Evidence for Intramolecular Triplex DNA Using a Methylation-Based Assay

Parniewski *et al.* (1990) presented the first evidence for intramolecular triplex DNA *in vivo* by showing that GATC *dam* methylation sites adjacent to or between $(GA)_n$ intramolecular-triplex-forming regions were undermethylated *in vivo*. By introducing a methylation site at the center of or adjacent to a homopurine·homopyrimidine mirror repeat, one expects the site to be methylated in double-stranded B-form DNA but not in an alternative non-B-DNA conformation. The result of Parniewski *et al.* (1990) was consistent with the existence of intramolecular triplexes *in vivo*. A problem with this interpretation, however, is that the *dam* methylase methylated the GATC sites *in vitro* under conditions in which the intramolecular triplexes existed (possibly when the sequence existed in the B-DNA form while in equilibrium with the triplex form). Parniewski *et al.* (1990) suggested that proteins might be involved in the stabilization of triplex structures *in vivo*, since the addition of chloramphenicol to cells (which stops protein synthesis) eliminated the undermethylation.

2. Chemical Probes Provide Evidence for Intramolecular Triplex DNA

a. Osmium Tetroxide

Karlovsky *et al.* (1990) developed an osmium tetroxide, 2,2′-dipyridine protocol to probe intramolecular triplex DNA *in situ*. In these experiments, *Escherichia coli* cells containing a plasmid with a potential intramolecular $(GA)_n$ triplex region were incubated in buffer at pH 4.5–5.4. The intracellular pH under these buffer conditions appeared to be about 0.7 pH units higher than that of the external buffer. At or below external pH 5.2 (intracellular pH 5.9), OsO_4 reacted with thymines at the center loop of the intramolecular triplex and, to some degree, at the ends of the mirror repeat (the triplex–duplex junctions). These experiments involved incubation of cells in a low pH buffer followed by a 30-minute exposure to OsO_4, 2,2′-bipyridine at 37°C. Under these conditions, the viability of *E. coli* is seriously compromised. This is a demonstration that intramolecular triplexes can form inside cells, and hopefully represents a snapshot of the *in vivo* situation.

b. Chloroacetaldehyde

Kohwi et al. (1992) utilized CAA as an intracellular probe of intramolecular triplex DNA formation at neutral pH. Using a $(dG)_{30}$ triplex-forming sequence, these researchers showed that the *in situ* CAA modification pattern was similar to that for intramolecular G·G·C triplex structures *in vitro*. CAA modified the unpaired pyrimidine strand and reacted at the unpaired duplex–triplex junction. Treatment with CAA, like OsO_4, compromises the viability of the *E. coli*. In these experiments, cells were grown in the presence of chloramphenicol prior to CAA treatment. Specific time points of chloramphenicol treatment resulted in an increase in the level of negative superhelical tension in DNA, which should promote triplex formation. More recently, Kohwi and Panchenko (1993) showed that a G·G·C triplex formed in *E. coli* following transcriptional activation of a downstream gene. Transcription presumably increases the local level of DNA supercoiling to promote intramolecular triplex formation.

3. Psoralen Photobinding as a Probe for Intramolecular Triplex DNA

Ussery and Sinden (1993) have presented evidence for the existence of intramolecular triplex DNA in living cells using a psoralen photobinding assay (similar to that described for Z-DNA). Their results demonstrated that the formation of the Hy3 intramolecular triplex *in vivo* was dependent on the level of DNA supercoiling and on intracellular pH. Psoralen photobinding to a central TA within an intramolecular-triplex-forming region was reduced *in vivo* relative to psoralen binding to a TA outside the triplex region, indicative of the triplex TA being single stranded. In addition, reactivity at a duplex–triplex junction was different from reactivity to the B-form of this sequence and matched the pattern observed for triplex DNA *in vitro*. Little triplex DNA was observed in wild-type cells incubated in stationary phase 36 hours in Luria broth at pH 8.0. However, a considerably higher level of intramolecular triplex was observed in topoisomerase I-deficient cells incubated 36 hours in K media. After long incubations in K media, the extracellular pH of the media decreased to about 5.0. Since this drop in pH would be reflected *in vivo,* it is not unexpected that a greater level of intramolecular triplex was found under these conditions. In addition, the presence of a mutation in topoisomerase I led to a higher level of supercoiling *in vivo,* which should support a higher level of triplex DNA formation. The incubation in stationary phase (24–36 hours) was required to allow the accumulation of intramolecular triplex structures *in vivo*. Presumably, the processes of DNA replication and transcription drive intramolecular triplexes back into the linear form. A distinct advantage of the psoralen photobinding assay for triplex DNA is that the measurements are made under conditions in which cell viability is not significantly compromised.

E. The Biology of Triple-Stranded DNA

1. Potential Biology of Polypurine·Polypyrimidine Tracts Existing as B-DNA Protein Binding Sites

The widespread occurrence of polypurine·polypyrimidine tracts in eukaryotic DNA suggests that these sequences may have a biological function. Analysis of eukaryotic sequence databases reveals thousands of polypurine·polypyrimidine tracts, many with the potential for triplex formation. Based on probability alone, polypurine·polypyrimidine sequences are found much more frequently than expected. Birnboim (1978) reported that polypurine tracts 25–250 nucleotides in length constitute 0.5% of the mouse genome. These polypurine regions of DNA can potentially influence biology in several ways. For example, they could provide binding sites for regulatory proteins, influence nucleosome positioning, or form intramolecular triplex structures that might have as yet unknown biological function.

Several examples of B-DNA polypurine·polypyrimidine binding proteins have been reported. Some of these proteins may be transcription factors, proteins that bind to a gene regulatory region and regulate the level of transcription from the gene (see Figure 6.12). In one well-studied system, $(CT)_n$·$(AG)_n$ regions upstream from the *Drosophila hsp* 26 gene are required for maximal heat-shock-induced expression (Glaser *et al.,* 1990; Lu *et al.,* 1993). However, chemical probe analysis suggests that this sequence exists in cells as B-DNA and not as an intramolecular triplex, which *can* form in this region *in vitro*. A protein that binds specifically to $(CT)_n \cdot (AG)_n$ regions of B-form DNA has been purified from *Drosophila* (Gilmour *et al.,* 1989) and identified as the GAGA factor (Lu *et al.,* 1993). Results suggest that binding of the GAGA factor to the $(CT)_n \cdot (AG)_n$ regions *in vivo* is required to establish the proper chromatin configuration necessary for transcription following heat shock. Moreover, Chen *et al.* (1993) have identified a nuclear protein(s) in extracts of rat cells that binds to $(CT)_n \cdot (AG)_n$ regions upstream from a gene encoding neural cell adhesion molecules (NCAMs). These researchers have cloned the $(CT)_n$ element 3′ and 5′ to a reporter gene. On transfection into several eukaryotic cell lines from different tissues, the $(CT)_n$ element can either repress or stimulate expression in an orientation-dependent fashion, thereby acting as a negative or positive enhancer.

Proteins that bind $(G)_n \cdot (C)_n$ regions have also been identified. Clark *et al.* (1990) described a protein that binds $(G)_n \cdot (C)_n$ sequences upstream from a chicken β-globin gene. Using band-shift analysis, Kohwi and Kohwi-Shigematsu (1991) also demonstrated the presence of $(G)_n \cdot (C)_n$ binding proteins in human cells. As discussed later, $(G)_n \cdot (C)_n$ sequences have been shown to influence gene expression in human cells. Kiyama and Camerini-Otero (1991) described the purification of a 55-kDa protein that has a high affinity for binding duplex $(AT)_n$ as wells as a *TAT* intermolecular triplex. The protein binds preferentially to triplex DNA; preliminary results suggested that the protein promotes triplex DNA formation.

The presence of long runs of poly(Pu)·poly(Py) may also affect the positioning of nucleosomes. Although this might be mediated by binding of $(GA)_n$-specific binding proteins (e.g., GAGA factor), polypurine·polypyrimidine sequences may not bind to nucleosomes because of the possible increased stiffness (decreased bendability) of poly(Pu)·poly(Py) tracts, formation of a non-B-form helix (e.g., A-DNA), or adoption of a helix repeat that is incompatible with histone interaction (Rhodes, 1979; Simpson and Künzler, 1979; Kunkel and Martinson, 1981; Prunell, 1982). Therefore, a polypurine region could act to phase nucleosomes by directing nucleosome formation away from the polypurine region.

2. Potential Biology of Polypurine·Polypyrimidine Sequences Existing as an Intramolecular Triplex *in Vivo:* Implications for Gene Expression

Intramolecular triplex DNA might influence the regulation of gene expression in a number of different ways. The formation of an intramolecular triplex could influence the local as well as the global structure of DNA. Intramolecular triplex DNA formation will affect the level of DNA supercoiling in the topological domain in which it forms. Triplex DNA formation could also affect nucleosome organization and nucleosome phasing from the triplex region.

The duplex–triplex transition could provide a molecular switch determining whether regulatory proteins bind or do not bind to DNA. The idea of a molecular switch is illustrated in Figure 6.12A, in which positive or negative transcription factors are bound to B-form DNA. On triplex formation, factors cannot bind and the gene is turned on or off. Figure 6.12B shows several ideas first presented by Lee *et al.* (1984) by which triplex DNA formation might turn off gene expression. A single-strand binding protein might stabilize or protect the single-stranded loop or the single-stranded loop might provide an entry point for RNA polymerase. Yet other possibilities are shown in Figure 6.12C in which a ribonucleoprotein binds to the triplex-forming region either by hybridization to the purine strand in duplex form (left) or by hybridizing to the purine or pyrimidine strand when it exists as the single unpaired strand associated with the intramolecular triplex (right).

An intramolecular triplex 200 bp upstream from the transcription initiation site for the human $^{G}\gamma$ and $^{A}\gamma$ globin genes may play a critical role in gene expression (Ulrich *et al.,* 1992). The sequence from positions -194 to -215 can form the triplex structure shown in Figure 6.13. The boxed bases represent four positions at which point mutations have been identified from patients with hereditary persistence of fetal hemoglobin. In this disease, the fetal hemoglobin is not turned off in adult individuals. There is a remarkable correlation between base changes that would destabilize the intramolecular triplex and the genetic disease. Further research will determine whether the formation of a stable intramolecular triplex is responsible for the developmental inactivation of this gene.

Kohwi and Kohwi-Shigematsu (1991) have provided evidence that $(G \cdot C)_n$ sequences might be forming an intramolecular triplex *in vivo* and influencing gene expression. These investigators showed a remarkable length dependence, under certain ionic conditions *in vitro,* of intramolecular triplex DNA formation. Under physiological ionic strengths with $n < 30$ the $(G \cdot C)_n$ sequence existed as B-DNA, whereas with $n \geq 32$ the region could form a G·G·C triplex. The effect of various lengths of $(G \cdot C)_n$ sequences on gene expression in eukaryotic cells paralleled the formation of the G·G·C intramolecular triplex *in vitro.* Lengths of $(G \cdot C)_n$ that did not form intramolecular triplex structures *in vitro* enhanced gene expression, whereas lengths that formed intramolecular triplex DNA *in vitro* inhibited gene expression *in vivo.* These results suggest that transcription could be repressed in the presence of the triplex structure. Kohwi and Kohwi-Shigematsu suggest, based on the results of an *in vivo* competition assay, that a $(G \cdot C)_n$ binding protein might act as a positive activator of gene expression. On intramolecular triplex DNA formation, binding of the protein is prevented, resulting in the decrease in gene expression.

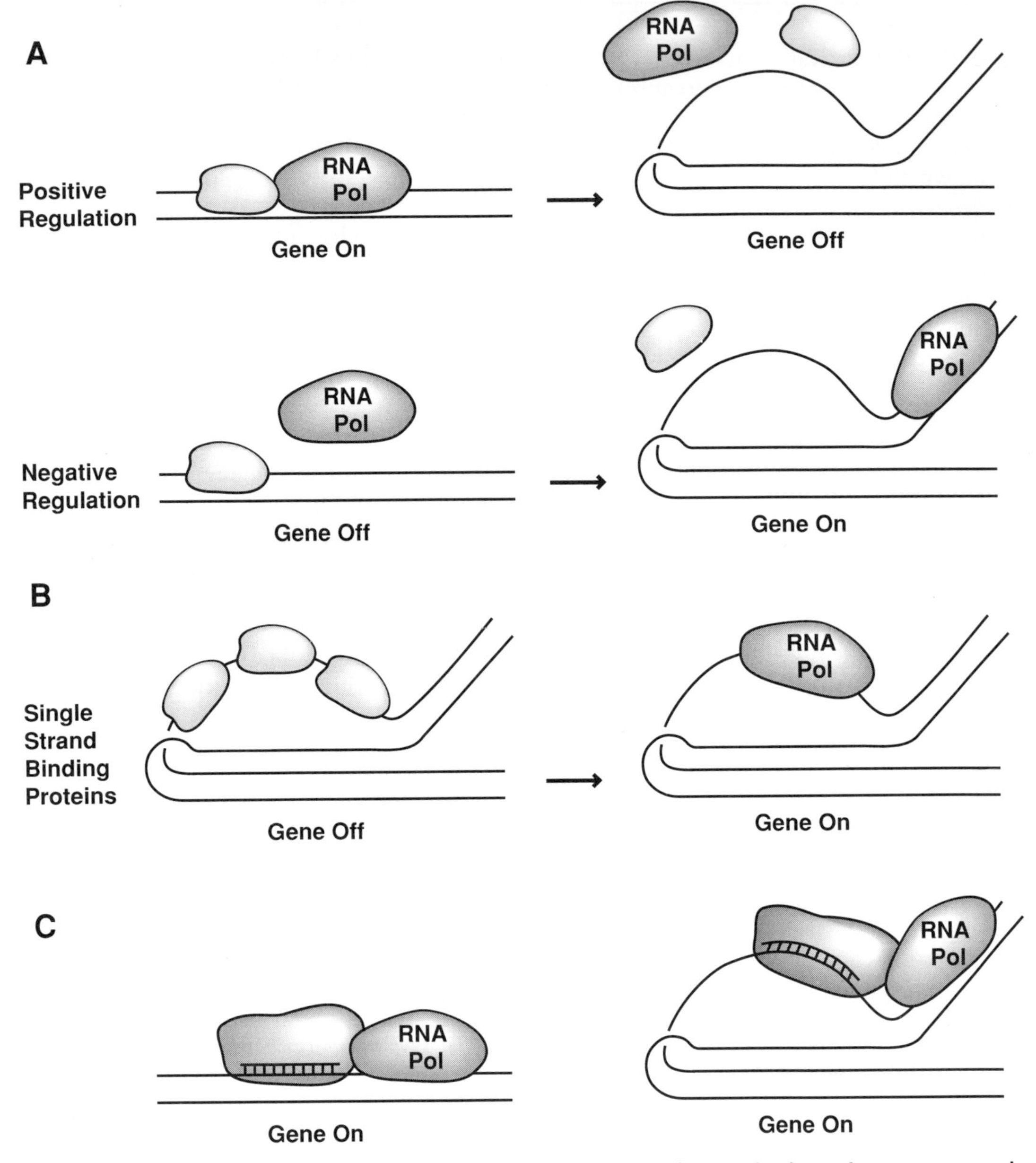

Figure 6.12 Hypothetical models for the involvement of intramolecular triplexes in gene regulation. (A) The potential involvement of B-DNA binding proteins in positive or negative gene regulation. In positive gene regulation, a B-DNA binding protein would bind to a gene and facilitate the binding of RNA polymerase ("Gene On," *left*). The formation of the intramolecular triplex structure would prevent binding of the positive transcription factor, which would prevent binding of RNA polymerase ("Gene Off," *right*). In the negative regulatory scheme, a B-DNA binding protein may bind to the promoter sequence acting as a repressor ("Gene Off," *left*); when an intramolecular triplex forms, this repressor protein can no longer bind to its recognition sequence ("Gene On,"

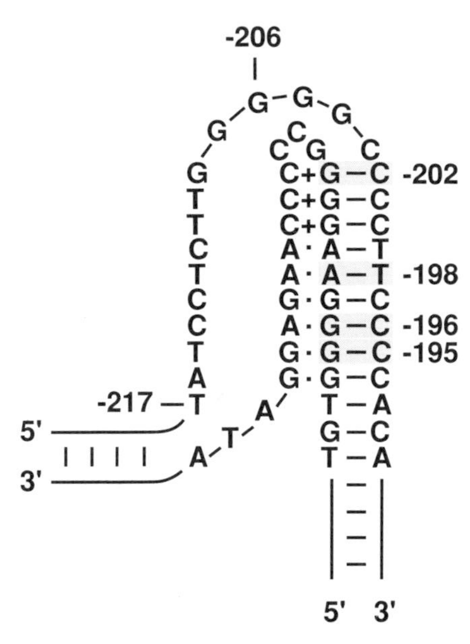

Figure 6.13 Potential intramolecular triplex structure associated with hereditary persistence of fetal hemoglobin. This intramolecular triplex structure forms *in vitro* in this DNA sequence found 200 bp upstream of the transcription start for two human globin genes. Mutations in the base pairs that are boxed (positions −195, −196, −198, and −202) lead to hereditary persistence of fetal hemoglobin in which the fetal hemoglobin gene is not turned off after birth. The correspondence between this genetic disease and mutations that would destabilize this intramolecular triplex structure strongly implicate the involvement of this intramolecular triplex in the developmental regulation of expression from these genes. Figure modified with permission from Ulrich *et al.* (1992).

---

right). The unwound single-strand region or the duplex–triplex junction may provide an entry site for RNA polymerase to the promoter. Transcription then may occur once RNA polymerase has entered the DNA double helix. (B) Models for the intramolecular triplex DNA structure as an entry site for RNA polymerase. RNA polymerase may use the unwound nature of the intramolecular triplex structure as an entry site for binding DNA and initiating transcription of the DNA double helix. There may be single-strand binding proteins that bind the single-stranded loop of the intramolecular triplex structure and prevent the entry of RNA polymerase ("Gene Off," *left*). Alternatively, specific triplex recognition proteins may bind the triplex structure (model not shown). If proteins are not bound, RNA polymerase may enter and begin transcription ("Gene On," *right*). Some regulatory signal or activity may act to remove the single-strand binding proteins when two different intramolecular triplex isomers could form. Binding proteins may be specific for one of the two isomers. (C) The potential interaction of ribonucleoprotein transcription factors with an intramolecular triplex structure. These proteins contain a single strand of RNA that may bind the DNA in one of two ways, forming an RNA–DNA double helix. In the model shown on the left, a ribonucleoprotein may bind double-stranded B-form DNA using its RNA molecule as a third strand in a recognition of a specific DNA sequences. Once bound, this transcription factor may interact with RNA polymerase and allow it to bind and initiate transcription. Alternatively, a ribonucleoprotein may use its RNA molecule to form a double-stranded RNA–DNA hybrid within a loop of an intramolecular triplex structure. This transcription factor may then recruit RNA polymerase and initiate transcription from the junction of a triplex structure. All these models are, at this point, hypothetical and are presented to illustrate the range of possibilities for the involvement of triplex DNA in gene regulation.

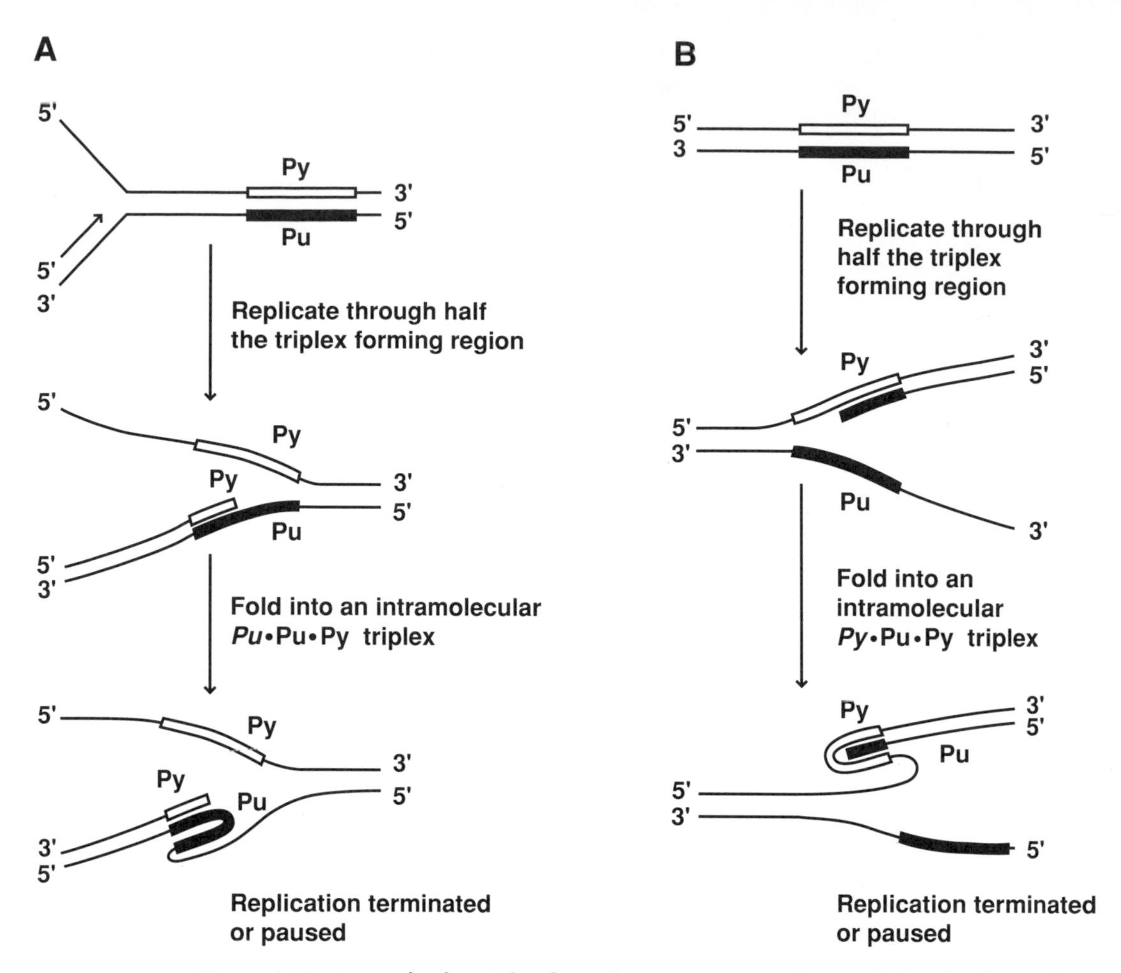

Figure 6.14 Intramolecular triplex formation may pause replication forks. This figure shows leading-strand DNA synthesis through a polypurine·polypyrimidine tract that can participate in formation of an intramolecular triplex structure. (A) Replication would occur through half of a purine-rich sequence resulting in synthesis of half of the pyrimidine-rich strand. The 5′ end of the purine-rich strand folds back and pairs in a Hoogsteen fashion to the 3′ end of the purine-rich strand. This would result in a Hu5 intramolecular triplex structure. The stability of the intramolecular triplex may prevent continued DNA polymerization and result in a temporary blockage or termination of DNA synthesis of the progeny strand. (B) Replication through half of a pyrimidine-rich sequence results in formation of a Hy5 intramolecular triplex structure. (C) A hypothetical situation in which a preformed intramolecular triplex structure may block DNA replication. Replication may approach the intramolecular triplex from two different directions. From one direction, a single-strand loop would be encountered first (*bottom*). DNA polymerase may unwind the third strand from the triplex structure if it first encounters the end of DNA containing the unpaired strand, or is blocked by this structure. DNA synthesis from the "duplex end" intramolecular triplex structure (*top*) would encounter a different structure in which the third strand is wrapped around the Watson–Crick duplex (see Figure 6.4B). *In vitro,* replication can be blocked from either direction (Dayn *et al.* 1992). It is not known how the accessory proteins associated with DNA polymerase forks, which include helicases and various clamp proteins, would respond to intramolecular triplex structures.

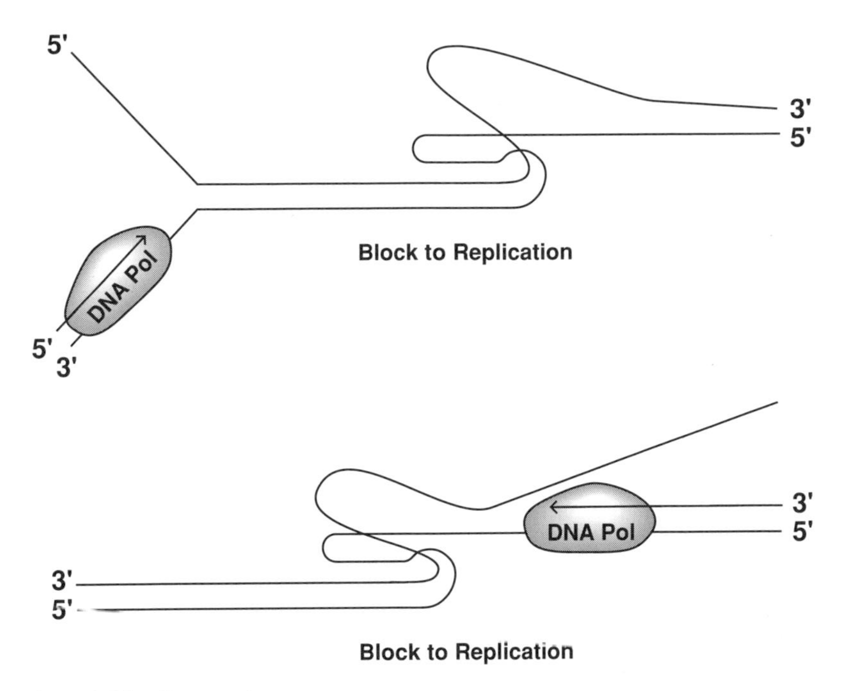

Figure 6.14 *Continued.*

3. Potential Role of Intramolecular Triplexes in Replication: Are Triplex Regions Replication Termination Regions?

Potential intramolecular-triplex-forming sequences may act as replication terminators. Baran *et al.* (1983) identified a region of DNA, 2 kb from the site of integration of polyoma virus, that acted as a strong terminator of DNA replication in rat cells. This region contains a $(GA)_{27}\cdot(TC)_{27}$ tract (Baran *et al.*, 1987) that pauses several DNA replication enzymes *in vitro* under conditions in which triplex DNA can form (Figure 6.14; Lapidot *et al.*, 1989; Baran *et al.*, 1991). Polymerization pauses near the center of the template polypurine tracts where, following synthesis of a $(CT)_n$ region, a (*GA*)·(GA)·(CT) triplex could form. Likewise, polymerization of half of a polypyrimidine tract could allow formation of a (*CT*)·(GA)·(CT) triplex at low pH (Figure 6.14). Since the pH requirements and thermal stabilities of these two triplexes are different, there might be a difference in the termination capacity depending on the direction of replication of the polypurine·polypyrimidine sequence.

Brinton *et al.* (1991) demonstrated that a polypurine·polypyrimidine sequence from the *dhfr* locus of Chinese hamster ovary cells has profound effects on replication of a plasmid shuttle vector containing the sequence in eukaryotic cells. This interesting DNA sequence

$$(GC)_5(AC)_{21}(G)_9(GAGA)GAGGGAGAGAGGCAGAGAGGG(AG)_{27}$$

contains a Z-DNA-forming region, $(GC)_5(AC)_{21}$, followed by a polypurine region that can form intramolecular triplex structures. *In vitro* replication in the Z-DNA-to-triplex direction occurs more easily than replication in the triplex-to-Z-DNA direction. Brinton *et al.* (1991) showed that *in vivo* replication in COS cells paused when going in the triplex-to-Z-DNA direction more than when the sequence was in the other direction. One copy of the Z–triplex motif reduced the amount of plasmid purified from the COS cells by 20–50%. Two copies of the Z–triplex motif, cloned on either side of the SV40 origin of replication, reduced the amount of DNA replicated in COS cells as much as 95% compared with replication of a shuttle vector without the Z–triplex motif (see box). Brinton *et al.* (1991) suggested that this Z–triplex DNA region acted as an "orientation-specific replication fork gate *in vivo*."

Intramolecular triplexes could play a role in the initiation of DNA replication, which requires an unwinding of the DNA double helix to permit binding of DNA helicases. Helicases are required to unwind the DNA so DNA polymerases can move down the double helix replicating the individual strands of the DNA. Since an intramolecular triplex provides a stable unpaired strand, this structure might provide a site of entry for proteins involved in DNA replication (in a fashion analogous to the suggested entry of RNA polymerase shown in Figure 6.12B).

Shuttle Vectors

A shuttle vector is a plasmid containing an origin of replication for two different organisms. In the case discussed in Section D,3, the plasmid can grow in *E. coli* as well as in monkey cells, the host for SV40. Replication of SV40 requires a SV40 virus-encoded protein called the large T antigen. This protein binds to specific sites at the replication origin and has DNA helicase activity. Since the shuttle vector does not encode the large T antigen but only contains the origin region of SV40, the large T antigen must be provided to allow replication of this shuttle vector. The large T antigen is provided by COS cells. COS cells are CV-1 monkey kidney cells that have been transformed by a defective SV40 virus that can neither autonomously replicate nor form infective virus particles. The viral genome is integrated into the chromosome of the COS cells where it produces the large T antigen, which can support replication of a shuttle vector containing the SV40 origin.

4. Potential Role of Intramolecular Triplexes in Recombination

Several polypurine·polypyrimidine triplex-forming regions have been identified near sites involved in genetic recombination. Collier *et al.* (1988) showed that the sequence $(AGGAG)_{28}$, which has mirror repeat symmetry, can readily form an intramolecular triplex *in vitro*. This sequence is found in the switch region of the mouse immunoglobulin Cα switch region encoding an IgA heavy chain. Antibody genes in chromosomes are spliced or recombined from a number of individual gene segments. The possibility exists that the formation of an intramolecular triplex at this region provides a single strand that could initiate genetic recombination by pairing with a homologous region of a second chromosome (or a region of the same chromosome). Another example of a potential intermolecular triplex that might be involved in a genetic recombination event comes from analysis of an unequal sister chromatid exchange event in which part of one heavy chain gene is duplicated on one chromosome and deleted from the other chromosome (Weinreb *et al.*, 1990). The unequal crossing over occurred between potential $(GA)_n$ intramolecular-triplex-forming regions that were adjacent to potential $(GT)_n$ Z-DNA forming regions. Kohwi and Panchenko (1993) have shown that the formation of intramolecular triplex structures *in vivo* can induce genetic recombination between two direct repeats flanking the triplex-forming sequence. In this case the intramolecular triplex structure may not act as a site for the initiation of a genetic exchange, but may act to bring two DNA sequences into close proximity to increase the probability of a RecA-independent recombination event.

There are several models for the involvement of intramolecular triplexes in genetic recombination. At this point these models are *purely speculative*. In one model this pairing could happen as the displaced strand of an intramolecular triplex invades a duplex region, forming a D loop (Figure 6.15A). Alternatively, the unpaired strand of an Hu5 *Pu*·Pu·Py triplex could Watson–Crick base pair with the unpaired strand of an Hy3 *Py*·Pu·Py triplex (Figure 6.15B). There are topological constraints on pairing two single-stranded looped regions. In the absence of topoisomerase activity, every 10 bp of B-DNA would need to be offset by 10 bp of left-handed DNA (this is a form V DNA situation; Brahmachari *et al.*, 1987; Stettler *et al.*, 1979).

5. Potential Role of Triplex Structures at Telomeres

The ends of telomeres of linear eukaryotic chromosomes may exist as triplex structures. Telomeres consist of long regions of tandem short repeats with a short run of guanines (for example G_4T_2 or G_4T_4). As discussed in Chapter 7, repetitive telomeric sequence elements can form a variety

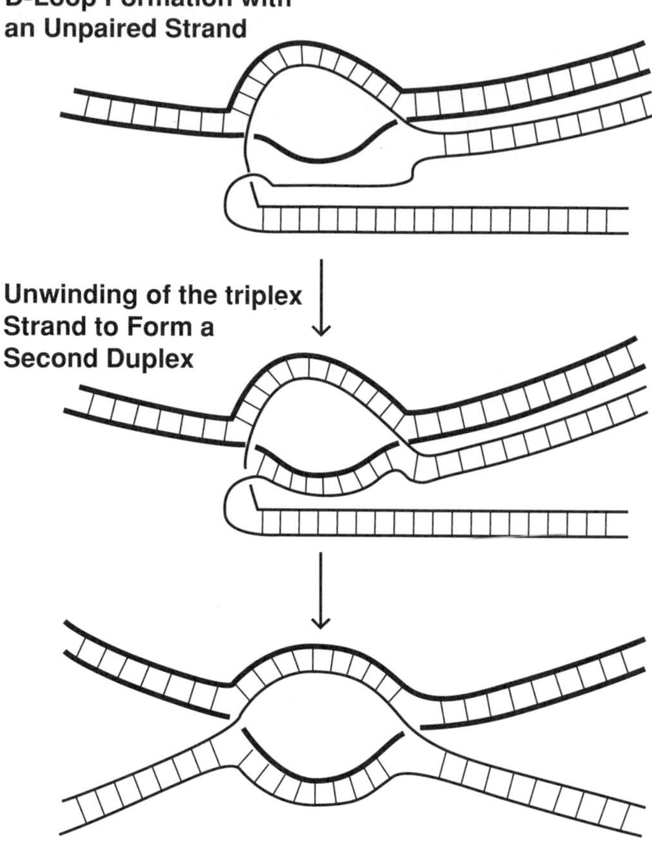

Figure 6.15 Possible involvement of an *intra*molecular triplex in genetic recombination. One step in genetic recombination involves pairing a single DNA strand with a complementary strand of a homologous DNA double helix. Genetic recombination is stimulated by the existence of free single strands of DNA. The single-strand region of an intramolecular triplex may promote genetic recombination. (A) The free single strand of an intramolecular triplex structure may pair with a complementary strand of a second homologous DNA duplex, resulting in D-loop formation. A D loop is a loop of DNA protruding out from a double-stranded DNA molecule because of the hybridization of a single-stranded DNA to its complementary strand in the double helix. Following D-loop formation with the intramolecular triplex, the third strand of the triplex structure could unpair. Since it is complementary to the displaced D-loop strand, these two strands can form Watson–Crick hydrogen-bonded base pairs as shown in the middle structure. Rotating the bottom strand from the right to the left results in the formation of a classic recombinational intermediate (*bottom*).

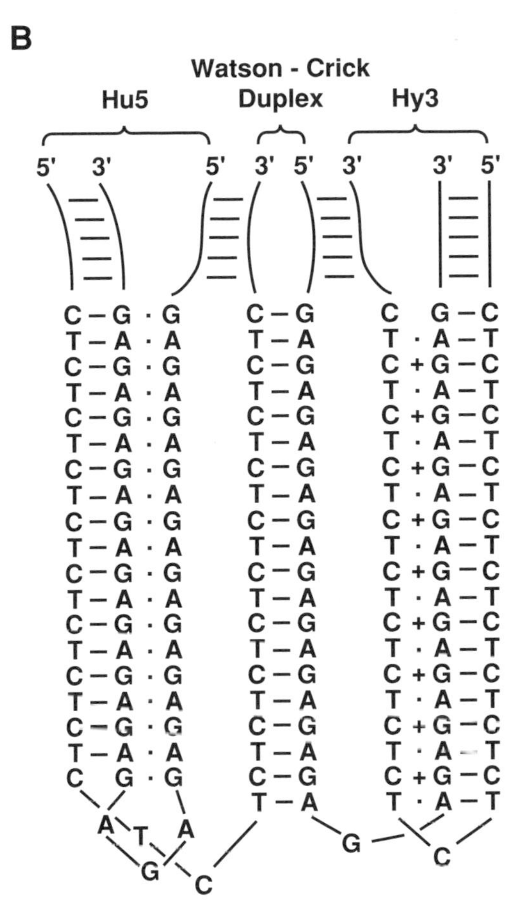

Figure 6.15 *Continued.* (B) Recombination involving two different intramolecular triplex structures. If Hu5 and Hy3 isomers formed in two different molecules containing the same polypurine·polypyrimidine sequence, the free single strands in Hu5 and Hy3 would be complementary. These two single strands might form a Watson–Crick duplex that would hold the two different molecules together at the position of the polypurine·polypyrimidine tract. Unwinding of the strands involved in the intramolecular triplex structures could result in further formation of Watson–Crick hydrogen bonds and the formation of a structure like that shown at the bottom of A. A structure similar to that in B could also be formed pairing the free strands of the Hu5 and Hy5 intramolecular triplex structures. In this isomeric structure, because of polarity differences of the free single strands, the direction of the DNA double helix extending from the Hy5 isomer would be pointing down compared with the helixes both pointing up in the Hy3 and Hy5 structure shown here. There are topological problems with forming right-handed Watson–Crick duplex turns within two loops of single-stranded DNA. A nick would be needed to allow one strand to wrap around another. A topoisomerase or specific protein involved in nicking synaptic recombination events would be required. In the absence of a nick, for every 10 bp of right-handed DNA formed, 10 bp of left-handed DNA would also have to form.

of duplex and quadruplex structures. The quartets are held together by G-quartet base pairing (Figures 7.9 and 7.10). However, Veselkov *et al.* (1993) showed that the single-stranded end of *Tetrahymena* chromosomes (5′ TTGGGGTTGGGG) can fold back forming a *Pu*·Pu·Py triple helix with G·G Hoogsteen base pairs *in vitro*. A triplex at the end of chromosomes may contribute to the stability of ends, preventing digestion by exonucleases or participation in genetic recombination.

F. Intermolecular Triplex DNA

Although most of this chapter has dealt with *intra*molecular triplexes in polypurine·polypyrimidine mirror repeats, triplexes were first discovered as *inter*molecular triplexes (Felsenfeld *et al.*, 1957) formed between homopolymers of A and T or G and C. Because of the third strand pairing rule (Table 6.1), a polypurine·polypyrimidine region will define a unique third strand pairing sequence. A third strand can be designed to Hoogsteen base pair with any polypurine·polypyrimidine region of DNA. This has significant biological implications. A fragment of DNA that Hoogsteen pairs to duplex DNA could affect DNA replication or gene expression. In addition, chemical nucleases or DNA damaging reagents can be covalently linked to the third strand, thus directing scission, damage, or covalent modification of a defined region of DNA. This has great potential for therapeutic applications for the treatment of certain diseases, and viral infections.

1. Application of Intermolecular Triplexes to Modify Gene Expression

Gene expression is a very precisely regulated process. In eukaryotic cells there are often many transcription factors that bind to DNA in the promoter region upstream from the start of transcription. As discussed in Chapter 8, the binding of proteins to DNA occurs by interaction of α helices, β sheets, or other protein domains with the major or minor grooves of DNA. The binding of a third strand to duplex DNA could prevent binding of a regulatory protein. Since polypurine·polypyrimidine regions are frequently found in the control regions of genes, there will be cases where gene regulation can be affected by addition of a third strand complementary to the purine strand.

Several examples of the modulation of gene expression by intermolecular triplex formation have been reported. There is a polypurine·polypyrimidine region upstream of the human c-*myc* oncogene at base pair positions −142 to −115 that is sensitive to S1 nuclease and can adopt an intramolecular triplex form *in vitro* (Boles and Hogan, 1987; Kinniburgh, 1989). This region interacts with transcription factors including one ribonucleoprotein complex (Davis *et al.*, 1989; Postel *et al.*, 1989). The ribonucleoprotein could bind to this region by formation of an intermolecular triplex or by formation of an RNA–DNA hybrid to the unpaired strand of an intramolecular triplex that could form in the polypurine·polypyrimidine region (Figure 6.12). Although polypurine·polypyrimidine sequences are important in gene regulation, the nature of their involvement remains to be elucidated. Cooney *et al.* (1988) reported that addition of an oligonucleotide designed to form an intermolecular triplex with the polypurine·polypyrimidine control region resulted in inhibition of c-*myc* transcription *in vitro*. This result demonstrates the potential for gene regulation via formation of an intermolecular triplex. Given the great specificity inherent in a unique sequence of DNA, it should be possible to design therapeutic oligonucleotides to perhaps turn off (or on) certain genes.

2. Application of Intermolecular Triplexes to Covalently Modify DNA

Several laboratories have coupled artificial nucleases or covalent modifiers of DNA to oligonucleotides that can form intermolecular triplexes at specific sites in DNA. This allows endonucleolytic cutting or covalent modification of the DNA at specific sites. Such approaches may have therapeutic value for gene regulation, killing abnormal cells, or killing cells infected with a virus that contains a unique polypurine·polypyrimidine triplex-forming DNA sequence. Perrouault *et al.* (1990) coupled an ellipticine derivative to the 3′ end of an 11-bp polypyrimidine sequence which formed an intermolecular triplex with a 32-bp duplex DNA fragment. On the addition of >300-nm light, the ellipticine induced cleavage of the 32-bp duplex in both strands of the DNA. Pei *et al.* (1990) coupled a calcium-dependent nuclease to the 5′ end of polypyrimidine oligonucleotides. Following triple strand formation, the nuclease was activated by the addition of Ca^{2+}, resulting in cutting of both strands of the DNA at the 5′ side of the cognate polypurine·polypyrimidine sequence. Dervan and colleagues have used a polypyrimidine oligonucleotide with EDTA (ethylene diaminetetraacetic acid) at each end to cleave DNA in a site-specific fashion. In the presence of iron (Fe), an EDTA-Fe complex produces free radicals that result in the cleavage of the phosphodiester backbone at the site of the *Py*·Pu·Py intermolecular triplex (Strobel and Dervan, 1990). Cleavage can also be accomplished by EDTA–Fe coupled to a polypurine strand in a *Pu*·Pu·Py triple helix (Beal and Dervan, 1991). Triplex-forming oligonucleotides tethered to a psoralen molecule have also been used to deliver a psoralen cross-link to a specific site in DNA adjacent to the triple helix (Takasugi *et al.*, 1991). The potential for site-specific cleavage or DNA adduct formation is enormous. This technology has great potential for mapping specific regions of chromosomes, cleaving DNA at specific locations, or introducing potentially lethal DNA damage to a selected region of a eukaryotic or viral genome.

3. Generalized Genetic Recombination Involves an Intermolecular Triplex

Genetic recombination is the process by which DNA from one homologous duplex is transferred and incorporated into a second homologous duplex (see Figure 4.18). Recombination is often promoted by single-stranded DNA. If DNA is negatively supercoiled *in vitro,* a denaturation bubble can open and expose the Watson–Crick base-pairing surfaces (Figure 6.16). Hybridization of the single strand can occur in an initial nucleation event. This nucleation is followed by incorporation of more of the single strand into the supercoiled duplex as negative supercoils are unwound. The displaced strand of the originally double-stranded DNA forms a single-stranded loop called a D loop. The relaxation of negative supercoils stabilizes D-loop formation. D-Loop formation can occur spontaneously between negative supercoiled plasmid and a homologous single strand of DNA.

In living cells, specialized recombination proteins are required for the strand exchange process. The best studied of these proteins is the RecA protein from *E. coli* (Cox and Lehman, 1987). An initial view of how RecA would pair a homologous strand with a duplex was based on the understanding that DNA normally exists as a duplex. It seemed reasonable that recA might unwind the duplex allowing the third single strand to pair with its complementary strand. However, the initial pairing of a single strand may occur in triple-helical fashion (see West, 1992). Evidence suggests that the third

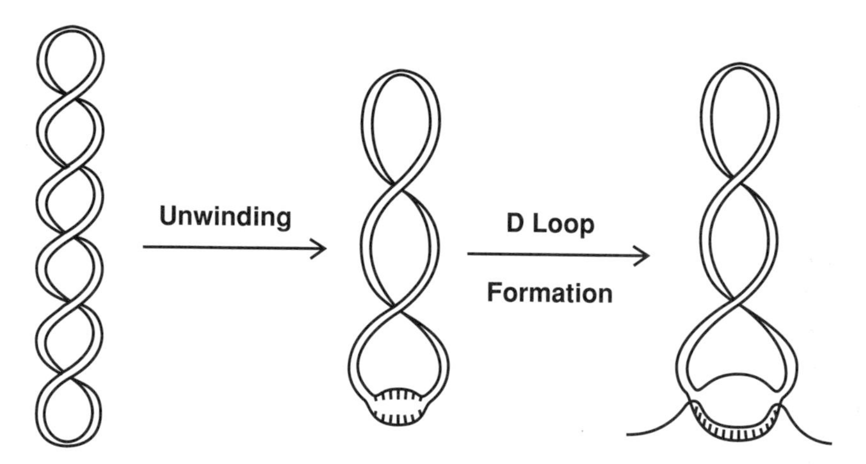

Figure 6.16 D-loop formation in generalized genetic recombinations. One model for the initiation of genetic recombination involves D-loop formation in which a denaturation bubble occurs, and a single complementary strand hydrogen bonds in a Watson–Crick fashion to bases within the denaturation bubble.

strand becomes associated with duplex DNA in the absence of extensive unwinding of duplex DNA (Hsieh and Camerini-Otero, 1989).

What kind of triplex DNA structure would be involved in *generalized* genetic recombination? Genetic recombination occurs throughout the chromosomes. Although a polypurine·polypyrimidine region might be an easy place to initiate pairing of a third strand, the polarity required for recombination is different when the third strand is paired in a Hoogsteen or reversed Hoogsteen fashion. For generalized recombination, the third strand that pairs with a complementary duplex region must do so when purines and pyrimidines are mixed on both strands. An early model for triplex pairing in genetic recombination proposed by Hsieh *et al.* (1990) involved Hoogsteen base pairing of the third strand to the purine in either the Watson or Crick strands. Significant rotations and contortions of the phosphate–sugar backbone would be required to allow the third strand to form Hoogsteen base pairs with either original duplex strand. More recent models involve the

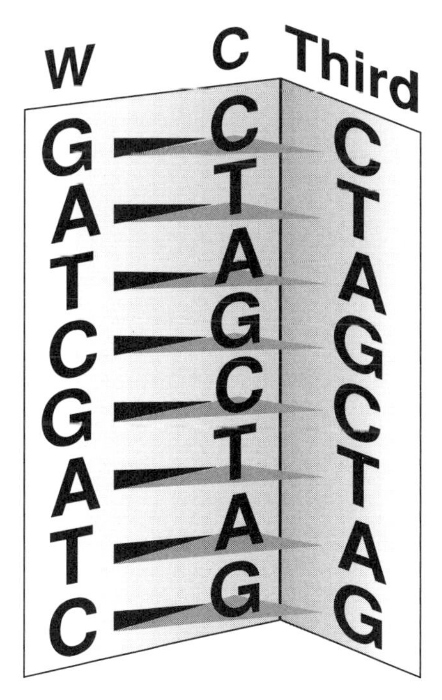

Figure 6.17 Triple-strand intermediate in general genetic recombination. In this model, a third strand binds in a parallel fashion with respect to its complementary strand. The third strand hydrogen bonds to both the Watson–Crick strands of the DNA duplex in a fashion that remains to be clearly demonstrated. See text for discussion.

RecA-mediated formation of an extended unwound triple helix in which the third strand binds on the major groove side in a parallel fashion with respect to its identical strand (Chiu *et al.*, 1993; Rao *et al.*, 1993; Zhurkin *et al.*, 1994). Although the validity of these models remains to be demonstrated conclusively, the unique pattern of hydrogen bond donor and acceptor sites inherent in each of the four Watson–Crick base-pair combinations (see Figure 8.1) may dictate binding an identical strand in a parallel orientation (Figure 6.17). To accommodate third strand pairing, the Watson–Crick duplex needs to be unwound and lengthened, and the single third strand may need to be organized into a uniform shape. These structural changes in the duplex and single strand are presumably mediated by the RecA protein (see Cox and Lehman, 1987; West, 1992, for review). Thus, general recombinase proteins may act by promoting the formation of a unique triple helix during genetic recombination. At some point during the process, unpairing of the original Watson–Crick helix occurs and the incoming third strand forms Watson–Crick base pairs.

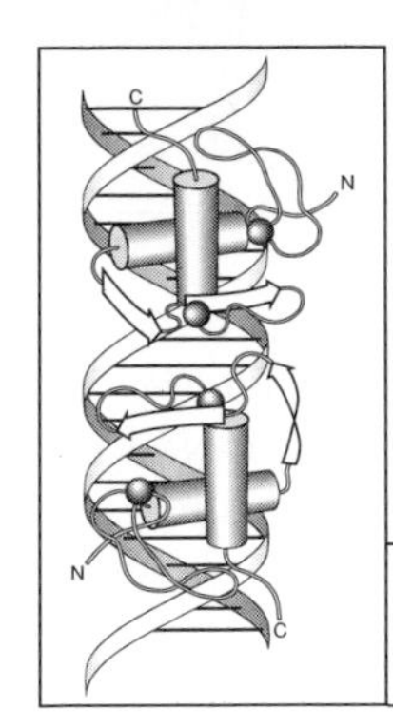

CHAPTER 7

Miscellaneous Alternative Conformations of DNA

A. Introduction

Several non-B forms of DNA (bent DNA, cruciforms, Z-DNA, and intramolecular triplex DNA) were described in previous chapters. There are many other alternative conformations of DNA. Some represent a slight variation of the B-DNA helix whereas others involve the formation of more elaborate structures. The conceptual organization of certain alternative helical forms to an "others" chapter by no means implies that the alternative conformations discussed here are less interesting or less important than those discussed separately in previous chapters. In fact, in the case of the DNA unwinding element, the importance in biology is very clearly understood. In time, our understanding of and appreciation for the structures included here will grow. No doubt other alternative conformations of DNA, in addition to those currently identified, will add to the list presented in this chapter.

B. Slipped, Mispaired DNA

Slipped, mispaired DNA structures can form in regions with direct repeat symmetry. To form a slipped, mispaired structure the entire region must unwind and one strand of one copy of the direct repeat must pair with the complementary strand of the other copy of the direct repeat. Figure 7.1 shows the two possible isomers of slipped, mispaired DNA (SMP-DNA) for a 20-bp direct repeat. One isomer has loops composed of the 5′ direct repeat in both strands whereas the other has loops composed of the 3′ direct repeats in both strands. Because this structure results in the unwinding of the DNA double helix, DNA supercoils will be lost on formation of SMP-DNA. This should make SMP-DNA structures thermodynamically favored. There is a loss of overall helix stability, however, from the loss of hydrogen bonding and base stacking interactions of DNA within the two loops.

The existence of slipped mispaired structures has been suggested by a number of groups who identified regions of eukaryotic DNA with direct repeat symmetry that were sensitive to S1 nuclease (Hentschel, 1982; Mace *et al.,* 1983; McKeon *et al.,* 1984). These groups appreciated that SMP-DNA structures could form within these regions. However, these sequences also contained mirror repeat symmetry, and intramolecular triplex structures could form at these sites. Williams and Müller (1987) cloned long repeats of two restriction enzyme sites into plasmids. These regions were unstable in *Escherichia coli* plasmids. The instability may have been the result of unusual secondary structures since the long repeats could form slipped mispaired as well as cruciform structures. To date, there remains no convincing physical evidence for the existence SMP-DNA structure in supercoiled DNA. It is likely to be just a matter of time before a different convincing physical demonstration is made of SMP-DNA stabilized by DNA supercoiling.

Biologically, SMP-DNA is very important in spontaneous frameshift mutagenesis. It is known that spontaneous deletion or addition mutations can occur within runs of a single base. In 1966, Streisinger *et al.* proposed a model to explain frameshift mutations within runs of a single base (Figure 7.2). (A frameshift mutation is one that changes the reading frame of DNA.) Since the genetic code is read as triplets, adding or deleting a single base shifts the reading frame of all bases downstream from the mutation. This event will result in an mRNA that encodes amino acids that are different from those present in the wild-type protein downstream from the frameshift mutation.

The mechanism of frameshift mutation within a homopolymeric run involves unpairing the template from the newly synthesized (nascent) DNA during the process of DNA replication and reforming hydrogen bonds with different bases. This slippage of hydrogen bonding between the template and nascent strand will result in the formation of an extrahelical base in either the template strand or the nascent strand. Continued replication results in dele-

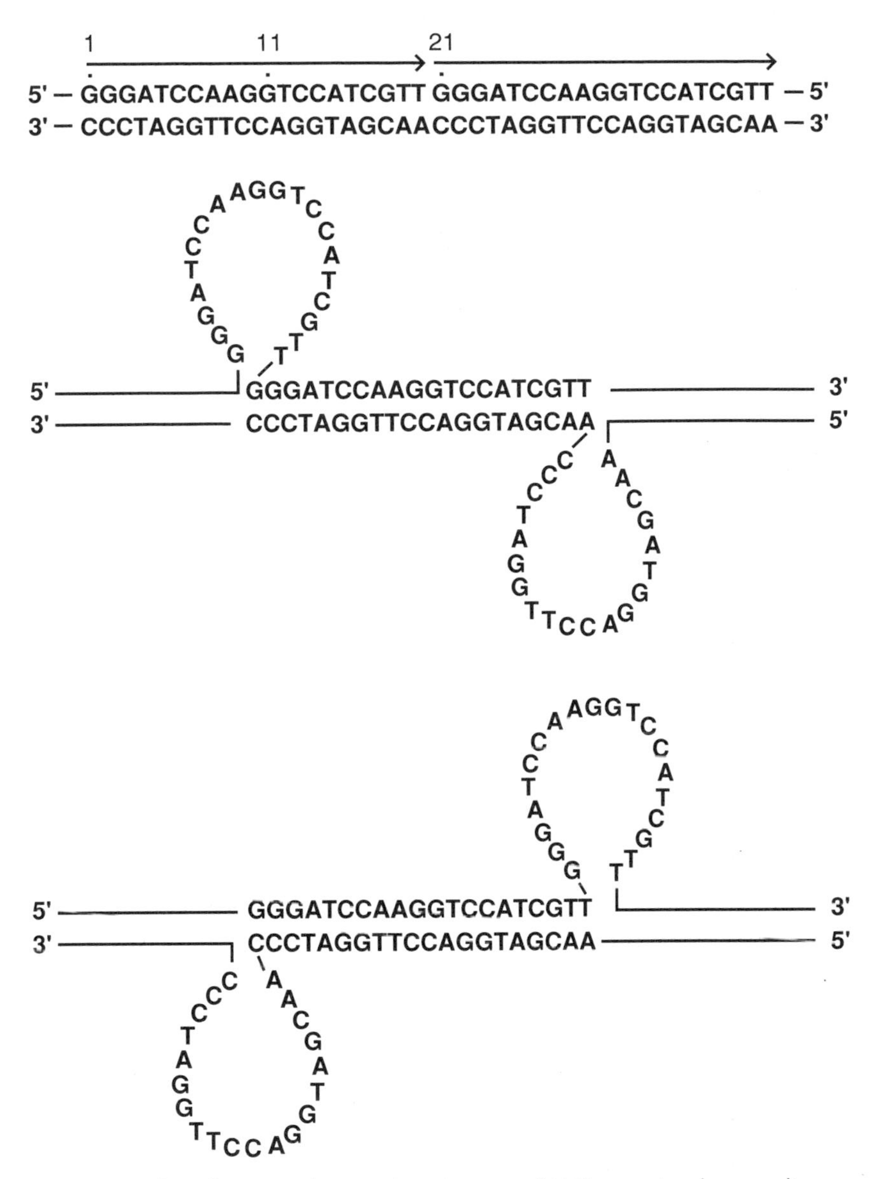

Figure 7.1 Slipped mispaired DNA. Slipped mispaired DNA can exist when two direct repeat sequences occur adjacent to each other. In this case, a 20-bp direct repeat is shown. The direct repeats are indicated by arrows above the DNA sequence. Two different slipped mispaired isomers of the sequence can exist. One involves pairing the second copy of the direct repeat in the top strand with the first copy of the direct repeat on the bottom strand. The other structure is an isomer in which the first copy of the direct repeat in the top strand pairs with the second copy of the direct repeat in the bottom strand. In each case, two single-strand loops in opposite strands would be extruded from the DNA.

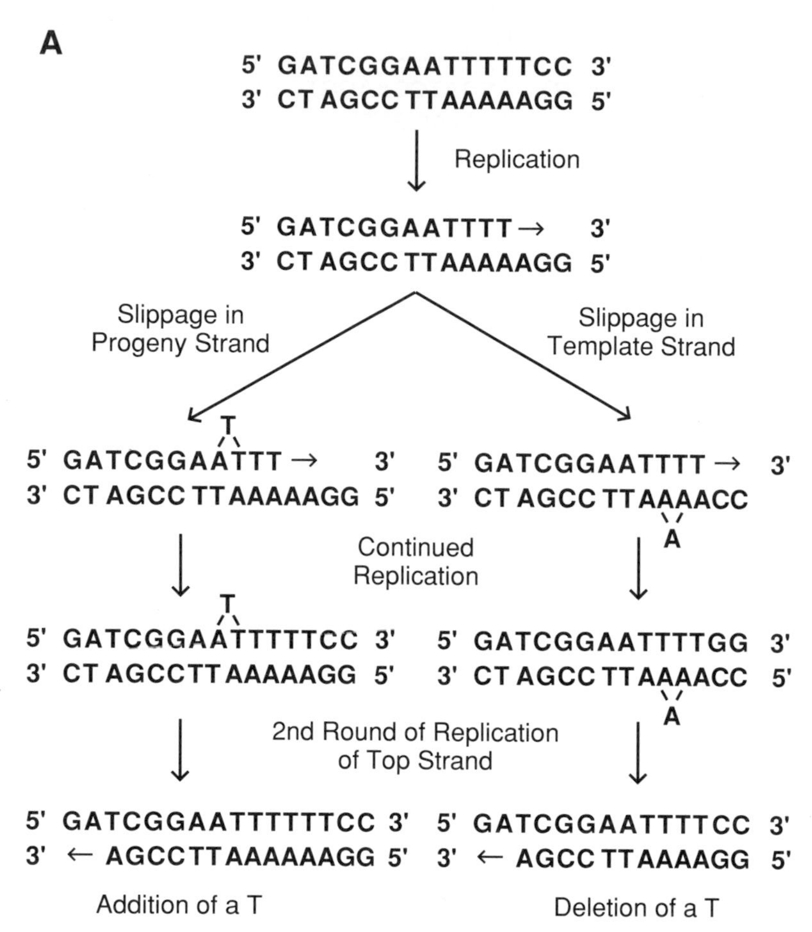

Figure 7.2 The Streisinger slipped misalignment model for frameshift mutagenesis. (A) The model for single-base frameshifts resulting from slipped mispairing. Replication occurs within a run of a single base, in this case copying a run of five As. (Left) A backward slippage in the progeny strand results in the formation of an extrahelical T in the progeny strand. Continued replication results in the addition of an extra T in the progeny strand. Following a second round of replication (in the absence of repair or editing), the duplex derived from the top strand contains the addition of a T·A base pair. (Right) A backward slippage in the template strand results in the formation of an extrahelical A in the template strand. Following replication, the duplex derived from the top strand contains a deletion of a T·A base pair.

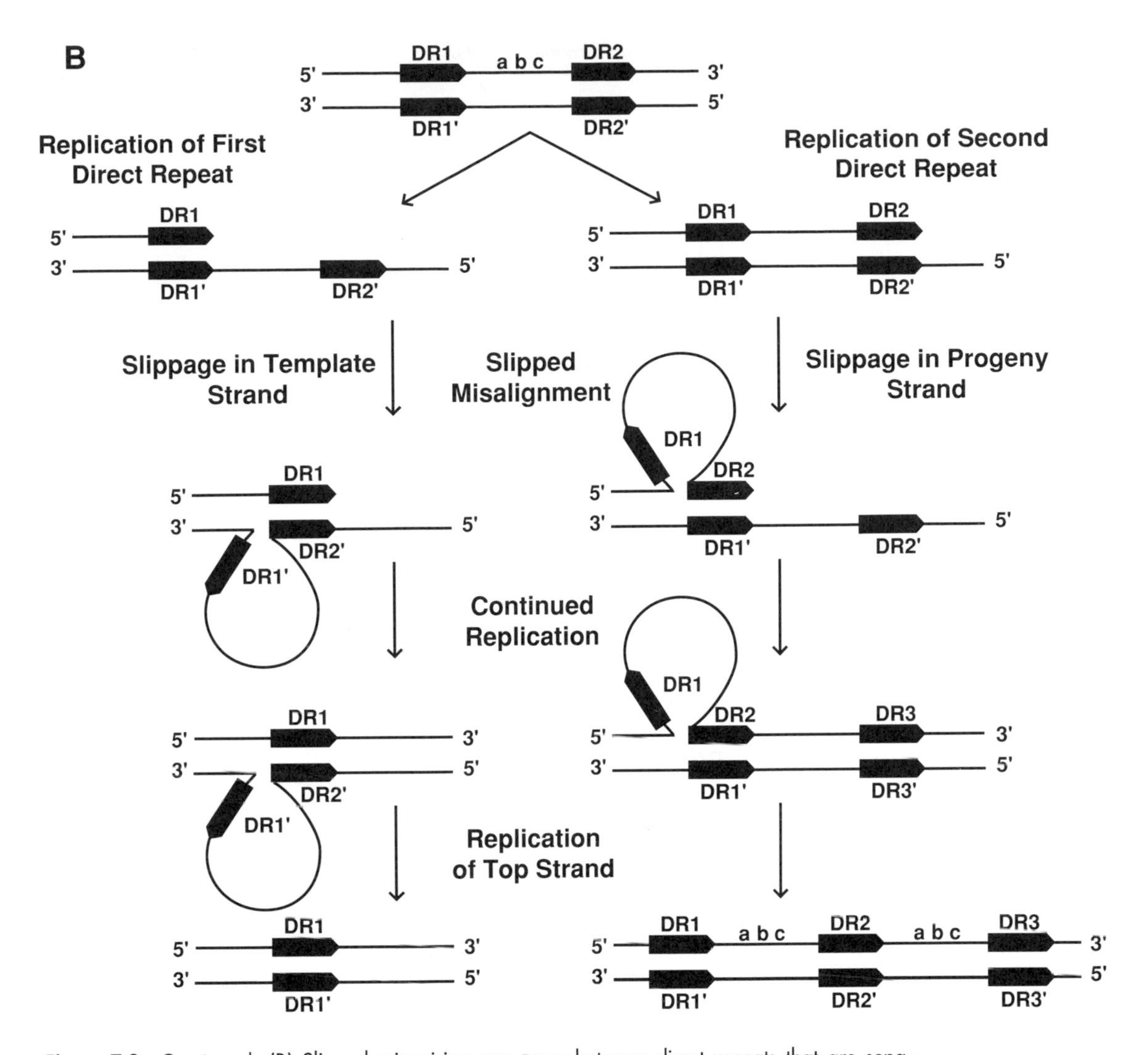

Figure 7.2 Continued. (B) Slipped mispairing can occur between direct repeats that are separated by intervening DNA. The DNA between the direct repeats with sequences a, b, and c could be several base pairs to hundreds of base pairs long. The length of the direct repeats between which deletion or duplication occurs can be from several to 10 or 20 bp in length. (Left) The pathway for deletion of one copy of the direct repeat and the intervening DNA is shown. A backward slippage in the template strand has occurred, with the progeny strand direct repeat number one (DR1) pairing with the template copy of direct repeat number two (DR2′). Continued replication following the second round of replication results in deletion of one copy of the direct repeat and the intervening DNA. (Right) The pathway for formation of a duplication event is shown. Following replication of the second copy of the direct repeat, the progeny strand slips back and the second copy of the direct repeat (DR2) in the progeny strand base pairs with the first copy of the direct repeat in the template strand (DR1′). Continued replication results in the resynthesis of a copy of the direct repeat and the intervening DNA, resulting in a duplication event in the progeny strand. It is very clear that mutation events consistent with the Streisinger model for misalignment occur in essentially all organisms that have been examined. This genetic evidence strongly suggests that loop-out structures exist and are maintained for extended periods of time in living cells.

tion (when the nascent strand slips forward) or in duplication (when the nascent strand slips backward).

Large deletion and duplication mutations occur between direct repeats that can form slipped structures during DNA replication (Figure 7.2B). These direct repeats can be adjacent to one another or separated by as many as hundreds of base pairs (Albertini *et al.,* 1982; Drake *et al.,* 1983; Ripley, 1991). The nature of the intervening DNA between the direct repeats is also important. If DNA between the direct repeats contains palindromic symmetry, the formation of a hairpin arm can stabilize the misalignment and increase the frequency of deletion (DasGupta *et al.,* 1987; Williams and Müller, 1987; Weston-Hafer and Berg, 1989; Sinden *et al.,* 1991; Trinh and Sinden, 1991,1993).

In conclusion, slipped mispaired structures are one alternative form of DNA that has not been physically identified *in vitro* or *in vivo.* Nevertheless, evidence from mutagenesis strongly suggests that slipped mispairing must occur during DNA replication. It is not certain if, like cruciforms, the formation of the alternative secondary structure occurs and presents a substrate for mutagenesis or if slippage simply occurs during the replication process.

C. DNA Unwinding Elements

A sequence element called a *DNA unwinding element* or DUE has been identified in DNA of both prokaryotes and eukaryotes. DUEs are A + T-rich regions of DNA that are commonly associated with replication origins and chromosomal matrix attachment sites. These A + T-rich sequences range from 30 to >100 bp in length. There appears to be little similarity in their DNA sequences (i.e., no "consensus" sequence) with the exception that the sequences are A + T rich. Several A + T-rich sequences are listed in Table 7.1. In the presence of DNA supercoiling, denaturation or melting of the DNA will occur first at A + T regions. The DUE *in vitro* can apparently unwind and be maintained as a *stably unwound structure* in the presence of negative DNA supercoiling.

Kowalski and co-workers were the first to show that large A + T-rich tracts appeared to have single-stranded character in supercoiled DNA (Sheflin and Kowalski, 1984). These investigators examined the pattern of cutting supercoiled DNA with mung bean nuclease, a single-strand-specific nuclease that works at neutral pH. Interestingly, the sites of nuclease sensitivity varied as a function of the Mg^{2+} concentration. In buffer containing no Mg^{2+}, the large A + T-rich tracts (DUEs) would melt and were sensitive to nuclease cutting. When Mg^{2+} was added, these A + T-rich regions remained double stranded whereas other regions within the supercoiled plasmid (such as cruci-

Table 7.1
DNA Unwinding Elements

Sequence	Organism	DNA sequence
3–13 mers[a]	*E. coli*	AGATCTATTTATTTAGAGATCTGTTCTATTGTGATCTCTTATTAGGATC
17 bp, A + T rich[b]	SV40	15 21 31 ATAAATAAAAAAAATTA
ARS[c]	Yeast	3711 3701 3691 3681 3671 3661 ATTTTTTATGTTTATCTCTAGTATTACTCTTTAGACAAAAAAATTGTAGTAAGA 3651 3641 3631 3621 3611 3601 ACTATTCATAGAGTGAATCGAAAACAATACGAAAATGTAAACATTTCCTATACGTAGTAT 3591 3581 3571 ATAGAGACAAAATAGAAGAAACCGTTCATAATTTTC
IgH (5′) enhancer[d]	Human	271 281 291 301 311 321 CTATCCAGAACTGACTTTTAACAATAATAAATTAAGTTTAAAATATTTTTAAATGAATTG
IgH (3′) enhancer[d]	Human	651 661 671 681 691 701 ACTTAAGTTTATCGACTTCTAAAATGTATTTAGAATTCATTTTCAAAATTAGGTTATGTA 711 721 731 741 751 761 AGAAATTGAAGGACTTTAGTGTCTTTAATTTCTAATATATTTAGAAAACTTCTTAAAATT 771 781 791 ACTCTATTATTCTTCCCTCTGATTATTGGT

[a]Kornberg and Baker (1992).
[b]Dean *et al.* (1987).
[c]Umek and Kowalski (1987).
[d]Kohwi-Shigematsu and Kohwi (1990)

forms) became sensitive to cutting. This dynamic variation in DNA structure is illustrated in Figure 7.3. The initial data of Kowalski suggested that, in the absence of Mg^{2+}, an entire large region of the A + T-rich DNA was maintained in a form that was melted or unwound in supercoiled DNA. This situation is likely to be similar to the C-type cruciform-inducing sequence discussed in Chapter 4.

When a large region of plasmid DNA unwinds, one supercoil should be removed for every 10.5 bp of DNA that is unwound. The two-dimensional agarose gel electrophoresis pattern of plasmid DNA containing an A + T-rich region is shown in Figure 7.4. The pattern was unlike any observed for cruci-

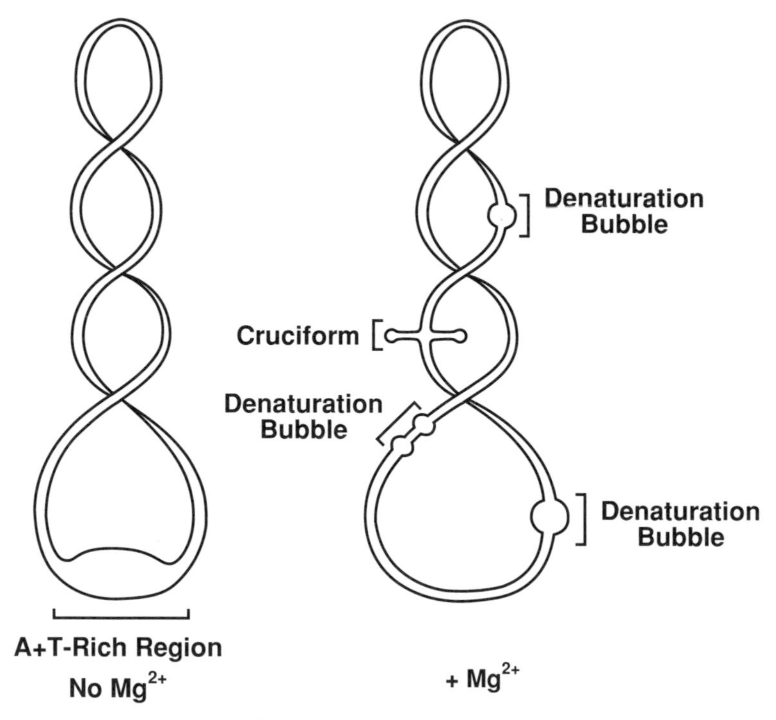

Figure 7.3 DNA supercoiling promotes melting at A + T-rich regions. A stably unwound region can be identified in some supercoiled DNA molecules when probed with a single-strand-specific nuclease. The region of DNA that unwinds is A + T rich. In supercoiled DNA, in the absence of divalent ions, a stably unwound region of DNA appears in which the two strands are not Watson–Crick hydrogen bonded (left). On addition of Mg^{2+}, which stabilizes the DNA double helix, this unwound region reforms a Watson–Crick double helix as evidenced by its insensitivity to a single-strand-specific nuclease. However, while this A + T-rich region forms a double helix, other regions of the supercoiled DNA molecule can form small denaturation bubbles or small cruciform structures that can be detected by cleavage at those specific locations by the single-strand-specific nuclease (right).

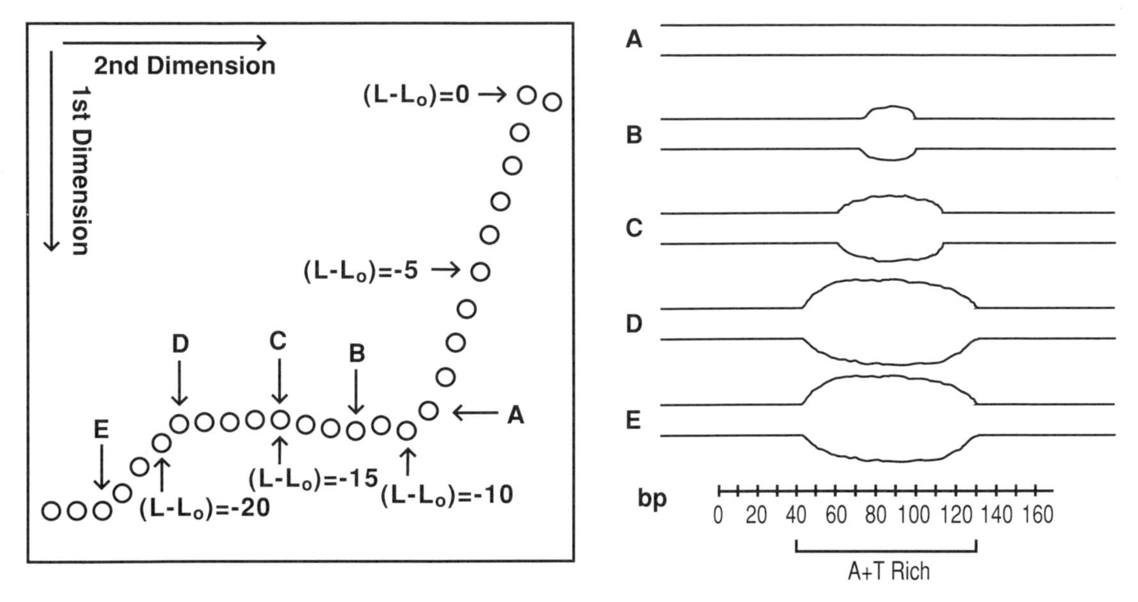

Figure 7.4 Identification of a DNA unwinding element by electrophoresis on a two-dimensional agarose gel. The pattern observed by stably unwound DNA within a plasmid DNA molecule on electrophoresis in a two-dimensional agarose gel is different from that observed for formation of a cruciform, Z-DNA, or triplex structure. Two-dimensional agarose gel electrophoresis was described at the end of Chapter 4 when the gel pattern for a cruciform transition was discussed. At some specific topoisomer number, here a linking number equal to L = −10, the subsequent topoisomers do not run more rapidly in the first dimension, as would be expected for molecules containing increasing numbers of supercoils. However, successive topoisomers migrate at a near constant rate representing a family of molecules in which the number of superhelical turns is identical in the first dimension. At some point, position D on the two-dimensional gel, the migration of topoisomers again increases in the first dimension. At the position of topoisomer E in the two-dimensional gel, molecules with increasing numbers of supercoils are no longer resolved by the gel. The results are consistent with the structures A–E shown to the right of the gel. The length of the unwound region, from structures B–D, increases in increments of 10.5 bp with an increase in the superhelical density of the plasmid of $\Delta L = -1$. Structure D would contain a 94-bp unwound A + T-rich region. The addition of four additional supercoils (structure E) results in no further unwinding (the entire A + T-rich region is unwound) but results in the introduction of four additional negative supercoils into the plasmid.

form, Z-DNA, or intramolecular-triplex-containing plasmids (for example, compare the 2-D gel for a cruciform transition in Figure 4.19). There is no sharp transition or break in the migration in the first dimension in which one topoisomer with more negative supercoiling migrates dramatically more slowly in the first dimension than the preceding (less negatively supercoiled) topoisomer. However, a "plateau" is observed in which many topoisomers migrate at the same position in the first dimension. All topoisomers migrating at the same position have the same number of negative supercoils even though the linking number (L) of each topoisomer is different. This means that each successive topoisomer migrating at the position of the plateau contains an unwound region 10 bp larger than the preceding topoisomer. This is illustrated in Figure 7.4.

DNA unwinding elements are required for the initiation of DNA replication at certain origins (Umek and Kowalski, 1987,1988). Progressive deletion of the A + T-rich region results in the eventual loss of function as an origin of DNA replication. When the region containing a deletion no longer functions as an origin, it is also no longer sensitive to mung bean nuclease. In addition, mutations within a DUE that reduce the calculated helix stability of the region result in the loss of the autonomously replicating sequence (ARS) function in yeast (Natale *et al.,* 1992). These results suggest that unwinding at an A + T-rich DNA region near the origin of DNA replication is required for the initiation of DNA replication. One of the best examples of just how the unwinding might be involved comes from studies of the *E. coli* origin of replication.

1. Unwinding at the Escherichia coli Origin of Replication

The events that occur during initiation at the origin of DNA replication in *E. coli* have been studied intensively by Kornberg and colleagues. These events are described in detail in an excellent text on DNA replication (Kornberg and Baker, 1992). The process of initiating a round of chromosomal DNA replication represents a major commitment for the cell. The cell must decide that the time and environment are sufficiently favorable to expend the energy needed to replicate its chromosome and then divide into two cells. Once begun, the process cannot be stopped. Mechanistically, the initiation of replication is a complicated process, with 20–30 different proteins involved. Since several copies of many individual proteins are needed, there may be as many as 50–100 total proteins working on DNA within a short region at the origin of replication. The steps involved consist of recognition of the origin by specific proteins, unwinding of the origin to allow helicase, primases, and DNA polymerase III to get onto the individual strands of DNA; and finally, the initiation of synthesis.

A current model for initiation of replication at the origin proposed by Kornberg is shown in Figure 7.5. Initiation-specific DnaA proteins first bind to specific sites (dnaA boxes) at the *E. coli* origin of replication in supercoiled DNA and organize DNA into a tight loop. Three A + T-rich direct repeat sequences then unwind. The unwinding of these regions can be detected by mung bean nuclease-induced cleavage or by the reactivity with single-strand-specific chemicals. A protein complex consisting of the DnaB and DnaC pro-

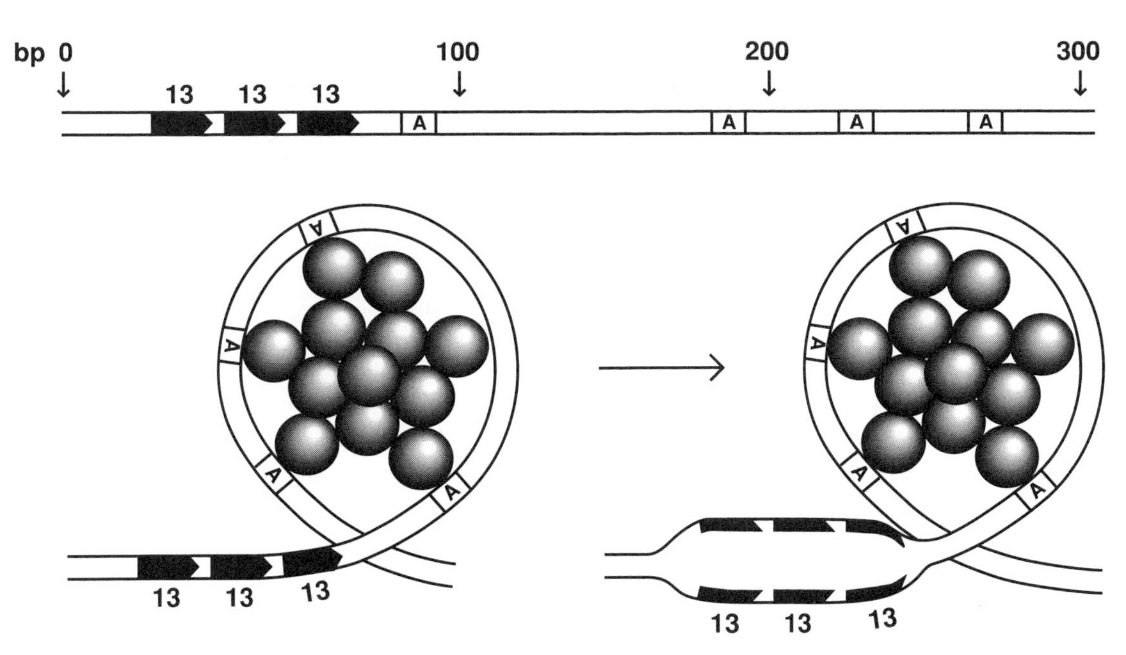

Figure 7.5 DNA unwinding element in oriC, the E. coli **DNA replication origin.** Four dnaA boxes indicated by A in a double helix are found in the 254-bp minimal oriC, the E. coli chromosomal replication origin. In addition, 5′ to the first dnaA box are three 13-bp A + T-rich direct repeats that bind DnaBC. The oriC region organizes itself into a three-dimensional structure in which DnaA protein binds to the four dnaA boxes and organizes the DNA into a loop. Following the organization into the loop, the A + T-rich region containing the three 13-bp A + T-rich direct repeats becomes stably unwound. This facilitates the entrance of the DnaBC helicase and primase (DnaG), which is needed to lay down the initital RNA primer for DNA replication. A common theme in many DNA replication origins appears to be the requirement for an A + T-rich region that can stably unwind to allow the entry of the multitude of proteins that must be associated with the replication fork.

teins then binds to this single-stranded region. The DnaB protein is a helicase that, once it has bound to DNA, begins unwinding the DNA double helix. When a larger bubble is formed by DnaB helicase activity, an RNA primase (DnaG) binds. Next the DNA polymerase holoenzyme (with its multiple subunits) binds, and the process of DNA replication begins.

D. Parallel-Stranded DNA

A parallel-stranded DNA molecule is one in which the polarity of the two DNA strands is the same (van de Sande *et al.,* 1988; Jovin *et al.,* 1990). A parallel orientation is opposite that found in the normal Watson–Crick double helix (Figure 7.6). Parallel-stranded DNA has even been engineered into a supercoiled plasmid molecule (Klysik *et al.,* 1991). Another example of base pairing in a parallel fashion was discussed in Chapter 6 in which the

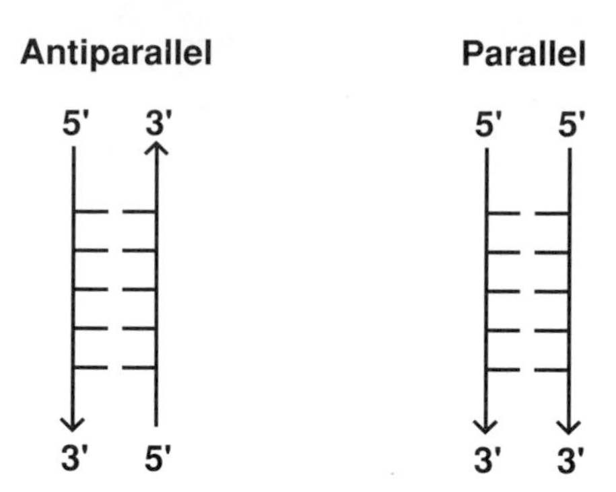

Figure 7.6 Antiparallel and parallel organization of complementary strands. DNA in the canonical Watson–Crick double helix is antiparallel. The polarity runs in one strand from the 5′→3′ direction and in the opposite complementary strand from the 3′→5′ direction. In parallel DNA, the polarity of the two complementary strands is identical. As shown in the next figure, under specialized conditions parallel-stranded DNA can exist.

third stand of a triple helix can pair in either a parallel or an antiparallel fashion with the central purine-rich strand.

A specific sequence arrangement is required for the formation of parallel-stranded DNA. Parallel-stranded DNA requires two complementary strands in which the polarity is the same. In parallel-stranded DNA, the bases are paired in a reverse Watson–Crick fashion. The reverse Watson–Crick base pair is one in which the glycosidic bonds and ribose sugars extend in a *trans* (or opposite) position from the bases compared with the *cis* configuration found in B-DNA. In the reverse Watson–Crick orientation, as shown in Figure 7.7, the base and ribose sugar (-R) have essentially flipped over 180°. (This is different from the *anti–syn* rotation that is found in Z-DNA.) In the reversed Watson–Crick base pair, thymine hydrogen bonds through the O2 and N3 positions whereas the N3 and O4 positions are hydrogen bonded in the normal Watson–Crick base-pairing scheme. Numerous polymers have been synthesized that can form parallel-stranded DNA, one of which is shown in Figure 7.7. Surprisingly, the stability of parallel-stranded DNA, with reverse Watson–Crick base pairs, is only slightly lower than that of a corresponding B-form antiparallel double helix.

Whether there is a biological role for parallel-stranded DNA is not known. It is possible that two single-stranded regions of DNA could interact in a parallel fashion, although large regions of chromosomal DNA are not typically single stranded. Single strand regions do occur, however, at telomeres as discussed later. Certain telomere sequences, specifically sequences containing runs of Gs, can form parallel helices. It is also possible that regions of RNA molecules or regions of two different RNA molecules may be organized into parallel helices.

Watson-Crick A•T Base Pair

Reverse Watson-Crick A•T Base Pair

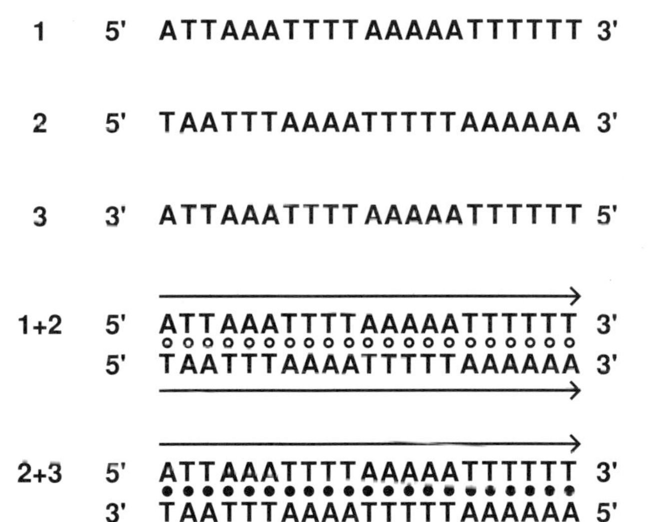

Figure 7.7 Parallel-stranded DNA. (Top) B-form DNA contains A·T base pairs (left) with the strands oriented in an antiparallel fashion. Parallel-strand DNA contains reverse Watson–Crick base pairs (right). In the reverse Watson–Crick base pair, the base is flipped 180° with respect to the Watson–Crick base pair. (Bottom) Ramsing and Jovin (1988) synthesized the sequences numbered 1, 2, and 3. Strands 1 and 2 pair in a parallel fashion using reverse Watson–Crick base pairing (indicated by ○). Strands 2 and 3 form traditional Watson–Crick base pairs (indicated by •) with antiparallel strands.

E. Four-Stranded DNA

There are a number of ways in which four strands of DNA may interact and pair into a stable structure. It was recognized by Morgan (1970) (as a model for the action of DNA polymerase) and McGavin (1971) that two pairs of complementary Watson–Crick hydrogen bonded bases could form a four-stranded hydrogen bonded structure (Figure 7.8). In this type of four-stranded structure, one base interacts with both its complementary Watson–Crick base pair and another "complementary" base of the second Watson–Crick base pair. As yet, there is no firm evidence for the existence of this type of four-stranded DNA. The interaction of two double helices by hydrogen bonding between the adenines has also been suggested (Figure 7.8; Borisova *et al.,* 1991).

1. Parallel Four-Stranded DNA, G-Quartet Structure

Sen and Gilbert (1988) demonstrated that DNA from the switch region in an immunoglobulin heavy chain gene could exist as a four-stranded, G-quartet structure in which all strands are parallel. (The immunoglobulin genes undergo specific recombination events resulting in extensive antibody diversity.) The DNA sequences involved in this alternative helical form are repeated motifs high in guanine, for example, 5′ GGGGAGCTGGG 3′. On incubation of single-stranded oligonucleotides containing this sequence element, the individual strands formed molecules that migrated as specific higher molecular weight structures during electrophoresis on polyacrylamide gels. A Hoogsteen base-pairing scheme was suggested from analysis of the chemical reactivity of the complex with dimethylsulfate (in which the N7 position of guanine was protected). In addition, the orientation of the four strands is parallel and the glycosidic bonds in all nucleotides are in the *anti* configuration. The base-pairing scheme and model for the parallel four-stranded G-quartet structure is shown in Figure 7.9.

This G-quartet DNA structure could have profound biological implications. As discussed by Sen and Gilbert (1988), runs of Gs at telomeres (see Section E,2) could form this parallel four-stranded structure. This might hold the four chromosomes together at meiosis. Runs of guanine within a chromosome might also interact and form this G-quartet structure. Single guanine-rich stands could come from the displaced fourth strand of a *Py*-Pu-Py intramolecular triplex. There are exciting structural possibilities inherent in a seemingly monotonous run of guanines.

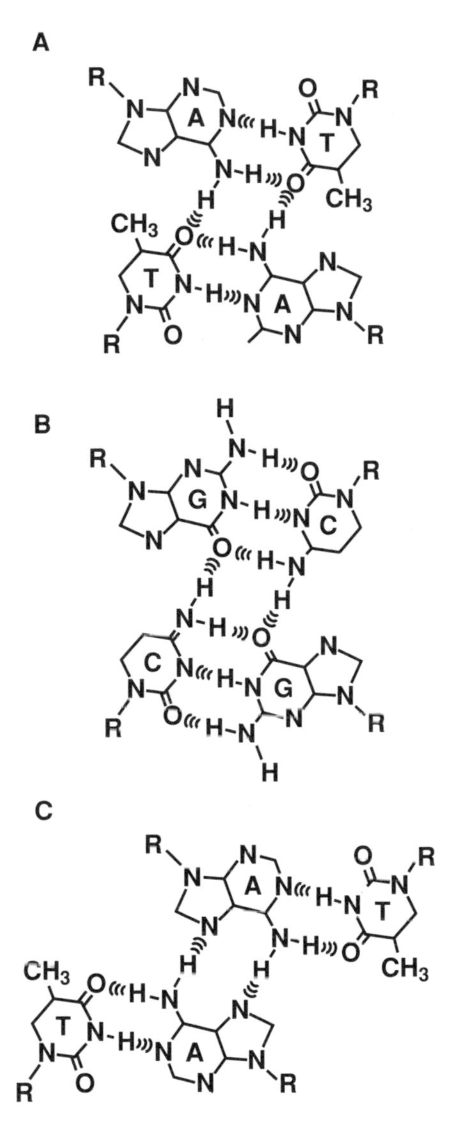

Figure 7.8 Quartet structures. There are several ways in which Watson–Crick A·T or G·C base pairs can form a quadruple-stranded helix by the interaction of two Watson–Crick helices. This interaction occurs through the major grooves of the two Watson–Crick helices. (A) The hypothetical interaction between two A·T Watson–Crick base pairs. This structure is stabilized by two hydrogen bonds between the N6 position of adenine and the O4 position of thymine. (B) The interaction between two G·C base pairs in which hydrogen bonding occurs between the O6 position of guanine and the N4 position of cytosine. (C) An additional base pairing scheme for a quadruple-stranded DNA structure between two A·T base pairs. In this case, base pairing can occur between the two purine nucleotides. Pairing would involve hydrogen bond formation between the N6 and N7 positions of the two adenines.

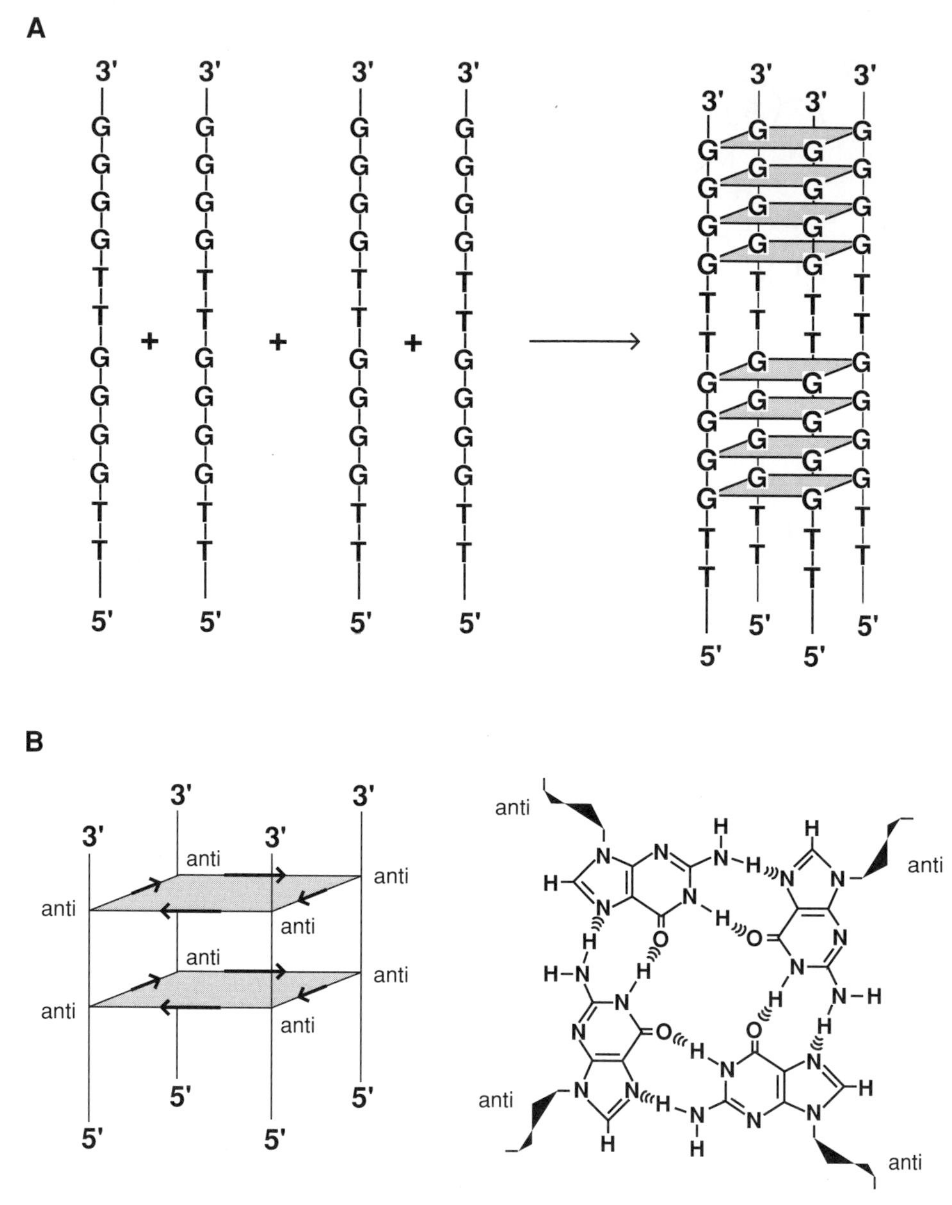

Figure 7.9 A G-quartet DNA structure. (A) Four single-strand tracts of polyguanine can form a quadruplex structure in which the polarity of all four strands is parallel. (B) These four poly(dG) strands are held together by Hoogsteen base pairs (right). All glycosidic bonds are in the anti conformation and the direction of hydrogen bonds are the same in each G quartet (left). The arrow shows the direction of the hydrogen bond from the hydrogen donor (H) to the acceptor (O or N).

2. Four-Stranded Structures at Telomeres

Telomeres are composed of repetitive simple sequences that are usually purine rich in one strand [for example, $(G_4T_2)_n$ or $(G_4T_4)_n$] and there are many alternative structures that can form from telomere DNA (Hardin *et al.*, 1991). The single-stranded *Tetrahymena* telomere DNA sequence (G_4T_2) can form a four-stranded structure or quadruplex consisting of a planar array of guanines held together by Hoogsteen base pairs (Williamson *et al.*, 1989). If association occurs as a single strand, the parallel structure (Figure 7.9) is more stable than an antiparallel orientation. Alternatively, the single strand overhanging telomere end has the potential to fold into a hairpin in which guanines are Hoogsteen base paired. In this fold-back structure, the two strands are antiparallel. Two fold-back structures can interact into four-stranded structures held together by Hoogsteen base pairs between guanines. Two possible structures are shown in Figures 7.10A and 7.10B.

The X-ray crystal structure of a four-stranded *Oxytricha* telomeric DNA sequence formed from two molecules of $G_4T_4G_4$ was determined by Kang *et al.* (1992). In this case, the individual $G_4T_4G_4$ molecules form hairpins, and two hairpins bind together with the loops at opposite ends of the quadruplex. Although there are two ways to join hairpins with opposite loops (Figures 7.10C and 7.10D), structure with alternating strands arranged in an antiparallel orientation was found in the crystal structure. The four guanines are hydrogen bonded as indicated in Figure 7.10F, although the base locations do not have the idealized spacing shown. The geometry of hydrogen bonding is different for two successive base quartets (Figures 7.10F, *top*, 7.10F, *bottom*). Alternate nucleotides in a quartet exist in the *anti* and *syn* conformation with the direction of hydrogen bonding (donor → acceptor) in one direction as indicated in Figure 7.10F, *top*. The pattern and direction of hydrogen bonding are reversed in adjacent quartets (Figure 7.10F, *bottom*).

A four-stranded structure formed from two *Oxytricha* $G_4T_4G_4$ molecules that do not form individual hairpins was characterized using nuclear magnetic resonance (NMR) spectroscopy by Smith and Feigon (1992). In this structure, the T_4 loops are on opposite ends of the quartet spanning the diagonals of the quartet at right angles to each other. The two isomers of this quartet arrangement (Figures 7.10G and 7.10H) are very similar, varying only in the polarity with which the two strands are associated. In these structures, the glycosidic bond angles are *syn–syn–anti–anti* around the quartet whereas the pattern in adjacent quartets is *anti–anti–syn–syn* (Figures 7.10I and 7.10J). The direction of hydrogen bonding also alternates between adjacent quartets. The isomer shown in Figure 7.10G was consistent with the NMR data of Smith and Feigon (1992).

The very different quartet structures identified for the same *Oxytricha* telomere sequence ($G_4T_4G_4$) by Smith and Feigon (1992) and Kang *et al.*

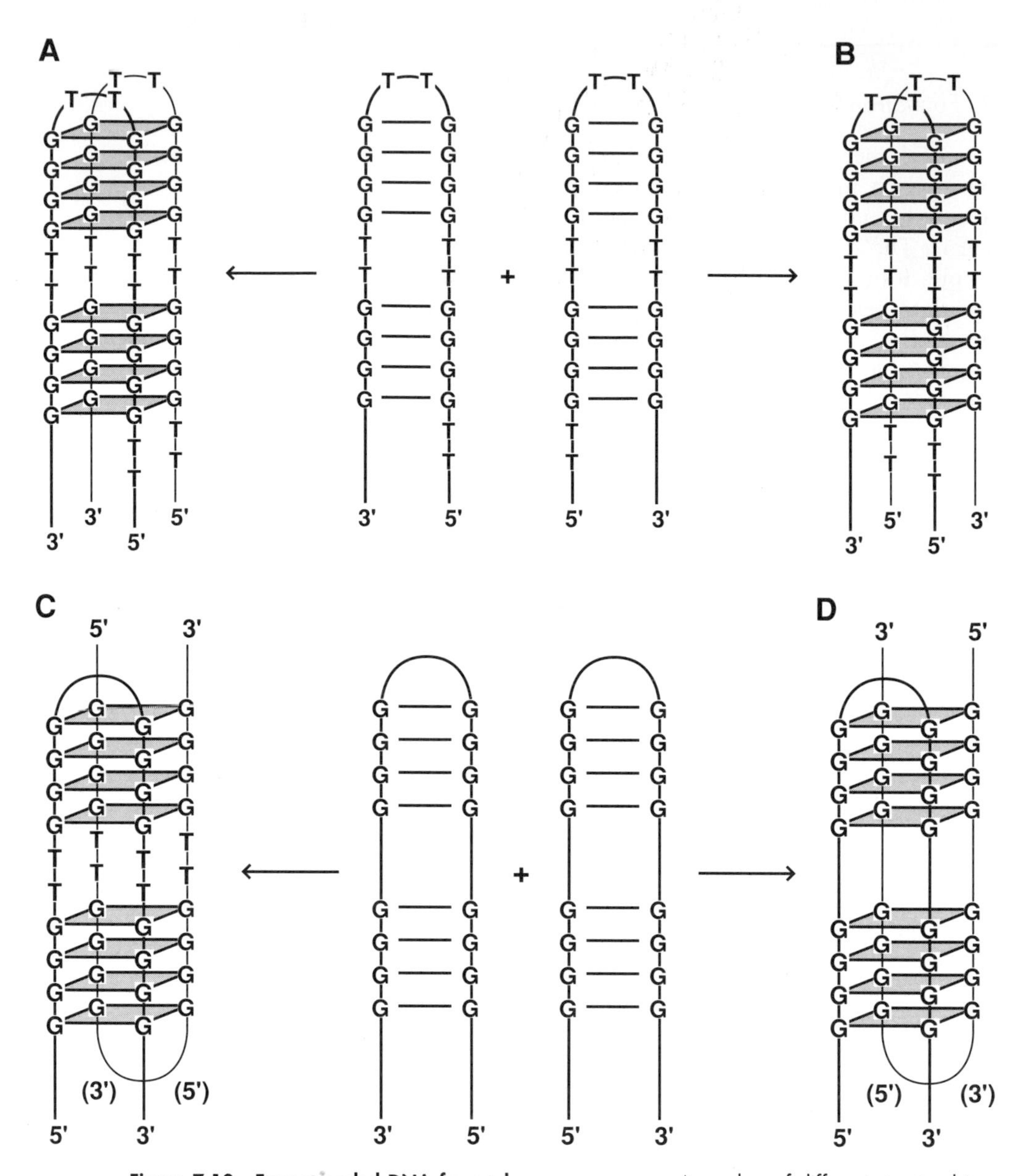

Figure 7.10 Four-stranded DNA from telomere sequences. A number of different structural isomers can form from sequences found at telomeres. Considering the Tetrahymena G_4T_2 sequence (also shown in Figure 7.9), if an antiparallel double strand forms, in which the guanines are hydrogen bonded in a Hoogsteen fashion, two duplexes can pair into four different quadruplex structures. In structure A, two antiparallel duplexes hydrogen bond in a parallel fashion with the hairpin loops at the same end, whereas in structure B they pair in an antiparallel fashion. In structures C and D, the hairpin loops are at opposite ends of the quadruplex. Structures C and D have antiparallel and parallel associations of the hairpin duplexes, respectively. The Oxytricha telomere sequence $G_4T_4G_4$ crystallized in the antiparallel quartet structure shown in C (Kang et al., 1992). As indicated by the representation in E and F within a quartet the glycosidic bonds alternate anti–syn–anti–syn. In addition, the direction of hydrogen bonding alternates in adjacent quartets. A central metal ion is believed to stabilize the quartet.

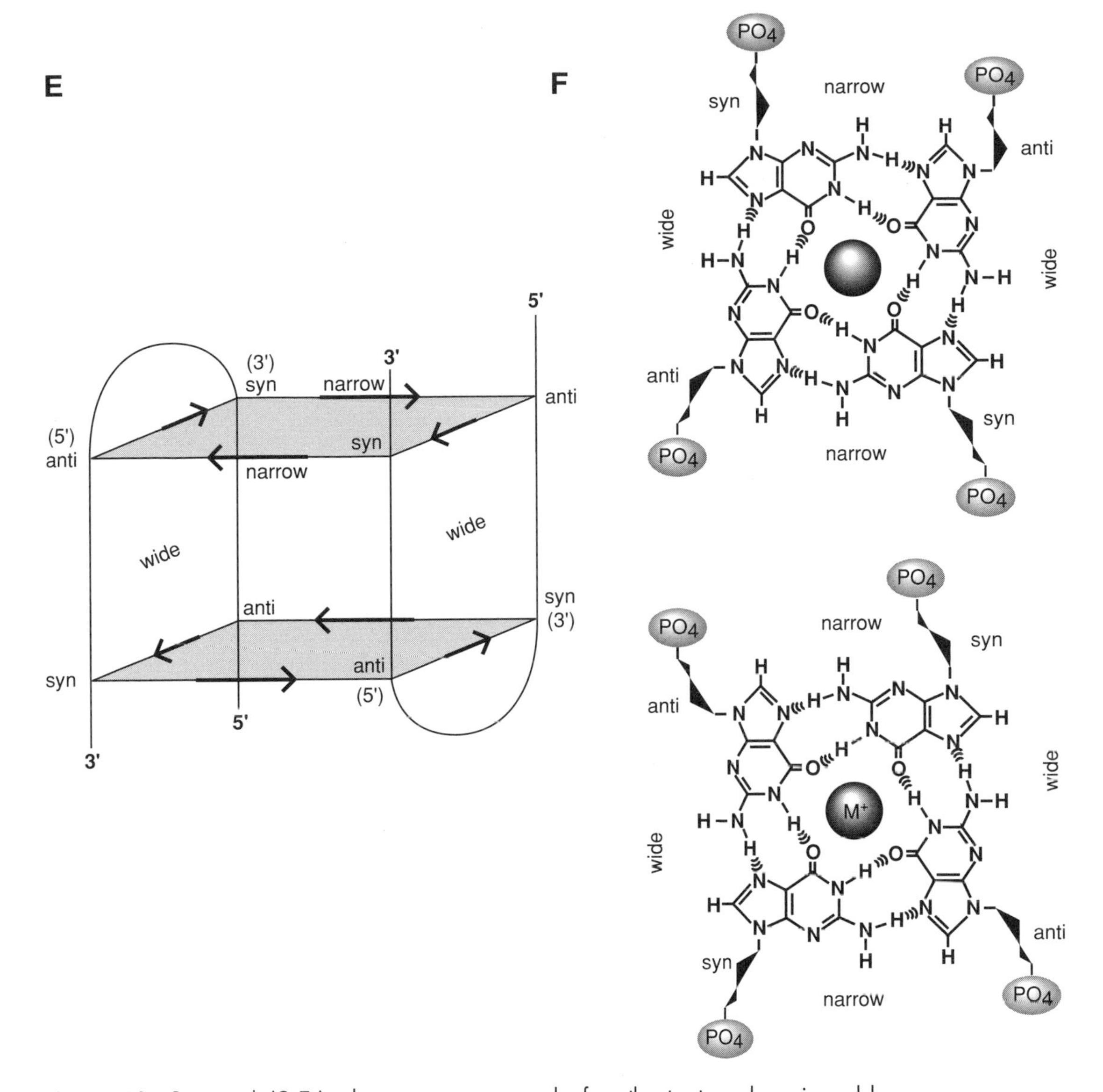

Figure 7.10 Continued. $(G_4T_4)_n$ telomere sequences can also form the structures shown in models G and H in which single strands fold forming the diagonals of a quartet. There is no intrastrand hydrogen bonding characteristic of a typical hairpin structure, but guanines from one strand hydrogen bond to two guanines from the second strand. The loops are at right angles to each other and at opposite ends of the quadruplex. Two different isomers can form (G and H) that differ in the polarity of the interaction. Both structures have two parallel and two antiparallel strand interactions. The model shown by structure G was identified in solution for $G_4T_4G_4$ using NMR spectroscopy (Smith and Feigon, 1992). Moreover, as indicated in I and J, the glycosidic bond configurations are anti–anti–syn–syn in one quartet and syn–syn–anti–anti in an adjacent quartet. Within one strand of the quartet, this gives an alternating anti–syn glycosidic configuration. The direction of hydrogen bonding also reverses between adjacent quartets.

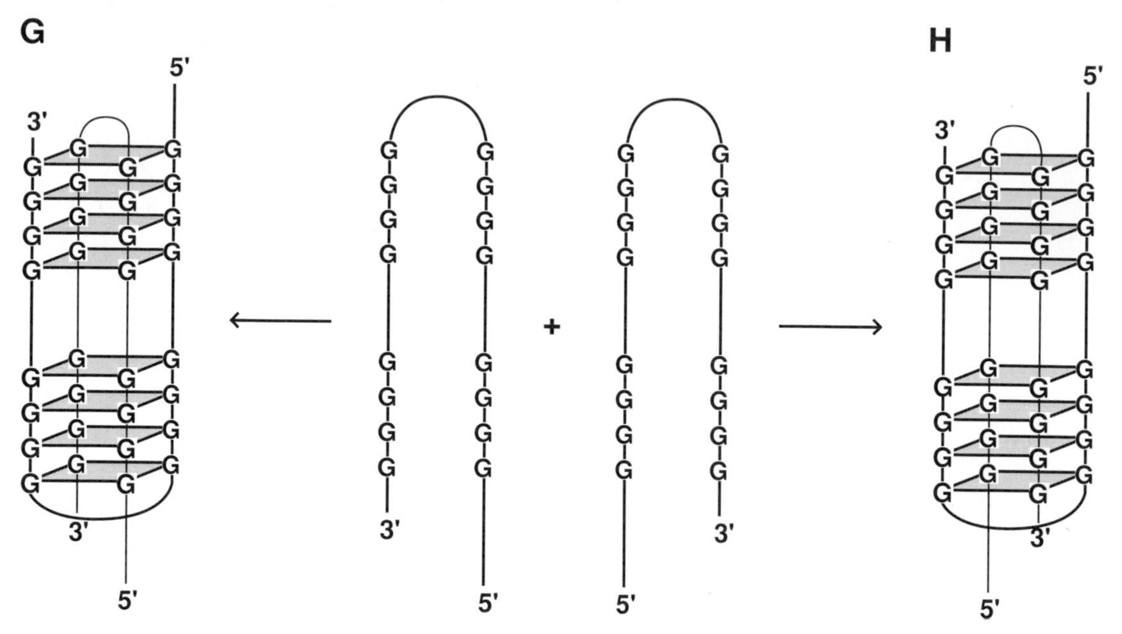

Figure 7.10 Continued.

(1992) demonstrate the variability inherent in the formation of quadruplex structures from poly(dG) sequences. The structures that form are sensitive to monovalent cations (Sen and Gilbert, 1990,1992; Hardin *et al.*, 1991; Scaria *et al.*, 1992); K^+ ion is believed to bind tightly within the center of the quartet. If quartet structures are forming in living cells, their stability might be controlled in part by the intracellular K^+ ion concentration.

Telomeres are critical for the maintenance and stability of linear chromosomes. Considering potential problems involved in replicating linear chromosomes, nature was wise to design circular chromosomes. DNA replication can occur around a circle in both directions. It is easy to see how the entire chromosome is replicated, since the leading strands can run up to or past each other. This ensures that all the genetic information is copied. The problem of replicating a linear chromosome is different. A leading strand can run off the end of the linear DNA molecule, the telomere. On the opposite strand, which represents a lagging strand, a polymerase would have to run backward, in the $3' \rightarrow 5'$ direction to replicate the very end of the linear molecule. To the best of our knowledge, this does not happen. At best, an RNA primer could be synthesized beginning at the exact $3'$ end of a linear molecule. DNA synthesis from this primer would still leave a short piece of RNA at the end of the chromosome. An incomplete length of DNA at each end would result

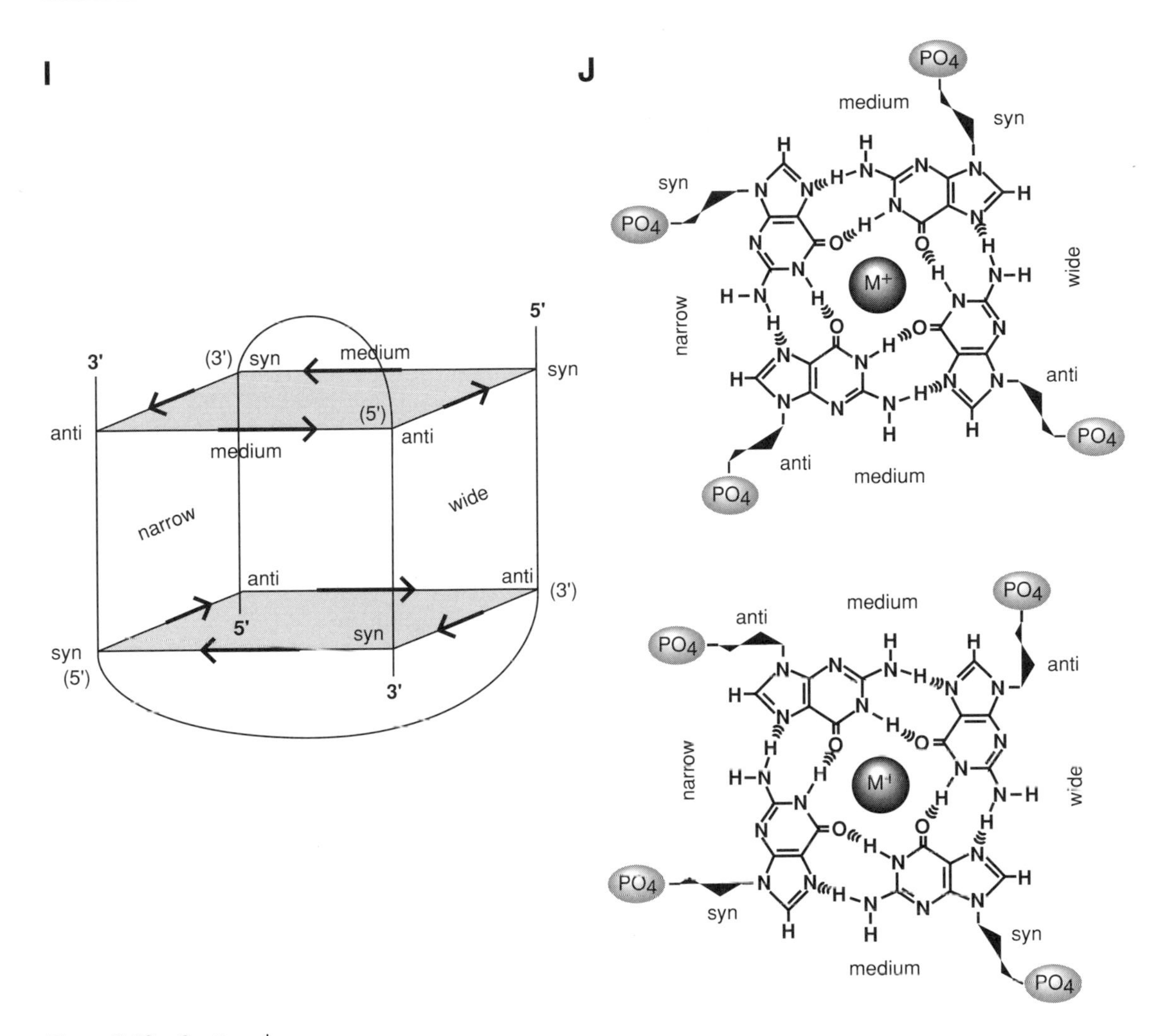

Figure 7.10 Continued.

in the synthesis of progressively shorter chromosomes following each round of DNA replication. This is a problem that primitive organisms with linear chromosomes must have had to solve rather quickly!

The problem of replicating the ends of chromosomes is solved by a special enzyme called *telomerase*. Telomerase is a ribonucleoprotein. The RNA that telomerase contains is a very simple short sequence that varies from organism to organism. Telomerase synthesizes repeating DNA sequences complementary to the RNA "template" within the telomerase. The telomerase is actually a specialized reverse transcriptase that synthesizes DNA from an RNA template. One telomere sequence in *Tetrahymena* is $(G_4T_2)_n$ encoded by an RNA molecule of sequence $(A_2C_4)_n$. The synthesis of the short repeating unit occurs multiple times producing chromosomes of slightly variable length. Finally, there *will* still be a 12- to 16-bp single-stranded purine-rich region protruding from the end of the chromosome. Since by this time the end is well away from unique or important genetic information, it really does not matter whether the ends are not completely double stranded. Estimates are that telomeres on human chromosomes are as long as 10 kb!

Specialized telomere proteins bind the unique repetitive DNA at the ends of the chromosomes. Certain proteins bind noncovalently to the single-stranded overhanging ends of the telomere. These proteins may protect the ends of chromosomes from digestion by nucleases. Once bound, they may also stop further elongation of the telomere by telomerase. Also, other structural proteins bind internal to the end of the chromosome to the repeating units of DNA composing the telomere. The function of the telomere binding proteins is not completely understood, but they may be involved in organizing and aligning interphase or metaphase chromosomes or they may be involved in chromosome segregation at cell division.

F. Anisomorphic DNA

Herpes simplex virus (HSV) has a region within its linear 100-kb chromosome that reverses or inverts its orientation when the DNA replicates. Near this site of inversion is the 12-bp direct repeat sequence, which is present in many copies (Figure 7.11). This sequence was cloned and the unusual structural properties of it were characterized (Wohlrab *et al.,* 1987; Wohlrab and Wells, 1989). The center of the direct repeat region was sensitive to S1 nuclease; the sensitivity depended on the length of the repeated region. If the number of the direct repeats was reduced below 5 copies, the direct repeat region lost S1 nuclease sensitivity. As the number of the direct repeats increased above 7 copies, S1 nuclease sensitivity returned. The region of nuclease sensitivity always mapped to the center of the direct repeats, regardless of the total number of the direct repeats.

The high G + C content and polypurine·polypyrimidine nature of the direct repeat is likely responsible for the unusual properties of this sequence element. This poly(Pu)·poly(Py) arrangement of bases may make a helix that is stiffer than usual, representing the opposite of bent DNA. A long, rather stiff region of DNA would not fit well into a supercoiled DNA molecule in which twisting, writhing, and compaction occur. One possibility is that, under pressure from DNA supercoiling, the region of anisomorphic DNA bends at the

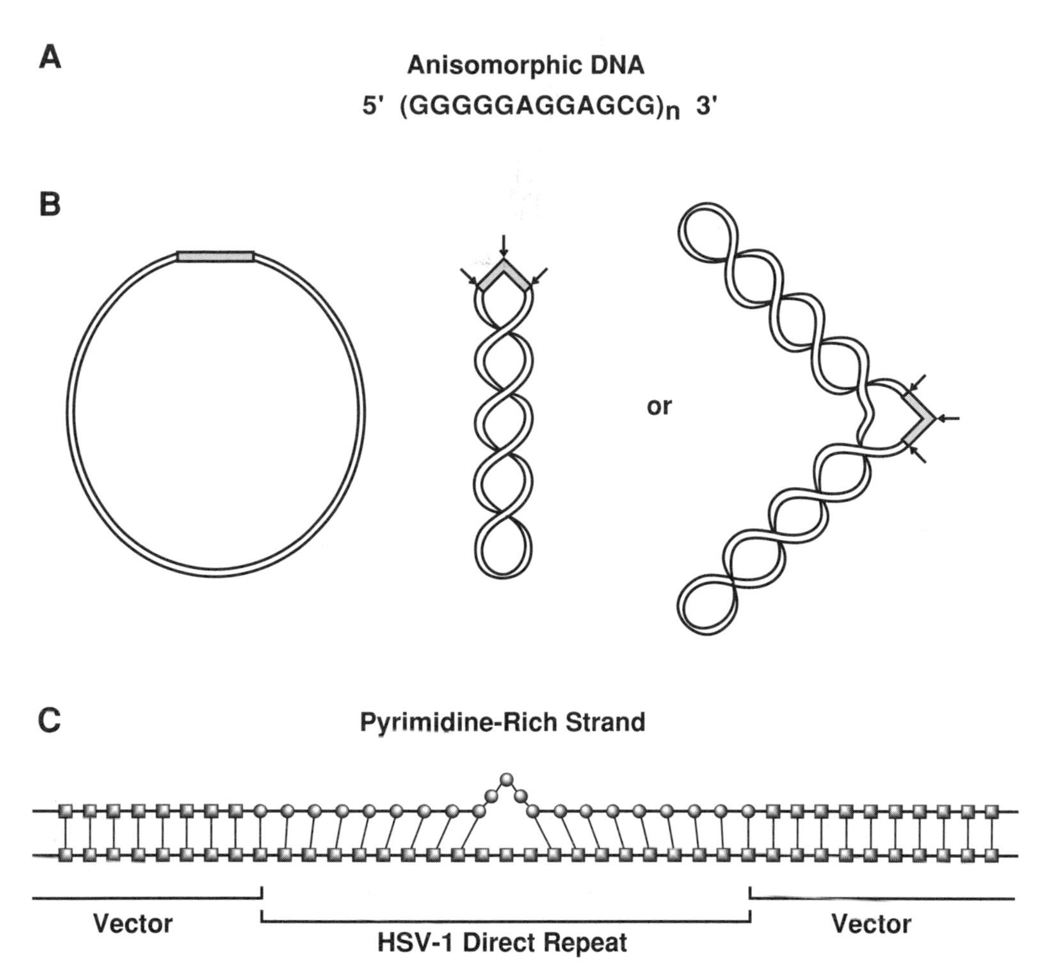

Figure 7.11 Anisomorphic DNA. The sequence of a 12-bp direct repeat present in multiple copies in the herpes simplex virus DNA. This 12-bp sequence is predominantly purine rich in one DNA strand. As the length of the direct repeat increases in plasmid DNA, eventually anomalous migration occurs on electrophoresis in an agarose gel, indicative of a family of several different DNA secondary structure transitions. To date, the structure of these unusual DNA conformations is not known. One possible model is shown in B, in which the polypurine·polypyrimidine stretch is unusually stiff so, on DNA supercoiling, it bends and kinks at the center of the direct repeat. This may affect the three-dimensional organization of the supercoiled DNA so there is the introduction of a branched, supercoiled DNA molecule. This kinking may be responsible for the anomalous electrophoretic mobility. An additional possibility is shown in C, in which alternative helical conformations are formed that result in the denaturation of several base pairs at the center of the directly repeated region. This may be due to different helical parameters present in each individual DNA strand. The anomalous electrophoretic migration may also be the result of partial unwinding or the formation of an alternative DNA conformation, perhaps involving a triplex or four-stranded structure that unwinds negative DNA supercoils (not shown). Figure modified with permission from Wohlrab et al., (1987) and Wells (1988).

center of the stiff region. This bend or kink could produce a phosphodiester bond that is sensitive to nucleases and a few bases that are susceptible to chemical attack. This model for anisomorphic DNA is shown in Figure 7.11.

Supercoiled plasmids containing several copies of these direct repeats show multiple DNA secondary structure transitions on two-dimensional agarose gels. Homopurine·homopyrimidine regions can form triplexes (Chapter 6) and other non-B-DNA forms (as described subsequently). It is possible that some as yet undefined secondary structural transition, an unusual triple-helical region or looped-back structure involving non-Watson–Crick pairing, forms in this region. It is also possible that slipped mispaired structures could be forming.

G. Nodule DNA and Higher-Order Pu·Py Structures

Long regions of $(dG)_n \cdot (dC)_n$ and $(GA)_n \cdot (CT)_n$ can form the "nodule DNA" structure shown in Figure 7.12 (Kohwi-Shigematsu and Kohwi, 1991; Panyutin and Wells, 1992). This structure forms from two smaller intramolecular triplex regions in which the single-stranded loop from the first intramolecular triplex interacts to form the third strand of an adjacent intramolecular triplex. In a poly(dG)·poly(dC) sequence longer than 40 bp, both a *dG*·dG·dC triplex and a dC^{+}·dG·dC triplex are formed in either half of the Pu·Py sequence.

Nodule DNA was identified by its unusual and characteristic pattern of reactivity with chemical probes (such as chloroacetaldehyde). As the length of a Pu·Py region increased, there was a transition from a single intramolecular triplex structure to the nodule structure, as evident from the changes in the pattern of chemical reactivity. In the poly(dG)·poly(dC) sequence, this structure forms in supercoiled DNA at low pH where the protonated cytosine can form the dC^{+}·dG·dC triplex structure.

When a homopurine·homopyrimidine region becomes very long, there are multiple opportunities for smaller triple helices to form. This is especially pronounced in very simple repetitive sequence elements such as poly(dG)·poly(dC) or poly(dGA)·poly(dTC). The work of Glover and Pulleyblank (1990) and Shimizu *et al.* (1990) demonstrated that considerable structural diversity can exist in long simple polypurine·polypyrimidine mirror repeats. The structural diversity is evident by the appearance of multiple bands indicative of DNA secondary transitions on two-dimensional gels. Some of these possible conformations are shown in Figure 7.13. Long simple repetitive polypurine regions, especially $(GA)_n$ repeats, exist widely in eukaryotic DNA. The biological significance of nodule DNA, if any, is not known.

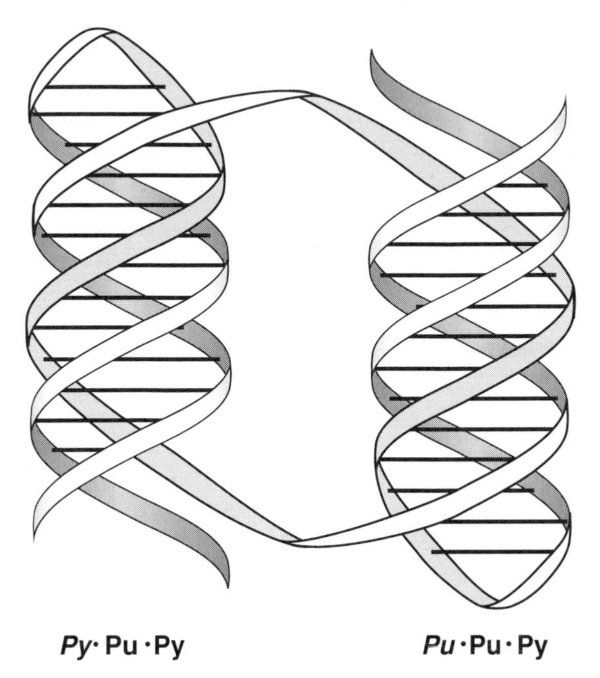

Figure 7.12 Nodule DNA. Within a long polypurine·polypyrimidine tract, the formation of two individual intramolecular triplexes is possible. At one end of the polypurine·polypyrimidine tract, a Pu·Pu·Py structure can form. In the second half of the polypurine·polypyrimidine tract, a Py·Pu·Py structure can form. The Py·Pu·Py structure (left) would contribute its displaced single purine strand to the other half of the Pu·Py region forming the Pu·Pu·Py intramolecular triplex structure (right). Figure modified with permission from Panyutin and Wells (1992).

H. Triplet Repeats, DNA Structure, and Human Genetic Disease

The molecular basis of several human genetic diseases that display unusual genetic properties has been linked with the expansion of simple triplet repeats. There are 10 different ways four bases can be organized into triplet repeats (Table 7.2). In the fragile X syndrome, a leading cause of mental retardation, an average of 29 copies of a CGG triplet repeat is found upstream from a gene associated with the disease, the *FMR*-1 gene. Males can carry the fragile X gene mutation and have no symptoms of the disease. In these individuals, the size of the CGG triplet repeat has expanded to 60–200 copies. Affected children of carrier males can have a massive expansion of the triplet repeat to several thousand base pairs in length. This expansion leads to methylation of CpG dinucleotides of the region encoding the triplet and to concomitant shut off of expression from the *FMR*-1 gene.

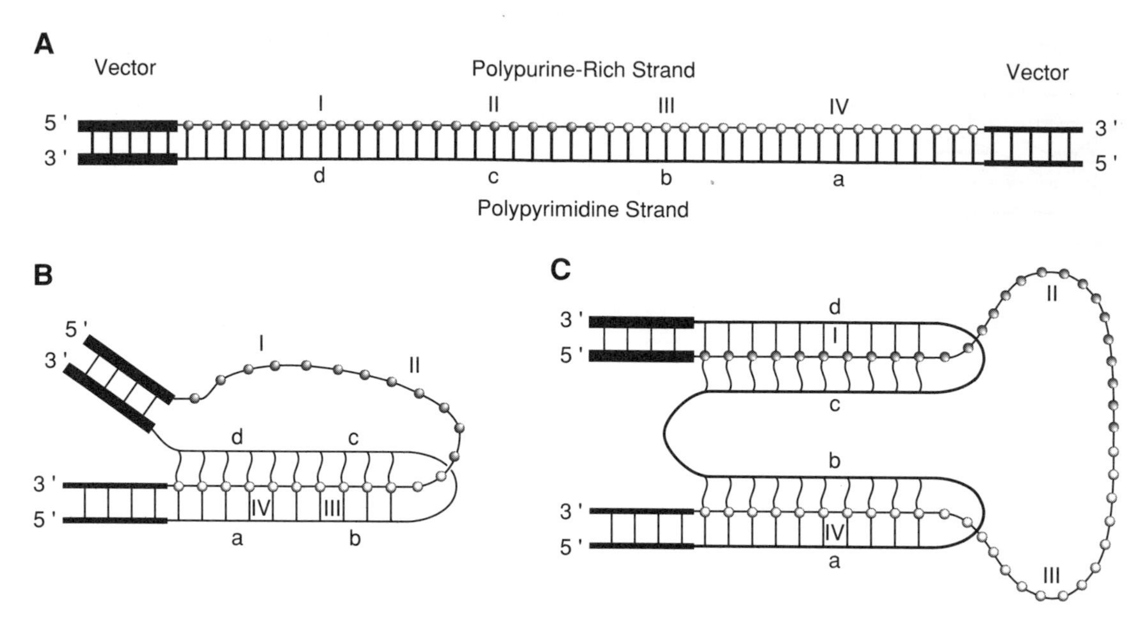

Figure 7.13 Higher order unusual DNA conformations in polypurine-polypyrimidine-rich DNA. In long polypurine-polypyrimidine regions, such as long regions of polyG·polyC or long stretches of $(GA)_n$, a number of different isomeric structures can form. (A) Different regions of the sequence are indicated as I, II, III, and IV in the purine-rich strand and a, b, c, and d in the pyrimidine-rich strand. A canonical intramolecular triplex structure can form, as shown in B, in which the c and d regions of the pyrimidine-rich strand form an intramolecular triplex, pairing with regions IV and III of the purine-rich strand. (C) Regions c and b of the pyrimidine-rich strand form a triplex structure with regions I and IV of the purine-rich strand.

Myotonic dystrophy, Kennedy's disease (X-linked spinal and bulbar muscular atrophy), Huntington's disease, spinocerebellar ataxia type-1, dentatorubral-pallidoluysian atrophy, and colon cancer have been associated with the expansion of CAG triplet repeats. As with the fragile X syndrome, these diseases can show the unusual pattern of inheritance called *anticipation* in which the severity of the condition worsens with successive generations. The severity of these diseases and their age of onset can be linked to the length of the triplet repeat.

1. The Mechanism of Mutation

The extremely interesting question related to triplet expansion is the one of mechanism. There are two mutational stages involved. First is a mutation from a stable length of triplet repeat (normal) into a range that is unstable (carrier), that can involve expansion by a factor of 2. In a second stage, expansion (to the affected stage) of the triplet repeat, in the case of fragile X and myotonic dystrophy, may occur by as much as a factor of 10. How can a length of the triplet repeat expand by more than a factor of 10! The answer

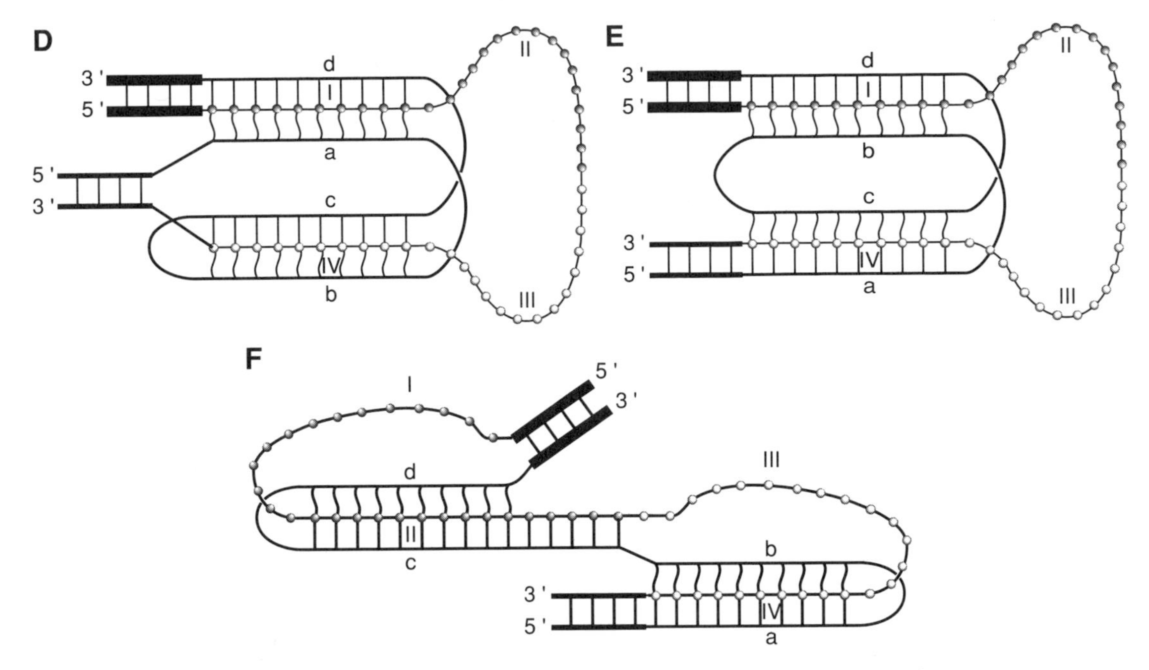

Figure 7.13 *Continued.* (D) Regions a and b form a triplex with regions I and IV of the purine-rich strand, respectively. (E) Regions b and c form an intramolecular triplex with regions I and IV, respectively. (F) Regions d and b form an intramolecular triplex with regions I and IV, respectively. It should be evident that there are myriad possibilities for the organization of multiple triplex structures in polypurine·polypyrimidine-rich regions of DNA. It is not known if any of these structures has biological significance. Figure modified with permission from Shimizu *et al.* (1990).

to this question is not yet known. A few ideas are presented here, but this topic is discussed in greater detail elsewhere by Sinden and Wells (1992; Wells and Sinden, 1993).

One of the first questions to consider is whether hypermutation involves a B-form or non-B-form DNA. The CGG and CAG repeats involved in these diseases have a high purine content on one strand (67%). Long regions of the CGG repeat have mirror repeat symmetry. These are classic requirements for formation of intramolecular triplex structures or the non-B-DNA structure (anisomorphic DNA) described by Wohlrab and Wells (1989). A number of possible structures might be formed by these repeats including the "classical" intramolecular triplex forms, looped intramolecular triplex forms, or looped four-stranded structures.

Genetic recombination does not appear to be a likely model to explain the massive triplet expansion, although recombination could lead to the Stage 1 expansion. DNA sequence analysis has shown that recombination of the DNA flanking the repeat has not occurred. In addition, repeated crossovers would have to occur to expand the repeat region by more than a factor of 10.

Table 7.2
The 10 Triplet Repeats[a]

poly(CGG)·poly(GCC)
poly(AAT)·poly(ATT)
poly(TAG)·poly(CTA)
poly(AGA)·poly(TCT)
poly(ACA)·poly(TGT)
poly(ATG)·poly(CAT)
poly(TCC)·poly(GGA)
poly(CCA)·poly(TGG)
poly(ACG)·poly(CGT)
poly(GCA)·poly(TGC)

[a]Note that poly(AAA)·poly(TTT) and poly(GGG)·poly(CCC) might also be considered triplet repeats although they qualify as lower and higher order repeats as well.

The triplet expansion may occur as a result of replication-based errors. Slipped mispairing can account for duplications and deletions from single bases to large regions of DNA between direct repeats. Repeated slipped misalignment during replication of short simple repeated sequence templates can lead to the production of long polymers, as first demonstrated by Kornberg *et al.* (1964). The rate of this "reiterative synthesis" slippage is dependent on repeat length, complexity, and stability, but independent of the length of the progeny strand (Schlotterer and Tautz, 1992). It is also possible that strand displacement synthesis (Lechner *et al.,* 1983) could lead to the expansion of triplet repeats. It is known that many DNA polymerases have great difficulty reading through the long CGG repeats associated with fragile X and the CAG repeats of myotonic dystrophy. A single slippage could not account for the massive amplification observed. Repeated or multiple slippage events would have to occur.

What could promote multiple slippage events? A strong block to replication might provide the opportunity for multiple slippage events. A strong block might be provided by an alternative DNA secondary structure that is stabilized by DNA supercoiling, by proteins, or by both. It is known, for example, that a potential triplex region provides a block to replication *in vitro* and *in vivo,* as discussed in Chapter 6. DNA polymerase may dissociate from the 3′ end, allowing breathing and potential slipped misalignment stabilized by poly(CGG)·poly(CGG) base pairs with a central G·G mismatch in a hairpin stem or in a structure stabilized by poly(CGG)·poly(GGC) base pairs that contain two G·G pairs and one C·C mispair. The latter structure may be similar to that found in some telomere fold-back structures, as discussed earlier.

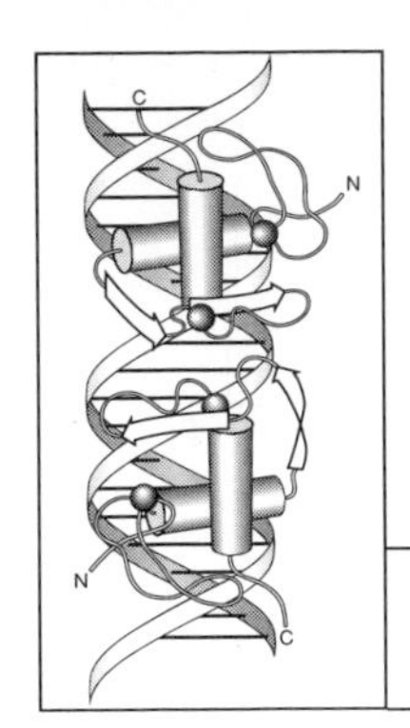

CHAPTER 8

DNA–Protein Interactions

A. Introduction

Consider the transition of interactions that must occur between DNA and proteins during the development from a human sperm and egg, which contain quiescently packaged DNA, to an adult with differential gene activity in a multitude of different tissues. Certain genes must be expressed at a precise time during development in a particular type of cell. Other genes must be expressed continually, or perhaps at one particular time during the cell cycle. The expression of the genetic information is controlled predominantly through the interaction of regulatory proteins with DNA. Regulatory proteins must have the ability to bind a short unique base sequence out of the billions of bases in the human genome. Other proteins involved in the structural organization of DNA, such as histones, must bind DNA in a sequence-independent fashion.

This chapter will enumerate some of the general principles involved in DNA–protein interactions, and then will discuss how these principles are utilized in several DNA–protein complexes. Although the discussion will generally be from the point of view of proteins interacting with a static uniform DNA template, keep in mind that the structure of the DNA itself can have profound influences on its interaction with regulatory proteins or proteins involved in DNA packaging, repair, recombination, or replication.

A·T

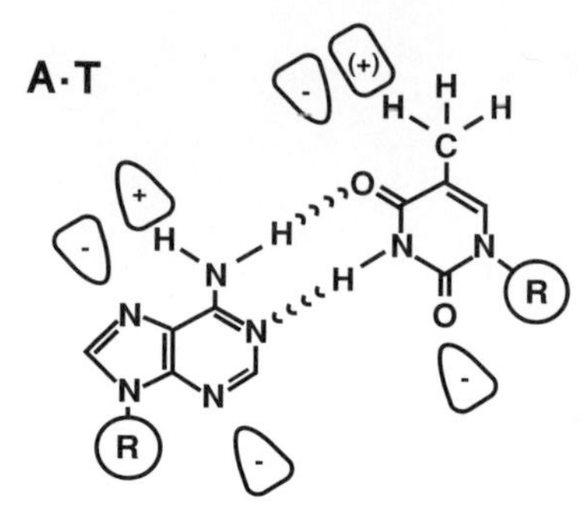

A T

G·C

G C

T·A

T A

C·G

C G

B. General Considerations on Protein Binding

How does a protein recognize DNA? How do some proteins, for example, histones and the bacterial HU protein, recognize DNA in a sequence-independent fashion? On the other hand, what contacts are responsible for a specific interaction, for example, between a repressor protein and its operator sequence on the DNA? Specific interactions are usually defined through hydrogen bonding interactions and salt bridges between functional groups of amino acid side chains or the peptide bonds and groups on the bases in the major or minor grooves. The phosphate backbone would, in general, provide a sequence-independent protein binding site (see Schleif, 1988; Pabo and Sauer, 1992, for review).

1. Bases in the Major and Minor Grooves Are Available for Specific Protein Recognition

Specificity in DNA–protein interactions comes from protein recognition of the linear order of base pairs through hydrogen bond and salt bridge contacts through the major and minor grooves. (A salt bridge is an interaction between oppositely charged groups. The interaction involves electrostatic interactions as well as some degree of hydrogen bonding.) Certain functional groups on the edges of the base pairs are available in the major and minor grooves. These groups present *hydrogen bonding donor* and *acceptor* sites that can participate in hydrogen bond formation with sites on one or more amino acids in the protein.

There are one hydrogen bond donor and two acceptor groups on the major groove surface of all four dinucleotide pairs (A·T, T·A, G·C, C·G; Figure 8.1). In addition, there is a methyl group at the C5 position of thymine that can participate in van der Waals interactions. Within the minor groove, there are two hydrogen bond acceptor sites in all four dinucleotide pairs; in the C·G and G·C dinucleotide pairs, the N2 position of G provides a hydrogen bond donor at the center of the minor groove. As shown in Table 8.1, the

Figure 8.1 Hydrogen bonding sites in the major and minor grooves. A·T, T·A, G·C, and C·G base pairs are shown with potential hydrogen bonding sites available to proteins in both the major and the minor grooves. The hydrogen bond donor positions are indicated by a triangle enclosing a "+" pointing away from the donor hydrogen. The hydrogen bond acceptors are indicated by a triangle enclosing a "−" pointing toward the electron-rich center. Non-hydrogen-bonding positions that can participate in electrostatic interactions are shown by rectangles. (Right) Linear representations of the order of potential hydrogen binding sites in the major groove (top) and the minor groove (bottom) for each base pair. Each of the four base pairs possesses a unique pattern of hydrogen binding potential. A defined sequence of DNA will dictate a unique three-dimensional protein surface that will be required for sequence-specific interaction.

Table 8.1
Potential Hydrogen Bonding Interactions at Ring Positions in the Major and Minor Groove[a]

		Purine				Pyrimidine					Pyrimidine			Purine			
Base pair	Groove	7	3	6	2	4	2	5	Base pair	Groove	5	2	4	2	6	3	7
A·T	Major	−		+		−		CH_3	T·A	Major	CH_3		−		+		−
	Minor		−				−			Minor		−				−	
G·C	Major	−		−		+			C·G	Major			+		−		−
	Minor		−		+		−			Minor		−		+		−	

[a]+, Hydrogen bond donor; −, hydrogen bond acceptor; CH_3, methyl group, capable of potential nonhydrogen-bonding electrostatic interaction.

pattern of hydrogen bond donors and acceptors is unique for each of the four dinucleotide pairs. Figure 8.2 lists the amino acids and shows potential hydrogen bond donors and acceptors on the amino acid side chains.

The unique arrangement of hydrogen bond donor and acceptor sites for each dinucleotide within the major or minor groove must provide the specificity utilized by proteins to discriminate specific regions of DNA sequence.[1] A few amino acids seem particularly well suited to form hydrogen bonds with dinucleotides. As recognized by Seeman *et al.* (1976a), glutamine (or asparagine) can form two hydrogen bonds with adenine in the major groove of an A·T or T·A base pair. Similarly, the guanidinium nitrogens of arginine can form hydrogen bonds with the N7 and O6 positions of guanine. In the minor groove, asparagine and glutamine can form two hydrogen bonds with the N3 and N2 positions of guanine. These amino acid–base pair interactions are shown in Figure 8.3. To date, a large number of specific interactions between these amino acids and base pairs has been identified. Many of these will be mentioned in the discussion of particular examples of protein–DNA interactions.

Water molecules play a very important role in specific interactions between DNA and proteins. DNA is always hydrated with a shell of well-ordered water molecules. These must be displaced to allow specific interaction with the functional groups of amino acids. In some cases, however, individual water molecules bridge a hydrogen bonding interaction between the DNA and the protein. In fact, in the first reported crystal structure of the *trp* repressor and a *trp* operator sequence *no direct contacts* were observed between the protein and the DNA (Otwinowski *et al.*, 1988). All hydrogen bond contacts were mediated through water molecules. (Figure 8.8 shows the participation of water molecules in the interaction between a glutamine and thymine in the bacteriophage λ repressor–operator.)

Proteins can also recognize particular regions of DNA through an *indirect readout* mechanism in which contacts are not made with the bases in the major or minor grooves, but with phosphate groups and sugar residues. There is a nonuniformity in the structure of the Watson–Crick double helix, since the primary DNA sequence defines a precise shape for the double helix. Primary sequence dictates local twist angles, the specific tilt and roll of bases, and bends in DNA, as well as the widths of the major and minor grooves. These structural parameters will precisely position in space the hydrogen

[1]This unique pattern of hydrogen bonding potential may also be utilized by a third DNA strand in the formation of a triplex molecule formed during RecA-dependent genetic recombination (see Figure 6.17).

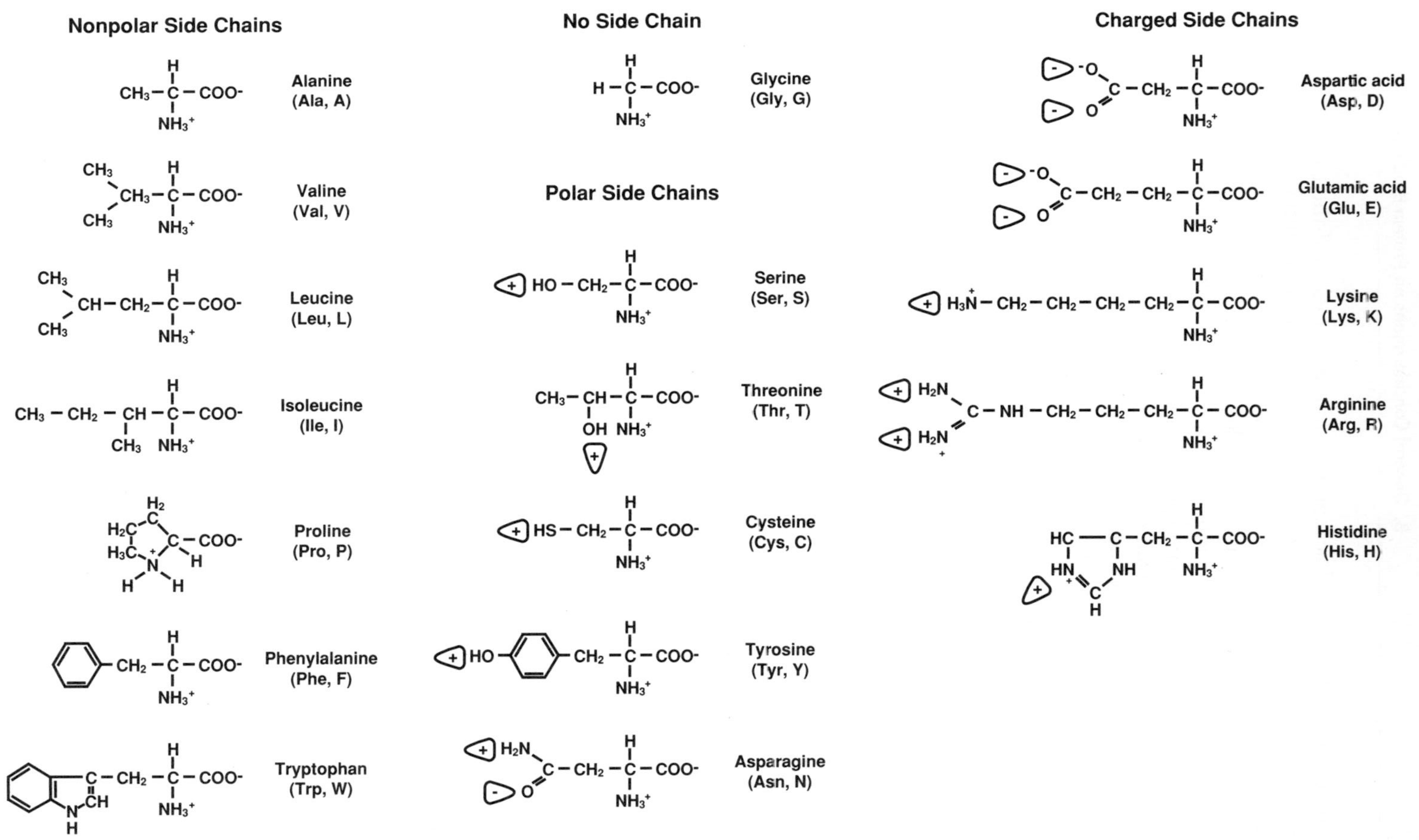

Nonpolar Side Chains

Alanine (Ala, A)

Valine (Val, V)

Leucine (Leu, L)

Isoleucine (Ile, I)

Proline (Pro, P)

Phenylalanine (Phe, F)

Tryptophan (Trp, W)

No Side Chain

Glycine (Gly, G)

Polar Side Chains

Serine (Ser, S)

Threonine (Thr, T)

Cysteine (Cys, C)

Tyrosine (Tyr, Y)

Asparagine (Asn, N)

Charged Side Chains

Aspartic acid (Asp, D)

Glutamic acid (Glu, E)

Lysine (Lys, K)

Arginine (Arg, R)

Histidine (His, H)

Figure 8.2 Structures of the amino acids are shown in three groups: those containing nonpolar side chains, those containing polar side chains, and those containing charged side chains. The name of the amino acid, its three letter code, and its single letter code are shown next to the chemical structures. The triangle designates a hydrogen bond donor or acceptor, as defined in the legend to Figure 8.1.

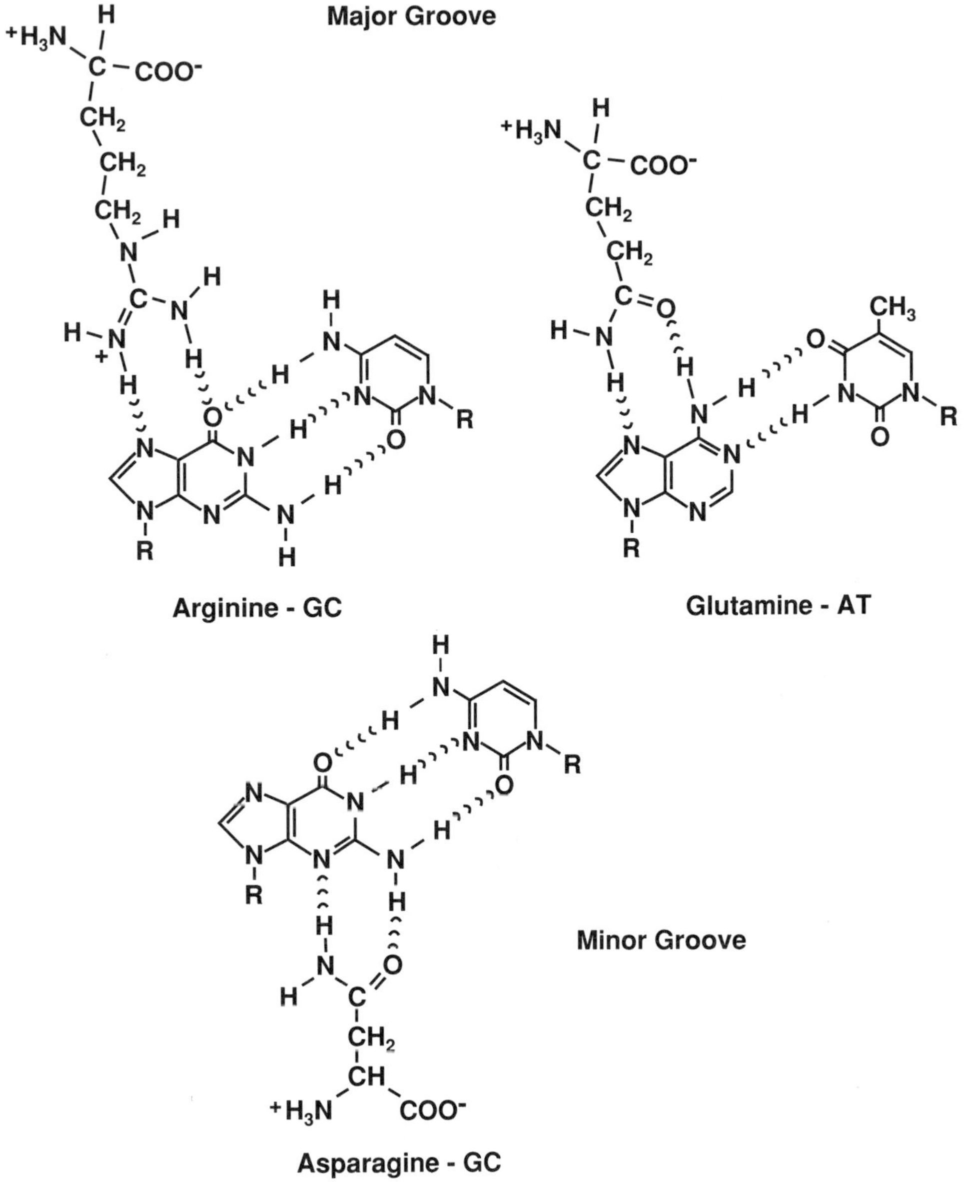

Major Groove

Arginine - GC

Glutamine - AT

Minor Groove

Asparagine - GC

Figure 8.3 Protein–DNA hydrogen bonding. Seeman et al. (1976a) described specific contacts that could be made by different amino acids to a C·G or an A·T base pair in the major groove. The structure of a C·G base pair bound to arginine is shown in which the guanidinium group of arginine binds to the N7 and 06 positions of guanine. Glutamine can hydrogen bond specifically to the adenine of an A·T base pair. The terminal amino group of glutamine binds to the N7 position of adenine whereas the carbonyl oxygen group binds to the N6 position of adenine. A specific amino acid–base pair hydrogen bond can be made in a minor groove between asparagine and a G·C base pair. Asparagine binds with the terminal amino group to the N3 position of guanine whereas the carbonyl oxygen hydrogen bonds to the N2 position of guanine. The hydrogen bonding potentials of glutamine and asparagine are identical and either could pair with G or A, through the minor or major groove, respectively. Examples of these hydrogen bond interactions have been found subsequently on resolution of the X-ray crystal structure of various protein–DNA complexes.

bonding donor and acceptor sites in the bases and the phosphate backbone. Theoretically, there should be sufficient specificity in the position of functional groups in the phosphate backbone to direct binding of a protein to a unique sequence of DNA. Most DNA binding proteins are designed to recognize a particular shape or flexibility of the double helix in addition to a direct readout of individual bases in the recognition site. In many cases, changes in the nucleotides within the binding site that do not make direct contact with the protein can strongly influence the binding affinity of a protein.

2. The Sugar–Phosphate Backbone Is Available for Nonspecific Protein Recognition

The phosphate backbone has a relatively uniform shape and negative charge available for nonspecific recognition by DNA binding proteins. Several examples are known in which protein interaction is sequence independent, utilizing interaction with the phosphate backbone. DNA polymerases constitute one class of proteins that must interact with DNA in a sequence-independent fashion. Polymerases must traverse the entire chromosome, contacting every individual base pair. In the X-ray crystallographic structure of *Escherichia coli* DNA polymerase I–DNA cocrystal, contacts between the enzyme and DNA are all made with the phosphate groups of DNA (Beese *et al.,* 1993). The interaction of the bacterial HU protein with DNA utilizes a different type of nonspecific interaction. In this case, an extended β sheet lies in the minor groove of DNA. It is stabilized by hydrogen bonding between the negatively charged phosphate backbone and positively charged groups on four arginines within the extended β sheet (Tanaka *et al.,* 1984; White *et al.,* 1989). DNase I provides a third example of nonspecific DNA–protein interactions. This nuclease cuts DNA in a sequence-independent fashion (although it does have preferred cleavage sites). In a cocrystal between DNA and DNase I, the contacts include one stacking interaction between a tyrosine ring and a pyrimidine and three hydrogen bonds between arginine and oxygen atoms of pyrimidine (Suck *et al.,* 1988). However, the majority of the hydrogen bonding contacts are made between 10 other amino acids and the phosphate backbone.

3. Dimeric Binding Sites in DNA

Many binding sites for regulatory proteins are dimeric, and are recognized by multimeric proteins, usually homotypic dimers or tetramers. A protein must select its small binding site, in which the contacts that provide the specificity are buried within the major groove, from a large amount of nonspecific sequence (the rest of the chromosome). The affinity of a protein for its DNA binding site is a function of the number and strength of electrostatic and hydrophobic interactions between the protein and the DNA. A larger binding site can provide a stronger interaction between the DNA and a protein than a smaller site. Placing two copies of a single site next to each other for recognition by a dimeric protein doubles the number of contacts relative to a protein monomer and increases the strength of the DNA–protein interaction.

Dimeric binding sites can also be used to vary the binding affinity between a protein and multiple binding sites in DNA. Groups of genes, controlled by the same regulatory protein but located at different positions on the chromosome are called regulons. One of the best studied examples is the SOS system in *E. coli*. At least 15 SOS genes are repressed under normal conditions by the LexA repressor and are induced by DNA damage. The timing and level of expression of the individual genes in response to SOS induction is different for each member of the regulon. Differential expression is possible because of different binding affinities of the LexA repressor for individual inverted repeat operator sites on the DNA. Sequence deviation from a preferred consensus DNA binding site creates operators with reduced binding affinity. The dimeric operator, in which the sequence of the two individual recognition sites can vary, allows variation of the affinity of the repressor for DNA. These changes in the affinity of the operator with the protein allow the differences in the timing and level of expression of the gene.

Variation in the affinity of a repressor for a DNA binding site can also be accomplished by utilizing heterotypic dimers. The *fos* and *jun* oncogene proteins are transcription factors that form a heterodimer. Cells frequently contain more than one of the *jun*- or *fos*-type transcription factors, each with a different binding affinity for the transcription factor binding site. The formation of different *jus–fos* heterotypic dimers creates a family of transcription factors with differential inherent binding affinities for a single binding site. The regulation of gene expression by these transcription factors could be modulated by differential expression of a type of *jun* or *fos* monomer, since this would affect the distribution of homo- or heterotypic dimers in the cell.

4. The Organization of Dimeric Binding Sites in DNA

There are a limited number of ways to organize dimeric binding sites including inverted repeats, mirror repeats, and direct repeats (see Figure 4.1). The *lac* operator and λ cI repressor binding site (discussed later) are examples of inverted repeat binding sites and represent the most common organization.

Two subunits of a dimeric protein would interact with an inverted repeat, as shown in Figure 8.4.

It is difficult to imagine how a dimeric protein could recognize a binding site that has mirror repeat symmetry because this sequence organization lacks symmetry within the B-DNA double helix. The relationship of functional groups in a protein domain that interact with sites on the DNA would not be positioned to make the same contacts within a major groove of a mirror repeat sequence. Moreover, the twist angles and stacking interactions be-

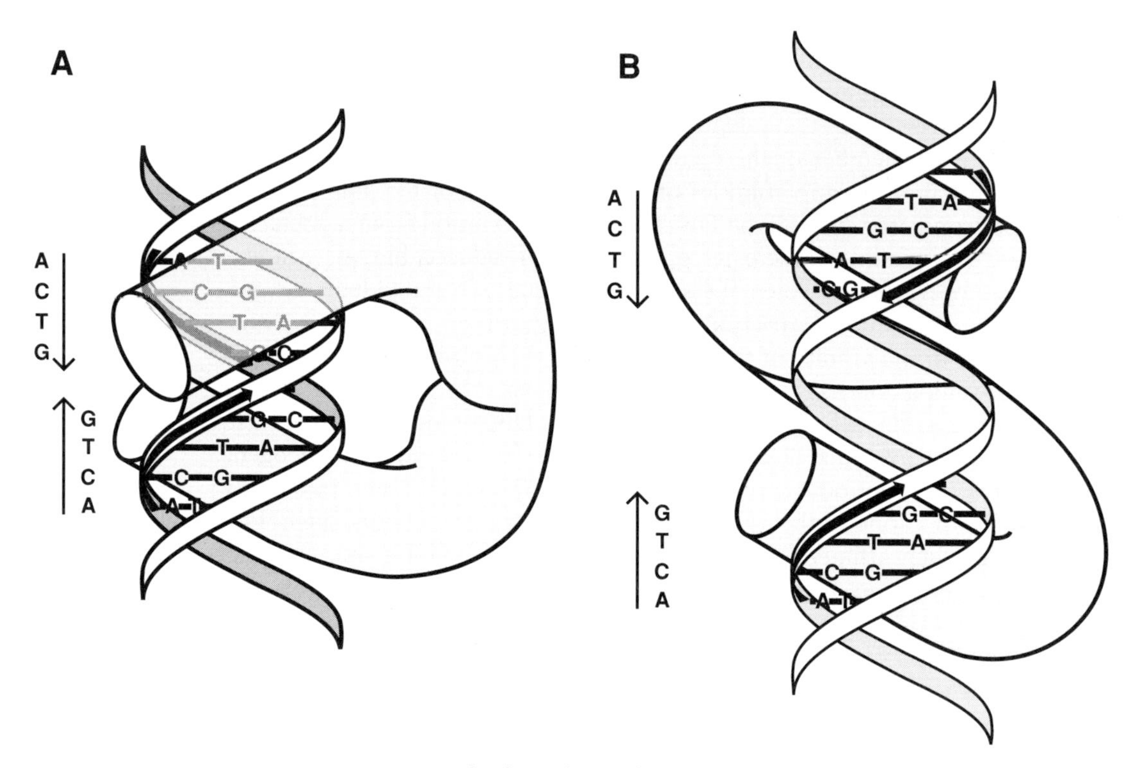

Figure 8.4 DNA–protein binding schemes for inverted repeats. (A) Inverted repeat, 5-bp spacing. The interaction of two α helices with an inverted repeat DNA binding site that occurs within one helical turn of DNA. The interaction of two α helices from identical subunits of a dimeric DNA binding protein with DNA requires the organization shown. One helix approaches and binds in the major groove from one side of the DNA helix. The second helix requires interaction in the major groove five bases away, on the opposite side of the DNA double helix. In this inverted repeat orientation, the two dimeric subunits would "scissor" or wrap around the DNA from both sides. (B) Inverted repeat, 10-bp spacing. If the inverted repeats were 10 bp apart, the two protein subunits approach the DNA from the same face.

tween dinucleotides within the two halves of a mirror repeat sequence may differ considerably and result in differences in the shape of the two mirror repeat halves. There are, however, several examples of mirror repeat elements in regulatory regions of genes, although it is not yet known if specific regulatory binding proteins recognize any of these regions.

Protein binding sites can also be organized in a direct repeat fashion. Binding sites for Zn-finger-containing proteins are often arranged in a tandem direct repeat orientation. For example the TFIIIA transcription factor (described later) has nine Zn fingers arranged in tandem that interact with repeated runs of guanines (Miller *et al.*, 1985).

C. Specific DNA–Protein Interactions

1. Interaction of the Helix–Turn–Helix Motif with DNA: General Considerations

A 20-amino-acid motif called a helix–turn–helix domain has been identified in a number of different DNA binding proteins. The structure of this protein domain is an 8-amino-acid α helix followed by a right "turn" consisting of 3 amino acids followed by another α helix of 9 amino acids. There are three positions in the helix–turn–helix motif that are highly conserved. An alanine is usually found at position 5 within the first α helix; a glycine is found at position 9, the first position of the turn; and either valine or isoleucine is usually found at position 15 within the second α helix. These critical positions are believed to be responsible for maintaining the structure of helix–turn–helix rather than specifying contacts with the DNA. With the exception of those conserved positions, there is a great deal of heterogeneity in the amino acid composition of other positions within this protein motif. In spite of differences in amino acid composition, the basic three-dimensional protein structure of this domain remains largely unchanged (see Pabo and Sauer, 1984,1992; Harrison and Aggarwal, 1990, for review). Since an α helix has a repeat length of 3.6 amino acid residues per turn (Figure 8.5), certain positions of a helix–turn–helix domain will face toward the body of the protein (a hydrophobic environment) whereas certain residues will face the solvent or the DNA (a hydrophilic environment). A similarity in the types of amino acids at these positions is observed within the family of helix–turn–helix proteins.

The helix–turn–helix motif has evolved to fit into the major groove of DNA. As discussed in detail by Pabo and Sauer (1984), the dimensions of the

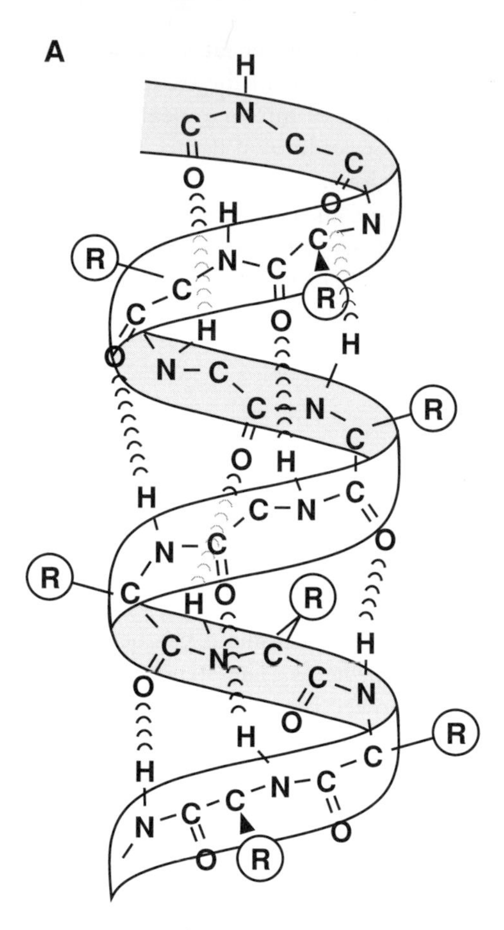

Figure 8.5 Fundamental protein structural motifs. (A) The α helix, a common structural motif of proteins, consists of a right-handed helix with a repeat length of 3.6 amino acid residues per helical turn. The α helix is stabilized by hydrogen bonds between an amide hydrogen of one amino acid and a carbonyl oxygen four amino acids away. The various side chains of the amino acids (designated as R) protrude from the surface of the cylindrical helix core, allowing interaction of these functional groups with other protein structural motifs or with hydrogen bonding sites in the major or minor grooves of the DNA. (B) The β sheet is another common structural motif of proteins in which two chains of amino acids are organized in a linear, side-by-side fashion. The two chains (which can be from different proteins or individual regions within a single polypeptide) are held together by hydrogen bonds between an amino hydrogen and a carbonyl oxygen. The β sheet is a relatively flat, but pleated structure. Two amino acid chains can be organized in either a parallel or an antiparallel fashion giving rise to two slightly different structures of a β sheet. (Left) A top view of an antiparallel β sheet. (Right) A side view, to show the planar, pleated structure of the β sheet. Note that the side chains of the amino acids are organized up and down with respect to the plane of the β sheet. (C) The interaction of an α helix and a β sheet with DNA. α Helices usually interact with DNA in the major groove and organize themselves with the long axis of the helix along the long

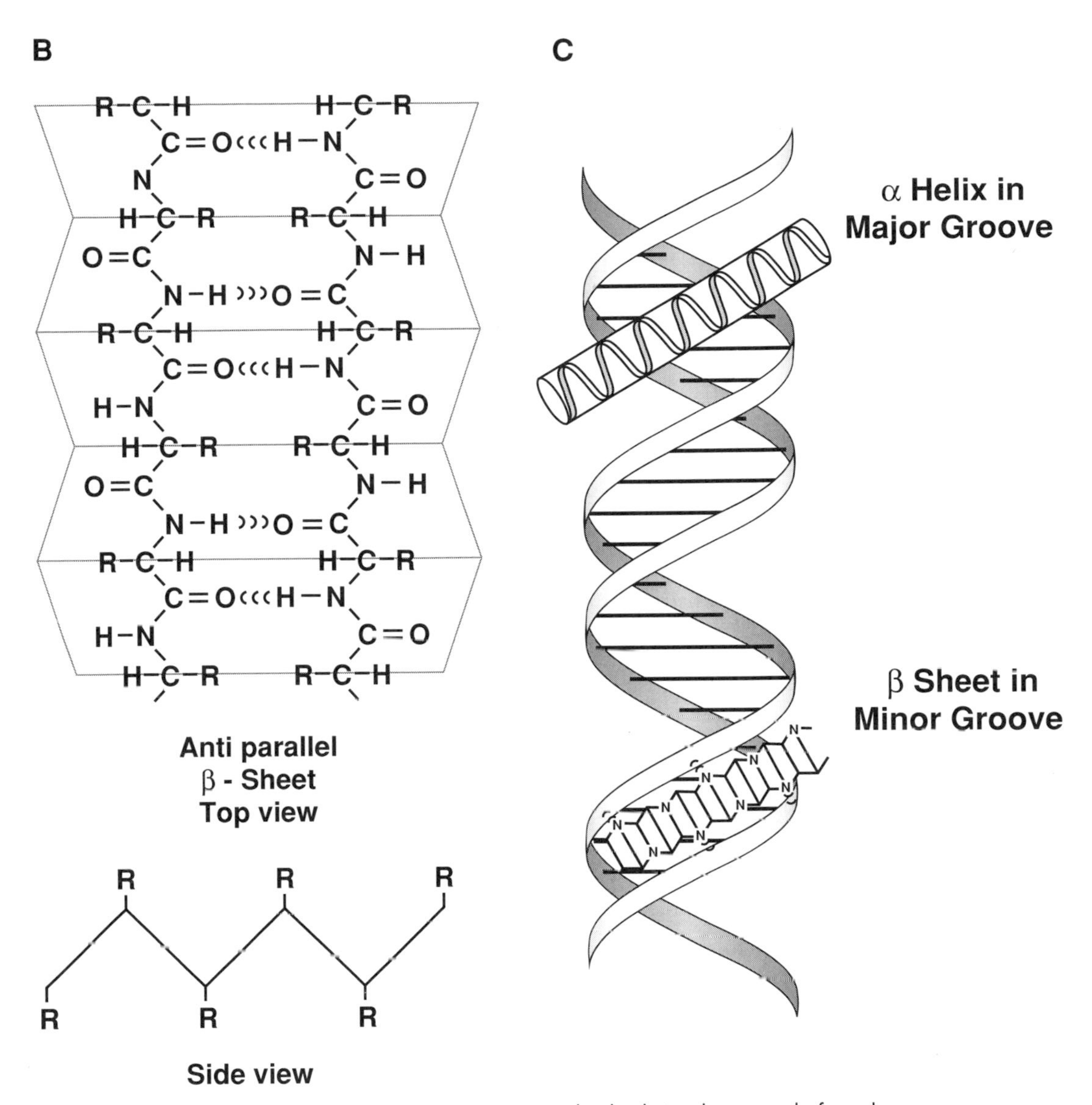

Figure 8.5 Continued. axis of the major groove. Functional side chains that protrude from the surface of the helix interact with phosphate groups from the backbone as well as with potential hydrogen bonding sites on the surfaces of the base pairs that are accessible from the major groove. A β sheet is believed to interact with DNA by lying along the long axis of the minor groove. A β sheet can bind nonspecifically to the minor groove of DNA, stabilized by hydrogen bond formation between the nitrogens of the amino group and the phosphate groups on the DNA backbone.

helix–turn–helix domain match those of the major groove. The diameter of a typical α helix is about 12 Å, which matches the 12 Å wide and 6–8 Å deep major groove of B-form DNA. If the α helix lies parallel to the direction of the major groove, the straight α helix can contact 4–6 bp before the bases curve out of the plane of the amino acids in the α helix. The α helix, however, can be positioned in the major groove in several different ways. For example, the α helix can be parallel to the major groove with either the NH_2 or COOH terminus in the groove. The NH_2 terminus of an α helix contains a slight positive charge that can interact with the negative charge of the phosphate backbone when positioned in the major groove. The short interaction with no more than 4–6 bp of DNA specifies the binding of a particular sequence of amino acids to a unique sequence of DNA.

α Helix and β Sheet

There are several structural features of polypeptides that define the overall shape of proteins. Figure 8.5 describes two major fundamental structural motifs of protein structure: the α helix and the β sheet. In addition to these well-defined structures, regions of a polypeptide can exist in an unordered configuration.

Amphipathic α Helices Can Sidle up to DNA

Certain proteins have been shown to adopt an α helix conformation where there is a marked polarity of charge distribution. If positive charges fall on one face of the helix and negative charges (or hydrophobic residues) on the opposite face, the α helix is said to be amphipathic. A potential role for basic residues on one side of a helix might be to bind the negatively charged phosphate groups on the DNA backbone. Many α helices are positioned within the major groove. It is important to position polar groups on one side of an α helix to interact with DNA. A hydrophobic surface of an α helix could interact with other hydrophobic environments within the protein.

2. Interaction of the Helix–Turn–Helix Motif with DNA: X-Ray Crystallographic Analysis

X-Ray crystallographic studies have revealed details of the specific interactions between DNA and helix–turn–helix DNA binding proteins. Two of the classic examples of DNA–protein interactions—the bacteriophage repressor–DNA and phage 434 repressor–DNA interactions—are described here.

a. The Lambda Repressor

The lambda repressor is a key regulatory protein responsible for controlling entry into the lytic or lysogenic bacteriophage λ life cycle. It controls expression of early genes that, when induced, signal the production of proteins that lead to λ replication, packaging into phage particles, and lysis of the bacterial host. The λ repressor binds the quasi-palindromic 17-bp operator sequence shown in Figure 8.6 which is one of six repressor binding sites in λ DNA. Each operator has a slightly different sequence in which the two quasi-palindromic halves of the operator flank a central C·G base pair. The slight differences in sequence of these operators are responsible for generating diversity in the affinity of the λ repressor binding to the sequences. This allows differential regulation of gene expression from genes controlled by the λ repressor. The left half of the operator is numbered 1–9, the right half 8′–1′. The structure of this short region of DNA is B-DNA like with an average helix twist of 34.4° (10.5 bp per turn). Individual twist angles, however, range from 21.5° to 47°. This region is also straight (it does not contain any stable bends) (Jordan and Pabo, 1988).

The λ repressor is a 236-amino-acid protein that exists as a dimer in solution. It has two functional domains. The NH_2-terminal domain consists of amino acids 1–92, which contain the DNA binding region, and the COOH-terminal domain consists of residues 93–236, which contain the protein dimerization sites. The NH_2-terminal domain is composed of five α helical regions: helix 1 (resides 9–23), helix 2 (residues 33–39), helix 3 (residues 44–52), helix 4 (residues 61–69), and helix 5 (residues 79–92). Helices 2 and 3 and the 5-amino-acid turn between them make up the 20-residue

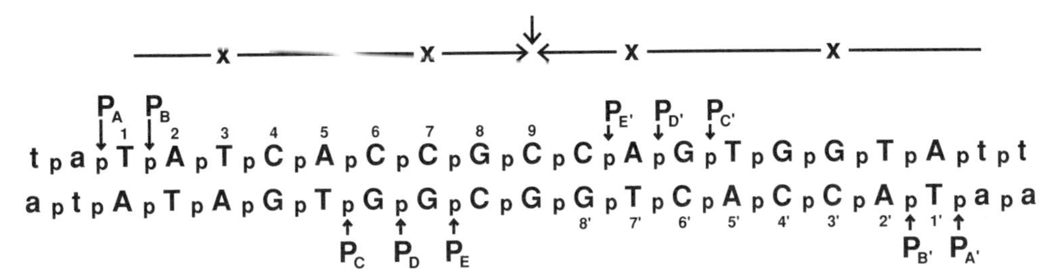

Figure 8.6 The 17-bp bacteriophage λ operator sequence is shown in capital letters. The inverted repeats are denoted by the arrows above the sequence. "x" represents a nonsymmetrical base within the inverted repeat. A few bases flanking the operator are shown in lowercase letters. The nucleotide positions are numbered 1–9 on the left side of the operator and 8′ through 1′ on the right side of the operator. The positions of several critical phosphate groups P_A, P_B, P_C, P_D, and P_E are indicated.

helix–turn–helix motif. The conserved critical residues required of the helix–turn–helix motif are Ala 37, Gly 41, and Val 47.

The interaction of the helix–turn–helix protein domain with the λ operator sequence is shown in Figure 8.7. Several features are worth mentioning about the interactions of the helix–turn–helix domain in the major groove. First, the long axis of α helix 3 lies parallel to the direction of the major groove. Second, there are a number of specific hydrogen bonding interactions between amino acids in the helix–turn–helix domain and bases in the λ operator. These contacts are listed in Table 8.2 and are shown in Figure 8.8. Specific hydrogen bond contacts with DNA are formed through glutamine, aspargine, serine, and lysine residues. These are some of the charged or polar amino acids that were predicted by Seeman *et al.* (1976a) to be capable of forming specific hydrogen bonds with individual base pairs (see Figure 8.3). The binding of Lys 4 in the so-called "N-terminal arm" is unusual because the arm reaches around the DNA and into the major groove, forming a salt bridge with the guanine of base pair 6. Third, there are a number of van der Waals contacts between the protein and the DNA. These specific contacts are also listed in Table 8.2. Typically, these contacts involve the interaction of the methyl group at the C5 position of thymine with other hydrophobic regions or "pockets" in the protein. These nonpolar groups interact with one another and exclude water molecules from the hydrophobic area.

Interactions between DNA and protein can occur through water molecules in addition to direct hydrogen bond and salt bridge contacts. In the λ repressor interaction, there are a number of well-ordered water molecules that are involved in interactions with bases within the major groove as well as interactions with the phosphate backbone (Table 8.2; Beamer and Pabo, 1992). For example, two water molecules participate in bridging an interaction between the carbonyl of Gln 44 and the O4 group of thymine in base pair 2 (Figure 8.8). The first dramatic demonstration of the importance of water molecules in the interaction of DNA and protein came from analysis of a *trp* repressor–operator cocrystal. Otwinowski *et al.* (1988) showed that no specific hydrogen bonds occurred in the *trp* repressor–operator complex. The few contacts with bases occurred through water molecules. (The majority of direct contacts in the *trp* repressor interaction are to the phosphate groups, an example of indirect readout.)

In addition to specific interactions within the major groove, there are a number of interactions between the protein and the phosphate backbone in the λ operator (Figure 8.9). In the first half of the inverted repeat, the phosphate groups are designated P_A, P_B, P_C, P_D, and P_E; in the other half they are called P_A', P_B', P_C', P_D', and P_E'. P_A and P_B precede bases 1 (T) and 2 (A) and P_C, P_D, and P_E precede bases 6 (G), 7(G), and 8(C) on the opposite strand (Figure 8.6). The 10.5-bp repeat length of the double helix will place the two

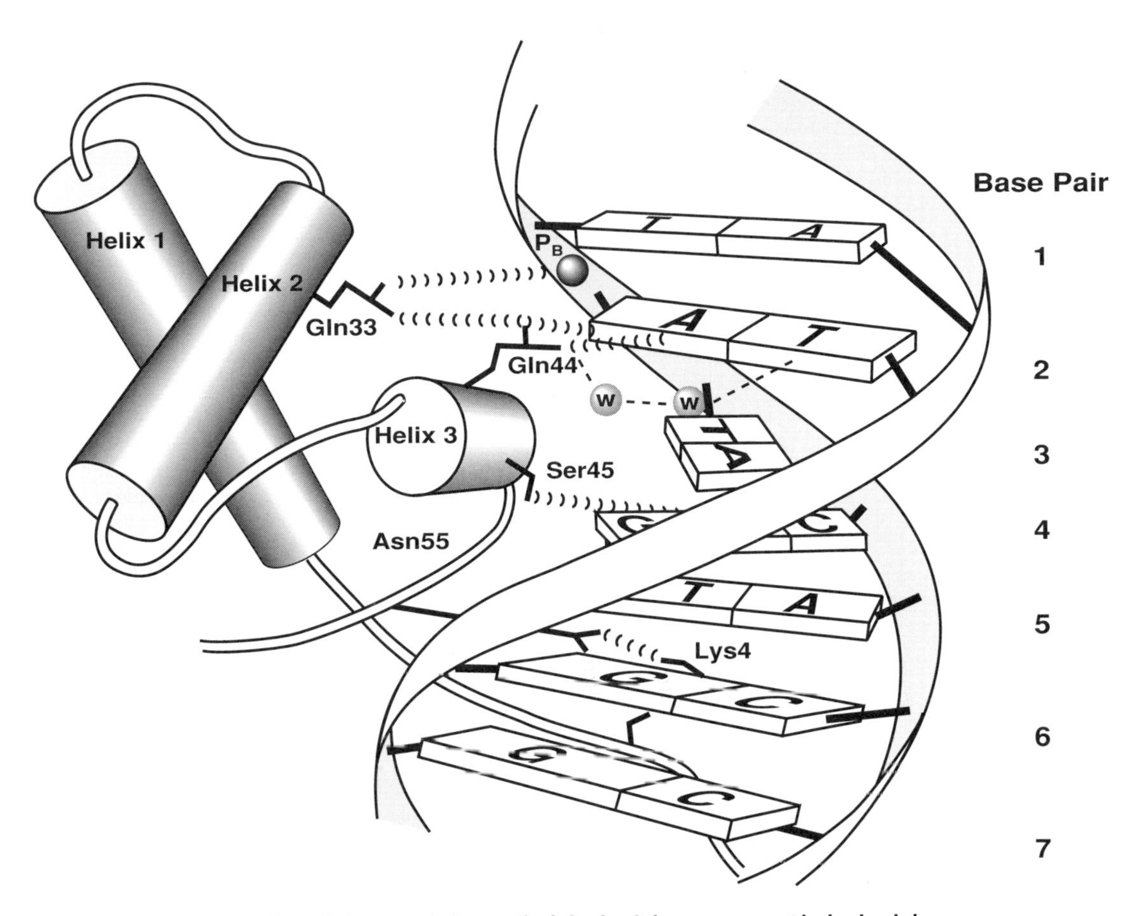

Figure 8.7 Interaction of the helix–turn–helix motif of the lambda repressor with the lambda operator. The helix–turn–helix motif of the λ repressor is composed of helix 2 and helix 3 with a 5-amino-acid turn between them. This motif interacts in the major groove of half of the lambda operator with α helix 3 lying parallel to the groove as it cuts across the long axis of the DNA double helix. The hydrogen bonding interactions between amino acid residues and α helix 3, α helix 2, and the amino acid turn between helices 3 and 4 (helix 4 is not shown) are shown. A detailed description of these interactions is presented in Table 8.2. Adapted with permission from Jordan Pabo (1988), Science **242.** Copyright 1988 by the AAAS.

groups of phosphates on the two strands within the same major groove. These phosphate groups are in contact with α helix 3 as it lies in the major groove, as shown in Figure 8.9. There are many hydrogen bonding contacts to the phosphate backbone that are important in making a tight complex of the protein and the DNA. In general, these contacts will not be sequence spe-

Table 8.2
Consensus Side Contacts between the λ Operator and λ Repressor

DNA position	Amino acid	Contact[a]
Major groove hydrogen bond contacts (shown in Figures 8.7 and 8.8)		
#2 A·T	Gln 33	Two hydrogen "bidentate" bonds are made between NH_2 of Gln 33 and N7 of A and the C=O of Gln 33 and N6 of A; the carbonyl of Gln 33 also contacts the O4 of T through a bridge of two water molecules
#2 A·T	Gln 44 (helix 3)	One hydrogen bond is made between NH_2 and the PO_4 group of A (P_A), and one hydrogen bond is made between the C=O and NH_2 of Gln 44
#4 C·G, #5 A·T	Ser 45	Hydrogen bonds are made between serine OH and N7 and O6 of G; the peptide carbonyl group contacts N4 of C 4 and O4 of T 5 through a single water molecule
#6 C·G, #7 C·G	Asn 55, Lys 4	One hydrogen bond is made between NH_2 of Asn 55 and N7 of G 6 (C·G) or G 6 and one hydrogen bond is made between C=O of Asn 55 and NH_3^+ of Lys 4; Lys 4 also makes hydrogen bonds between its NH_3^+ and the O6 of G 6 and the O6 of G 7
Major groove van der Waals contacts		
#3 T·A	Ala 49, Ile 54, Gly 48, Ser 45	The methyl on thymine (C5) contains a hydrophobic pocket made of the α and β carbon of Ser 45, the α carbons of Gly 48 and Ala 49, and the δ carbon of Ile 54.
#5 T·A	Gly 46, Ser 45	Methyl group on thymine (C5) contacts Gly 46 and the α and β carbons of Ser 45
Hydrogen bonds to the phosphate backbone		
P_A	Gln 33, Tyr 22, Gln 44	Inside: The peptide NH of Gln 33 forms a hydrogen bond with a P_A oxygen Outside: The OH of Tyr 22 forms a hydrogen bond with the oxygen on P_A farthest from the major groove; the peptide NH of Gln 44 makes contact with the P_A oxygen through a water molecule (not shown)
P_B	Asn 52, Lys 19, Gln 33	Inside: The NH_2 side chain of Asn 52 at the end of helix 3 forms a hydrogen bond with P_B Outside: The NH_3^+ of Lys 19 forms a hydrogen bond with P_B and the carbonyl side chain of Asn 52; the side chain NH_2 of Gln 33 also forms a hydrogen bond with P_B
P_C	Gly 43, Ser 45	Inside: The peptide NH of Gly 43 forms a hydrogen bond with a phosphate oxygen of P_C; the OH of Ser 45 forms a hydrogen bond through a water molecule to a P_C oxygen
P_D	Asn 61, Asn 58	Inside: The NH_2 Asn 61 side chain forms a hydrogen bond with an oxygen of the phosphate group of P_D; the NH_2 of Asn 58 forms a hydrogen bond with P_D through a water molecule
P_E	Asn 58	Inside: The NH_2 of Asn 58 forms a hydrogen bond with the phosphate oxygen of P_E; the Ala 56 backbone carbonyl contacts P_E through a water bridge

[a]Data from Jordan and Pabo (1988); Beamer and Pabo (1992).

cific. The local variation in twist, a consequence of primary base sequence, will influence the three-dimensional positioning of the phosphates. However, there is less binding specificity than is provided by the base-specific hydrogen bonding contacts.

The protein–phosphate hydrogen bonds fall into two classes: those located in the major groove between phosphate and the helix–turn–helix domain and those located outside the major groove between phosphates and α helix 1 and amino acids between helices 1 and 2 (Figure 8.9, Table 8.2). The strongest hydrogen bonding occurs to P_A and P_B through residues within the major groove at the ends of α helices 2 and 3. Residues within α helix 1 but outside the major groove also make contact with these phosphates. Hydrogen bonds to phosphates P_C, P_D, and P_E are made through amino acids in the helix–turn–helix domain. The two individual P_E phosphates, near the center of symmetry of the operator, are believed to make contact with different amino acids in the two protein subunits.

b. The 434 Repressor

The 434 repressor is produced by bacteriophage 434, which is similar to bacteriophage λ. As does the lambda repressor, the 434 repressor binds to a number of related operator sites on the 434 chromosome. The 434 repressor binds as a dimer to a 14-bp operator sequence using a helix–turn–helix motif (Figure 8.10). Moreover, like the λ repressor, α helix 3 of the 434 repressor lies in a parallel orientation in the major groove of the DNA. The DNA–protein interaction involves the hydrogen bonding and hydrophobic interactions between the 434 repressor and operator shown in Figure 8.11 and listed in Table 8.3 (Aggarwal *et al.*, 1988). However, that is where the similarities end. The sequences of the 434 operators, the amino acids composing the 20-residue helix–turn–helix motifs, and the specific contacts are different between the two systems. Moreover, unlike the λ operator, the 434 operator is bent when repressor binds.

The shape of the DNA in the 434 repressor–operator complex is altered relative to the structure of the protein-free DNA. The 14-bp 434 operator exists basically as B-DNA but is bent when the repressor is bound. The contacts between the protein and DNA between bases -2 to $+1$ require a bending of the DNA toward the NH_2 terminus of α helix 2. This requires a compression of the minor groove at the center of the bend. At the center of the operator, the minor groove width is compressed to 8.8 Å while the width opens to 14 Å at the ends of the 14-bp operator (the normal width of the minor groove in B-DNA is 11.5 Å). Near the center of the operator, between nucleotides 7L and 7R, Arg 43 extends from the loop between α helices 3 and 4 into the minor groove. Arg 43 from each repressor monomer forms *asymmet-*

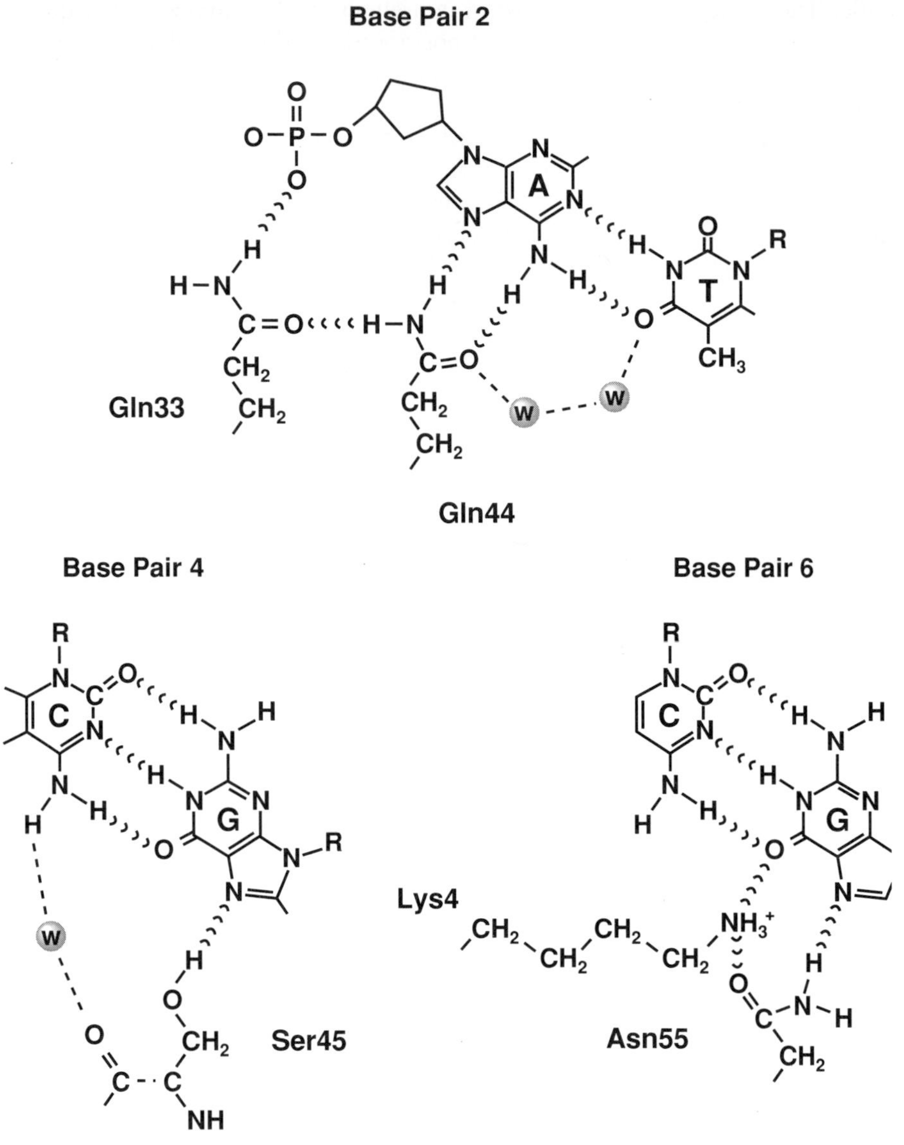

Base Pair 2

Gln33

Gln44

Base Pair 4

Ser45

Base Pair 6

Lys4

Asn55

Figure 8.8 Amino acid–base pair hydrogen bond contacts between the λ repressor and the λ operator. Contacts with base pair 2 (A·T) in the λ operator are made through Gln 33 of α helix 2 and Gln 44 of α helix 3. Two hydrogen bonds are made with between Gln 44 and A. In addition, the carbonyl of Gln 44 contacts O4 of T through a bridge of two water molecules. Gln 33 makes a single hydrogen bond with the phosphate group of A through an amino group hydrogen. A hydrogen bond is also made between the carbonyl group of Gln 33 and the terminal amino group of Gln 44. The interaction with DNA stabilizes the interaction between α helices 2 and 3. Base pair 4

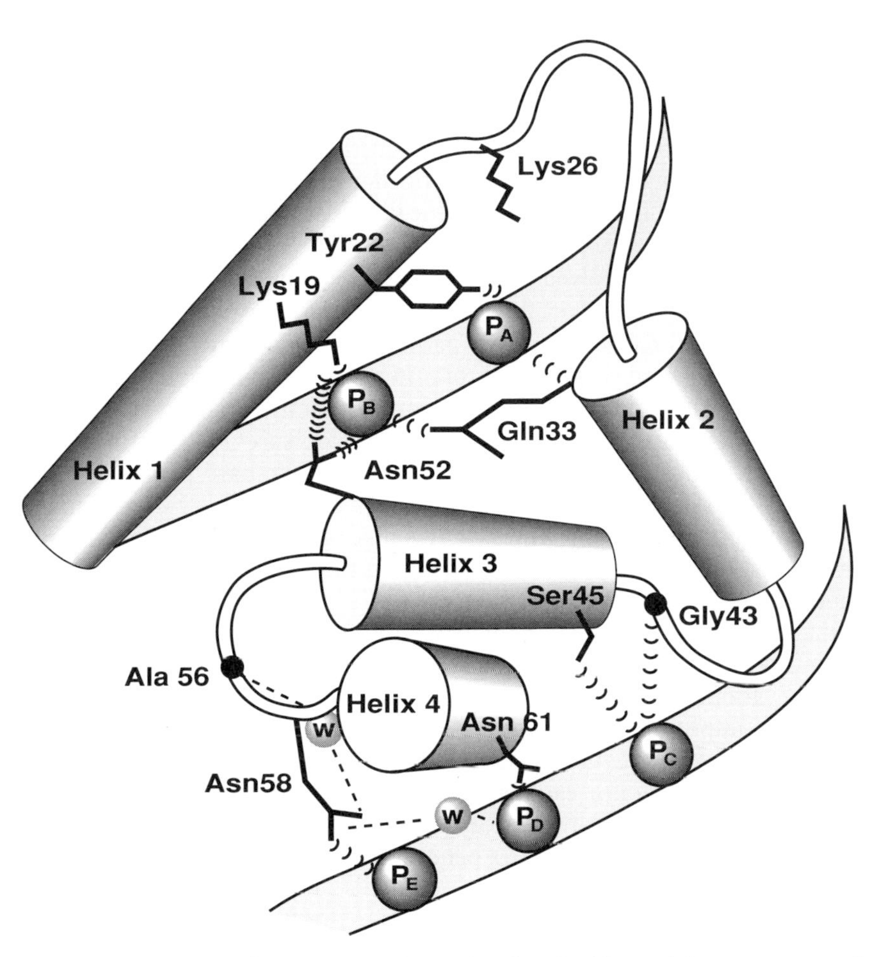

Figure 8.9 Interaction of λ repressor with the phosphate backbone of the λ operator. Hydrogen bond contacts are made from inside the major groove between phosphates P_A, P_B, P_C, P_D, and P_E and α helices 2, 3, and 4, as well as between the amino acid turns separating helices 2 and 3 and helices 3 and 4. In addition, contacts outside of the major groove are made through α helix 1 and outside the amino acid turn between the helices 1 and 2 to P_A and P_B. These hydrogen bond contacts are outlined in detail in Table 8.2. Adapted from Jordan and Pabo (1988), Science **242.** Copyright 1988 by the AAAS.

(C·G) contacts repressor through hydrogen bonds between the N7 and O6 of G and the hydroxyl group of Ser 45. The peptide carbonyl contacts the N4 of C through a single water molecule. Base pair 6 (C·G) makes hydrogen bond contacts with Lys 4 and Asn 55 through the G residue. The C4 carbonyl oxygen hydrogen bonds with the NH_3^+ group on Lys 4. The NH_3^+ of Lys 4 also hydrogen bonds to the carbonyl group of the terminal amino of Asn 55, which is hydrogen bonded through the amino group hydrogens to the N7 of G. (Only the side chains of the amino acids are shown.) Adapted from Jordan and Pabo (1988), Science **242.** Copyright 1988 by the AAAS.

ric contacts with the bases, sugars, and phosphates. The presence of the positive charges (from the guanidinium groups) in the minor groove is thought to neutralize the repulsive negative charges on the phosphate backbone and facilitate the compression of the minor groove at the center of the operator. In addition, at the center of the bend, the propeller twist of two A·T base pairs and one C·G base pair is increased dramatically to about 29° (from just a few degrees), and there is a change in the twist angle The propeller twisting positions functional groups involved in Watson–Crick hydrogen bonding close to groups on an adjacent base pair. As a result, three *bifurcated* hydrogen bonds form between these three adjacent base pairs at the bend center (Figure 8.12).

The phage 434 repressor–operator interaction provides an interesting demonstration of the importance of bases in the operator that are not specifically contacted by amino acids in the protein. Koudelka *et al.* (1987) showed that changing the base composition of the central four base pairs—6L, 6R, 7L, and 7R (AAAG in OR1 as shown in Figure 8.10)—which do not make hydrogen bond contacts with repressor, resulted in dramatic changes in the affinity of repressor for the operator. Changing the central four base pairs of a symmetrical 14-bp operator (ACAATATATATTGT) to ACGT or AGCT reduced the affinity of the operator for DNA more than 50-fold. Replacement of the central base pairs with AAAT increased the affinity of the repressor for the operator about 3-fold. The base composition at the center of symmetry is believed to be important in allowing compression of the minor groove to facilitate the bending of the 434 operator on binding repressor. G·C base pairs at the center of the operator cannot participate in bending as easily and consequently exhibit a reduced affinity for repressor. Adjacent A·T base pairs are readily compressible which facilitates bending.

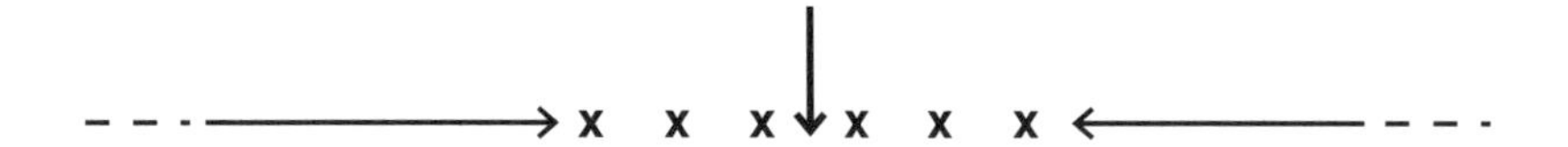

-3'L -2'L -1'L 1'L 2'L 3'L 4'L 5'L 6'L 7'L 7'R 6'R 5'R 4'R 3'R 2'R 1'R -1'R -2'R -3'R -4'R

t p a p t p A p C p A p A p G p A p A p A p G p T p T p T p G p T p a p c p t p t

a p t p a p T p G p T p T p C p T p T p T p C p A p A p A p C p A p t p g p a p a

-3'L -2'L -1'L 1'L 2'L 3'L 4'L 5'L 6'L 7'L 7R 6R 5R 4R 3R 2R 1R -1R -2R -3R -4R

Figure 8.10 Sequence and numbering system of the 434 operator. The 14-bp operator (uppercase letters) has inverted repeated symmetry in which the central 6 bp are nonpalindromic (indicated by x above the base). The vertical arrow denotes the center of symmetry. The left half of the operator is numbered 1L–7L whereas the right half of the operator is labeled 7′R–1′R. Several bases flanking the operator are shown in lowercase letters.

c. The CAP–DNA Complex: DNA Bending around Proteins

Schultz *et al.* (1991) described the structure of the CAP (catabolite activator protein) interaction with DNA. The binding of this protein introduces a 90° bend in DNA. In the CAP–DNA complex, the bend is composed of two sharp ≈40° kinks at dinucleotides 5 bp on either side of the center of the CAP binding site. These kinks, separated by 10 bp, act to curve the DNA around the body of the CAP protein. The specific interactions between the DNA and CAP are mediated through helix–turn–helix motifs positioned in adjacent major grooves. Hydrogen bonds are made in the major groove between guanidinium groups of two arginines and two guanines, and a carboxylate group of glutamic acid and a cytosine.

3. Interaction of the β Sheet with DNA

A β sheet is composed of two or more polypeptides held together by hydrogen bonding between alternating NH and carbonyl groups in the backbone of opposite strands. β Sheets can be either antiparallel, as shown in Figure 8.5B, or parallel, in which the polarity of the two amino acid chains is the same (not shown). This structural organization leaves one free NH group on every other amino acid on each strand. The distance between these potential hydrogen bonding donors matches the potential hydrogen bonding acceptor distances on the C3′ oxygens in the sugar–phosphate backbone.

In 1974 Carter and Kraut proposed a model for the interaction of a β sheet with the minor groove of double-stranded RNA. Shortly thereafter Church *et al.* (1977) proposed a model for the interaction of a β sheet with the minor groove of B-form DNA. A β sheet could lie in the minor groove and form hydrogen bonds with DNA in one of two orientations. Church *et al.* (1977) suggested that the best fit occurred when the NH_2 terminus of the β sheet was oriented in the 3′→5′ direction with respect to the DNA strand to which it hydrogen bonded. Since the hydrogen bonding contacts are with oxygens of the phosphate backbone, this type of interaction should lack sequence specificity.

a. Interaction of the HU Family of Proteins with DNA

The bacterial HU protein is a small, basic, abundant "histone-like" protein found in bacteria. It binds DNA with no sequence specificity, bends DNA, and restrains negative supercoils in DNA (Drlica and Rouviere-Yaniv, 1987). The biological properties of the HU protein as well as the topology of the HU–DNA interaction are discussed in Chapter 9.

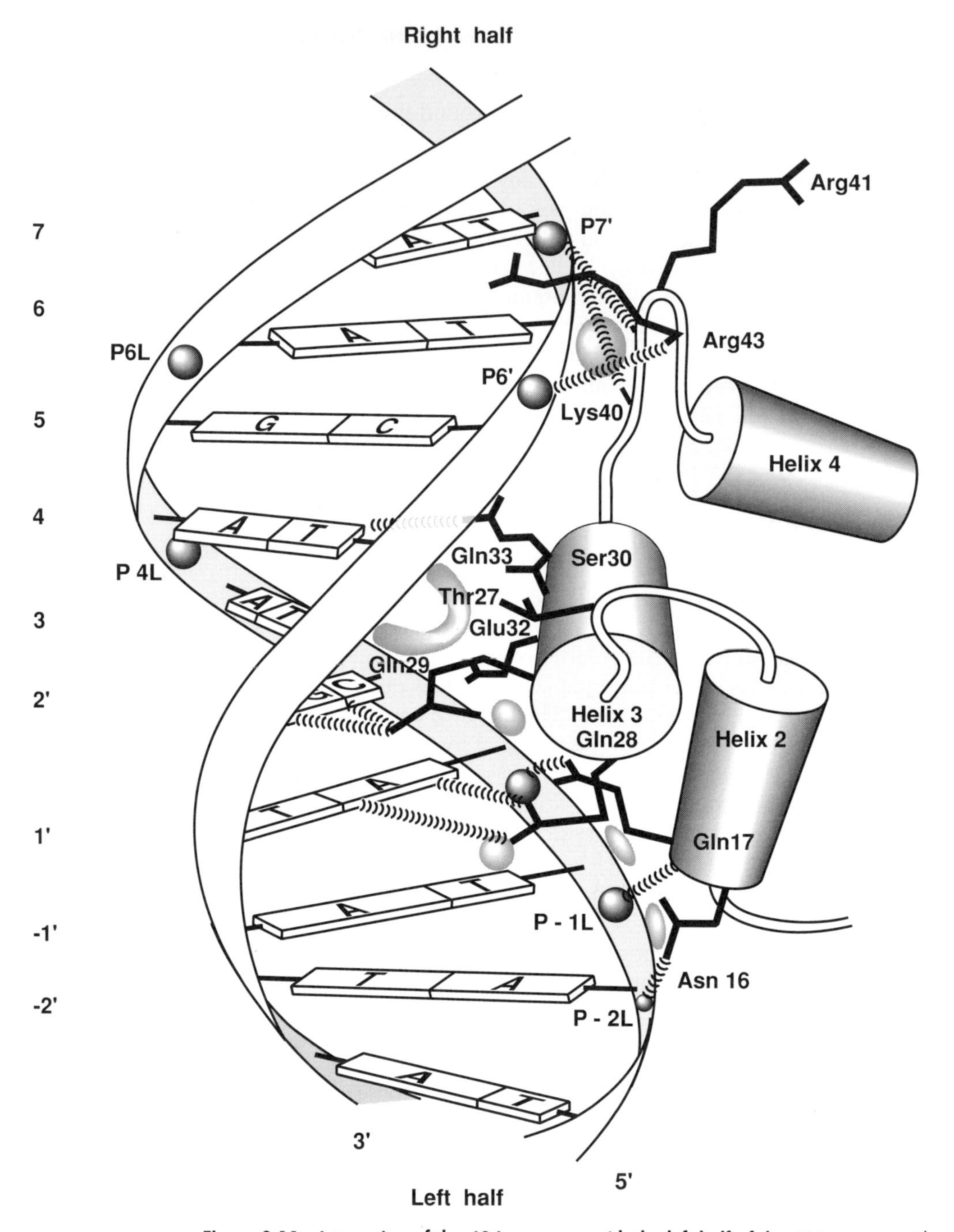

Figure 8.11 Interaction of the 434 repressor with the left half of the 434 operator. The orientation of helix 3 of the helix–turn–helix motif (helix 2 and helix 3) within the major groove of the left half of the 434 operator is shown. Helix 3 lies in the major groove and makes hydrogen bonding contacts with functional groups on base pairs 2 and 4. Helix 2 makes hydrogen bond contacts with

Table 8.3
Contacts between the 434 Repressor and Operator

DNA position	Amino acid	Contact[a]
Major groove hydrogen bond contacts (shown in Figure 8.11)		
#1 A·T	Gln 28	Bidentate hydrogen bonds are formed between the NH_2 of Gln 28 and N7 of A, and between the C=O of Gln 28 and the N6 of A
#2 C·G	Gln 29	The terminal NH_2 of Gln 29 makes bidentate contacts with the O6 carbonyl group and the N7 position of guanine
#4 A·T	Gln 33	The O4 carbonyl of T forms a hydrogen bond with the NH_2 of Gln 33
Major groove van der Waals contacts		
#−1 T·A	Gln 28	The methyl group of thymine at position −1 (outside the 14-bp operator) contacts the methyl groups on the side chain of Gln 28
#3 A·T	Thr 27, Gln 29	Side chain methyl groups of Thr 27 and Gln 29 form a van der Waals pocket to bind the methyl group of thymine
#4 A·T	Gln 29, Ser 30	The methyl group on thymine contacts the side chain methyl groups of Gln 29 and Ser 30
P1, Sugar 1	Asn 16, Gln 17	van der Waals contacts made between the PO_4, and sugar and the CH_2 side chains of the amino acids
Hydrogen bonds to the phosphate backbone		
P2	Asn 16	The P2 phosphate (on the 5′ side of A 2, between T 3 and A 2) forms a hydrogen bond with the terminal NH_2 of Asn 16
P1	Gln 17, Arg 10	The P1 phosphate (between T 1 and A 2) hydrogen bonds to the NH_3 of Arg 10 (not shown) and the main chain NH of Gln 17, at the end of helix 2
P1	Gln 17, Asn 36	The terminal NH_2 of Gln 17 forms a hydrogen bond to P1; Asn 36 also forms a hydrogen bond with P1 (not shown)
P6′	Arg 43	The main chain NH group of Arg 43 forms a hydrogen bond with P6′
P7′	Lys 40, Arg 41	The main chain NH groups of Lys 40 and Arg 41 form hydrogen bonds with P7′

[a]Data from Aggarwal *et al.* (1988).

The HU protein exists in solution primarily as a heterotypic dimer formed by the intermeshing of the two monomeric subunits. In 1984 Tanaka *et al.* published an X-ray crystallographic analysis of the HU protein from *Bacillus stearothermophilus* (see Figure 8.13A). There are three α helices: (1) residues 3–13; (2) residues 21–39; and (3) residues 84–89. The important

base pairs 1 and 2 whereas the amino acid turn between helices 3 and 4 makes hydrogen bonding contacts near the center of the 434 operator. Hydrogen bonds are shown as short curves. Hydrophobic interactions are shown as shaded areas between specific amino acids and the bases or phosphate backbone. These hydrogen bonding and hydrophobic interactions are described in detail in Table 8.3. Adapted from Aggarwal et al. (1988), Science **242.** Copyright 1988 by the AAAS.

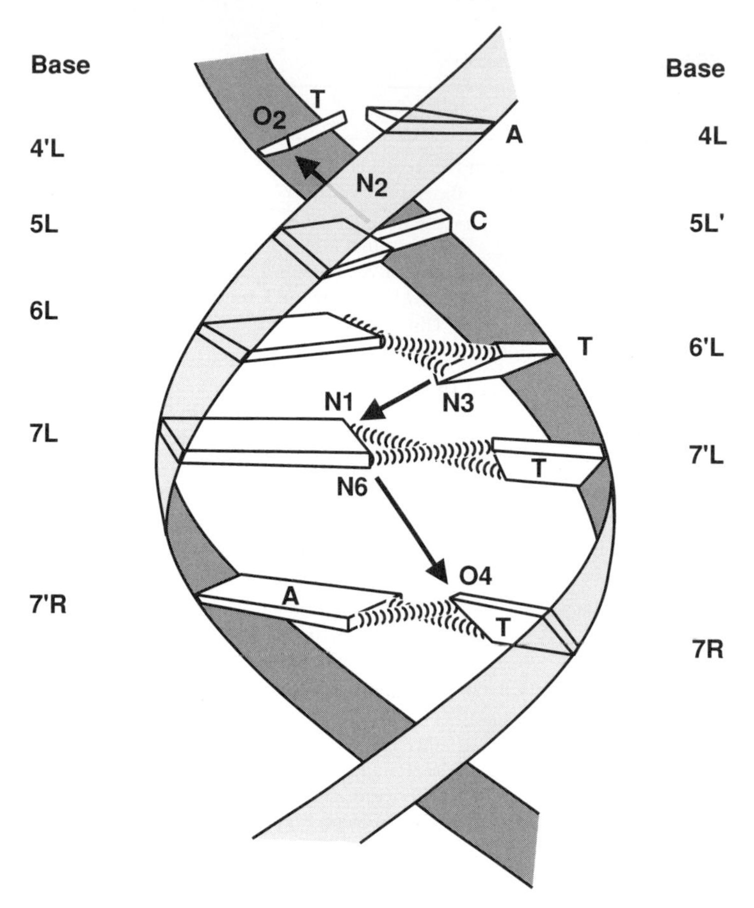

Figure 8.12 Interdinucleotide hydrogen bonds in the 434 operator. When 434 repressor binds the 434 operator, a bend is introduced at the center of the operator. This bend results in the significant propeller twisting of one C·G and two A·T base pairs at the center of the operator. Bifurcated hydrogen bonds form between adjacent dinucleotide pairs as a result of this increase in propeller twist. Watson–Crick hydrogen bonds between A·T and C·G base pairs in the operator are shown using the standard short curves. The hydrogen bonds between the adjacent dinucleotide pairs are shown as bold arrows. The atom positions from which these hydrogen bonds originate and terminate are indicated at the ends of the arrows. Hydrogen bonds are formed between the N3 position of T 6′L and the N1 position of A 7L; N6 of A 7L and O4 T 7R; and N2 of C 5′L and O2 of T 4′L. Adapted from Aggarwal et al. Science **242.** Copyright 1988 by the AAAS.

area, as far as the interaction with DNA is concerned, is the β sheet, which is composed of three strands. Strand 1 comprises residues 40–44; strand 2, residues 48–51; and strand 3, residues 78–83. The β ribbon arms of the two subunits stick out like "lobster claws," forming a pocket that matches closely the diameter of the B-form double helix. The section of the arm next to the body of the protein adopts a two-stranded antiparallel β ribbon conformation that, as discussed earlier, can interact in the minor groove of B-form DNA (Figure 8.13B). The ribbon is twisted and bent, which may be important for allowing the ribbon to bind DNA as it follows the shape of the minor groove (White *et al.*, 1989). Amino acids in this arm region are conserved in HU proteins from a number of different organisms, suggesting that the structure of the arm is critical for the interaction with DNA. Positive charges on Arg 53, Arg 55, Arg 58, Lys 80, and Lys 86 within the arm region can make reasonable interactions with the phosphate groups of five adjacent nucleotides. Although the position of Arg 61 is not resolved in the X-ray crystal structure, it is believed to make contact with an adjacent sixth phosphate. The exclusive contacts with the negatively charged phosphate backbone insure a nonspecific binding to DNA.

b. The Specific Binding of Integration Host Factor

Integration Host Factor (IHF) is closely related to the HU protein. It is an *E. coli* protein that, among other things, participates in the integration of bacteriophage λ into DNA (see Figures 2.19 and 2.22). The two similar subunits of IHF are α-IHF and β-IHF. These are present at lower concentrations than HU and, in contrast to HU, bind to specific sites on the DNA. What are the differences between IHF and HU that confer sequence-specific recognition to IHF? Changes in the β ribbon might suffice. A number of groups in the β ribbon of HU have positive charges that are used to bind the negative phosphate groups of DNA. These charged amino acids are also found in IHF and provide the general electrostatic affinity of IHF for DNA. Within the region of the arm that interacts with DNA, Ala 56, Ala 57, and Pro 72 of HU are replaced by Asn 56, Gln 57, and Thr 72 in the α-IHF protein and by Asn 56, Gln 57, and Glu 72 in the β-IHF subunit. These amino acids have the potential to form hydrogen bonds with bases in the minor groove. Also, positively charged arginine and lysine residues are present in the α and β subunits, respectively, at position 74. In a variety of HU proteins, position 74 is not conserved. The exact details of the interactions between specific amino acids and functional groups must await an X-ray crystallographic analysis of an IHF–DNA cocrystal.

c. Sequence-Specific β-Sheet DNA Major Groove Binding Proteins

There are also examples of proteins in which a β sheet interacts with specific bases in the major groove (Phillips, 1991). The MetJ repressor from *E. coli* is a dimer of two identical 104-amino-acid subunits in which two 10-amino-acid β strands interact to form an antiparallel β sheet on the surface of

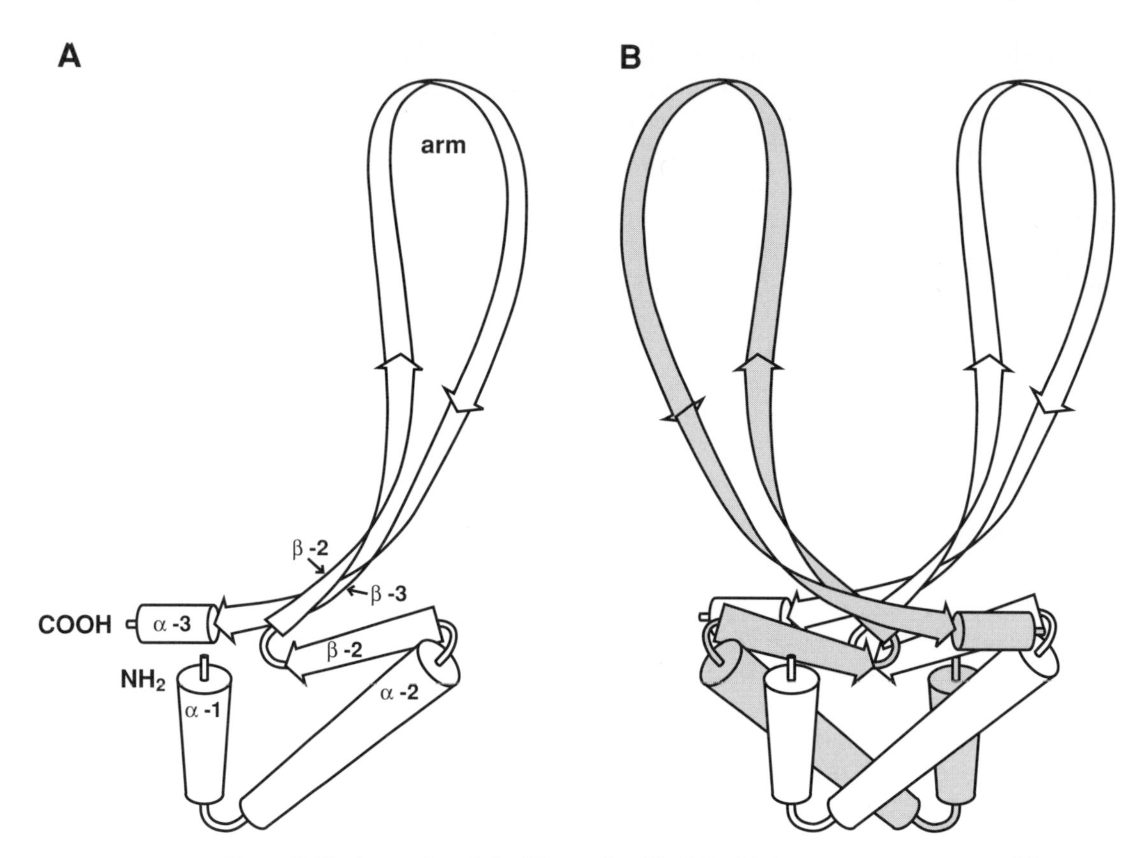

Figure 8.13 Interaction of the HU protein with DNA. (A) An HU monomer consists of three α helices represented by the cylinders and several extended β sheets represented by the ribbons. The top loop of this structure is not resolved in the x-ray analysis and therefore its position represents a "best guess" as to the position of the amino acids in the extended β loops. Positions of the COOH and NH_2 termini are indicated. (B) The HU dimer is composed of two monomeric subunits of HU in which the two α domains at the NH_2 termini interact forming the base of the HU dimer. The two extended β sheets stick out from the top of the HU dimer and these extended β sheets lay in the minor grooves of the DNA double helix. (C) HU (shown 90° to the representation in A) binds the DNA by laying its two β pleated sheet "lobster claw" arms into the minor grooves of DNA. On doing this, it is believed to tightly bend the DNA into a loop of <99 bp, or perhaps as small as 39 bp (Figure 9.2). The interaction of the arm regions with the minor groove occurs in a nonspecific fashion. Hydrogen bonds form between multiple arginine and lysine groups on the HU arms and phosphate groups of the DNA backbone. Adapted with permission from Nature (Tanaka et al., 1984). Copyright (1984) Macmillan Magazines Limited.

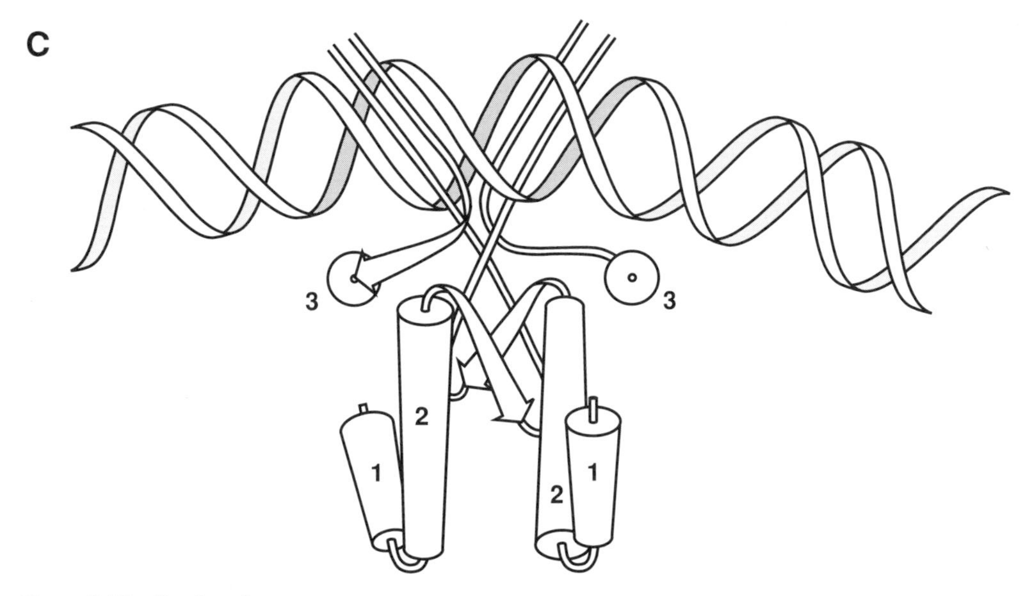

Figure 8.13 Continued.

the protein (Rafferty *et al.*, 1989). The β sheet makes specific contacts with an 8-bp consensus palindromic recognition sequence (AGACGTCT) that is tandemly repeated two to five times in the operator sequences in six *Met* genes in the *E. coli* chromosomes (Phillips *et al.*, 1989). The DNA is slightly bent around the protein when the β sheet interacts in the major groove. A Lys 23 and Thr 25 in each β sheet make contact via hydrogen bonding to the second and third bases of the operator (usually G and A, respectively). Similar β sheet DNA binding motifs are utilized by the Arc and Mnt repressors from *Salmonella* bacteriophage P22 to recognize their cognate operators in P22 phage DNA (Knight *et al.*, 1989; Phillips, 1991).

4. Interaction of Zn-DNA Binding Domains with DNA

A large number of proteins containing Zn binding domains have been identified (Miller *et al.*, 1985; Klug and Rhodes, 1987; Kaptein, 1991a,b; Krizek *et al.*, 1991; Pabo and Sauer, 1992; Berg, 1993; Giedroc, 1994). Many of these are transcription factors that bind to regulatory regions upstream from genes. In the case of transcription factors, the Zn binding domain must recognize specific base pairs. Other proteins containing Zn binding domains, such as the *E. coli* protein UvrA, which is involved in the excision of dam-

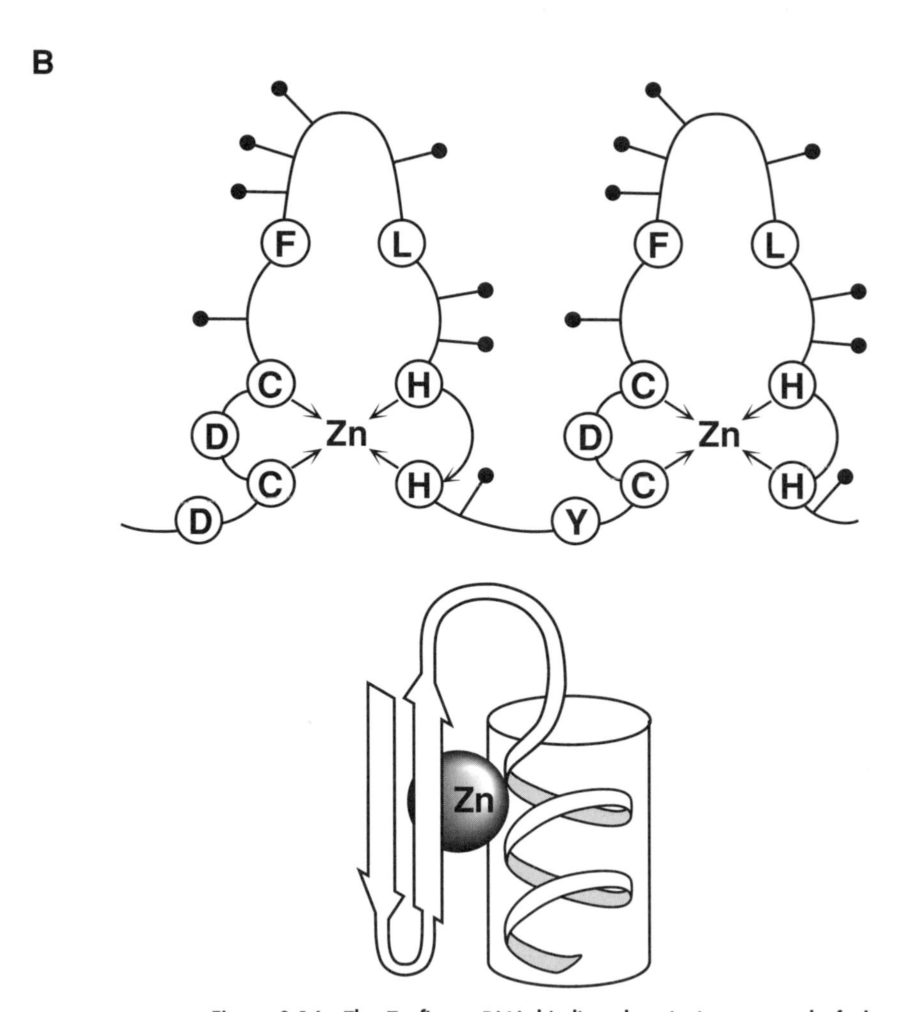

Figure 8.14 The Zn finger DNA binding domain is composed of about 30 conserved amino acids. (A) The conserved residues include two cysteines (C), two histidines (H), a tyrosine (Y), a phenylalanine (F), and a leucine (L) at the positions indicated (the conserved positions are circled). The stars in the sequence indicate positions at which an extra amino acid is found in some fingers. The dots indicate nonconserved amino acid positions. The amino acids composing the β sheets are indicated by the zigzag lines (positions 2–4 and 11–13). The α helix is represented by the coiled line (positions 16–27). (B) The first suggested organization of a Zn finger domain around a Zn

aged bases from DNA, and the bacteriophage T_4 gene 32 protein, which is a multifunctional single-stranded DNA binding protein, must bind in a sequence-independent manner. To date, there are three types of Zn DNA binding domains that have been identified: the Zn finger or $Zn(Cys_2–His_2)$ domain, the Zn twist or $[Zn(Cys_4)]_2$ domain, and a bidentate Zn cluster or $Zn_2(Cys_6)$ domain (Vallee *et al.*, 1991).

a. The Zn-Finger, $Zn(Cys_2–His_2)$ Domain

In 1985 Miller, McLachlen, and Klug reported that nine 30-amino-acid repeating units were present in the DNA binding protein *t*ranscription *f*actor IIIA (TFIIIA). In addition, the protein could be proteolytically digested into a large number of 3-kDa subunits believed to represent the 30-amino-acid repeating units. Finally, 9–11 Zn atoms were associated with the TFIIIA protein. The arrangement of conserved amino acids within the 30-amino-acid residue region is shown in Figure 8.14A. Conserved residues include two cysteines (C) and two histidines (H) as well as tyrosine (Y), phenylalanine (F), and leucine (L). Miller *et al.* (1985) proposed that each 30-residue unit folded into a structural domain in which the cysteines and histidines were coordinated tetrahedrally to an atom of Zn.

The Zn finger folds into a short antiparallel β sheet composed of residues 2–4 and 11–13 (Berg, 1988; Lee *et al.*, 1989). The cysteines at positions 4 and 9 are bound to Zn on one side of the sheet. Amino acid residues 14 and 15 then form a loop between the β sheet and an α helix. Residues 16–27 compose an α helix, which is held next to the β sheet by coordination through the histidines to the Zn atom (Figure 8.14B). Amino acid residues 28–30 and 1 provide a linker region to the next Zn finger.

The nine tandem Zn fingers in the TFIIIA gene were initially believed to interact with about 50 bp of the 5S RNA gene (Rhodes and Klug, 1986). Thus, each Zn finger could interact with about 5 bp of DNA, that is, half a helical turn of DNA. Fairall *et al.* (1986) suggested two models for the interaction of multiple tandemly repeated binding domains with DNA (Figures 8.15A and 8.15B). In Model I, the Zn fingers would essentially wrap around the DNA following the path of the major groove. In Model II, the Zn fingers interdigitate with the major groove from one side (face) of the DNA helix. The protein would dock with the DNA, sliding its fingers into grooves.

atom. The conserved nucleotides are shown as small open circles. The two cysteine and two histidine residues are coordinated to Zn, organizing the rest of the domain into a finger-like projection of amino acids. Berg (1988) suggested the structure of a Zn finger shown (bottom) in which the two cysteines in the β sheets and the two histidines in the α helix are coordinated through Zn, resulting in the finger-like projection containing the loop between the α helix and the β sheets. Figure modified with permission from Miller et al. (1985) and Berg (1988).

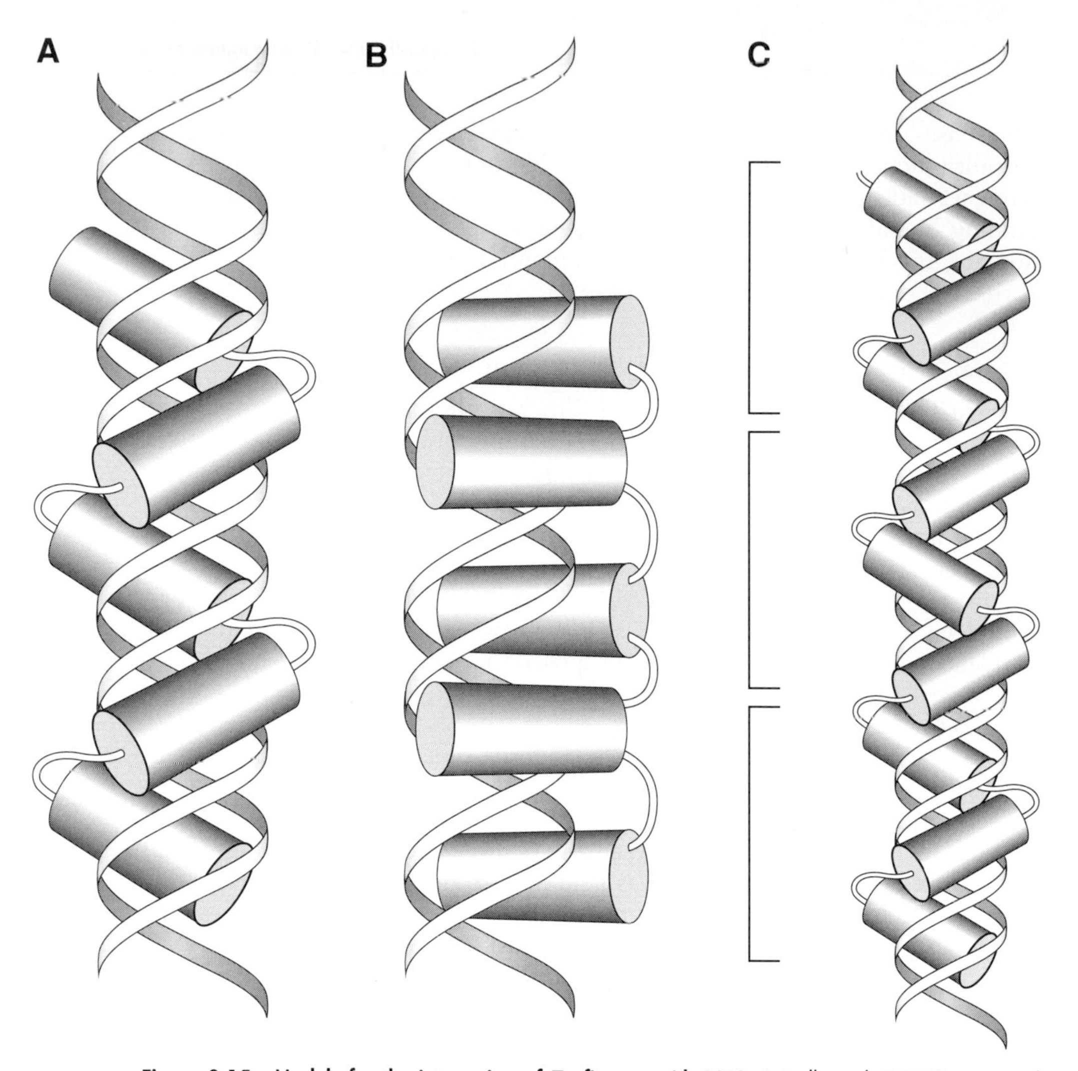

Figure 8.15 Models for the interaction of Zn fingers with DNA. Fairall *et al.* (1986) suggested two different models for the interaction of Zn fingers with DNA. (A) In Model I the Zn fingers wrap around the DNA lying in the major grooves on alternating faces of the DNA helix. For example, one cylinder lies on top of the helix while the next cylinder lies perpendicular to the first in the major groove on the bottom of the DNA molecule (as shown on the plane of the paper). This would require wrapping the Zn fingers around the DNA. (B) Model II involves positioning the Zn fingers into the major groove from one side or face of the helix. The Zn fingers would bind alternatively to the top or the bottom of the DNA helix with respect to the plane of the paper. In this configuration, part of the Zn finger region crosses the major groove. (C) A schematic model for the interaction of the nine Zn fingers of TFIIIA with its recognition site. The three outermost fingers wrap around the groove whereas the central three fingers interact across the minor groove. The interaction of individual Zn fingers is not drawn to scale for TFIIIA. The actual interaction involves three Zn fingers with about 10 bp of DNA, as shown for Zif268 (Figure 8.16). Figure modified with permission from Fairall *et al.* (1986).

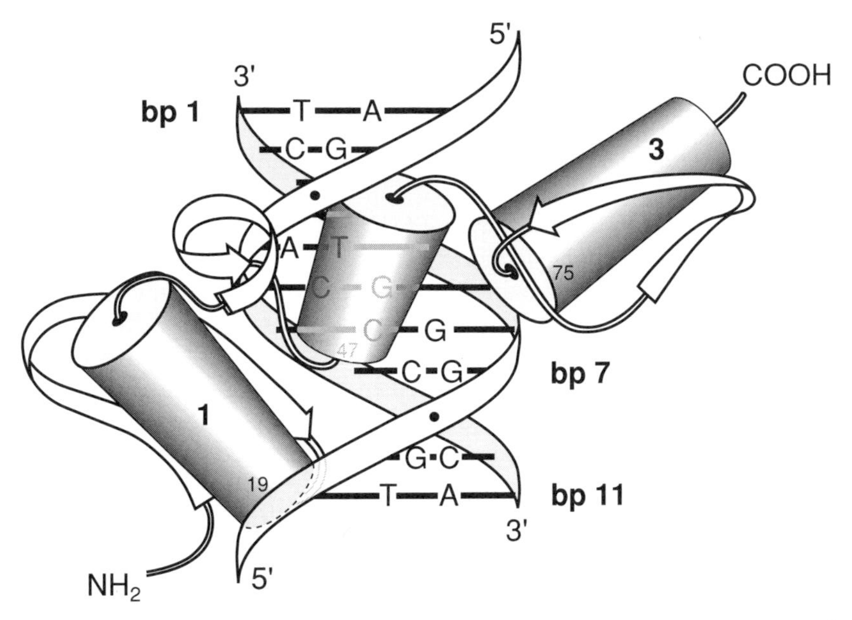

Figure 8.16 The X-ray crystal structure of Zif268 protein interaction with DNA. The Zif268 protein interacts with an 11–bp recognition site. Each of the three Zn fingers interacts with a 3-bp G + C-rich binding subsite. The three Zn fingers wrap around the major groove making contact with the recognition sites. Adapted from Pavletich and Pabo (1991), *Science* **252.** Copyright 1991 by the AAAS.

Model II necessitates that the protein cross the minor groove. As shown subsequently, the first X-ray cocrystal of a Zn finger showed interaction with DNA by Model I.

X-Ray Crystal Structure of Zif268 The X-ray crystal structure of a Zn finger–DNA complex was solved by Pavletich and Pabo (1991). The Zif268 protein from mouse contains three Zn fingers that bind to an 11-bp recognition sequence A GCG TGG GCG T. Each Zn finger binds to a 3-bp subsite (underlined) in the 11-bp recognition site. The Zif268 protein binds DNA by Model I (Figure 8.16), in which the three Zn fingers wrap around the DNA and make contact with the six guanines (bold) in the major groove. The NH_2-terminal end of the α helix is directed into the major groove. In each finger a hydrogen bond is made between the last guanine of the subsite and an arginine that precedes the α helix of each finger (finger 1, Arg 18; finger 2, Arg 46; finger 3, Arg 74). Hydrogen bonds are made in fingers 1 and 3 with the first guanine of the subsite and an arginine near the middle of the α helix, six amino acids away (finger 1, Arg 24; finger 3, Arg 80). In finger 2 the guanine adjacent to the terminal guanine hydrogen bonds to His 49 three amino acids away from the arginine before the α helix. There are also a number of hydrogen bond contacts with the phosphate backbone. This is a very different type of interaction than the helix–turn–helix motif discussed earlier.

The Organization of Zn Fingers in TFIIIA In contrast to the binding of Zif286, initial experimental evidence by Fairall *et al.* (1986) and Churchill *et al.* (1990) suggested an interdigitated model for the interaction of DNA with TFIIIA in which Zn fingers crossed the minor groove. Moreover, Tramtrack, a regulatory protein from *Drosophila* with two Zn fingers may bind across the minor groove, as determined by footprinting studies (Fairall *et al.*, 1992). The most recent picture to emerge regarding the binding of TFIIIA is that it interacts with a 27-bp site (Hayes and Tullius, 1992). The terminal three Zn fingers appear to wrap around the major groove like the Zif286 protein, interacting with about 10 bp each. The central three Zn fingers span two minor grooves laying on one side of the DNA helix (Figure 8.15C). Therefore, there appears to be more than one mode of interaction of Zn fingers with the DNA double helix.

b. The Zn Twist, $[Zn(Cys_4)]_2$ Domain

Certain nuclear receptors bind hormones, including glucocorticoids and estrogens, and then bind specific hormone response elements in the DNA to activate gene expression. The glucocorticoid receptor contains a 60-amino-acid DNA binding domain with two regions containing four cysteines each. The positions of these cysteines are highly conserved among this large family of hormone receptors. The $Zn(Cys_4)$ domains fold into a globular domain containing two α helices at right angles to each other. A Zn atom is located near the NH_2 terminus of each α helix (Härd *et al.*, 1990; Schwabe *et al.*, 1990). The NH_2-terminal amphipathic α helix contains a number of basic residues on one side of the helix that can make specific contacts to the phosphate backbone from within the major groove of DNA (Luisi *et al.*, 1991), as shown in Figure 8.17A.

c. The Binuclear Zn Cluster, $Zn_2(Cys_6)$ Domain

A 30-amino-acid domain found in the GAL4 transcription factor from yeast contains six cysteine residues. The spacing of these cysteines is conserved around many transcription activators found in a variety of fungi. Two cysteines bind each atom of Zn while the two Zn atoms are bridged through the third pair of cysteines (Cys_2–Zn–Cys_2–Zn–Cys_2). These cysteines are organized around the binuclear Zn cluster with two short α helices joined by a short loop on opposite sides of the Zn atoms. One of these helices within this

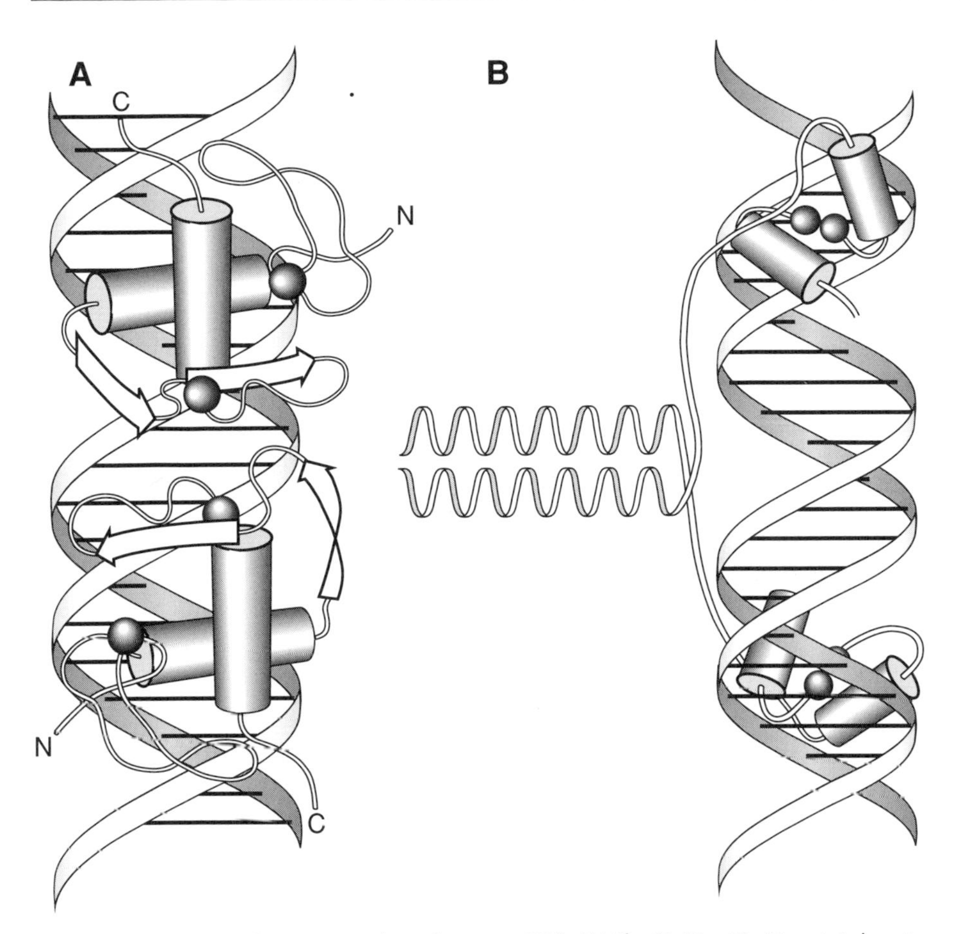

Figure 8.17 Binding of Zn twist and Zn clusters to DNA. (A) The $[Zn(Cys4)]_2$ (Zn twist) domains of the dimeric glucocorticoid receptor bind to an inverted repeat recognition site in which the symmetrical halves are separated by 3 bp. A representation of the crystal structure of a glucocorticoid receptor bound to a completely symmetrical DNA binding site with the half sites separated by 4 bp is shown (Luisi *et al.,* 1991). The α helices are shown as cylinders and the Zn atoms as dark spheres. Adapted with permission from Nature (Luisi *et al.,* 1991). Copyright (1991) Macmillan Magazines Limited. (B) The Zn_2Cys_6 (Zn cluster) domains of a dimeric GAL4 yeast transcriptional activator bind to a 17-bp inverted repeat recognition site. The two α helices of the Zn_2Cys_6 domain are held together by two Zn atoms. The dimerization of two GAL4 monomers by a coiled coil allows contact in the major grooves to conserved CCG tri-nucleotides separated by 1.5 helical turns. This spacing requires interaction from opposite sides of the double helix. Adapted with permission from *Nature* (Marmorstein *et al.,* 1992). Copyright (1992) Macmillan Magazines Limited.

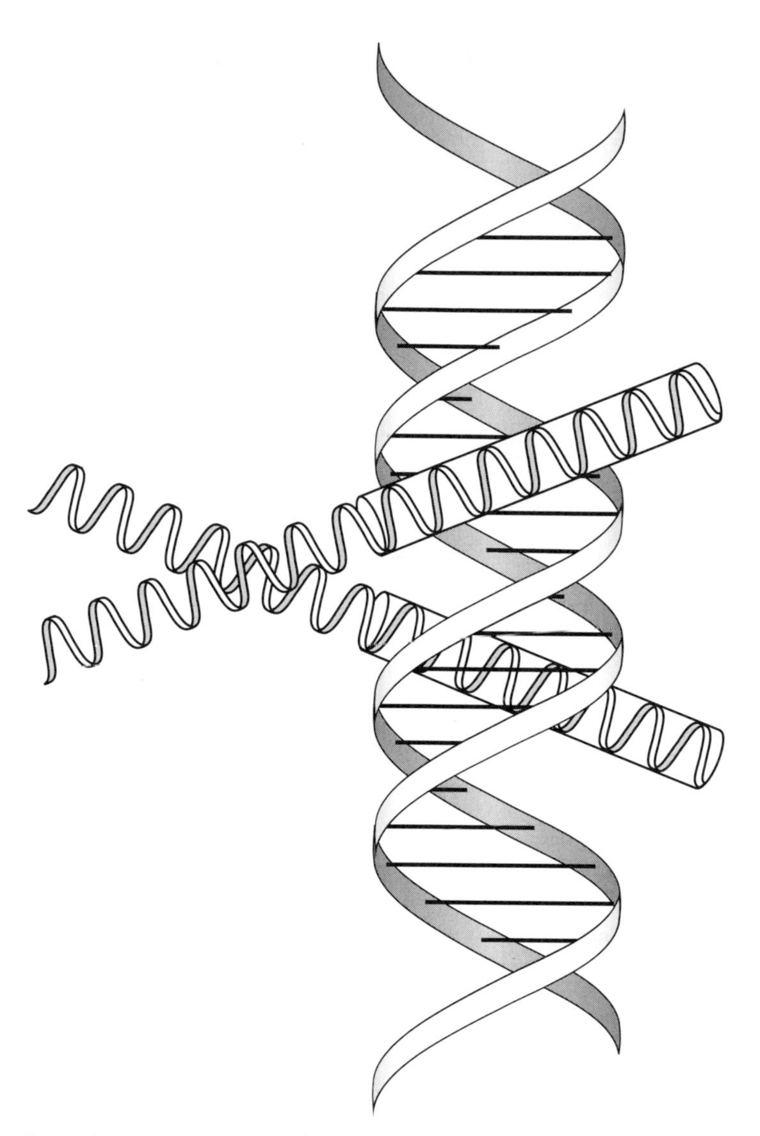

Figure 8.18 Leucine zipper–basic domain interaction. A schematic representation of a bZip leucine zipper–basic motif DNA–protein interaction is shown. A long α helix region is composed of the leucine-rich domain and the basic region. Hydrophobic interactions between the leucine-rich helices hold the two subunits together. The relatively straight α helical basic regions lie in the major grooves in scissor-like fashion making contact with functional groups of several bases and with several phosphate groups. Figure modified with permission from Ellenberger *et al.* (1992).

recognition molecule interacts through the major groove with DNA (Figure 8.17B; Marmorstein *et al.*, 1992). The two subunits of the GAL4 activator protein are held together by a parallel coiled coil extending at right angles from the DNA. A dimer of the binuclear Zn recognition module recognizes a 17-bp DNA binding site, making specific contacts with 3 bp in each half-site separated by about 1.5 helical turns.

5. An Assortment of Other DNA Binding Motifs

a. The bZip Family of Transcription Factors

bZip proteins are dimeric transcription factors containing a dimerization domain and a basic motif that is the DNA binding domain. Examples of bZip proteins include the yeast transcription factor GCN4 and oncogene proteins of the *fos* and *jun* families (Landschultz *et al.*, 1988; Kerppola and Curran, 1991; Pathak and Sigler, 1992; Wolberger, 1993). The dimerization domain, called a leucine zipper, is an extended, slightly curved amphipathic α helix in which leucines are regularly spaced every 7 amino acids (O'Shea *et al.*, 1991). With an α helix repeat of 3.6 amino acids, a repeat length of seven places the leucines on one side of the α helix. The two α helices containing the leucines coil together into a coiled coil in which the helices are held together predominantly by hydrophobic interactions between the leucines. The interaction between the two subunits forms a DNA binding domain made up of the basic motif region of each subunit. The DNA binding domain contains many basic residues that can interact with DNA.

The structure of the GCN4–DNA cocrystal shows a dimer of two large subunits containing a 56-amino-acid region that forms a continuous α helix (Ellenberger *et al.*, 1992). One end of the α helix consists of a 30-amino-acid leucine zipper region whereas the other end consists of a 25-amino-acid basic region. The leucine zipper end forms a coiled coil holding the two α helices together. The coiled coil is oriented perpendicular to the DNA, positioning each basic α helical region in a major groove on opposite sides of the DNA double helix (Figure 8.18). As the α helix passes through the major groove, specific hydrogen bond contacts are made between asparagine, alanine, serine, arginine, and lysine residues and hydrogen bonding sites on the bases and the phosphate backbone.

b. The Amazing Binding of TBP: The TATA-Box Binding Protein

The simultaneous reports of the DNA cocrystal structures of two TATA-box binding proteins (TBP) from different sources with two different TATA-box-containing promotor regions has revealed an astounding interaction of DNA with a protein (Kim, J. L., *et al.*, 1993; Kim, Y., *et al.*, 1993).

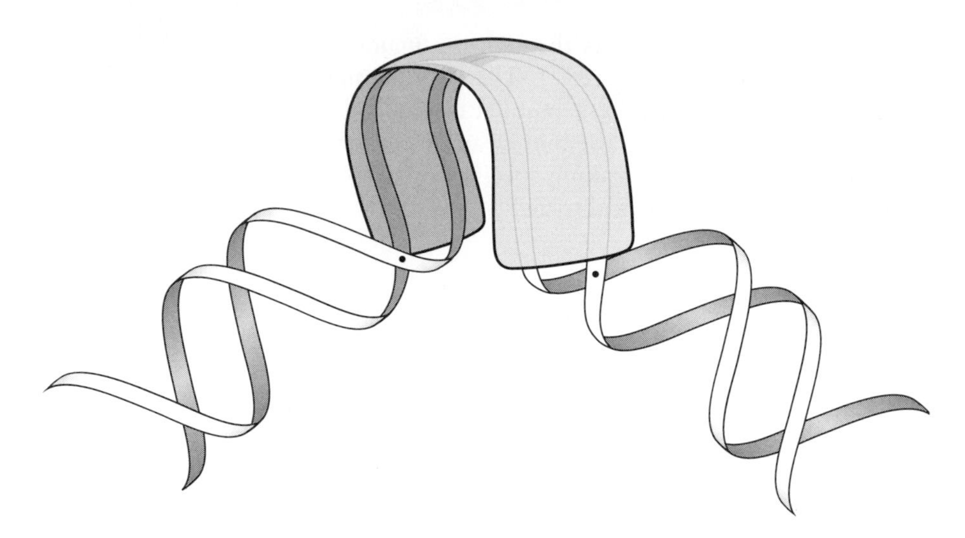

Figure 8.19 DNA bending by the TATA-box binding protein. The TATA-box binding protein (TBP) (also known as TFIID) binds to an 8-bp T + A-rich region just upstream of the transcription start site in RNA polymerase II-dependent genes. The saddle-like protein drastically alters the structure of the 8-bp recognition site. DNA is unwound to an average of 18° per base pair. Moreover, within this 8-bp region the DNA is sharply bent by 100°. Immediately adjacent to the TBP binding site, the DNA remains as typical B-form DNA. The position of the TBP is schematically indicated by the saddle-like structure.

TFIID is a large multisubunit protein that recognizes the TATA box of eukaryotic genes. The TATA-box binding protein is the DNA binding subunit of TFIID. Binding of TBP is required to form a preinitiation complex as a prerequisite to RNA polymerase binding and subsequent transcription. TBP contains a curved β sheet made of 10 antiparallel strands. The β sheet is curved into a saddle shape with four α helices on top. The TBP saddle does not straddle a linear DNA duplex as might be expected; rather the DNA helix axis runs up and down the curve of the saddle bending by 100°. This sharp bend in DNA is accompanied by an unwinding of the TATA box by 110°. The unwinding of the DNA should relax DNA supercoils, but the sharp bend forms a compensating writhe that topologically cancels the effect of unwinding. Thus, in terms of superhelical energy, these two dramatic changes in DNA structure are neutral (or offset each other). The 8-bp TATA-box element in DNA comprises the entire sharp 100° bend in the DNA–TBP complex. The TATA element is bent toward the major groove. This bending flares open the minor groove, exposing the bases within to the underside of the TBP saddle. The average twist angle is 18° within the TATA element with a 3° angle at the center of the bend. Bases within the bend also have unusually large roll angles. The dinucleotide base pairs at the ends of the TATA element (bp 1, 2, 7, and 8) are kinked and buckled (the plane of the bases is bent at the center) because of the aromatic rings of two pairs of phenylalanines being thrust into the remainder of the minor groove between these terminal dinucleotides. Astonishingly, the base pairs immediately adjacent to the TATA element are in a typical B-DNA form (Figure 8.19).

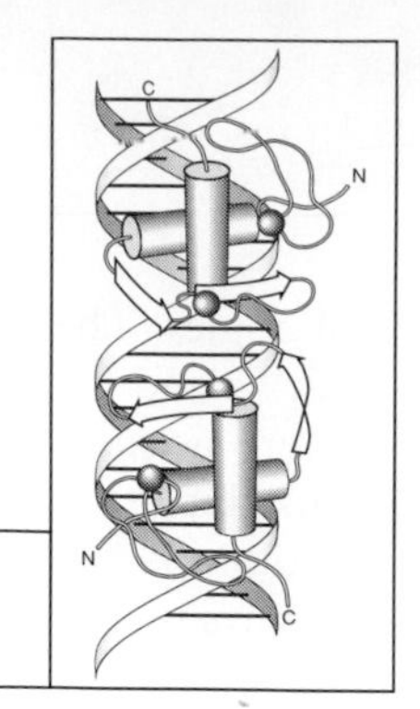

9 CHAPTER

The Organization of DNA into Chromosomes

A. Introduction

Have you ever unpacked something new and found that, on putting the item back into its original box, no matter how hard you tried the box bulged, and the ends of the box would not stay tucked in? You were certain that the item you were repacking could not have fit into its original package! DNA is a lot like that. The cell is faced with an enormous problem in packaging its chromosome into a single cell. In the case of *Escherichia coli,* the DNA is about 1 mm long, about the size of the period at the end of this sentence. The packaging for this chromosome, the *E. coli* cell, is only 1 μ long—about 1000 times shorter in length than the DNA. Human cells are much larger than the *E. coli* cell but the DNA in the 23 human chromosomes lined up end to end is nearly 1 m long! Not only does the DNA have to fit into the cell, but the DNA must be packaged so its information can be efficiently accessed by RNA polymerases, DNA polymerases, repair enzymes, and the myriad regulatory proteins of the cell.

The basic organizational rationales for packaging DNA in bacterial or higher eukaryotic cells are quite different. In bacterial cells, a large fraction of the genome is important for normal growth and maintenance of the cell during either exponential growth or stationary phase. There may be few regions of the chromosome that are inactive for very long periods of time. Most regions of the chromosome must be ready at a moment's notice to participate in transcription or regulatory activities. In the human genome, on the other hand, there is a great deal of "junk DNA" or DNA of undefined function. A large fraction of the genome contains repetitive elements (Alu sequences, for example) that do not appear to encode RNA with any informational or structural function critical to the cell. (See the box entitled "Junk DNA and Mismatch Repair.") Moreover, in a multicellular organism with complex developmental regulatory schemes, there are large blocks of genetic information that are not used in particular cell types. Certain genes may be utilized only during one short period during development. Eukaryotic cells have mechanisms for packaging regions of their chromosomes into configurations that do not become involved in transcription. This chapter will present a brief description of DNA supercoiling and the DNA–protein interactions that are responsible for the organization and packaging of DNA in prokaryotic and eukaryotic cells.

Junk DNA and Mismatch Repair

"Junk" DNA is defined as the repetitive sequences that make up much of the chromosome in many organisms. Junk DNA does not apparently encode proteins or RNA molecules needed for the immediate survival of the cell. Why would "junk" DNA be so prevalent if it were not important? Recently it has been suggested that "junk DNA" may be involved in preventing excessive genetic recombination betweel closely related species (Radman, 1991; Kricker *et al.*, 1992). One does not want recombination to tinker with DNA encoding a beautifully designed protein that may have required hundreds of thousands of years to develop. The DNA of different species can be distinguished by its repetitive DNA which, since it does not contain information that encodes proteins or RNA, can vary greatly from one species to the next. In the presence of enzymes that carry out genetic recombination very efficiently, divergent repetitive DNA would prevent genetic recombination between closely related species by stimulating the destruction of recombination events containing mismatches from recombining "junk" DNA. This is done by the mismatch repair system which is designed to detect and then correct regions of DNA containing mismatches.

B. The Packaging of DNA in Bacterial Cells

1. The Bacterial HU Protein

Griffith (1976) published pictures of DNA spilling out of an *E. coli* cell very quickly after the cell was broken open, in which the DNA complexed with proteins looked like "beads on a string." Griffith's pictures were remarkably similar to those observed for eukaryotic chromatin. There was about a six- to sevenfold compaction of DNA in the beaded structure. However, although eukaryotic nucleosomes are very stable, the *E. coli* nucleosomal structure was unstable and could only be observed immediately following lysis of the cell.

The protein responsible for the bacterial nucleosomes observed by Griffith may be the abundant HU protein. The bacterial HU protein (Rouviere-Yaniv and Gros, 1975) is a small basic protein of 18,000 daltons. It exists as a heterodimer of two nearly identical subunits (HU-1 and HU-2) (see Figure 8.13). HU binds to DNA and changes the shape and supercoiling of the DNA. There are also a number of other small abundant DNA binding proteins including H-NS (a histone-like protein previously called H1) in *E. coli* that may be involved in chromosome organization. These may act alone or they may interact with the HU protein to organize the chromosome *in vivo* (Drlica and Rouviere-Yaniv, 1987).

Rouviere-Yaniv *et al.* (1979) and Broyles and Pettijohn (1986) showed that the bacterial HU protein can act in a fashion consistent with the requirements for a DNA packaging protein. A packaging protein must compact the DNA or shorten its end-to-end distance. This can be accomplished by folding DNA back and forth or by coiling. The HU protein (and the histones described subsequently) utilize coiling as a mechanism of DNA compaction. The DNA is coiled around the body of the protein, thereby greatly reducing its overall length. In addition, coiling DNA defines a path or writhe of DNA in space. As discussed in Chapter 3, the negative superhelical energy of supercoiled DNA is manifest as changes in the twist of the double helix (T) and the writhe of DNA in space (W). Therefore, wrapping DNA around HU proteins in prokaryotes or around histones in eukaryotes influences the level of DNA supercoiling and the amount of free energy potentially available for biological reactions.

2. DNA Supercoiling and DNA Packaging

DNA supercoiling is a very important component in packaging DNA inside cells. Chromosomes in prokaryotes are frequently circular, which reduces the end-to-end length by a factor of at least 2. In addition, most circular DNAs isolated from cells are negatively supercoiled. The formation of interwound supercoils compacts DNA, as shown in the electron micrograph in Figure 3.2. The energy from negative DNA supercoiling can also be used to drive the wrapping of DNA into a toroidal coil around a protein (Figure 3.5). The writhe associated with wrapping around a protein will affect the superhelical topology of the rest of the DNA molecule. One 360° wrap around a protein will introduce a writhe of +1 or −1 depending on whether the wrap is clockwise (+1) or counterclockwise (−1). One toroidal 360° wrap is equivalent to one interwound supercoil *provided* there is no change in the twist (T) of the helix (see Figure 3.5).

3. Restrained and Unrestrained Supercoils

A negative supercoil can exist in two forms, *restrained* or *unrestrained.* A circular DNA molecule with one negative supercoil that exists as an interwound negative supercoil or as a toroidal coil that is not stably wrapped around a protein is considered *unrestrained.* The deficiency in linking number ($L - L_0$) is distributed in part as the negative supercoil $W = -1$ and in part as a deviation from the preferred twist (where $T_{\text{Preferred}} = L_0$). In the unrestrained supercoil, torsional strain or torsional stress is felt in the winding of the DNA double helix *over the entire DNA molecule.* If a nick were introduced anywhere in the circular molecule, the linking number would increase (to $L = L_0$) as negative supercoils were relaxed by the rotation of one strand about the other. Resealing the nick by the action of DNA ligase would result in the formation of a relaxed DNA molecule (Figure 9.1).

A negative supercoil is said to be *restrained* when the linking number deficit ($L - L_0$) is restrained by the stable coiling or writhing (W) of the DNA around a protein or by the stable interaction with a denaturing protein or a protein that changes twist. In the example shown in Figure 9.1, the single supercoil is restrained by wrapping around the protein. The part of the DNA molecule that is not physically associated with the protein is effectively relaxed since there is no linking number deficit distributed over this part of the DNA molecule. If this molecule is nicked, no unwinding occurs. When this molecule is resealed by DNA ligase, it retains its original linking number deficit ($\Delta L = -1$).

4. Are Supercoils Restrained in *Escherichia coli*?

Pettijohn and Pfenninger (1980) showed that supercoils were restrained in living *E. coli* cells. To demonstrate this, they introduced nicks into DNA *in vivo* by irradiating cells with X-rays. Following incubation in media, the breaks were repaired in cells by the action of DNA repair enzymes and DNA ligase. Before being sealed, however, the breaks provided a swivel through which supercoils could be lost by the rotation of one strand around the other. *In vitro,* following the introduction of nicks into purified protein-free DNA,

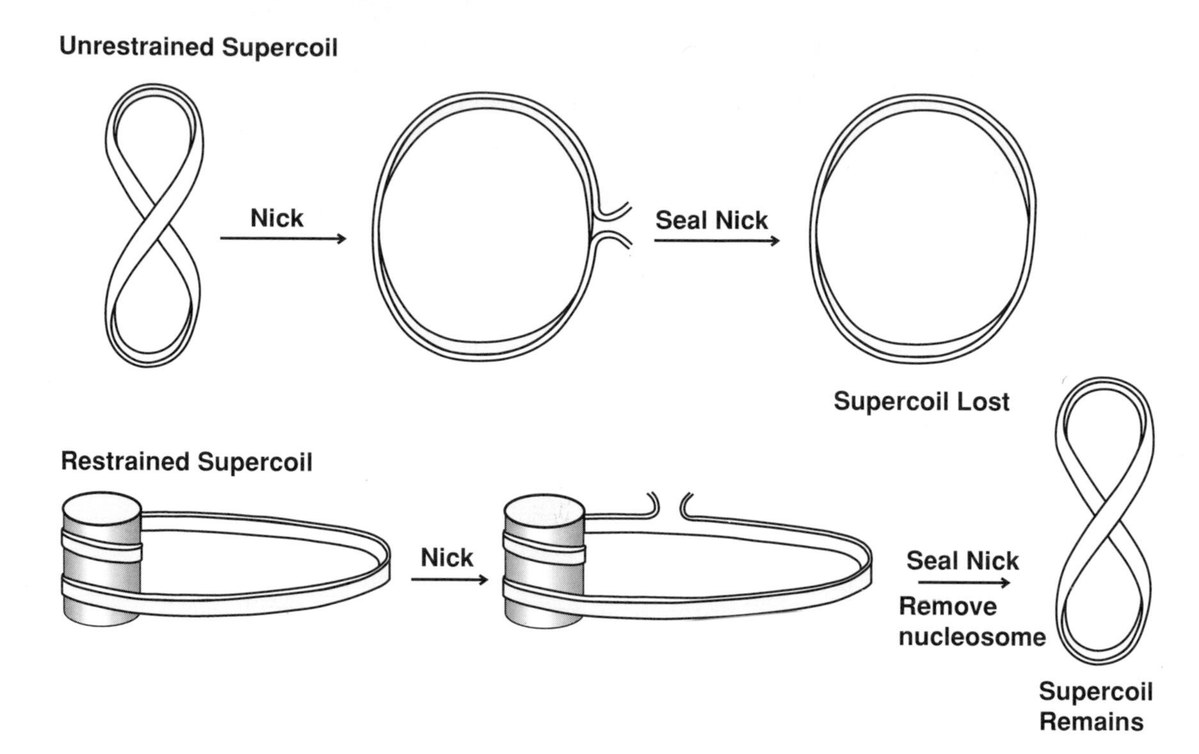

Figure 9.1 Restrained and unrestrained supercoils. (Top) On the introduction of nick into a circular DNA molecule, a swivel is introduced from which one strand can rotate around the other (shown by breaking one edge of the rubber band model). This allows the deficiency in the helical winding inherent in negatively supercoiled DNA ($L < L_0$) to be restored (an increase in L), resulting in the loss of the negative supercoil. (Bottom) DNA containing one negative supercoil in which the supercoil is restrained by the stable winding of the DNA in a left-handed fashion around a protein. In eukaryotes, supercoils are restrained by nucleosomes. On introduction of a nick into the DNA, there is no change in the linking number when this nick is sealed because the single negative supercoil is organized in a stable fashion by winding DNA around the protein. The DNA that is not associated with the protein is not wound with negative unrestrained superhelical tension.

all supercoils are rapidly lost. To determine if all supercoils would be lost from a large circular DNA (an F plasmid) packaged in *E. coli,* it was necessary to prevent the introduction of new supercoils into the DNA. Thus, during repair, cells were incubated in media containing coumermycin, an inhibitor of DNA gyrase (the enzyme responsible for supercoiling DNA). Consequently, if all supercoils were lost by the introduction of a nick in the DNA, no new supercoils could be reintroduced into DNA by the action of DNA gyrase. When DNA was purified from cells in which repair occurred in the presence of coumermycin, the F plasmid DNA contained about half the number of supercoils as DNA purified before nicking *in vivo*. This result demonstrated that about half the supercoils present in DNA in living *E. coli* cells are restrained, presumably by association with specific proteins in cells.

5. The HU Protein Can Restrain Supercoils *in Vitro*

Rouviere-Yaniv *et al.* (1979) and Broyles and Pettijohn (1986) demonstrated that, when bound to DNA *in vitro*, HU compacts DNA and restrains DNA supercoils. Topoisomerases can relax a protein-free supercoiled DNA molecule. If supercoils are restrained by the interaction of DNA with a protein, they will not be lost through the action of a topoisomerase. Moreover, if relaxed DNA wraps around a protein to restrain a negative supercoil, a positive supercoil must be introduced into another part of the molecule (to satisfy the relationship $L = T + W$). If topoisomerase activity is present to relax the positive supercoil introduced by the negative superhelical wrapping, then there will be a net *introduction* of one negative supercoil into DNA (see Figure 3.10B). By incubating DNA in the presence of HU protein and topoisomerase, HU can be shown to restrain supercoils in negatively supercoiled DNA and introduce restrained supercoils in relaxed DNA.

The number of supercoils that could be restrained by HU was linearly dependent on the HU concentration. At a 1 : 1 weight ratio of HU protein to DNA, a maximum number of supercoils was restrained by HU protein. The addition of more HU did not result in the restraint of additional supercoils. The maximum number of supercoils HU restrained *in vitro* was $\sigma \approx -0.03$ or about half that found in DNA purified from cells. Perhaps not surprisingly, this was the same level of restraint that Pettijohn and Pfenninger (1980) found *in vivo* after nicking DNA with X-rays. (See the box entitled "The Topology of the DNA–Hu Interaction.")

6. HU Binds Transiently to DNA

Griffith found that, following lysis of cells, the DNA–protein complexes rapidly dissociated, indicating that the DNA packaging that existed in *E. coli* was not very stable. The *in vitro* results of Broyles and Pettijohn supported the idea that HU binds weakly to DNA. In physiological levels of salt, the half-time ($t_{1/2}$) for the dissociation of DNA was too rapid to measure. At a lower salt concentration (0.05 *M*), $t_{1/2}$ of HU binding was longer and kinetic measurements could be made. This rapidly dissociable binding may be advantageous for *E. coli* in which most of the genome must be accessible to regulatory proteins and RNA polymerase.

The Topology of the DNA–HU Interaction

The maximum number of supercoils restrained by HU corresponds to a superhelical density of $\sigma = -0.03$. Superhelical density (σ) equals $[(10.5 \times W) \div N]$, where W is the number of supercoils and N is the number of base pairs in the DNA molecule. At a density of $\sigma = -0.03$ there is one negative supercoil per 290 bp of DNA. Although one negative supercoil is restrained, this provides no information about the path of the DNA around the HU protein. From the equation $L = T + W$, if L and T are known then W, the path or shape of the DNA, could be determined for the HU–DNA complex. The interaction of 290 bp of DNA with HU restrains one negative supercoil, or $\Delta L = -1$ in 290 bp of DNA. Can the twist of DNA in a DNA–HU complex be determined?

The twist or helical pitch of DNA can be determined by nuclease digestion experiments (Klug and Lutter, 1981). If DNA with a helical repeat of 10.5 bp lays on a flat surface, a nuclease has access only to the phosphodiester bonds away from the surface. This will result in cutting the DNA at intervals of 10.5 bp. If the helix twist of DNA on a surface is 8 or 12 bp, then the nuclease cutting pattern will reflect this 8- or 12-bp pattern, respectively. The surface protecting regions of the double helix from nuclease digestion need not be flat. If DNA is wrapped around a protein, the region of the DNA phosphodiester backbone on the inside, next to the protein, will not be accessible to nuclease exchange. The regions of the phosphodiester backbone on the outside of the surface will be susceptible to nuclease attack. Broyles and Pettijohn (1986) utilized nuclease digestion of DNA–HU complexes to determine that the helical twist of DNA associated with the protein was 8.5 bp. This is an astounding change from the helical repeat in solution of 10.5 bp and represents an overwinding of the helix in terms of the number of helical turns in a DNA molecule. This situation, with $L > L_0$ and fewer base pairs per turn, creates an energetically unfavorable, *positively supercoiled* situation. If the helical periodicity of DNA in the HU complex is in the direction of positive supercoils, how can negative supercoils be introduced by HU? The answer must lie in the path or writhe of the DNA in the HU complex.

To calculate the ΔW component of DNA in an HU complex, consider that 290 bp of DNA restrains one negative supercoil ($\Delta L = -1$). The twist number (or number of helical turns) of a relaxed DNA fragment of 290 bp is $T_{relaxed} = 290 \div 10.5 = 27.62$. The twist number of DNA in the HU complex is $T_{HU} = 290 \div 8.5 = 34.12$. This means that, on organization around HU, there are 6.5 more helical turns introduced into the 290 bp of DNA. Since $\Delta L = \Delta T + \Delta W$, with $\Delta L = -1$ and $\Delta T = +6.5$, ΔW must be -7.5. This means that the DNA must be forming 7.5 left-handed toroidal coils in the 290 bp of DNA. One toroidal coil must therefore oc-

continues

continued

cur in 38–39 bp of DNA. At a 1:1 mass ratio there are about 4 HU monomers per 60 bp of DNA (or about 19 monomers of HU for the 290 bp). These considerations suggested the DNA–HU organization shown by Drlica and Rouviere-Yaniv (1987; Figure 9.2). DNA wraps around an HU tetramer in about 39 bp and a 19-bp linker gene goes to the next HU tetramer. The expected effect of the toroidal coiling on supercoiling is negated for the most part, by a change in the helical twist of the DNA. The organization shown in Figure 9.2 would contribute to the compaction of the DNA in prokaryotic cells.

There is a second model for the organization of HU in which HU simply kinks DNA every 8–9 bp (the size of the binding site for one dimer) as the DNA becomes organized into a coil (White *et al.*, 1989).

7. HU Bends DNA

From the topology of DNA supercoiling on interaction of HU with DNA, it is expected that HU will bend DNA sharply into a tight circle. Hodges-Garcia *et al.* (1989) tested this possibility by measuring the circularization rate of small DNA fragments (<200 bp). Typically, it is very difficult to ligate DNA fragments shorter than the persistence length of DNA into a circle. For example, 80-bp fragments could not easily bend or be ligated into a circle in the absence of the HU protein. However, in the presence of HU, 80-bp fragments of DNA were efficiently ligated into a circle.

HU has been implicated in a variety of reactions involving DNA. In some cases the increased bendability or flexibility of DNA in the presence of HU is important. For example, the HU protein may facilitate wrapping the *E. coli* chromosome replication origin, *oriC*, into a complex with the initiation specific DnaA protein; see Figure 7.5.

C. The Packaging of DNA in the Eukaryotic Nucleosome

DNA in eukaryotic cells is organized into nucleosomes, in which the DNA is wrapped into two left-handed coils around a histone octamer. A core nucleosome consists of two subunits each of small basic histone proteins H2A, H2B, H3, and H4. Histones bind tightly to DNA, forming very stable complexes. Because histones must be intimately involved with the organization and expression of eukaryotic DNA, their structure and properties have

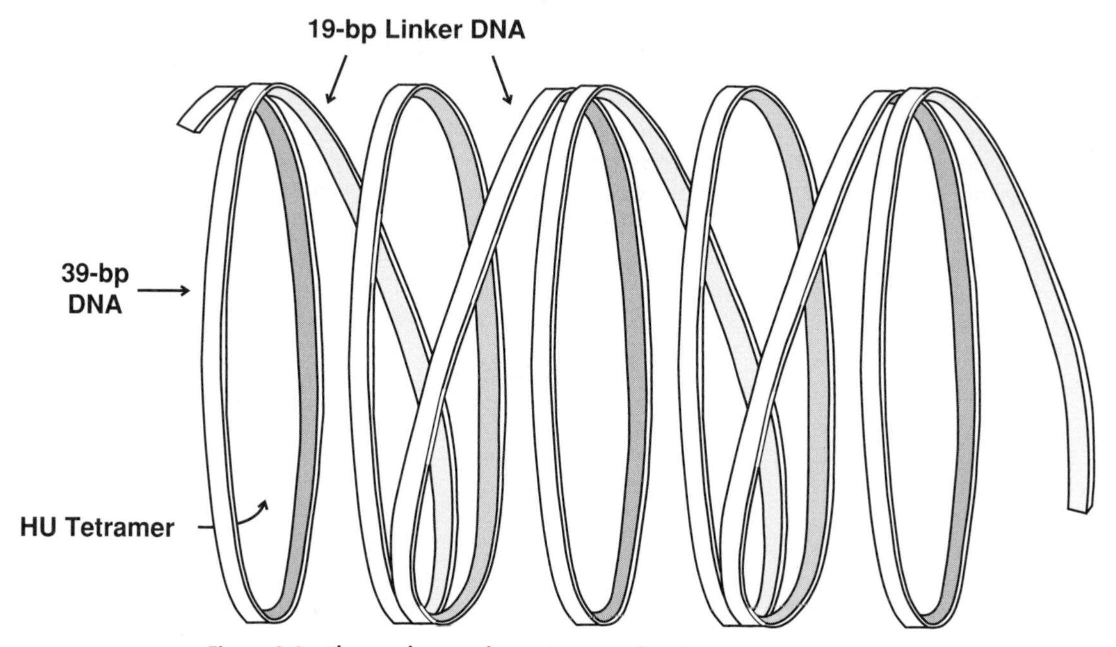

Figure 9.2 The topology and organization of DNA around an HU tetramer. The experiments of Broyles and Pettijohn (1986) suggested that 290 bp of DNA were associated with 20 HU tetramers. It was known that one negative supercoil was restrained in 290 bp and that a change twist to 8.5 bp per helical turn occurred in the DNA associated with HU. As described in detail in the text, seven negative supercoils would need to be introduced within the 290 bp. In the model presented by Drlica and Rouviere-Yaniv (1987), 39 bp of DNA are associated in a tight stable left-handed toroidal coil with around 4 HU monomers (proteins not shown). Between the five 39-bp toriodal coils, 19-bp linkers occur. They represent one half of a toroidal turn each. Consequently, the 290 bp were made up of five 39-bp coils around HU tetramers, linked together by four 19-bp spacers. This accounts for the required seven negative superhelical turns associated with the organization of 290 bp of DNA around 20 HU monomers. Figure modified with permission from Drlica and Rouviere-Yaniv (1987).

been studied extensively for a number of years (for review, see van Holde, 1989; Thoma, 1991; Felsenfeld, 1992; Morse, 1992; Wolffe, 1992).

1. The Structure of the Nucleosome

In 1984, Richmond *et al.* reported the structure of the nucleosome determined by X-ray crystallography. The nucleosome is a disk 57 Å thick and 110 Å across around which 145 bp of DNA wraps in a left-handed coil. The two H3 and H4 subunits form a symmetrical unit at the center of the disk. Two pairs of H2A–H2B dimers then bind to the outside of the H3–H4 complex completing the octameric histone core. The H3–H4 central protein complex is believed to contact the central 70–80 bp of the DNA. The outer regions of the DNA are associated with the H2A–H2B subunits as well as with the H3 subunit. A schematic model for the organization of the nucleosome is shown in Figure 9.3. The DNA completes 1.6–1.8 turns around the nucleosome within 145 bp. Arents and Moudrianakis (1993) have described a detailed view of the organization of DNA around a nucleosome.

A single fifth histone called H1 binds to the outside of the nucleosome and is believed to contact the DNA where it enters and exits the nucleosome. With H1 present, about 167 bp of DNA are associated with the nucleosome (H1 is not shown in Figure 9.3). The 167 bp of DNA form two complete helical turns. The nucleosome with an H1 histone bound is called a *chromatosome*. DNA in a chromatosome is organized in a more stable and more compact form than in the nucleosome. Regions of DNA that are transcriptionally inactive are believed to be organized as chromatosomes.

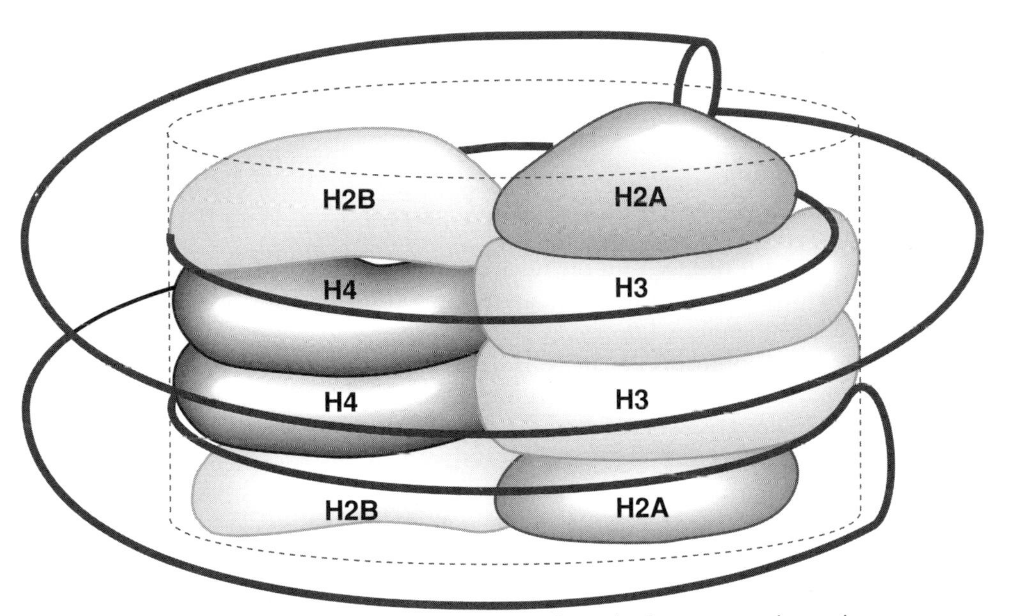

Figure 9.3 The organization of DNA in nucleosomes. The path of DNA around a nucleosome core is shown. Nucleosomes are composed of two subunits each of H4, H3, H2B, and H2A histones. Two H4, H3 dimers are associated at the center of the nucleosome and H2B, H2A dimers are positioned on top of H4 and H3, respectively. The nucleosome makes a cylinder roughly 57 Å in length by 110 Å in diameter. The path of the DNA is shown as the tube running around the surface of the nucleosome core. DNA makes about 1.75 left-handed rotations around the nucleosomes in a space of about 145 bp. The outer edges of these 145 bp are associated predominantly with the H2A and H2B subunits. The central region of the 145 bp is more tightly associated with the H3 and H4 subunits.

Other proteins called HMG (*high mobility group*) proteins are often found with nucleosomes in regions of chromosomes that are transcriptionally active. These proteins have recently been shown to bind to cruciform structures (see Lilley, 1992, for review). Biologically, they may bind two helices of DNA together as they cross in space, stabilizing an X structure of DNA. Specialized texts discuss other chromatin-associated proteins (van Holde, 1989; Wolffe, 1992).

Histone subunits are also subject to covalent modifications such as phosphorylation and acetylation. These modifications can influence the protein–protein–DNA interactions and may affect the biology of the histone–DNA interaction. For example, the acetylation of histones H3 and H4 on the ε amino group of lysines can destabilize the nucleosome (Oliva *et al.*, 1990). The level of restraint of negative superhelical tension may be altered when histones are acetylated (Norton *et al.*, 1989,1990). Acetylation is associated with regions of chromatin that are active in transcription. The destabilization of the nucleosome should facilitate the passage of RNA polymerase through the nucleosome during transcription. In addition, Lee *et al.* (1993) have shown that histone acetylation can lead to binding of transcription factor TFIIIA to a *Xenopus borealis* 5S RNA gene organized in a nucleosome. Without acetylation, the transcription factor cannot bind its recognition sequence.

2. Nucleosomes and the Restraint of Supercoiling

The path of DNA on the nucleosome constitutes about 1.75 left-handed turns. However, a nucleosome appears to restrain only one negative supercoil, a phenomenon known as *the linking number paradox*. For example, SV40 minichromosomes can be isolated as circular molecules with about 25 nucleosomes on the DNA. There is no torsional stress in the winding of the helix in linker DNA between the nucleosomes and, as seen using the electron microscope, the circles appear relaxed. When the histones are removed by treatment with a very high salt concentration, detergent, or protease, 25 negative supercoils are detected in the DNA. The numbers may seem about 2-fold low considering the topological rules for DNA supercoiling where $L = T + W$ (see Chapter 3). Although there are 25 negative supercoils in SV40 DNA ($L - L_0 = -25$), the writhe of the DNA is described by $W = 25 \times -1.75 = -43.75$. As discussed in Chapter 3, one writhe is equivalent to one interwound supercoil. Why do 1.75 turns not restrain 1.75 supercoils? To maintain the relationship $L = T + W$, where $L = -25$ and $W = -43.75$, T must increase relative to a situation where $L = W$. If the helical twist of the DNA wrapped around the nucleosome changes from 10.5 bp per turn to about 10 bp per turn, the overall increase in T will compensate for the 1.75 writhes in the path of DNA around the nucleosome (see Hayes *et al.*, 1991).

3. Nucleosome Positioning

The organization of DNA into nucleosomes can have dramatic consequences for gene expression and DNA repair. Knowledge of the factors that govern the placement and level of association of nucleosomes on DNA is critical to understanding the regulation of gene expression. This section briefly summarizes the biological implications and significance of nucleosome positioning.

a. Models for Positioning

Nucleosomes can be *randomly* or *precisely positioned* on a DNA molecule (Figure 9.4). A random organization would exist if all DNA molecules had an equal ability to wrap around histones. With the immense variability of helical structure and deformability inherent in different sequence organizations, a totally random organization is unlikely. In fact, sequences of DNA that resist bending are not usually associated with nucleosomes, whereas certain bent regions strongly associate with nucleosomes. The term "*nucleosome phasing*" has been used to describe the relationship between nucleosomes on DNA. Phased nucleosomes have a uniform spacing, resulting in a uniform length of the linker DNA. Phased nucleosomes can be random or positioned as shown in Figure 9.4A.

Translational and Rotational Positioning. In 1985, Drew and Travers described a subtle elegance to the way DNA can orient itself in solution. A 179-bp piece of DNA from a *tyrT* gene (encoding a tryptophan tRNA) was known to be bent since it contained several phased A tracts. When circularized, the 179 bp of DNA oriented themselves in space in a precise, nonrandom configuration with the bends directed toward the center of the circle. In this configuration, the phased A tracts were oriented with the minor grooves of the A tracts on the inside of the circle and minor grooves of the G + C-rich regions on the outside of the circle. This was determined from the preferential DNase I cutting of the A tracts in the circular molecule. There was a periodicity of cutting at 10.6-bp intervals, in phase with the A tracts. If individual DNA molecules were oriented randomly in space, then DNase I would not have shown preferential cutting sites within the A tracts, nor would a periodicity of cutting have been observed. (A caveat of this experiment is that the nuclease, DNase I, must be too large to get inside and act on the strands in the interior of the small circle of DNA. Otherwise cutting would have occurred randomly throughout the DNA molecule.) This simple experiment demonstrated a *rotational* preference for DNA when organized into a circle (Drew and Travers, 1985).

When the linear *tyrT* DNA was associated with a nucleosome, the same rotational organization was observed. DNase I showed preferential cutting at 10.6-bp intervals, in phase with the A tracts. This result demonstrated a pre-

A

Randomly Positioned Nucleosomes

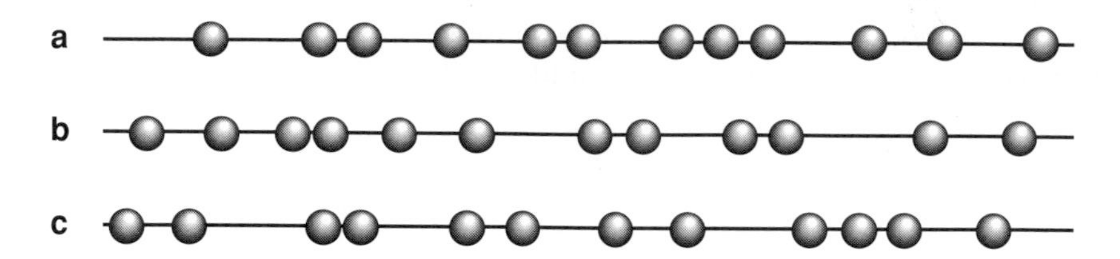

Phased Nucleosomes, Randomly Positioned

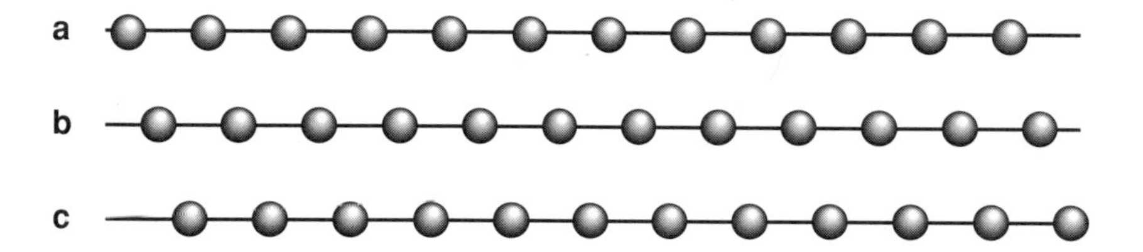

Phased Positioned Nucleosomes

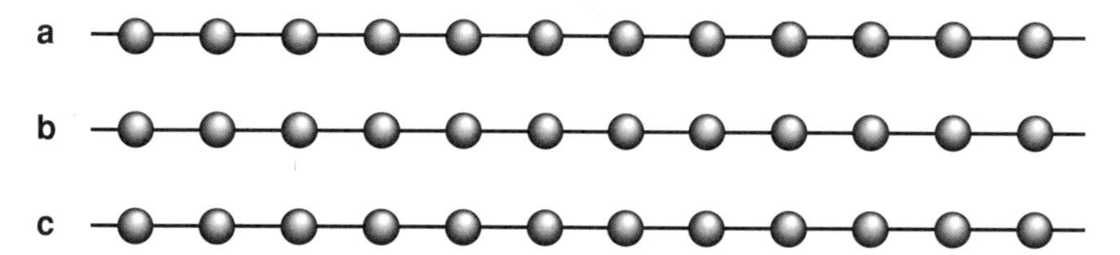

Figure 9.4 Nucleosome phasing. Nucleosomes can associate with DNA in a number of different ways. (A) (*Top*) On three identical DNA molecules, the position of the nucleosomes is different. (*Middle*) Nucleosomes can be evenly spaced. This is referred to as "phasing," but in this example the phased nucleosomes are randomly positioned on identical pieces of DNA. DNA molecules a, b, and c have evenly spaced nucleosomes but the exact position of the nucleosome within a gene or a particular piece of DNA varies from molecule to molecule. (*Bottom*) Three molecules in which nucleosomes are evenly spaced or phased, and in which they are positioned on the same regions of each individual DNA molecule. (B) Nucleosomes can be rotationally phased and *translationally positioned,* as shown. Rotational phasing refers to the organization of sequences of DNA on the outside or inside of the DNA that is wrapped around the nucleosome. Since many DNA molecules will organize themselves into a stable bend or curve, it may seem reasonable that as DNA coils around nucleosome it may do so using the natural curvature inherent in a particular DNA sequence. A naturally curved piece of DNA can associate with nucleosomes in a number of different *translational positions.* At the same time, the DNA can use its natural curvature to wrap around the nucleosome

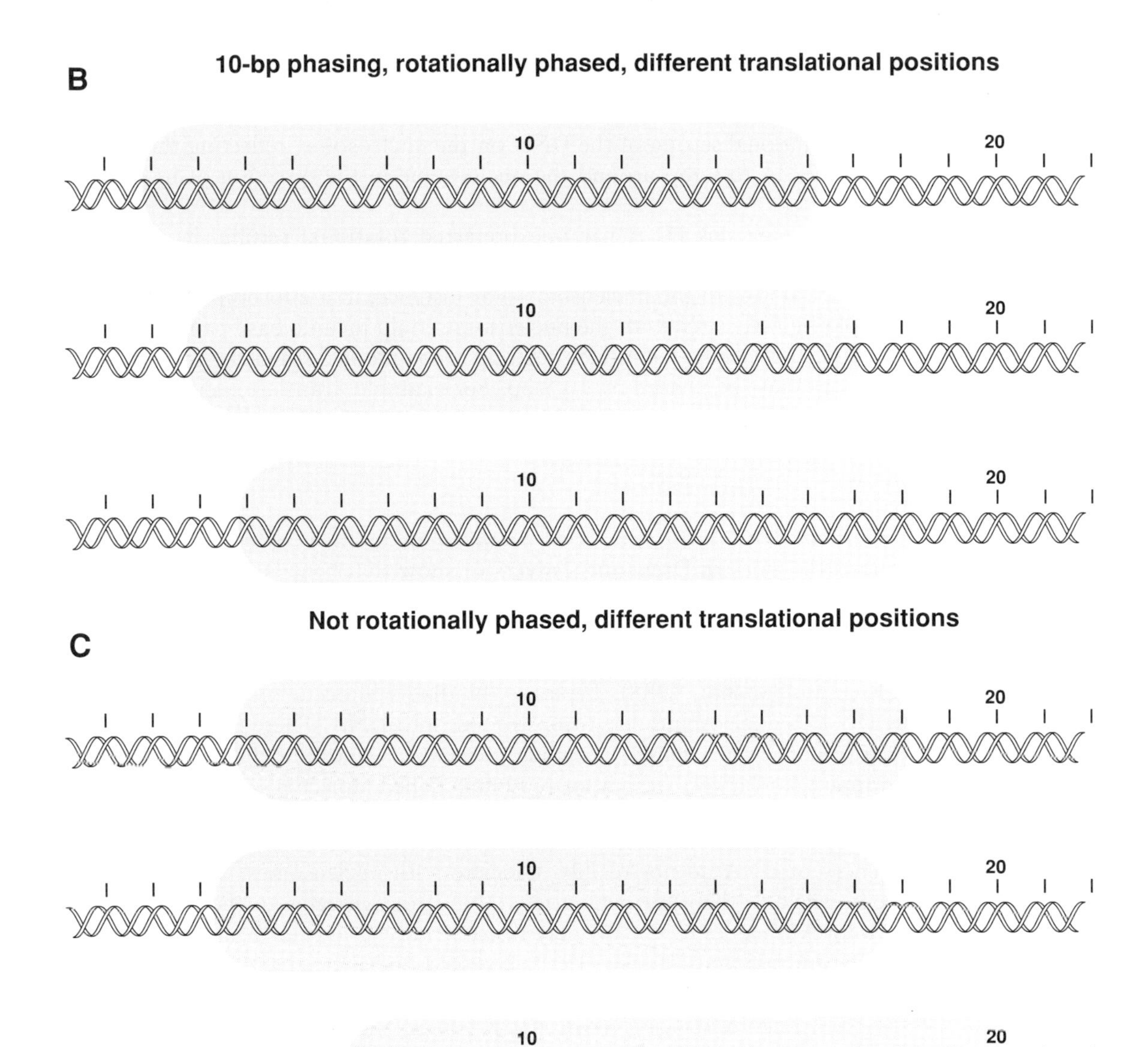

Figure 9.4 *Continued.* with the same *rotational positioning.* The three individual molecules contain a nucleosome shown by the shaded area which covers 145 bp of DNA. Each nucleosome is associated with the DNA helix at 10-bp intervals. This means that the same base pairs spaced at 10-bp intervals will be on the inside of the curve and make contact with nucleosome. However, these bases will make contact with different regions of the core histone proteins within the nucleosome. (C) This set of three molecules shows nucleosomes that are neither rotationally phased nor differently positioned translationally. In this case, different regions of the DNA will be in contact with the nucleosomal core proteins. In one molecule, particular base pairs may be on the inside of the curvature in contact with histones whereas in another these bases may be on the outside of the helix in contact with the solvent.

ferred rotational setting of the DNA on the nucleosome, reflecting the preference of DNA to wrap around the nucleosome in the direction of its natural bend.

Although this DNA has one preferred rotational setting, it may have several translational positions. A translational position refers to which bases are associated with the nucleosome. For instance, in a 200-bp piece of DNA, the 145 bp associated with the nucleosome could include base pairs 1–145 or 55–200. If the same rotational position is maintained, then the translational positions must be phased by 10.5 bp. For example, translational positions of base pairs 1–145, 11.5–155.5, 22–166, 32.5–176.5, 43–187, or 53.5–197.5 are rotationally phased. Different patterns of nucleosome phasing are shown in Figures 9.4B and 9.4C.

Bent DNA Positions Nucleosomes. A number of studies have extended the work of Drew and Travers to show that bent DNA strongly positions nucleosomes (Shrader and Crothers, 1990). DNA sequences from the bacterial plasmids can position nucleosomes. There is a strong bend near the terminus of SV40 DNA replication that binds nucleosomes very tightly (Hsieh and Griffith, 1988). Typically, binding will occur first at bent regions of DNA as nucleosomes are reconstituted onto DNA. There must be a lower energy barrier for tightly wrapping curved DNA around histones than for wrapping a relatively straight or randomly coiled molecule.

Sequences That Resist Organization into Nucleosomes. Certain sequences of DNA do not readily associate with nucleosomes. Sequences that are less flexible than "garden variety DNA" (or average sequence DNA) such as long A tracts or G·C runs may resist wrapping around histones (Rhodes, 1979; Simpson and Kunzler, 1979; Jayasena and Behe, 1989). Long A tracts are rarely found near the center of nucleosomes but tend to be organized at the ends of DNA associated with nucleosomes (Satchwell *et al.*, 1986). At the region where DNA enters and exits the nucleosome, the DNA may be more flexible and may be less tightly wrapped than DNA at the center of the nucleosome.

Nucleosomes and Gene Expression. This section will present a brief overview of the potential effects of nucleosomal organization on gene expression (for review, see Gross and Garrard, 1988; Grunstein, 1990; Felsenfeld, 1992; Pina *et al.* 1991; Freeman and Garrard, 1992; Garrard, 1990; Wolffe, 1992). The organization of DNA into nucleosomes generally represses gene expression. The level of nucleosomal organization in terms of higher order packing (which is described later) affects the level or degree of repression. Nucleosomal DNA has been divided into two classes: *heterochromatin* and *euchromatin.* Heterochromatin has historically been defined as densely staining regions of chromosomes that condense early in prophase and that replicate late in S phase. These regions include centromeres and telomeres, and represent regions of the genome that may be inert transcriptionally. Females contain two X chromosomes, one of which is inactive. The inactive X chromosome is maintained as a heterochromatic structure called a Barr body. If a recombinant DNA construct containing a gene that can be expressed in the eukaryotic genome is introduced into a heterochromatic region it will become transcriptionally inactive, a situation known as a *position effect.* Euchromatin refers to the rest of the genome that does not stain as intensely. Euchromatin contains regions of DNA that are transcriptionally active or that may need to be transcriptionally active at some point during the growth and development of the cell.

Gene expression in euchromatin is not controlled by the level of DNA accessibility alone, but involves the presence and action of transcription factors. The deposition of nucleosomes on promoters can lead to an inability of RNA polymerase II to initiate transcription of the gene. Knezetic and Luse (1986) showed an inverse correlation between the level of deposition of nucleosomes on DNA and transcriptional initiation activity. As more nucleosomes were reconstituted onto DNA, transcription initiation was inhibited. RNA polymerase is apparently unable to bind to a promoter if it is wrapped around a nucleosome. If nucleosomes are reconstituted onto a DNA template that has one or more transcription factors bound to the promoter, the transcription factor-bound promoter will not become organized into a repressive nucleosome. Even after the heavy deposition of nucleosomes, transcription initiation can occur from these templates when relevant transcription factors and RNA polymerase are supplied. (Workman and Roeder, 1987; Knezetic *et al.,* 1988).

b. What Happens When RNA Polymerase Encounters a Nucleosome?

When RNA polymerase encounters a nucleosome, it can either stop or transcribe through the nucleosome (for review, see Thoma 1991; Morse, 1992; van Holde, 1992). Nucleosomes may be inhibitory and prevent movement of RNA polymerase. Under certain experimental conditions nucleosomes can prevent transcription elongation. Alternatively, RNA polymerase may transcribe through nucleosomes. For a gene to be transcribed, the DNA must presumably unwrap from the core histones, although the nucleosome may not entirely dissociate from the DNA.

There are several possible mechanisms by which transcription could occur through a nucleosome (see Wolffe, 1992). First, DNA could unwrap one turn from the nucleosome, allowing transcription of 70 bp of DNA. Once transcribed, the DNA might rewrap around the histone core while the second 70-bp turn of DNA begins to unwrap and be transcribed. Second, the nucleosome may dissociate and unfold into two tetrameric subunits (half nucleosomes) each consisting of one subunit of H2A, H2B, H3, and H4 (Weintraub *et al.,* 1976). There is chemical and biophysical evidence that the nucleosomes on transcriptionally active chromatin are somewhat unfolded or less tightly associated into a core octamer (Prior *et al.,* 1983; Cartwright and Elgin, 1986). Each of these half nucleosomes may be loosely bound to the DNA. Third, positive supercoiling generated ahead of transcription by RNA polymerase may weaken the nucleosome organization, leading to dissociation from the DNA (Clark and Felsenfeld, 1991; Lee and Garrard, 1991a,b). Nucleosomes may rapidly reform behind RNA polymerase in a localized region of negative supercoiling.

Biochemically, what distinguishes a nucleosome on inactive DNA from a nucleosome on DNA that is transcriptionally active? The answer to this question is not yet known with certainty, but chemical modifications of histones may be associated with transcriptionally active nucleosomes. There are a number of sites on histones that can be either phosphorylated or acetylated. Most of these sites are within the NH_2-terminal ends of the core histones. The amino group of lysine, which has a positive charge that may be involved in binding the histone to the negatively charged DNA phosphate backbone, is the site of acetylation. The addition of the acetyl group ($-COCH_3$) to the amino group removes the positive charge on the lysine. If the tenacity of histone binding to DNA involves electrostatic interaction, acetylation will weaken histone–DNA binding. Thus, acetylation may "loosen up" the nucleosome and facilitate movement of RNA polymerase through DNA. In fact, when histones are purified from transcriptionally active chromatin, histones H3 and H4 are hyperacetylated. Moreover, histone acetylation can result in a change in the amount of supercoils restrained by nucleosomes suggesting that some supercoiling might be introduced by histone modification (Norton *et al.,* 1990).

Transcriptionally Active Genes Can Contain Precisely Positioned Nucleosomes in Vivo. Several examples are known in which nucleosomes *in vivo* are precisely positioned across a gene. Certain sequences in DNA can strongly position a nucleosome because of the intrinsic curvature or flexibility of a particular sequence. Sequence information, which influences the structure and shape of DNA, may direct nucleosomes to a preferred rotational and translational setting. For example, the precise positioning of nucleosomes in the *his4* and *pho5* genes in yeast leaves the promoter regions available for the binding of transcription factors and RNA polymerase (see Simpson, 1990, for review).

D. The Organization of Chromosomes in Cells

1. The Bacterial Nucleoid

The *E. coli* chromosome is a large circular molecule in which the DNA is negatively supercoiled. It would seem advantageous to partition the chromosome into independent topological regions or domains in which supercoiling might be differentially regulated. Since certain genes are optimally transcribed at a precise level of supercoiling, topological domain organization might provide one mechanism for global gene regulation. Independent domain organization also conserves the chemical energy required to supercoil DNA on introduction of a nick into DNA. Without multiple topological domains, a single break in the chromosomes could provide a swivel through which all supercoils could be lost.

In the 1970s the bacterial chromosome was characterized *in vitro* (for review, see Drlica and Riley, 1990). Methods were developed to gently break open the bacterial cell and to characterize the *entire* circular *E. coli* chromosome using gentle physical chemical methods (Stonington and Pettijohn 1971). The purified *E. coli* chromosome is divided into about 45 independent *topological domains* or loops of DNA (Figure 9.5; for review, see Pettijohn and Sinden, 1985). A topological domain is defined as a region of DNA in which the rotation of the DNA strands at the ends of the domain is prevented. Newly synthesized (nascent) RNA molecules defined domains in the purified chromosome, presumably by binding to two different regions of the DNA. However, nascent RNA molecules are not responsible for defining domains in living cells (Sinden and Pettijohn, 1981). Although it is possible that type II topoisomerases, or specific proteins, may be responsible for defining domains in living cells, definitive *in vivo* experiments addressing this question are lacking.

The level of supercoiling *in vivo* appears to be between $\sigma = -0.025$ and $\sigma = -0.05$. Estimates of the level of supercoiling *in vivo* come from experiments employing various assays for negative superhelical tension. One assay involves measuring the rate of psoralen photobinding to DNA (see Figure 1.22). The intercalative binding of psoralen to DNA, which results in unwinding of the helix and the relaxation of supercoils, is thermodynamically favored in negatively supercoiled DNA (Sinden *et al.*, 1980; Sinden and Pettijohn, 1981). The estimation of supercoiling in living cells involved measuring the rate of psoralen photobinding to DNA in *E. coli* under conditions in which the DNA should be supercoiled. This rate was compared with that measured in cells following the relaxation of supercoils. Supercoils are relaxed in cells following the introduction of nicks in DNA by X-ray irradiation or following relaxation by treatment with DNA gyrase inhibitors.

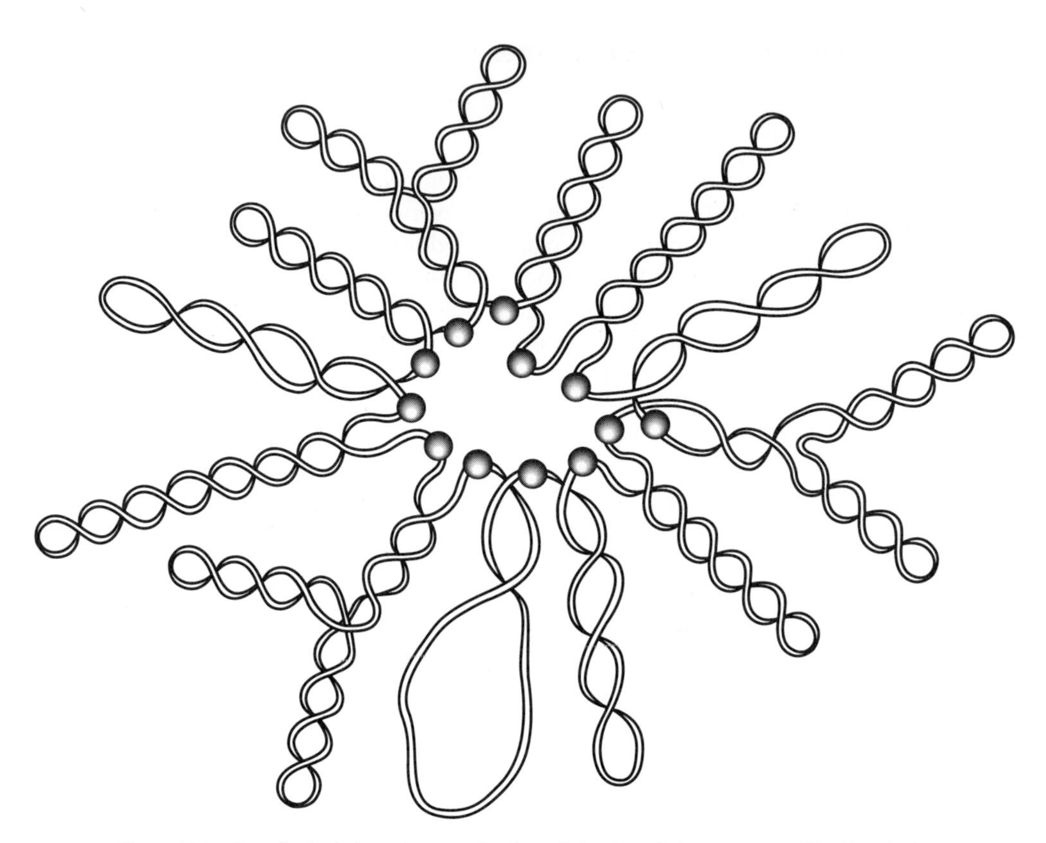

Figure 9.5 Topological domain organization of the *E. coli* chromosome. The *E. coli* chromosome is organized into about 45 independent topological domains or loops of DNA. The nature of the interactions responsible for organizing these topological domains *in vivo* is not yet fully understood. The domain organization may involve the binding of DNA to topoisomerases such as *E. coli* DNA gyrase or the *par* gene products, or binding to other unidentified domain organizing proteins. (Quinolone antibiotics will induce DNA gyrase to cleave the *E. coli* chromosome into 50 pieces, suggesting an equal domain-sized spacing of DNA gyrase.) The shaded spheres represent the domain boundaries. These may represent sites of stable coiling of DNA around a protein that has been bound to the membrane or otherwise prevented from rotation inside a living cell. This structure contains 12 independent topological domains in which DNA is negatively supercoiled. By organizing DNA into independent topological domains, the level of supercoiling can theoretically vary independently within individual domains. For example, the domain near the bottom of the structure is nearly relaxed, containing only two negative superhelical turns. The adjacent domain in a counterclockwise position contains four negative supercoils whereas the next domain contains five negative supercoils. Two domains contain a branched supercoiled structure of DNA. It is known that the regulation of gene expression from certain promoters in *E. coli* is dependent on the level of negative supercoiling. Certain genes are stimulated and certain genes are inhibited by increases or decreases in the levels of supercoiling. This topological domain organization would theoretically allow the maintenance of differences in the levels of supercoiling in different parts of the chromosome. Consequently, certain families of genes could be turned on or off independently.

Following either treatment, the rate of psoralen photobinding decreased, due to the lower binding constant of psoralen to relaxed rather than supercoiled DNA. Other estimates of *in vivo* supercoiling in plasmids involve the detection of certain superhelical-density-dependent conformations of DNA including cruciforms and Z-DNA. A precise level of supercoiling is required to drive the formation of alternative DNA conformations (cruciforms, Z-DNA, intramolecular triplexes). Therefore, measuring the amount of the alternative conformation of one of these "torsionally tuned DNA probes" provides an estimate of the level of energy in the DNA helix in the living cell (see relevant boxes in Chapters 4 and 5). The level of supercoiling in cells has also been inferred by analyzing products of recombination, which are characteristic for certain energy levels in DNA (Bliska and Cozzarelli, 1987) or by measuring the level of transcription from genes that are sensitive to the level of supercoiling (Borowiec and Gralla, 1987).

Many questions about the organization of the bacterial chromosome *in vivo* remain unanswered. Is the region of the genome that constitutes a topological domain defined or fluid? If domains were defined, a particular length of the *E. coli* chromosome would always constitute a single domain. A single gene, for example the gene for topoisomerase I, might be at the center or at the base of a domain. Alternatively, domains may be fluid and constantly changing. DNA may slide through domain anchorage points on the membrane. It is also possible that regions of DNA could constitute defined domains under certain growth conditions or temperatures, or when cells enter stationary phase. The domain organization may be influenced by differences in the proteins present in cells in different growth stages.

Another question is whether the level of supercoiling may be different in individual domains of the chromosome. This may be important for the regulation of gene expression from supercoil-dependent promotors. It is known that different parts of one domain (a plasmid molecule) can be differentially supercoiled by transcriptional activity. It is not known, however, whether different domains of the *E. coli* chromosome are maintained at different average levels of negative supercoiling.

2. The Organization of Chromosomes in Eukaryotic Cells

The enormous length of the chromosomes in most higher eukaryotes may be part of the reason that the genetic information is divided into numerous independent chromosomes. Table 9.1 compares the length and complexity of genomes from several organisms. The number and size of chromosomes can vary. For example, the chromosomes of yeast are on average about 5 times smaller than that of *E. coli*, whereas each human chromosome contains

Table 9.1
Chromosome Size and Organization

Organism	Base pairs	Length	Number of chromosomes	Chromosome form
Bacteriophage T7	3.8×10^4	12.5 μm	1	Linear
Bacteriophage T4	2×10^5	68 μm	1	Linear
E. coli	4×10^6	1.36 mm	1	Circular
Yeast	1.35×10^7	4.6 mm	17	Linear
Drosophila	1.65×10^8	5.6 mm	4	Linear
Human	2.9×10^9	990 mm	23	Linear

an average of more than 150 million base pairs. Another difference in the organization of DNA into eukaryotic chromosomes and those of bacteria is that eukaryotic chromosomes are linear.

a. Linear Genomes in Eukaryotes Are Organized into Topological Domains

As discussed in Chapter 3, a topological domain can be defined in linear DNA by preventing the rotation of the DNA double helix at the ends of the molecule. For example, shortly after infection of *E. coli* with bacteriophage T4, the linear genome becomes organized into a single, large, moderately supercoiled domain. An individual human chromosome, which may be 50 times longer than the *E. coli* chromosome, could be supercoiled by fixing the ends to a solid surface within the cell. However, a single chromosome is a very large molecule, and if the topology and superhelical-dependent variation in DNA structure are going to be important in the regulation of gene expression, the chromosome must be subdivided into smaller, potentially variable regions.

Chromosomes of higher eukaryotes are organized into topological domains (Figure 9.6; for review, see Gasser and Laemmli, 1987; Gross and Garrard, 1987; Garrard, 1990; Freeman and Garrard, 1992). Specific A + T-rich regions of DNA preferentially associate with the nuclear matrix, a proteinaceous scaffold. This periodic looping of the DNA on the nuclear matrix forms the topological domains of eukaryotic chromosomes. The sizes of topological domains vary in different organisms. Transcription and DNA replication may take place on the matrix.

The A + T-rich regions of DNA at the base of the loops that attach to the nuclear matrix have been called MARs (matrix attachment regions) or SARs (scaffold attachment regions). These A + T-rich regions typically contain topoisomerase II DNA binding sites. Moreover, DNA topoisomerase II has been identified as a major structural protein of the nuclear matrix. One model for the formation of topological domains is that topoisomerases in the nuclear matrix bind DNA and define a domain by preventing rotation of the double helix. The superhelicity and topological integrity of DNA within a topological domain may be important in gene regulation (for review, see Esposito and Sinden, 1988; Freeman and Garrard, 1992). The association of type II topoisomerases at the base of the topological domains suggests that the ability to supercoil DNA or the capacity to relax local regions of negative or positive supercoils is important biologically.

MARs Contain DUEs (DNA Unwinding Elements)

Kohwi-Shigematsu and Kohwi demonstrated that a MAR flanking an enhancer for an immunoglobulin heavy chain gene can act as a DNA unwinding element (see Chapter 7). Within the MAR, a 200-bp region containing repeats of the sequence AATATATTT can stably unwind in supercoiled DNA. Mutational changes within this sequence prevented unwinding within this entire region (Kohwi-Shigematsu and Kohwi, 1990). Bode *et al.* (1992) chemically synthesized 25-bp A + T-rich regions containing AATATATTT elements and showed that multimers of these fragments would remain unwound in supercoiled DNA. Mutations in the AATATATTT element prevented the stable unwinding of oligomers of this sequence. DNA molecules containing the unwinding element bound to the nuclear matrix, whereas molecules containing the mutated sequence bound less well to the matrix. Moreover, the sequence that could unwind acted as a transcriptional enhancer when cloned upstream from a reporter gene and stably transfected into mouse cells. The mutated sequence reduced expression of a reporter gene when transfected into mouse cells. These results suggest that the ability of A + T-rich MAR regions to stably unwind may be an important feature of DNA sequences involved in organizing chromosomes into topological domains. Unwinding may be required for the initiation of DNA replication that occurs at the matrix. Unwinding may also be involved in transcription, perhaps by acting as a "torsional sink" absorbing the local negative supercoiling generated by transcribing RNA polymerase pulling away from the nuclear matrix.

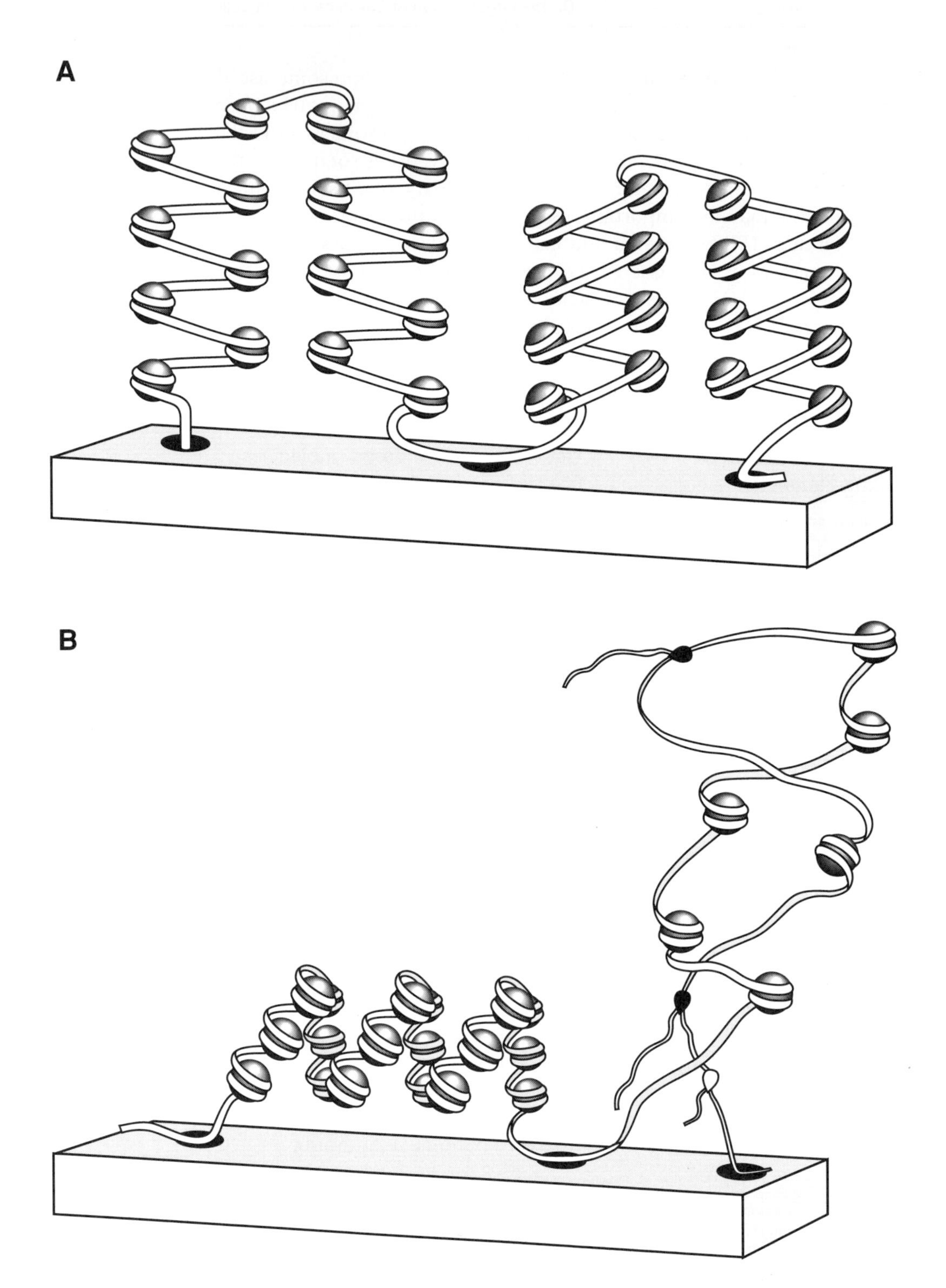
A
B

C

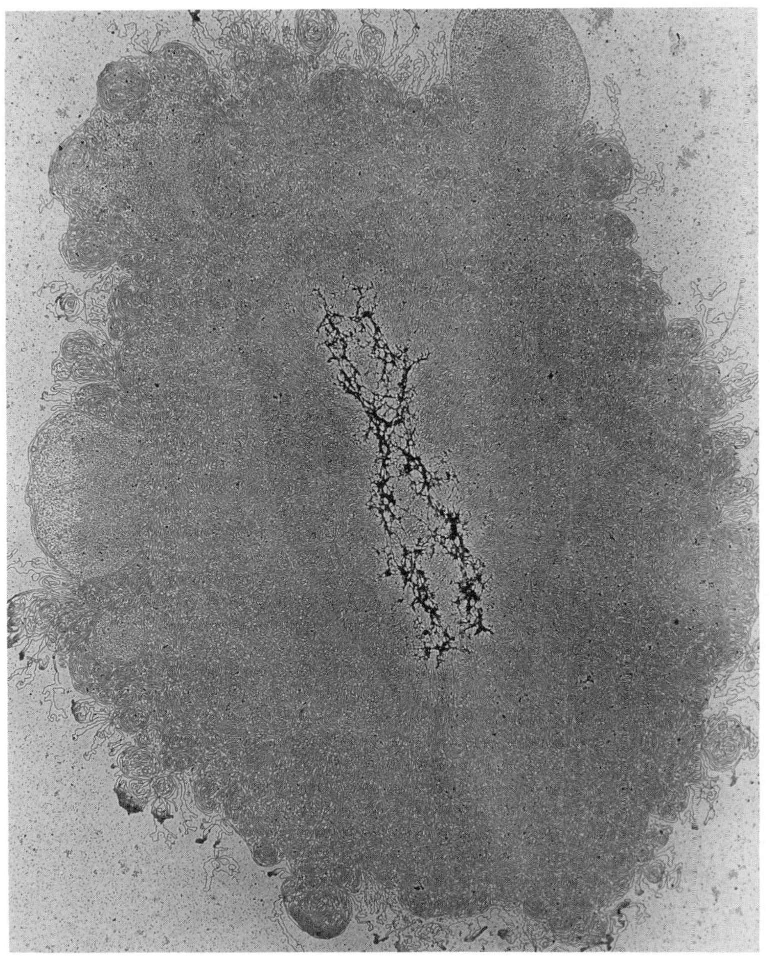

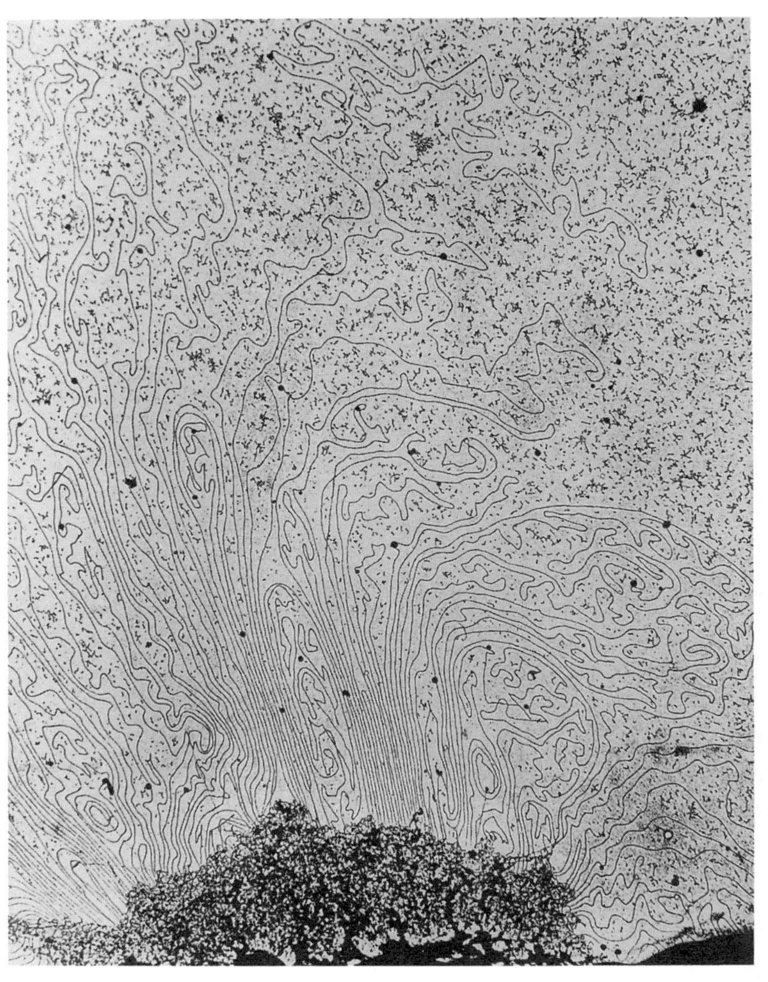

b. Torsional Tension Exists over Coding Regions in Eukaryotic Cells

Initial studies of unrestrained supercoiling in eukaryotes, using the psoralen photobinding approach, failed to detect a high energy supercoiled state of the DNA (Sinden *et al.*, 1980). This result suggested that the bulk of DNA in eukaryotic chromosomes was wrapped into nucleosomes that effectively restrained supercoils. These experiments, however, were averaging measurements that would not have detected short, local regions of negative supercoiling. The psoralen photobinding approach has been modified to look at the rate of cross-linking of specific regions of the genome. Specifically, restriction fragments corresponding to coding regions of genes or their 3′ and 5′ regions can be examined by hybridization analysis following electrophoretic separation of the cross-linked and non-cross-linked fractions of the sequence in question. Jupe *et al.* (1993) used this approach to carefully examine the state of unrestrained supercoiling over the *Drosophila hsp70* and 18S ribosomal genes. Ljungman and Hanawalt (1992) have demonstrated unrestrained tor-

Figure 9.6 Models for higher order structure of eukaryotic DNA. The first order of packaging of DNA in chromosomes is the formation of a string of nucleosomes along the DNA. These nucleosomes can then organize themselves into a 10-nm fiber. A number of different models have been proposed for the organization of nucleosomes into higher order structures. (A) (*Left*) A model for nucleosome packing is shown in which there is a zigzag appearance of the linker between adjacent nucleosomes. (*Right*) A topological domain with a slightly different organization of nucleosomes called the crossed linker model. Note that the linker between adjacent nucleosomes crosses the path of the previous linker, giving a zigzag appearance to the linkers between adjacent nucleosomes. The organization of nucleosomes in the two conformations shown here would lead to the appearance of rather flat ribbon structures that can then undergo higher orders of coiling into a 30-nm fiber (See Worcel *et al.* (1981), Woodcock *et al.* (1984), and Freeman and Garrard (1992) for a discussion of these and other models.) (B) (*Left*) A third organization of nucleosomes is a model involving a toroidal coiling of nucleosomes in a left-handed coil with about 6 nucleosomes per helical turn (see Worcel and Benyajati, 1977; Finch and Klug, 1976). (*Right*) On activation of a topological domain, so genes within this topological domain could be expressed and transcribed into messenger RNA, the higher order compact structure of chromatin unfolds making nucleosomes and the DNA more accessible to the environment. Certain nucleosomes may come off the DNA or weaken their association with the DNA. In addition, regions of DNA around active genes become negatively supercoiled in transcriptionally active chromatin. The 6 nucleosomes in this domain are less closely spaced than in the three other domains. Three RNA polymerases and their nascent RNA transcripts are also shown in the topological domain. The processes of transcription and DNA replication may occur at the base of the topological domain where particular DNA sequences called matrix attachment regions or scaffold attachment regions are involved in the association of the DNA with a nuclear matrix. (C) Electron micrographs of a histone-depleted metaphase chromosome. (*Left*) The central dense region is the protein scaffold. This is surrounded by a halo of individual DNA loops emanating from the scaffold. (*Right*) A magnified view of a region showing the individual DNA strands emerging from adjacent regions on the scaffold. In these figures the DNA has been nicked to relax the DNA and allow visualization of the extended DNA strands. Reproduced with permission from Paulson and Laemmli (1977).

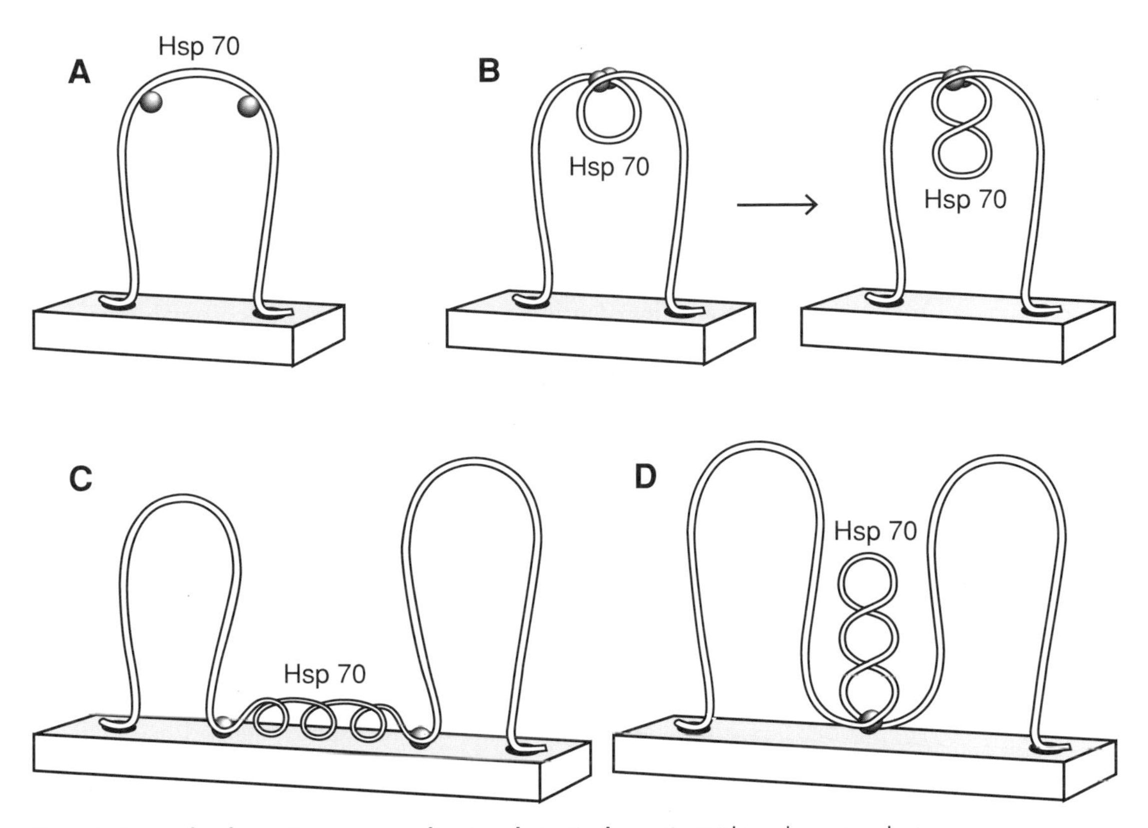

Figure 9.7 Localized constitutive supercoiling in eukaryotic chromatin. Within a larger topological domain in eukaryotic chromatin, a microdomain encompassing the 10-kb *hsp70* genes at locus 87A7 exists under a state of unrestrained negative supercoiling. Although the mechamism by which this organization is accomplished is unknown, a few examples are shown. (A) The *hsp70* region, delineated by round balls that represent scs sequences (specialized chromatin structures that define functional domain boundaries), exists within a larger topological domain and might not be organized differently with respect to sequences flanking the scs sites. (B) A microdomain has been defined by the association of the ends of the *hsp70* functional domain. Within this microdomain, DNA is organized with unrestrained supercoiling. (C, D) Two other ways in which a microdomain could be defined by organization of domain boundaries at separate sites (C) or a unique site (D) on the nuclear matrix. Once the topological domain is defined, the negative supercoiling could be introduced and maintained by differential topoisomerase I and II activity or by an alteration in nucleosome organization of the DNA.

sional tension at the 5′ end of the amplified dihydrofolate reductase gene in human cells. Negative superhelical tension was maintained over the coding regions of these genes.

Could this supercoiling be introduced by movement of RNA polymerase? As discussed in Chapter 3, the movement of RNA polymerase can generate a local transient differential in supercoiling (see Figure 3.10C). However, the supercoiling observed over the *Drosophila hsp70* gene was not dependent on transcription. The *hsp70* gene encodes a *h*eat *s*hock *p*rotein 70-kDa in size. On elevation of the growth temperature, transcription from heat shock genes is induced to very high levels. There is little expression during growth at lower temperatures. Jupe *et al.* found an equal level of negative supercoiling whether or not the gene was being transcribed. Although the level of supercoiling was unchanged, there is significant reorganization in the chromatin structure on transcription, as evidenced by a change in the accessibility to psoralen photobinding. During transcription, the nucleosomal organization of DNA is less pronounced, concomitant with the requirement for the transient dissociation or unfolding of the nucleosomes needed to allow passage of RNA polymerase.

Negative supercoiling occurs in a localized microdomain at the *hsp70* coding region. The level of negative supercoiling decreases significantly in sequences flanking the coding region. This microdomain occurs within the larger context of a chromosome loop or "domain." How is negative superhelical tension maintained in the presence of active topoisomerases? There may be physical organization of a microdomain by attaching the ends of the coding region to the nuclear matrix or to some large protein complex. Could the binding of topoisomerase molecules at the ends of the coding region and their association with the nuclear matrix define the domain? Alternatively, could topoisomerases attach at either end of the coding region and associate to form a small loop of DNA that could contain a few supercoils? Several possibilities are presented in Figure 9.7.

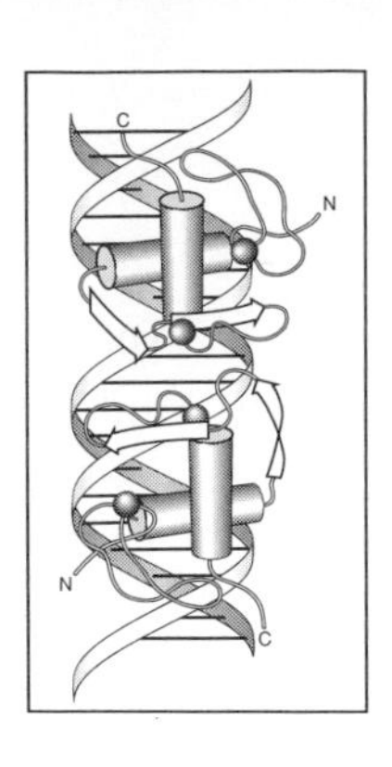

Bibliography

Adams, D. E., Shekhtman, E. M., Zechiedrich, E. L., Schmid, M. B., and Cozzarelli, N. R. (1992). The role of topoisomerase IV in partitioning bacterial replicons and the structure of catenated intermediates in DNA replication. *Cell* **71,** 277–288.

Aggarwal, A. K., Rodgers, D. W., Drottar, M., Ptashne, M., and Harrison, S. C. (1988). Recognition of a DNA operator by the repressor of phage 434: A view at high resolution. *Science* **242,** 899–907.

Albertini, A. M., Hofer, M., Calos, M. P., and Miller, J. H. (1982). On the formation of spontaneous mutations: The importance of short sequence homologies in generation of large deletion. *Cell* **29,** 319–328.

Ansevin, A. T., and Wang, A. H. (1990). Evidence for a new Z-type left-handed DNA helix: Properties of Z[WC]-DNA. *Nucleic Acids Res.* **18,** 6119–6126.

Arents, G., and Moudrianakis, E. N. (1993). Topography of the histone octamer surface: Repeating structural motifs utilized in the docking of nucleosomal DNA. *Proc. Natl. Acad. Sci. USA* **90,** 10489–10493.

Arndt–Jovin, D. J., Uduardy, A., Garner, M. M., Ritter, S., and Jovin, T. M. (1993). Z-DNA binding and inhibition by GTP of *Drosophila* topoisomerase II. *Biochemistry* **32,** 4862–4872.

Arnott, S., and Selsing, E. (1974). Structures for the polynucleotide complexes poly(dA) · poly(dT) and poly(dT) · poly(dA) · poly(dT). *J. Mol. Biol.* **88,** 509–521.

Arnott, S., Wilkins, M. H. F., Hamilton, L. D., and Langridge, R. (1965). Fourier synthesis studies of lithium DNA. III. Hoogsteen Models. *J. Mol. Biol.* **11,** 391–402.

Arnott, S., Chandrasekaran, R., Hukins, D. W. L., Smith, P. J. C., and Watts, L. (1974). Structural details of a double-helix observed for DNAs containing alternating purine–pyrimidine sequences. *J. Mol. Biol.* **88,** 523–533.

Arnott, S., Chandrasekaran, R., Hall, I. H., and Puigjaner, L. C. (1983). Heteronomous DNA. *Nucleic Acids Res.* **11,** 4141–4156.

Azorin, F., and Rich, A. (1985). Isolation of Z-DNA binding proteins from SV40 minichromosomes: Evidence for binding to the viral control region. *Cell* **41,** 365–374.

Azorin, F., Hahn, R., and Rich, A. (1984). Restriction endonucleases can be used to study B–Z junction in supercoiled DNA. *Biochemistry* **81,** 5714–5718.

Balbinder, E., Mac Vean, C., and Williams, R. E. (1989). Overlapping direct repeats stimulate deletions in specifically designed derivatives of plasmid pBR325 in *Escherichia coli. Mutat. Res.* **214,** 233–252.

Balke, V. L., and Gralla, J. D. (1987). Changes in the linking number of supercoiled DNA accompany growth transitions in *Escherichia coli. J. Bacteriol.* **169,** 4499–4506.

Banerjee, R., and Grunberger, D. (1986). Enhanced expression of the bacterial chloramphenicol acetyltransferase gene in mouse cells cotransfected with synthetic polynucleotides able to form Z-DNA. *Proc. Natl. Acad. Sci. USA* **83,** 4988–4992.

Banerjee, R., Carothers, A. M., and Grunberger, D. (1985). Inhibition of the herpes simplex virus thymidine kinase gene transfection in Ltk^- cells by potential Z-DNA forming polymers. *Nucleic Acids Res.* **13,** 5111–5126.

Baran, N., Neer, A., and Manor, H. (1983). "Onion skin" replication of integrated polyoma virus DNA and flanking sequences in polyoma-transformed rat cells: Termination within a specific cellular DNA segment. *Proc. Natl. Acad. Sci. USA* **80,** 105–109.

Baran, N., Lapidot, A., and Manor, H. (1991). Formation of DNA triplexes accounts for arrests of DNA synthesis at $d(TC)_n$ and $d(GA)_n$ tracts. *Proc. Natl. Acad. Sci. USA* **88,** 507–511.

Barton, J. K., and Raphael, A. L. (1985). Site-specific cleavage of left-handed DNA in pBR322 by lambda-tris(diphenylphenanthroline)cobalt(III). *Proc. Natl. Acad. Sci. USA* **82,** 6460–6464.

Bauer, W. R. (1978). Structure and reactions of closed duplex DNA. *Annu. Rev. Biophys. Bioeng.* **7,** 287–313.

Beal, P. A., and Dervan, P. B. (1991). Second structural motif for recognition of DNA by oligonucleotide-directed triple-helix formation. *Science* **251**, 1360–1363.

Beamer, L. J., and Pabo, C. O. (1992). Refined 1.8 Å crystal structure of the lambda repressor–operator complex. *J. Mol. Biol.* **227**, 177–196.

Becker, M. M., and Grossmann, G. (1992). Photofootprinting DNA *in vitro*. *Meth. Enzymol.* **212**, 262–272.

Becker, M. M., and Wang, J. C. (1984). Use of light for footprinting DNA *in vivo*. *Nature (London)* **309**, 682–687.

Beerman, T. A., and Lebowitz, J. (1973). Further analysis of the altered secondary structure of superhelical DNA. Sensitivity to methylmercuric hydroxide, a chemical probe for unpaired bases. *J. Mol. Biol.* **79**, 451–470.

Beese, L. S., Derbyshire, V., and Steitz, T. A. (1993). Structure of DNA polymerase I Klenow fragment bound to duplex DNA. *Science* **260**, 352–355.

Belotserkovskii, B. P., Veselkov, A. G., Filippov, S. A., Dobrynin, V. N., Mirkin, S. M., and Frank-Kamenetskii, M. D. (1990). Formation of intramolecular triplex in homopurine–homopyrimidine mirror repeats with point substitutions. *Nucleic Acids Res.* **18**, 6621–6624.

Benham, C. J. (1982). Stable cruciform formation at inverted repeat sequences in supercoiled DNA. *Biopolymers* **21**, 679–696.

Berg, J. M. (1988). Proposed structure for the zinc-binding domains from transcription factor IIIA and related proteins. *Proc. Natl. Acad. Sci. USA* **85**, 99–102.

Berg, J. M. (1993). Zinc-finger proteins. *Curr. Opin. Struct. Biol.* **3**, 11–16.

Bhattacharyya, A., Murchie, A. I., von Kitzing, E., Diekmann, S., Kemper, B., and Lilley, D. M. J. (1991). A model for the interaction of DNA junctions and resolving enzymes. *J. Mol. Biol.* **221**, 1191–1207.

Birnboim, H. C. (1978). Spacing of polypyrimidine regions in mouse DNA as determined by poly(adenylate,guanylate) binding. *J. Mol. Biol.* **121**, 541–559.

Blaho, J. A., and Wells, R. D. (1987). Left-handed Z-DNA binding by the recA protein of *Escherichia coli* [published erratum appears in *J. Biol. Chem.* **263(22)**, 11015]. *J. Biol. Chem.* **262**, 6082–6088.

Blaho, J. A., and Wells, R. D. (1989). Left-handed Z-DNA and genetic recombination. *Prog. Nucleic Acids Res. Mol. Biol.* **37**, 107–126.

Blaho, J. A., Larson, J. E., McLean, M. J., and Wells, R. D. (1988). Multiple DNA secondary structures in perfect inverted repeat inserts in plasmids: Right-handed B-DNA, cruciforms, and left-handed Z-DNA. *J. Biol. Chem.* **263**, 14446–14455.

Bliska, J. R., and Cozzarelli, N. R. (1987). Use of site-specific recombination as a probe of DNA structure and metabolism. *J. Mol. Biol.* **194**, 205–218.

Bode, J., Kohwi, Y., Dickinson, L., Joh, T., Klehr, D., Mielke, C., and Kohwi-Shigematsu, T. (1992). Biological significance of unwinding capability of nuclear matrix-associating DNAs. *Science* **255**, 195–197.

Boehm, T., Mengle-Gaw, L., Kees, U. R., Spurr, N., Lavenir, I., Forster, A., and Rabbitts, T. H. (1989). Alternating purine–pyrimidine tracts may promote chromosomal translocations seen in a variety of human lymphoid tumours. *EMBO J.* **8**, 2621–2631.

Boles, T. C., and Hogan, M. E. (1987) DNA structure equilibria in the human c-*myc* gene. *Biochemistry* **26**, 367–376.

Boles, T. C., White, J. H., and Cozzarelli, N. R. (1990). Structure of plectonemically supercoiled DNA. *J. Mol. Biol.* **213**, 931–951.

Borisova, O. F., Golova, Y. B., Gottikh, B. P., Zibrov, A. S., Il'icheva, I. A., Lysov, Y. P., Mamayeva, O. K., Chernov, B. K., Chernyi, A. A., Shchyolkina, A. K., and Florentiev, V. L. (1991). Parallel double stranded helices and the tertiary structure of nucleic acids. *J. Biomol. Struct. Dynam.* **8**, 1187–1210.

Borowiec, J. A., and Gralla, J. D. (1987). All three elements of the *lac ps* promoter mediate its transcriptional response to DNA supercoiling. *J. Mol. Biol.* **195**, 89–97.

Bowater, R., Aboul-ela, F., and Lilley, D. M. J. (1991). Large-scale stable opening of supercoiled DNA in response to temperature and supercoiling in (A+T)-rich regions that promote low-salt cruciform extrusion. *Biochemistry* **30**, 11495–11506.

Brahmachari, S. K., Shouche, Y. S., Cantor, C. R., and McClelland, M. (1987). Sequences that adopt non-B-DNA conformation in form V DNA as probed by enzymic methylation. *J. Mol. Biol.* **193**, 201–211.

Brinton, B. T., Caddle, M. S., and Heintz, N. H. (1991). Position and orientation-dependent effects of a eukaryotic Z-triplex DNA motif on episomal DNA replication in COS-7 cells. *J. Biol. Chem.* **266**, 5153–5161.

Brown, D. M. (1974). Chemical reactions of polynucleotides and nucleic acids. *In* "Basic Principles in Nucleic Acid Chemistry" (P. O. P. Ts'o, ed.), Vol. II, pp. 1–91. Academic Press, New York.

Broyles, S., and Pettijohn, D. E. (1986). Interaction of the *E. coli* HU protein with DNA: Evidence for formation of nucleosome-like structures with altered DNA helical pitch. *J. Mol. Biol.* **187**, 47–60.

Burkhoff, A. M., and Tullius, T. D. (1988). Structural details of an adenine tract that does not cause DNA to bend. *Nature (London)* **331**, 455–457.

Calladine, C. R., and H. R. Drew. (1992). "Understanding DNA—The Molecule and How It Works." Academic Press, San Diego, CA.

Cantor, C. R., and Efstratiadis, A. (1984). Possible structures of homopurine–homopyrimidine S1-hypersensitive sites. *Nucleic Acids Res.* **12,** 8059–8072.

Carter, C. W., Jr., and Kraut, J. (1974). A proposed model for interaction of polypeptides with RNA. *Proc. Natl. Acad. Sci. USA* **71,** 283–287.

Cartwright, I. L., and Elgin, S. C. R. (1986). Nucleosomal instability and induction of new upstream protein-DNA associations accompany activation of four small heat shock protein genes in Drosophila melanogaster. *Mol. Cell. Biol.* **6,** 779–791.

Castleman, H., Hannau, L. H., and Erlanger, B. F. (1983). Stabilization of (dG–dC)n · $(dG–dC)_n$ in the Z conformation by cross-linking reaction. *Nucleic Acids Res.* **11,** 8421–8429.

Chamberlin, M. J. (1965). Comparative properties of DNA, RNA, and hybrid homopolymer pairs. *Fed. Proc.* **24,** 1446–1457.

Chamberlin, M. J., and Patterson, D. L. (1965). Physical and chemical characterization of the ordered complexes formed between polyinosinic acid, polycytidylic acid and their deoxyribo-analogues. *J. Mol. Biol.* **12,** 410–428.

Chargaff, E. (1951). Structure and function of nucleic acids as cell constituents. *Fed. Proc.* **10,** 654–659.

Chen, A., Reyes, A., and Akeson, R. (1993). A homopurine:homopyrimidine sequence derived from the rat neuronal cell adhesion molecule-encoding gene alters expression in transient transfections. *Gene* **128,** 211–218.

Chiu, S. K., Rao, B. J., Story, R. M., and Radding, C. M. (1993). Interactions of three strands in joints made by RecA protein. *Biochemistry* **32,** 13146–13155.

Choo, K. B., Lee, H. H., Liew, L. N., Chong, K. Y., and Chou, H. F. (1990). Analysis of the unoccupied site of an integrated human papillomavirus 16 sequence in a cervical carcinoma. *Virology* **178,** 621–625.

Church, G. M., Sussman, J. L., and Kim, S. H. (1977). Secondary structural complementarity between DNA and proteins. *Proc. Natl. Acad. Sci. USA* **74,** 1458–1462.

Churchill, M. E. A., Tullius, T. D., and Klug, A. (1990). Mode of interaction of the zinc finger protein TFIIIA with a 5S RNA gene of *Xenopus. Proc. Natl. Acad. Sci. USA* **87,** 5528–5532.

Cimino, G. D., Gamper, H. B., Isaacs, S. T., and Hearst, J. E. (1985). Psoralens as photoactive probes of nucleic acid structure and function: Organic chemistry, photochemistry, and biochemistry. *Annu. Rev. Biochem.* **54,** 1151–1193.

Clark, D. J., and Felsenfeld, G. (1991). Formation of nucleosomes on positively supercoiled DNA. *EMBO J.* **10**, 387–395.

Clark, S. P., Lewis, C. D., and Felsenfeld, G. (1990). Properties of BPG1, a poly(dG)-binding protein from chicken erythrocytes. *Nucleic Acids Res.* **18**, 5119–5126.

Collier, D. A., and Wells, R. D. (1990). Effect of length, supercoiling, and pH on intramolecular triplex formation: Multiple conformers at pur.pyr mirror repeats. *J. Biol. Chem.* **265**, 10652–10658.

Collier, D. A., Griffin, J. A., and Wells, R. D. (1988). Non-B right-handed DNA conformations of homopurine · homopyrimidine sequences in the murine immunoglobulin C alpha switch region. *J. Biol. Chem.* **263**, 7397–7405.

Collins, J. (1980). Instability of palindromic DNA in *Escherichia coli. Cold Spring Harbor Symp. Quant. Biol.* **45**, 409–416.

Connolly, B., Parsons, C. A., Benson, F. E., Dunderdale, H. J., Sharples, G. J., Lloyd, R. G., and West, S. C. (1991). Resolution of Holliday junctions *in vitro* requires the *Escherichia coli ruv*C gene product. *Proc. Natl. Acad. Sci. USA* **88**, 6063–6067.

Cook, D. N., Ma, D., Pon, N. G., and Hearst, J. E. (1992). Dynamics of DNA supercoiling by transcription in *Escherichia coli. Proc. Natl. Acad. Sci. USA* **89**, 10603–10607.

Cooney, M., Czernuszewicz, G., Postel, E. H., Flint, S. J., and Hogan, M. E. (1988). Site-specific oligonucleotide binding represses transcription of the human c-*myc* gene *in vitro. Science* **241**, 456–459.

Cooper, J. P., and Hagerman, P. J. (1987). Gel electrophoretic analysis of the geometry of a DNA four-way junction. *J. Biol. Chem.* **198**, 711–719.

Cooper, J. P., and Hagerman, P. J. (1989). Geometry of a branched DNA structure in solution. *Proc. Natl. Acad. Sci. USA* **86**, 7336–7340.

Courey, A. J., and Wang, J. C. (1988). Influence of DNA sequence and supercoiling on the process of cruciform formation. *J. Mol. Biol.* **202**, 35–43.

Cox, M. M., and Lehman, I. R. (1987). Enzymes of general recombination. *Annu. Rev. Biochem.* **56**, 229–262.

Cozzarelli, N. R., Boles, T. C., and White, J. H. (1990). Primer on the topology and geometry of DNA supercoiling. *In* "DNA Topology and Its Biological Effects" (N. R. Cozzarelli and J. C. Wang, eds.), pp. 139–184. Cold Spring Harbor Laboratory Press, Cold Spring Harbor, New York.

Dasgupta, U., Weston-Hafer, K., and Berg, D. E. (1987). Local DNA sequence control of deletion formation in *Escherichia coli* plasmid pBR322. *Genetics* **115**, 41–49.

Davis, T. L., Firulli, A. B., and Kinniburgh, A. J. (1989). Ribonucleoprotein and protein factors bind to an H-DNA-forming c-*myc* DNA element:

Possible regulators of the c-*myc* gene (transcription element/DNA conformer/gene regulation). *Proc. Natl. Acad. Sci. USA* **86,** 9682–9686.

Dayn, A., Samadashwily, G. M., and Mirkin, S. (1992). Intramolecular DNA triplexes: Unusual sequence requirements and influence on DNA polymerization. *Proc. Natl. Acad. Sci. USA* **89,** 11406–11410.

Dean, F. B., Borowiec, J. A., Ishimi, Y., Deb, Sumitra, Tegtmeyer, P., and Hurwitz, J. (1987). Simian virus 40 large tumor antigen requires three core replication origin domains for DNA unwinding and replication *in vitro*. *Proc. Natl. Acad. Sci. USA* **84,** 8267–8271.

Dean, W. W., and Lebowitz, J. (1971). Partial alteration of secondary structure in native superhelical DNA. *Nature (London)* **231,** 5–8.

deMassey, B., Weisberg, R. A., and Studier, F. W. (1987). Gene 3 endonuclease of bacteriophage T7 resolves conformationally branched structures in double-stranded DNA [published erratum appears in *J. Mol. Biol.*, 1987, **196**(3), following 742]. *J. Mol. Biol.* **193,** 359–376.

deMassey, B., Studier, F. W., Dorgai, L., Appelbaum, F., and Weisberg, R. A. (1984). Enzymes and the sites of genetic recombination: Studies with gene-3 endonuclease of phage T7 and with site-affinity mutants of lambda phage. *Cold Spring Harbor Symp. Quant. Biol.* **49,** 715–726.

Depew, R. E., and Wang, J. C. (1975). Conformational fluctuations of the DNA helix. *Proc. Natl. Acad. Sci. USA* **72,** 4275–4279.

Dickerson, R. E. (1983). Base sequence and helix structure variation in B and A DNA. *J. Mol. Biol.* **166,** 419–441.

Dickerson, R. E., and Drew, H. R. (1981). Structure of a B-DNA dodecamer II. Influence of base sequence on helix structure. *J. Mol. Biol.* **149,** 761–786.

Dickerson, R. E., Bansal, M., Calladine, C. R., Diekmann, S., Hunter, W. N., *et al.* (1989). Definitions and nomenclature of nucleic acid structure parameters. *J. Mol. Biol.* **205,** 787–791.

Drake, J. W. (1991). A constant rate of spontaneous mutation in DNA-based microbes. *Proc. Natl. Acad. Sci. USA* **88,** 7160–7164.

Drake, J. W., Glickman, B. W., and Ripley, L. S. (1983). Updating the theory of mutation. *Am. Scientist* **71,** 621–630.

Drew, H. R., and Travers, A. A. (1985). DNA bending and its relation to nucleosome positioning. *J. Mol. Biol.* **186,** 773–790.

Drlica, K. (1984). Biology of bacterial deoxyribonucleic acid topoisomerases. *Microbiol. Rev.* **48,** 273–289.

Drlica, K. (1990). Bacterial topoisomerases and the control of DNA supercoiling. *Trends Genet.* **6,** 433–437.

Drlica, K. (1992). Control of bacterial DNA supercoiling. *Mol. Microbiol.* **6,** 425–433.

Drlica, K. and Riley, M. (1990). "The Bacterial Chromosome." American Society for Microbiology, Washington, D.C.

Drlica, K., and Rouviere-Yaniv, J. (1987). Histonelike proteins of bacteria. *Microbiol. Rev.* **51**, 301–319.

Duckett, D. R., Murchie, A. I. H., Diekmann, S., von Kitzing, E., Kemper, B., and Lilley, D. M. J. (1988). The structure of the Holliday junction, and its resolution. *Cell* **55**, 79–89.

Durand, R., Job, C., Zarling, D. A., Teissere, M., Jovin, T. M., and Job, D. (1983). Comparative transcription of right- and left-handed poly[d(G-C)] by wheat germ RNA polymerase II. *EMBO J.* **2**, 1707–1714.

Elborough, K. M., and West, S. C. (1990). Resolution of synthetic Holliday junctions in DNA by an endonuclease from calf thymus. *EMBO J.* **9**, 2931–2936.

Elias, P., and Lehman, I. R. (1988). Interaction of origin binding protein with an origin of replication of herpes simplex virus 1. *Proc. Natl. Acad. Sci. USA* **85**, 2959–2963.

Ellenberger, T. E., Brandl, C. J., Struhl, K., and Harrison, S. C. (1992). The GCN4 basic region leucine zipper binds DNA as a dimer of uninterrupted α-helices: Crystal structure of the protein–DNA complex. *Cell* **71**, 1223–1237.

Ellison, M. J., Kelleher III, R. J., Wang, A. H.-J., Habener, J. F., and Rich, A. (1985). Sequence-dependent energetics of the B-Z transition in supercoiled DNA containing nonalternating purine-pyrimidine sequences. *Proc. Natl. Acad. Sci. USA* **82**, 8320–8324.

Esposito, F., and Sinden, R. R. (1987). Supercoiling in eukaryotic and prokaryotic DNA: Changes in response to topological pertubation of SV40 in CV-1 cells and of plasmids in *E. coli. Nucleic Acids Res.* **15**, 5105–5124.

Esposito, F., and Sinden, R. R. (1988). DNA supercoiling and eukaryotic gene expression. *Oxford Surv. Euk. Genes* **5**, 1–50.

Fairall, L., Rhodes, D., and Klug, A. (1986). Mapping of the sites of protection on a 5S RNA gene by the *Xenopus* transcription factor IIIA. A model for the interaction. *J. Mol. Biol.* **192**, 577–591.

Fairall, L., Martin, S., and Rhodes, D. (1989). The DNA binding site of the *Xenopus* transcription factor IIIA has a non-B-form structure. *EMBO J.* **8**, 1809–1817.

Fairall, L., Harrison, S. D., Travers, A. A., and Rhodes, D. (1992). Sequence-specific DNA binding by a two zinc-finger peptide from the *Drosophila melanogaster* tramtrack protein. *J. Mol. Biol.* **226**, 349–366.

Felsenfeld, G. (1992). Chromatin as an essential part of the transcriptional mechanism. *Nature (London)* **355**, 219–283.

Felsenfeld, G., and Miles, H. T. (1967). The physical and chemical properties of nucleic acids. *Annu. Rev. Biochem.* **26**, 407–468.

Felsenfeld, G., Davies, D. R., and Rich, A. (1957). Formation of a three-stranded polynucleotide molecule. *J. Am. Chem. Soc.* **79**, 2023–2024.

Finch, J. T., and Klug, A. (1976). Solenoidal model for superstructure in chromatin. *Proc. Natl. Acad. Sci. USA* **73**, 1897–1901.

Fishel, R. A., Detmer, K., and Rich, A. (1988). Identification of homologous pairing and strand-exchange activity from a human tumor cell line based on Z-DNA affinity chromatography. *Proc. Natl. Acad. Sci. USA* **85**, 36–40.

Fox, K. R. (1990). Long (dA)n · (dT)n tracts can form intramolecular triplexes under superhelical stress. *Nucleic Acids Res.* **18**, 5387–5391.

Frappier, L., Price, G. B., Martin, R. G., and Zannis-Hadjopoulos, M. (1987). Monoclonal antibodies to cruciform DNA structures. *J. Mol. Biol.* **193**, 751–758.

Freeman, L. A., and Garrard, W. T. (1992). DNA supercoiling in chromatin structure and gene expression. *Crit. Rev. Euk. Gene Exp.* **2**, 165–209.

Friedberg, E. C. (1985). "DNA Repair." Freeman, New York.

Fuller, W., Wilkins, M. H. F., Wilson, H. R., Hamilton, L. D., and Arnott, S. (1965). The molecular configuration of deoxyribonucleic acid. IV. X-Ray diffraction study of the A form. *J. Mol. Biol.* **12**, 60–80.

Garner, M. M., and Felsenfeld, G. (1987). Effect of Z-DNA on nucleosome placement. *J. Mol. Biol.* **196**, 581–590.

Garrard, W. T. (1990). Chromosomal loop organization in eukaryotic genomes. *Nucleic Acids Mol. Biol.* **4**, 163–175.

Gasser, S. M., and Laemmli, U. K. (1987). A glimpse at chromosomal order. *Trends Genet.* **3**, 16.

Gellert, M. (1981). DNA topoisomerases. *Annu. Rev. Biochem.* **50**, 879–910.

Gellert, M., O'Dea, M. H., and Mizuuchi, K. (1983). Slow cruciform transitions in palindromic DNA. *Proc. Natl. Acad. Sci. USA* **80**, 5545–5549.

Giedroc, D. P. (1994). Zinc: DNA-binding proteins. *In* "Encylopedia of Inorganic Chemistry" (R. B. King, ed.). Wiley, Sussex, U.K.

Gierer, A. (1966). Model for DNA and protein interactions and the function of the operator. *Nature (London)* **212**, 1480–1481.

Gilmour, D. S., Thomas, G. H., and Elgin, S. C. R. (1989). Droshophila nuclear proteins bind to regions of alternating C and T residues in gene promoters. *Science* **241**, 1487–1490.

Glaser, R. L., Thomas, G. H., Siegfried, E., Elgin, S. C., and Lis, J. T. (1990). Optimal heat-induced expression of the *Drosophila hsp*26 gene requires a promoter sequence containing $(CT)_n \cdot (GA)_n$ repeats. *J. Mol. Biol.* **211**, 751–761.

Glover, J. N., and Pulleyblank, D. E. (1990). Protonated polypurine/polypyrimidine DNA tracts that appear to lack the single-stranded pyrimidine loop predicted by the "H" model. *J. Mol. Biol.* **215**, 653–663.

Glover, J. N., Farah, C. S., and Pulleyblank, D. E. (1990). Structural characterization of separated H DNA conformers. *Biochemistry* **29,** 11110–11115.

Glucksmann, M. A., Markiewicz, P., Malone, C., and Rothman-Denes, L. B. (1992). Specific sequences and a hairpin structure in the template strand are required for N4 virion RNA polymerase promoter recognition. *Cell* **70,** 491–500.

Goldstein, E., and Drlica, K. (1984). Regulation of bacterial DNA supercoiling: Plasmid linking numbers vary with growth temperature. *Proc. Natl. Acad. Sci. USA* **81,** 4046–4050.

Gough, G. W., and Lilley, D. M. (1985). DNA bending induced by cruciform formation. *Nature (London)* **313,** 154–156.

Griffith, J. D. (1976). Visualization of prokaryotic DNA in a regularly condensed chromatin-like fiber. *Proc. Natl. Acad. Sci. USA* **73,** 563–567.

Gross, D. S., and Garrard, W. T. (1987). Poising chromatin for transcription. *Trends Biochem. Sci.* **12,** 293–297.

Gross, D. S., and Garrard, W. T. (1988). Nuclease hypersensitive sites in chromatin. *Annu. Rev. Biochem.* **57,** 159–197.

Grunberger, D., and Santella, R. M. (1981). Alternative conformations of DNA modified by *N*-2-acetylaminofluorene. *J. Supramol. Struct.* **17,** 231–244.

Grunstein, M. (1990). Histone function in transcription. *Annu. Rev. Cell Biol.* **6,** 643–678.

Gut, S. H., Bischoff, M., Hobi, R., and Kuenzle, C. C. (1987). Z-DNA-binding proteins from bull testis. *Nucleic Acids Res.* **15,** 9691–9705.

Hagerman, P. J. (1985). Sequence dependence of the curvature of DNA: A test of the phasing hypothesis. *Biochemistry* **24,** 7033–7037.

Hagerman, P. J. (1986). Sequence-directed curvature of DNA. *Nature (London)* **321,** 449–450.

Hagerman, P. J. (1988). Flexibility of DNA. *Annu. Rev. Biophys. Biophys. Chem.* **17,** 265–286.

Hagerman, P. J. (1990). Sequence-directed curvature of DNA. *Annu. Rev. Biochem.* **59,** 755–781.

Haniford, D. B., and Pulleyblank, D. E. (1983a). Facile transition of poly[d(TG) · d(CA)] into a left-handed helix in physiological conditions. *Nature (London)* **302,** 632–634.

Haniford, D. B., and Pulleyblank, D. E. (1983b). The *in-vivo* occurrence of Z DNA. *J. Biomol. Struct. Dynam.* **1,** 593–609.

Hayes, J. J., Clark, D. J., and Wolffe, A. P. (1991). Histone contributions to the structure of DNA in the nucleosome. *Proc. Natl. Acad. Sci. USA* **88,** 6829–6833.

Hanvey, J. C., Klysik, J., and Wells, R. D. (1988). Influence of DNA sequence on the formation of non-B right-handed helices in oligopurine · oligopyrimidine inserts in plasmids. *J. Biol. Chem.* **263,** 7386–7396.

Hanvey, J. C., Shimizu, M., and Wells, R. D. (1989). Intramolecular DNA triplexes in supercoiled plasmids. II. Effect of base composition and noncentral interruptions on formation and stability. *J. Biol. Chem.* **264,** 5950–5956.

Härd, T., Kellenbach, E., Boelens, R., Maler, B. A., Dahlman, K., Freedman, L. P., Carlstedt-Duke, J., Yamamoto, K. R., Gustafsson, J., and Kapstein, R. (1990). Solution structure of the glucocorticoid receptor DNA-binding domain. *Science* **249,** 157–160.

Hardin, C. C., Henderson, E., Watson, T., and Prosser, J. K. (1991). Monovalent cation induced structural transitions in telomeric DNAs: G-DNA folding intermediates. *Biochemistry* **30,** 4460–4472.

Harrison, S. C., and Aggarwal, A. K. (1990). DNA recognition by proteins with the helix–turn–helix motif. *Annu. Rev. Biochem.* **59,** 933–969.

Hayes, J. J., and Tullius, T. D. (1992). Structure of the TFIIIA-5S DNA complex. *J. Mol. Biol.* **227,** 407–417.

Hentschel, C. C. (1982). Homocopolymer sequences in the spacer of a sea urchin histone gene repeat are sensitive to S1 nuclease. *Nature (London)* **295,** 714–716.

Herbert, A. G., Spitzner, J. R., Lowenhaupt, K., and Rich, A. (1993). Z-DNA binding protein from chicken blood nuclei. *Proc. Natl. Acad. Sci. USA* **90,** 3339–3342.

Hill, R. J., and Stollar, B. D. (1983). Dependence of Z-DNA antibody binding to polytene chromosomes on acid fixation and DNA torsional strain. *Nature (London)* **305,** 338–340.

Hodges-Garcia, Y., Hagerman, P. J., and Pettijohn, D. E. (1989). DNA ring closure mediated by protein HU. *J. Biol. Chem.* **264,** 14621–14623.

Hoepfner, R. W., and Sinden, R. R. (1993). Amplified primer extension assay for psoralen photoproducts provides a sensitive assay for a $(CG)_6TA(CG)_2(TG)_8$ Z-DNA torsionally tuned probe: Preferential psoralen photobinding to one strand of a B-Z junction. *Biochemistry* **32,** 7542–7548.

Hoogsteen, K. (1963). The crystal and molecular structure of a hydrogen-bonded complex between 1-methylthymine and 9-methyladenine. *Acta Crystallogr.* **16,** 907–916.

Horwitz, M. S. Z., and Loeb, L. A. (1988). An *E. coli* promoter that regulates transcription by DNA superhelix-induced cruciform extrusion. *Science* **241,** 703–705.

Hsieh, C. H., and Griffith, J. D. (1988). The terminus of SV40 DNA replication and transcription contains a sharp sequence-directed curve. *Cell* **52,** 535–544.

Hsieh, C. H., and Griffith, J. D. (1989). Deletions of bases in one strand of duplex DNA, in contrast to single-base mismatches, produce highly kinked molecules: Possible relevance to the folding of single-stranded nucleic acids. *Proc. Natl. Acad. Sci. USA* **86,** 4833–4837.

Hsieh, L. S., Burger, R. M., and Drlica, K. (1991). Bacterial DNA supercoiling and [ATP]/[ADP] changes associated with a transition to anaerobic growth. *J. Mol. Biol.* **219,** 1–8.

Hsieh, P., and Camerini-Otero, R. D. (1989). Formation of joint DNA molecules by two eukaryotic strand exchange proteins does not require melting of a DNA duplex. *J. Biol. Chem.* **264,** 5089–5097.

Hsieh, P., Camerini-Otero, C. S., and Camerini-Otero, R. D. (1990). Pairing of homologous DNA sequences by proteins: Evidence for three-stranded DNA. *Genes Dev.* **4,** 1951–1963.

Htun, H., and Dahlberg, J. E. (1988). Single strands, triple strands, and kinks in H-DNA. *Science* **241,** 1791–1796.

Htun, H., and Dahlberg, J. E. (1989). Topology and formation of triple-stranded H-DNA. *Science* **243,** 1571–1576.

Htun, H., Lund, E., and Dahlberg, J. E. (1984). Human U1 RNA genes contain an unusually sensitive nuclease S1 cleavage site within the conserved 3′ flanking region. *Proc. Natl. Acad. Sci. USA* **81,** 7288–7292.

Hudson, G. S., Bankier, A. T., Satchwell, S. C., and Barrell, B. G. (1985). The short unique region of the B95-8 Epstein–Barr virus genome. *J. Virol.* **147,** 81–98.

Husain, I., Griffith, J., and Sancar, A. (1988). Thymine dimers bend DNA. *Proc. Natl. Acad. Sci. USA* **85,** 2558–2562.

Jackson, D. A., and Cook, P. R. (1985). A general method for preparing chromatin containing intact DNA. *EMBO J.* **4,** 913–918.

Jackson, D. A., Yuan, J., and Cook, P. R. (1988). A gentle method for preparing cyto- and nucleo-skeletons and associated chromatin. *J. Cell Sci.* **90,** 365–378.

Jackson, D. A., Dickinson, P., and Cook, P. R. (1990). The size of chromatin loops in HeLa cells. *EMBO J.* **9,** 567–571.

Jaworski, A., Hsieh, W. T., Blaho, J. A., Larson, J. E., and Wells, R. D. (1987). Left-handed DNA *in vivo*. *Science* **238,** 773–777.

Jayasena, S. D., and Behe, M. J. (1989). Competitive nucleosome reconstitution of polydeoxynucleotides containing oligoguanosine tracts. *J. Mol. Biol.* **208,** 297–306.

Job, D., Marmillot, P., Job, C., and Jovin, T. M. (1988). Transcription of left-handed Z-DNA templates: Increased rate of single-step addition reactions catalyzed by wheat germ RNA polymerase II. *Biochemistry* **27,** 6371–6378.

Johnston, B. H. (1988). The S1-sensitive form of $d(C\text{-}T)_n \cdot d(A\text{-}G)_n$: Chemical evidence for a three-stranded structure in plasmids. *Science* **241**, 1800–1804.

Johnston, B. H., and Rich, A. (1985). Chemical probes of DNA conformation: Detection of Z-DNA at nucleotide resolution. *Cell* **42**, 713–724.

Jordan, S. R., and Pabo, C. O. (1988). Structure of the lambda complex at 2.5 Å resolution: Details of the repressor–operator interactions. *Science* **242**, 893–899.

Jovin, T. M., Rippe, K., Ramsing, N. B., Klement, R., Elhorst, W., and Vojtiskova, M. (1990). Parallel stranded DNA. *In* "Structure and Methods, Vol.3: DNA and RNA" (R. H. Sarma and M. H. Sarma, eds.), pp. 155–174. Adenine Press, Schenectady, New York.

Jupe, E. R., Sinden, R. R., and Cartwright, I. L. (1993). Stably maintained microdomain of localized unrestrained supercoiling at a *Drosophila* heat shock gene locus. *EMBO J.* **12**, 1067–1075.

Kabsch, W., Sander, C., and Trifonov, E. N. (1982). The ten helical twist angles of B-DNA. *Nucleic Acids Res.* **10**, 1097–1104.

Kang, C. K., Zhang, X., Ratliff, R., Moyzis, R., and Rich, A. (1992). Crystal structure of four-stranded *Oxytricha* telomeric DNA. *Nature (London)* **356**, 126–131.

Kang, D. S., Harvey, S. C., and Wells, R. D. (1985). Diepoxybutane forms a monoadduct with B-form $(dG\text{-}dC)_n \cdot (dG\text{-}dC)_n$ and a cross-linked diadduct with the left-handed Z-form. *Nucleic Acids Res.* **13**, 5645–5656.

Kaptein, R. (1991a). Zinc-finger structures. *Curr. Opin. Struct. Biol.* **2**, 109–115.

Kaptein, R. (1991b). Zinc fingers. *Curr. Opin. Struct. Biol.* **1**, 63–70.

Karlovsky, P., Pecinka, P., Vojtiskova, M., Makaturova, E., and Palecek, E. (1990). Protonated triplex DNA in *E. coli* cells as detected by chemical probing. *FEBS Lett.* **274**, 39–42.

Kerppola, T. K., and Curran, T. (1991). Transcription factor interactions: Basics on zippers. *Curr. Opin. Struct. Biol.* **1**, 71–79.

Kikuchi, A. (1990). Reverse gyrase and other archaebacterial topoisomerases. *In* "DNA Topology and Its Biological Effects" (N. R. Cozzarelli and J. C. Wang, eds.), pp. 285–298. Cold Spring Harbor Laboratory Press, Cold Spring Harbor, New York.

Kim, J. L., Nikolov, D. B., and Burley, S. K. (1993). Co-crystal structure of TBP recognizing the minor groove of a TATA element. *Nature (London)* **365**, 520–527.

Kim, Y., Geiger, J. H., Hahn, S., and Sigler, P. B. (1993). Crystal structure of a yeast TBP/TATA-box complex. *Nature (London)* **365**, 512–520.

Kinniburgh, A. J. (1989). A cis-acting transcription element of the c-*myc* gene can assume an H-DNA conformation. *Nucleic Acids Res.* **17**, 7771–7778.

Kirchhausen, T., Want, J. C., and Harrison, S. C. (1985). DNA gyrase and its complexes with DNA: Direct observation by electron microscopy. *Cell* **41**, 933–943.

Kiyama, R., and Camerini-Otero, R. D. (1991). A triplex DNA-binding protein from human cells: Purification and characterization. *Proc. Natl. Acad. Sci. USA* **88**, 10450–10454.

Klug, A., and Lutter, L. C. (1981). The helical periodicity of DNA on the nucleosome. *Nucleic Acids Res.* **9**, 4267–4283.

Klug, A., and Rhodes, D. (1987). "Zinc fingers": A novel protein motif for nucleic acid recognition. *Trends. Biochem. Sci.* **12**, 464–469.

Klysik, J., Rippe, K., and Jovin, T. M. (1991). Parallel-stranded DNA under topological stress: Rearrangement of $(dA)_{15} \cdot (dT)_{15}$ to a $d(A \cdot A \cdot T)_n$ triplex. *Nucleic Acids Res.* **19**, 7145–7154.

Klysik, J., Stirdivant, S. M., Larson, J. E., Hart, P. A., and Wells, R. D. (1981). Left-handed DNA in restriction fragments and a recombinant plasmid. *Nature (London)* **290**, 672–677.

Klysik, J., Zacharias, W., Galazka, G., Kwinkowski, M., Uznanski, B., and Okruszek, A. (1988). Structural interconversion of alternating purine–pyrimidine inverted repeats cloned in supercoiled plasmids. *Nucleic Acids Res.* **16**, 6915–6933.

Kmiec, E. B., Angelides, K. J., and Holloman, W. K. (1985). Left-handed DNA and the synaptic pairing reaction promoted by *Ustilago* rec1 protein. *Cell* **40**, 139–145.

Knezetic, J. A., and Luse, D. S. (1986). The presence of nucleosomes on a DNA template prevents initiation by RNA polymerase II *in vitro*. *Cell* **45**, 95–104.

Knezetic, J. A., Jacob, G. A., and Luse, D. S. (1988). Assembly of RNA polymerase II preinitiation complexes before assembly of nucleosomes allows efficient initiation of transcription on nucleosomal templates. *Mol. Cell. Biol.* **8**, 3114–3121.

Knight, K. L., Bowie, J. U., Vershon, A. K., Kelly, R. D., and Saver, R. T. (1989). The Arc and Mnt repressors: A new class of sequence-specific DNA-binding proteins. *J. Biol. Chem.* **264**, 3639–3642.

Kochel, T. J., and Sinden, R. R. (1988). Analysis of trimethylpsoralen photoreactivity to Z-DNA provides a general *in vivo* assay for Z-DNA: Analysis of the hypersensitivity of $(GT)_n$ B-Z junctions. *BioTechniques* **6**, 532–543.

Kochel, T. J., and Sinden, R. R. (1989). Hyperreactivity of B-Z junctions to 4,5′,8-trimethylpsoralen photobinding assayed by an exonuclease III/photoreversal mapping procedure. *J. Mol. Biol.* **205**, 91–102.

Kohwi, Y., and Kohwi-Shigematsu, T. (1988). Magnesium ion-dependent triple-helix structure formed by homopurine–homopyrimidine sequences in supercoiled plasmid DNA. *Proc. Natl. Acad. Sci. USA* **85**, 3781–3785.

Kohwi, Y., and Kohwi-Shigematsu, T. (1991). Altered gene expression correlates with DNA structure. *Genes Dev.* **5**, 2547–2554.

Kohwi, Y., and Panchenko, Y. (1993). Transcription-dependent recombination induced by triple-helix formation. *Genes Dev.* **7**, 1766–1778.

Kohwi, Y., Malkhosyan, S. R., and Kohwi-Shigematsu, T. (1992). Intramolecular dG · dG · dC triplex detected in *Escherichia coli* cells. *J. Mol. Biol.* **223**, 817–822.

Kohwi-Shigematsu, T., and Kohwi, Y. (1985). Poly(dG)–poly(dC) sequences, under torsional stress, induce an altered DNA conformation upon neighboring DNA sequences. *Cell* **43**, 199–206.

Kohwi-Shigematsu, T., and Kohwi, Y. (1990). Torsional stress stabilizes extended base unpairing in suppressor sites flanking immunoglobulin heavy chain enhancer. *Biochemistry* **29**, 9551–9560.

Kohwi-Shigematsu, T., and Kohwi, Y. (1991). Detection of triple-helix related structures adopted by poly(dG) · poly(dC) sequences in supercoiled plasmid DNA. *Nucleic Acids Res.* **19**, 4267–4271.

Koo, H.-S., and Crothers, D. M. (1988). Calibration of DNA curvature and a unified description of sequence-directed bending. *Proc. Natl. Acad. Sci. USA* **85**, 1763–1767.

Koo, H.-S., Wu, H. M., and Crothers, D. (1986). DNA bending at adenine–thymine tracts. *Nature (London)* **320**, 501–506.

Kornberg, A., and Baker, T. A. (1992). "DNA Replication." Freeman, New York.

Kornberg, A., Bertsch, L. L., Jackson, J. F., and Khorana, H. G. (1964). Enzymatic synthesis of deoxyribonucluc acid, XVI. Oligonucleotides as templates and the mechanism of their replication. *Proc. Natl. Acad. Sci. USA* **51**, 315.

Koudelka, G. B., Harrison, S. C., and Ptashne, M. (1987). Effect of non-contacted bases on the affinity of 434 operator for 434 repressor and Cro. *Nature (London)* **326**, 886–888.

Kricker, M. C., Drake, J. W., and Radman, M. (1992). Duplication-targeted DNA methylation and mutagenesis in the evolution of eukaryotic chromosomes. *Proc. Natl. Acad. Sci. USA* **89**, 1075–1079.

Krishna, P., Kennedy, B. P., van-de-Sande, J. H., and McGhee, J. D. (1988). Yolk proteins from nematodes, chickens, and frogs bind strongly and preferentially to left-handed Z-DNA. *J. Biol. Chem.* **263**, 19066–19070.

Krishna, P., Kennedy, B. P., Waisman, D. M., van de Sande, J. H., and McGhee, J. D. (1990). Are many Z-DNA binding proteins actually phospholipid-binding proteins? *Proc. Natl. Acad. Sci. USA* **87**, 1292–1295.

Krizek, B. A., Amann, B. T., Kilfoil, V. J., Merkle, D. L., and Berg, J. M. (1991). A consensus zinc finger peptide: Design, high-affinity metal binding, a pH-dependent structure, and a His to Cys sequence variant. *J. Am. Chem. Soc.* **113,** 4518–4523.

Kunkel, G. R., and Martinson, H. G. (1981). Nucleosomes will not form on double-stranded RNA or over poly(dA) · poly(dT) tracts in recombinant DNA. *Nucleic Acids Res.* **9,** 6869–6888.

Lafer, E. M., Sousa, R., and Rich, A. (1985a). Anti-Z-DNA antibody binding can stabilize Z-DNA in relaxed and linear plasmids under physiological conditions. *EMBO J.* **4,** 3655–3660.

Lafer, E. M., Sousa, R., Rosen, B., Hsu, A., and Rich, A. (1985b). Isolation and characterization of Z-DNA binding proteins from wheat germ. *Biochemistry* **24,** 5070–5076.

Lafer, E. M., Sousa, R. J., and Rich, A. (1988). Z-DNA-binding proteins in *Escherichia coli:* Purification, generation of monoclonal antibodies and gene isolation. *J. Mol. Biol.* **203,** 511–516.

Landschulz, W. H., Johnson, P. F., and McKnight, S. L. (1988). The leucine zipper: A hypothetical structure common to a new class of DNA binding proteins. *Science* **240,** 1759–1764.

Langridge, R., Wilson, H. R., Hooper, C. W., Wilkins, M. H. F., and Hamilton, L. D. (1960a). The molecular configuration of deoxyribonucleic acid. I. X-Ray diffraction study of a crystalline form of the lithium salt. *J. Mol. Biol.* **2,** 19–37.

Langridge, R., Marvin, D. A., Seeds, W. E., Wilson, H. R., and Hamilton, L. D. (1960b). The molecular configuration of deoxyribonucleic acid. II. Molecular models and their Fourier transforms. *J. Mol. Biol.* **2,** 38–64.

Lapidot, A., Baran, N., and Manor, H. (1989). (dT-dC)n and (dG-dA)n tracts arrest single stranded DNA replication *in vitro*. *Nucleic Acids Res.* **17,** 883–900.

Leach, D. R. F., and Stahl, R. W. (1983). Viability of lambda phages carrying a perfect palindrome in the absence of recombination nucleases. *Nature (London)* **305,** 448–451.

Lechner, R. L., Engler, M. J., and Richardson, C. C. (1983). Characterization of strand displacement synthesis catalyzed by bacteriophage T7 DNA polymerase. *J. Biol. Chem.* **258,** 11174–11184.

Lee, D. H., and Schleif, R. F. (1989). *In vivo* DNA loops in *ara*CBAD: Size limits and helical repeat. *Proc. Natl. Acad. Sci. USA* **86,** 476–480.

Lee, D. Y., Hayes, J. J., Pruss, D., and Wolffe, A. P. (1993). A positive role for histone acetylation in transcription factor access to DNA. *Cell* **72,** 73–84.

Lee, J. S., Woodsworth, M. L., Latimer, J. P., and Morgan, A. R. (1984). Poly(pyrimidine) · poly(purine) synthetic DNAs containing 5-methylcytosine form stable triplexes at neutral pH. *Nucleic Acids Res.* **12,** 6603–6614.

Lee, M. S., and Garrard, W. T. (1991a). Transcription-induced nucleosome "splitting": An underlying structure for DNase I sensitive chromatin. *EMBO J.* **10,** 607–615.

Lee, M. S., and Garrard, W. T. (1991b). Positive DNA supercoiling generates a chromatin conformation characteristic of highly active genes. *Proc. Natl. Acad. Sci. USA* **88,** 9675.

Lee, M. S., Gippert, G. P., Soman, K. V., Case, D. A., and Wright, P. E. (1989). Three-dimensional solution structure of a single zinc finger DNA-binding domain. *Science* **245,** 635–637.

Leroy, J. L., Charretier, E., Kochoyan, M., and Gueron, M. (1988). Evidence from base-pair kinetices for two types of adenine tract structures in solution: Their relation to DNA curvature. *Biochemistry* **27,** 8894–8898.

Letai, A. G., Palladino, M. A., Fromm, E., Rizzo, V., and Fresco, J. R. (1988). Specificity in formation of triple-stranded nucleic acid helical complexes: Studies with agarose-linked polyribonucleotide affinity columns. *Biochemistry* **27,** 9108–9112.

Lilley, D. M. (1980). The inverted repeat as a recognizable structural feature in supercoiled DNA. *Proc. Natl. Acad. Sci. USA* **77,** 6468–6472.

Lilley, D. M. (1981). *In vivo* consequences of plasmid topology. *Nature (London)* **292,** 380–382.

Lilley, D. M. (1985). The kinetic properties of cruciform extrusion are determined by DNA. *Nucleic Acids Res.* **13,** 1443–1465.

Lilley, D. M. J. (1989). Structural isomerization in DNA: The formation of cruciform structures in supercoiled DNA molecules. *Chem. Soc. Rev.* **18,** 53–83.

Lilley, D. M. J. (1992). HMG has DNA wrapped up. *Nature (London)* **357,** 282–283.

Lilley, D. M. J., and Clegg, R. M. (1993). The structure of the four-way junction in DNA. *Annu. Rev. Biophys. Biomol. Struct.* **22,** 299–328.

Liu, L. F., and Wang, J. C. (1987). Supercoiling of the DNA template during transcription. *Proc. Natl. Acad. Sci. USA* **84,** 7024–7027.

Ljungman, M., and Hanawalt, P. C. (1992). Localized torsional tension in the DNA of human cells. *Proc. Natl. Acad. Sci. USA* **13,** 6055–6059.

Lockshon, D., and Galloway, D. A. (1986). Cloning and characterization of *ori*L2, a large palindromic DNA replication origin of herpes simplex virus type 2. *J. Virol.* **58,** 513–521.

Lockshon, D., and Morris, D. R. (1983). Positively supercoiled plasmid DNA is produced by treatment of *Escherichia coli* with DNA gyrase inhibitors. *Nucleic Acids Res.* **11,** 2999–3018.

Lu, Q., Wallrath, L. L., Granok, H., and Elgin, S. C. R. (1993). $(CT)_n \cdot (GA)_n$

repeats and heat shock elements have distinct roles in chromatin structure and transcriptional activation of the *Drosophila hsp*26 gene. *Mol. Cell. Biol.* **13,** 2802–2814.

Luisi, B. F., Zu, W. X., Otwinowski, Z., Freedman, L. P., Yamamoto, K. R., and Sigler, P. B. (1991). Crystallographic analysis of the interaction of the glucocorticoid receptor with DNA. *Nature* (*London*) **352,** 497–505.

Lyamichev, V. I., Mirkin, S. M., and Frank-Kamenetskii, M. D. (1985). A pH-dependent structural transition in the homopurine–homopyrimidine tract in superhelical DNA. *J. Biomol. Struct. Dynam.* **3,** 327–338.

Lyamichev, V. I., Mirkin, S. M., and Frank-Kamenetskii, M. D. (1986). Structures of homopurine–homopyrimidine tract in superhelical DNA. *J. Biomol. Struct. Dynam.* **3,** 667–669.

Lyamichev, V. I., Frank-Kamenetskii, M. D., and Soyfer, V. N. (1990). Protection against UV-induced pyrimidine dimerization in DNA by triplex formation. *Nature* (*London*) **344,** 568–570.

Mace, H. A. F., Pelham, H. R. B., and Travers, A. A. (1983). Association of an S1 nuclease-sensitive structure with short direct repeats 5′ of *Drosophila* heat shock genes. *Nature* (*London*) **304,** 555–557.

Marini, J. C., Levene, S. D., Crothers, D. M., and Englund, P. T. (1982). Bent helical structure in kinetoplast DNA. *Proc. Natl. Acad. Sci. USA* **79,** 7664–7668.

Marini, J. C., Levene, S. D., Crothers, D. M., and Englund, P. T. (1983). Bent helical structure in kinetoplast DNA. *Proc. Natl. Acad. Sci. USA* **80,** 7678.

Markiewicz, P., Malone, C., Chase, J. W., and Rothman-Denes, L. B. (1992). *Escherichia coli* single-stranded DNA-binding protein is a supercoiled template-dependent transcriptional activator of N4 virion RNA polymerase. *Genes Dev.* **6,** 2010–2019.

Marmorstein, R., Carey, M., Ptashne, M., and Harrison, S. C. (1992). DNA recognition by GAL4: Structure of a protein–DNA complex. *Nature* (*London*) **356,** 408–414.

Marmur, J., and Doty, P. (1962). Determination of the base composition of deoxyribonucleic acid from its thermal denaturation temperature. *J. Mol. Biol.* **5,** 109–118.

Marvin, D. A., Spencer, M., Wilkins, M. H. F., and Hamilton, L. D. (1961). The molecular configuration of deoxyribonucleic acid. III. X-Ray diffraction study of the C form of the lithium salt. *J. Mol. Biol.* **3,** 547–565.

Maxam, A. M., and Gilbert, W. (1980). Sequencing end-labeled DNA with base-specific chemical cleavages. *Meth. Enzymol.* **65,** 499–557.

McClellan, J. A., Boublikova, P., Palecek, E., and Lilley, D. M. (1990).

Superhelical torsion in cellular DNA responds directly to environmental and genetic factors. *Proc. Natl. Acad. Sci. USA* **87**, 8373–8377.

McClure, W. P. (1985). Mechanism and control of transcription in prokaryotes. *Annu. Rev. Biochem.* **54**, 171–204.

McGavin, S. (1971). Models of specifically paired like (homologous) nucleic acid structures. *J. Mol. Biol.* **55**, 293–298.

McKeon, C., Schmidt, A., and de Crombrugghe, B. (1984). A sequence conserved in both the chicken and mouse α2(I) collagen promoter contains sites sensitive to S1 nucleases. *J. Biol. Chem.* **259**, 6636–6640.

McLean, M. J., Lee, J. W., and Wells, R. D. (1988). Characteristics of Z-DNA helices formed by imperfect (purine–pyrimidine) sequences in plasmids. *J. Biol. Chem.* **263**, 7378–7385.

Menzel, R., and Gellert, M. (1983). Regulation of the genes for *E. coli* DNA gyrase: Homeostatic control of DNA supercoiling. *Cell* **34**, 105–113.

Michel, D., Chatelain, G., Herault, Y., and Brun, G. (1992). The long repetitive polypurine/polypyrimidine sequence $(TTCCC)_{48}$ forms DNA triplex with Pu-Pu-Py base triples *in vivo*. *Nucleic Acids Res.* **20**, 439–443.

Miller, J., McLachlan, A. D., and Klug, A. (1985). Repetitive zinc-binding domains in the protein transcription factor IIIA from *Xenopus* oocytes. *EMBO J.* **4**, 1609–1614.

Mirkin, S. M., Lyamichev, V. I., Drushlyak, K. N., Dobrynin, V. N., Filippov, S. A., and Frank-Kamenetskii, M. D. (1987). DNA H-form requires a homopurine–homopyrimidine mirror repeat. *Nature (London)* **330**, 495–497.

Mitsui, Y., Langridge, R., Grant, R. C., Kodama, M., Wells, R. D., Shortle, B. E., and Cantor, C. R. (1970). Physical and enzymatic studies on poly(dI-dC) · poly(dI-dC), an unusual double-helical DNA. *Nature (London)* **228**, 1166–1169.

Mizuuchi, K., Mizuuchi, M., and Gellert, M. (1982). Cruciform structures in palindromic DNA are favored by DNA supercoiling. *J. Mol. Biol.* **156**, 229–243.

Molineaux, S. M., Engh, H., de-Ferra, F., Hudson, L., and Lazzarini, R. A. (1986). Recombination within the myelin basic protein gene created the dysmyelinating shiverer mouse mutation. *Proc. Natl. Acad. Sci. USA* **83**, 7542–7546.

Moller, A., Gabriels, J. E., Lafer, E. M., Nordheim, A., Rich, A., and Stollar, D. B. (1982). Monoclonal antibodies recognize diffrent parts of Z-DNA. *J. Biol. Chem.* **257**, 12081–12085.

Morales, N. M., Cobourn, S. D., and Muller, U. R. (1990). Effect of *in vitro* transcription on cruciform stability. *Nucleic Acids Res.* **18**, 2777–2782.

Morgan, A. R. (1970). Model for DNA replication by Kornberg's DNA polymerase. *Nature (London)* **227,** 1310–1313.

Morgan, A. R., and Wells, R. D. (1968). Specificity of the three-stranded complex formation between double-stranded DNA and single-stranded RNA containing repeating nucleotide sequences. *J. Mol. Biol.* **37,** 63–80.

Morse, R. H. (1992). Transcribed chromatin. *Trends Biochem. Sci.* **17,** 23–26.

Murchie, A. I., and Lilley, D. M. (1987). The mechanism of cruciform formation in supercoiled DNA: Initial opening of central basepairs in salt-dependent extrusion. *Nucleic Acids Res.* **15,** 9641–9654.

Murchie, A. I., Clegg, R. M., von Kitzing, E., Duckett, D. R., Diekmann, S., and Lilley, D. M. (1989). Fluorescence energy transfer shows that the four-way DNA junction is a right-handed cross of antiparallel molecules. *Nature (London)* **341,** 763–766.

Murphy, K. E., and Stringer, J. R. (1986). RecA independent recombination of poly[d(GT)·d(CA)] in pBR322. *Nucleic Acids Res.* **14,** 7325–7340.

Nadal, M., Mirambeau, G., Forterre, P., Reiter, W.-D., and Duguet, M. (1986). Positively supercoiled DNA in a virus-like particle of an archaebacterium. *Nature (London)* **321,** 256–258.

Nadeau, J. G., and Crothers, D. M. (1989). Structural basis for DNA bending. *Proc. Natl. Acad. Sci. USA* **86,** 2622–2626.

Natale, D. A., Schubert, A. E., and Kowalski, D. (1992). DNA helical stability accounts for mutational defects in a yeast replication origin. *Proc. Natl. Acad. Sci. USA* **89,** 2654–2658.

Naylor, L. H., and Clark, E. M. (1990). $d(TG)_n \cdot d(CA)_n$ sequences upstream of the rat prolactin gene form Z-DNA and inhibit gene transcription. *Nucleic Acids Res.* **18,** 1595–1601.

Nejedly, K., Klysik, J., and Palecek, E. (1989). Supercoil-stabilized left handed DNA in the plasmid $(dA\text{-}dT)_{16}$ insert formed in the presence of Ni^{2+}. *FEBS Lett.* **243,** 313–317.

Noirot, P., Bargonetti, J., and Novick, R. P. (1990). Initiation of rolling-circle replication in pT181 plasmid: Initiator protein enhances cruciform extrusion at the origin. *Proc. Natl. Acad. Sci. USA* **87,** 8560–8564.

Nordheim, A., and Rich, A. (1983). The sequence $(dC\text{-}dA)_n \cdot dG\text{-}dT)_n$ forms left-handed Z-DNA in negatively supercoiled plasmids. *Proc. Natl. Acad. Sci. USA* **80,** 1821–1825.

Nordheim, A., Pardue, M. L., Lafer, E. M., Moller, A., Stollar, B. D., and Rich, A. (1981). Antibodies to left-handed Z-DNA bind to interband regions of *Drosophila* polytene chromosomes. *Nature (London)* **294,** 417–422.

Nordheim, A., Lafer, E. M., Peck, L. J., Wang, J. C., Stollar, B. D., and Rich, A. (1982a). Negatively supercoiled plasmids contain left-handed Z-DNA segments as detected by specific antibody binding. *Cell* **31**, 309–318.

Nordheim, A., Tesser, P., Azorin, F., Kwon, Y. H., Moller, A., and Rich, A. (1982b). Isolation of *Drosophila* proteins that bind selectively to left-handed Z-DNA. *Biochemistry* **79**, 7729–7733.

Norton, V. G., Marvin, K. W., Yau, P., and Bradbury, E. M. (1990). Nucleosome linking number change controlled by acetylation of histones H3 and H4. *J. Biol. Chem.* **265**, 19848–19852.

Norton, V. G., Imai, B. S., Yau, P., and Bradbury, E. M. (1989). Histone acetylation reduces nucleosome core particle linking number change. *Cell* **57**, 449–457.

Oliva, R., Bazett-Jones, D. P., Locklear, L., and Dixon, G. H. (1990). Histone hyperacetylation can induce unfolding of the nucleosome core particle. *Nucleic Acid Res.* **18**, 2739–2747.

Ornstein, R. L., Rein, R., Breen, D. L., and MacElroy, R. D. (1978). An optimized potential function for the calculation of nucleic acid interaction energies. I. Base stacking. *Biopolymers* **17**, 2341–2360.

O'Shea, E. K., Kleimm, J. D., Kim, P. S., and Alber, T. (1991). X-Ray structure of the GCN4 leucine zipper, a two-stranded, parallel coiled coil. *Science* **254**, 539–544.

Otwinowski, Z., Schevitz, R. W., Zhang, R.-G., Lawson, C. L., Joachimiak, A., Marmorstein, R. Q., Luisi, B. F., and Sigler, P. B. (1988). Crystal structure of *trp* repressor/operator complex at atomic resolution. *Nature (London)* **335**, 321–329.

Pabo, C. O., and Sauer, R. T. (1984). Protein–DNA recognition. *Annu. Rev. Biochem.* **53**, 293–321.

Pabo, C. O., and Sauer, R. T. (1992). Transcription factors: Structural families and principles of DNA recognition. *Annu. Rev. Biochem.* **61**, 1053–1095.

Paleček, E. (1991). Local supercoil-stabilized DNA structures. *CRC Crit. Rev. Biochem. Mol. Biol.* **26**, 151–226.

Paleček, E. (1992). Probing of DNA structure in cells with osmium tetroxide-2,2′-bipyridine. *Meth. Enzymol.* **212**, 305–318.

Panayotatos, N., and Fontaine, A. (1987). A native cruciform DNA structure probed in bacteria by recombinant T7 endonuclease. *J. Biol. Chem.* **262**, 11364–11368.

Panayotatos, N., and Wells, R. D. (1981). Cruciform structures in supercoiled DNA. *Nature (London)* **289**, 466–470.

Panyutin, I. G., and Wells, R. D. (1992). Nodule DNA in the $(GA)_{37} \cdot (CT)_{37}$ insert in superhelical plasmids. *J. Biol. Chem.* **267**, 5495–5501.

Pardue, M. L., Nordheim, A., Lafer, E. M., Stollar, B. D., and Rich, A. (1983). Z-DNA and the polytene chromosome. *Cold Spring Harbor Symp. Quant. Biol.* **47,** 171–176.

Parniewski, P., Kwinkowski, M., Wilk, A., and Klysik, J. (1990). Dam methyltransferase sites located within the loop region of the oligopurine–oligopyrimidine sequences capable of forming H-DNA are undermethylated *in vivo. Nucleic Acids Res.* **18,** 605–611.

Pathak, D., and Sigler, P. B. (1992). Updating structure–function relationships in the bZip family of transcription factors. *Curr. Opin. Struct. Biol.* **2,** 116–123.

Paulson, J. R., and Laemmli, U. K. (1977). The structure of histone-depleted metaphase chromosomes. *Cell* **12,** 817–828.

Pavletich, N. P., and Pabo, C. O. (1991). Zinc finger-DNA recognition: Crystal structure of a Zif268–DNA complex at 2.1 Å. *Science* **252,** 809–817.

Peck, L. J., and Wang, J. C. (1985). Transcriptional block caused by a negative supercoiling-induced structural change in an alternating CG sequence. *Cell* **40,** 129–137.

Peck, L. J., Nordheim, A., Rich, A., and Wang, J. C. (1982). Flipping of cloned $d(pCpG)_n \cdot d(pG\text{-}pC)_n$ DNA sequences from right- to left-handed helical structure by salt, Co(III), or negative supercoiling. *Proc. Natl. Acad. Sci. USA* **79,** 4560–4564.

Pei, D., Corey, D. R., and Schultz, P. G. (1990). Site-specific cleavage of duplex DNA by a semisynthetic nuclease via triple-helix formation. *Proc. Natl. Acad. Sci. USA* **87,** 9858–9862.

Perrouault, L., Asseline, U., Rivalle, C., Thuong, N. T., Bisagni, E., Giovannangeli, C., Le Doan, T., and Hélène, C. (1990). Sequence-specific artificial photo-induced endonucleases based on triple helix-forming oligonucleotides. *Nature (London)* **344,** 358–360.

Pettijohn, D. E., and Pfenninger, O. (1980). Supercoils in prokaryotic DNA restrained *in vivo. Proc. Natl. Acad. Sci. USA* **77,** 1331–1335.

Pettijohn, D. E., and Sinden, R. R. (1985). Structure of the isolated nucleoid. *In* "Molecular Cytology of *E. coli.*" (N. Nanninga, ed.), pp. 199–227. Academic Press, London.

Phillips, S. E. V. (1991). Specific β-sheet interactions. *Curr. Opin. Struct. Biol.* **1,** 89–98.

Phillips, S. E. V., Manfield, I., Parsons, I., Rafferty, J. B., Somers, W. S., Margarita, D., Cohen, G. N., Santi-Girons, I., and Stockely, P. G. (1989). Cooperative tandem binding of the *E. coli* methionine repressor. *Nature (London)* **341,** 711–714.

Pina, B., Barettino, D., and Beato, M. (1991). Nucleosome positioning and regulated gene expression. *Oxford Surv. Euk. Genes* **7,** 83–117.

Platt, J. R. (1955). Possible separation of inter-twisted nucleic acid chains by transfer-twist. *Proc. Natl. Acad. Sci. USA* **71,** 181–183.

Pohl, F. M., and Jovin, T. M. (1972). Salt-induced co-operative conformational change of a synthetic DNA: Equilibrium and kinetic studies with poly (dG-dC). *J. Mol. Biol.* **67,** 375–396.

Postel, E. H., Mango, S. E., and Flint, S. J. (1989). A nuclease-hypersensitive element of the human c-*myc* promoter interacts with a transcription initiation factor. *Mol. Cell. Biol.* **9,** 5123–5133.

Prior, C. P., Cantor, C. R., Johnson, E. M., Littau, V. C., and Allfrey, V. G. (1983). Reversible changes in nucleosome structure and histone H3 accessibility in transcriptionally active and inactive states of rDNA chromatin. *Cell* **34,** 1033–1042.

Prunell, A. (1982). Nucleosome reconstitution on plasmid-inserted poly(dA) · poly(dT). *EMBO J.* **1,** 173–179.

Pruss, G. J., and Drlica, K. (1986). Topoisomerase I mutants: The gene on pBR322 that encodes resistance to tetracycline affects plasmid DNA supercoiling. *Proc. Natl. Acad. Sci. USA* **83,** 8952–8956.

Pruss, G. J., Manes, S. H., and Drlica, K. (1982). *Escherichia coli* DNA topoisomerase I mutants: Increased supercoiling is corrected by mutations near gyrase genes. *Cell* **31,** 35–42.

Pulleyblank, D. E., Shure, M., Tang, D., Vinograd, J., and Vosberg, H. (1975). Action of nick-closing enzyme on supercoiled and nonsupercoiled closed circular DNA: Formation of a Boltzmann distribution of topological isomers. *Proc. Natl. Acad. Sci. USA* **72,** 4280–4284.

Pulleyblank, D. E., Haniford, D. B., and Morgan, A. R. (1985). A structural basis for S1 nuclease sensitivity of double-stranded DNA. *Cell* **42,** 271–280.

Radman, M. (1991). Avoidance of inter-repeat recombination by sequence divergence and a mechanism of neutral evolution. *Biochimie* **73,** 357–361.

Rafferty J.B., , Somers, W. S., Saint-Girons, I., and Phillips, S. E. V. (1989). Three-dimensional crystal structures of *E. coli* Met repressor with and without corepressor. *Nature (London)* **341,** 705–710.

Rahmouni, A. R., and Wells, R. D. (1989). Stabilization of Z DNA *in vivo* by localized supercoiling. *Science* **246,** 358–363.

Rahmouni, A. R., and Wells, R. D. (1992). Direct evidence for the effect of transcription on local DNA supercoiling *in vivo*. *J. Mol. Biol.* **223,** 131–144.

Ramsing, N. B., and Jovin, T. M. (1988). Parallel stranded duplex DNA. *Nucleic Acids Res.* **16,** 6659–6676.

Rao, B. J., Dutreix, M., and Radding, C. M. (1991). Stable three-stranded DNA made by RecA protein. *Proc. Natl. Acad. Sci. USA* **88,** 2984–2988.

Rao, B. J., Chiu, S. K., and Radding, C. M. (1993). Homologous recognition and triplex formation promoted by RecA protein between duplex oligonucleotides and single-stranded DNA. *J. Mol. Biol.* **229,** 328–343.

Rhodes, D. (1979). Nucleosome cores reconstituted from poly (dA · dT) and the octamer of histones. *Nucleic Acids Res.* **6,** 1805–1816.

Rhodes, D., and Klug, A. (1986). An underlying repeat in some transcriptional control sequences corresponding to half a double helical turn of DNA. *Cell* **46,** 123–132.

Rice, J. A., and Crothers, D. M. (1989). DNA bending by the bulge defect. *Biochemistry* **28,** 4512–4516.

Rich, A., Nordheim, A., and Wang, A. H.-J. (1984). The chemistry and biology of left-handed Z-DNA. *Annu. Rev. Biochem.* **53,** 791–846.

Richmond, T. J., Finch, J. T., Rushton, B., Rhodes, D., and Klug, A. (1984). Structure of the nucleosome core particle at 7 Å resolution. *Nature (London)* **311,** 532–537.

Rio, P., and Leng, M. (1986). *N*-Hydroxyaminofluorene: A chemical probe for DNA conformation. *J. Mol. Biol.* **191,** 569–572.

Ripley, L. S. (1991). Frameshift mutation: Determinants of specificity. *Annu. Rev. Genet.* **24,** 189–213.

Rohner, K. J., Hobi, R., and Kuenzle, C. C. (1990). Z-DNA-binding proteins. Identification critically depends on the proper choice of ligands. *J. Biol. Chem.* **265,** 19112–19115.

Rosenberg, J. M., Seeman, N. C., Day, R. O., and Rich, A. (1976). RNA double helical fragment at the atomic resolution. II. The crystal structure of sodium guanylyl-3′,5′-cytidine nonahydrate. *J. Mol. Biol.* **104,** 145–167.

Rouviere-Yaniv, J. and Gros, F. (1975). Characterization of a novel, low molecular weight DNA-binding protein from *Escherichia coli. Proc. Natl. Acad. Sci. USA* **72,** 3428–3432.

Rouviere-Yaniv, J., Yaniv, M., and Germond, J. E. (1979). *E. coli* DNA binding protein HU forms nucleosome-like structure with circular double-stranded DNA. *Cell* **17,** 265–274.

Ryder, K., Silver, S., DeLucia, A. L., Fanning, E., and Tegtmeyer, P. (1986). An altered DNA conformation in origin region I is a determinant for the binding of SV40 large T antigen. *Cell* **44,** 719–725.

Saenger, W. (1984). "Principles of Nucleic Acid Structure." Springer-Verlag, New York.

Sanger, F., and Coulson, A. R. (1975). A rapid method for determining sequences in DNA by primed synthesis with DNA polymerase. *J. Mol. Biol.* **94,** 441–448.

Santella, R. M., Grunberger, D., Weinstein, I. B., and Rich, A. (1981). Induction of the Z conformation in poly(dG-dC)–poly(dG-dC) by binding of *N*-2-acetylaminofluorene to guanine residues. *Proc. Natl. Acad. Sci. USA* **78,** 1451–1455.

Santoro, C., Costanzo, F., and Ciliberto, G. (1984). Inhibition of eukaryotic tRNA transcription by potential Z-DNA sequences. *EMBO J.* **3,** 1553–1559.

Satchwell, S. C., Drew, H. R., and Travers, A. A. (1986). Sequence periodicities in chicken nucleosome core DNA. *J. Mol. Biol.* **191,** 659–675.

Scaria, P. V., Shire, S. J., and Shafer, R. H. (1992). Quadruplex structure of d($G_3T_4G_3$) stabilized by K^+ or Na^+ is an asymmetric hairpin dimer. *Proc. Natl. Acad. Sci. USA* **89,** 10336–10340.

Schleif, R. (1988). DNA binding by proteins. *Science* **241,** 1182–1187.

Schleif, R. (1992). DNA looping. *Annu. Rev. Biochem.* **61,** 119–223.

Schlotterer, C., and Tautz, D. (1992). Slippage synthesis of simple sequence DNA. *Nucleic Acids Res.* **20,** 211–215.

Schultz, S. C., Shields, G. C., and Steitz, T. A. (1991). Crystal structure of a CAP-DNA complex: The DNA is bent by 90 degrees. *Science* **253,** 1001–1007.

Schwabe, J. W., Neuhaus, D., and Rhodes, D. (1990). Solution structure of the DNA-binding domain of the oestrogen receptor. *Nature (London)* **348,** 458–461.

Scovell, W. M. (1986). Supercoiled DNA. *J. Chem. Ed.* **63,** 562–565.

Seeman, N. C., Rosenberg, J. M., and Rich, A. (1976a). Sequence-specific recognition of double helical nucleic acids by proteins. *Proc. Natl. Acad. Sci. USA* **73,** 804–808.

Seeman, N. C., Rosenberg, J. M., Suddath, F. L., Kim, J. J. P., and Rich, A. (1976b). RNA double helical fragment at atomic resolution. I. The crystal and molecular structure of sodium adenylyl-3′,5′-uridine hexahydrate. *J. Mol. Biol.* **104,** 109–144.

Selsing, E., Wells, R. D., Alden, C. J., and Arnott, S. (1979). Bent DNA: Visualization of a base-paired and stacked A-B conformational junction. *J. Biol. Chem.* **254,** 5417–5422.

Sen, D., and Gilbert, W. (1988). Formation of parallel four-stranded complexes by guanine-rich motifs in DNA and its implications for meiosis. *Nature (London)* **334,** 364–366.

Sen, D., and Gilbert, W. (1990). A sodium–potassium switch in the formation of four-stranded G4-DNA. *Nature (London)* **344,** 410–414.

Sen, D., and Gilbert, W. (1992). Novel DNA superstructures formed by telomere-like oligomers. *Biochemistry* **31,** 65–70.

Sheflin, L. G., and Kowalski, D. (1984). Mung bean nuclease cleavage of dA + dT-rich sequence or an inverted repeat sequence in supercoiled PM2 DNA depends on ionic environment. *Nucleic Acids Res.* **12,** 7087–7104.

Shimizu, M., Hanvey, J. C., and Wells, R. D. (1989). Intramolecular DNA triplexes in supercoiled plasmids. I. Effect of loop size on formation and stability. *J. Biol. Chem.* **264,** 5944–5949.

Shimizu, M., Hanvey, J. C., and Wells, R. D. (1990). Multiple non-B-DNA conformations of polypurine · polypyrimidine sequences in plasmids. *Biochemistry* **29,** 4704–4713.

Shimizu, M., Kubo, K., Matsumoto, U., and Shindo, H. (1994). The loop sequence plays crucial roles for isomerization of intramolecular DNA triplexes in supercoiled plasmids. *J. Mol. Biol.* **235,** 185–197.

Shrader, T. E., and Crothers, D. M. (1990). Effects of DNA sequence and histone-histone interactions on nucleosome placement. *J. Mol. Biol.* **216,** 69–84.

Simpson, R. T. (1990). Nucleosome positioning: Occurrence, mechanisms, and functional consequences. *Prog. Nucleic Acid Res. Mol. Biol.* **40,** 143–184.

Simpson, R. T., and Kunzler, P. (1979). Chromatin and core particles formed from the inner histones and synthetic polydeoxyribonucleotides of defined sequence. *Nucleic Acids Res.* **6,** 1387–1415.

Sinden, R. R. (1987). Supercoiled DNA: Biological significance. *J. Chem. Ed.* **64,** 294–301.

Sinden, R. R., and Cole, R. S. (1978). Topography and kinetics of genetic recombination in *Escherichia coli* treated with psoralen and light. *Proc. Natl. Acad. Sci. USA* **75,** 2373–2377.

Sinden, R. R., and Kochel, T. J. (1987). Reduced 4,5′,8-trimethylpsoralen cross-linking of left-handed Z-DNA stabilized by DNA supercoiling. *Biochemistry* **26,** 1343–1350.

Sinden, R. R., and Pettijohn, D. E. (1981). Chromosomes in living *Escherichia coli* cells are segregated into domains of supercoiling. *Proc. Natl. Acad. Sci. USA* **78,** 224–228.

Sinden, R. R., and Pettijohn, D. E. (1984). Cruciform transitions in DNA. *J. Biol. Chem.* **259,** 6593–6600.

Sinden, R. R., and Ussery, D. W. (1992). Analysis of DNA structure *in vivo* using psoralen photobinding: Measurement of supercoiling, topological domains, and DNA–protein interactions. *Meth. Enzymol.* **212,** 319–335.

Sinden, R. R., and Wells, R. D. (1992). DNA structure, mutations, and human genetic disease. *Curr. Opin. Biotech.* **3,** 612–622.

Sinden, R. R., Broyles, S. S., and Pettijohn, D. E. (1983a). Perfect palindromic lac operator DNA sequence exists as a stable cruciform structure in supercoiled DNA *in vitro* but not *in vivo*. *Proc. Natl. Acad. Sci. USA* **80,** 1797–1801.

Sinden, R. R., Broyles, S. S., and Pettijohn, D. E. (1983b). Procedure using psoralen cross-linking to quantitate DNA cruciform structures shows that cruciforms can exist *in vitro* but not *in vivo*. *In* "Mechanisms of DNA Replication and Recombination" (N. R. Cozzarelli, ed.), pp. 19–28. Liss, New York.

Sinden, R. R., Carlson, J. O., and Pettijohn, D. E. (1980). Torsional tension in the DNA double helix measured with trimethylpsoralen in living *E. coli* cells: Analogous measurements in insect and human cells. *Cell* **21,** 773–783.

Sinden, R. R., Zheng, G., Brankamp, R. G., and Allen, K. N. (1991). On the deletion of inverted repeated DNA in *Escherichia coli*: Effects of length, thermal stability, and cruciform formation *in vivo*. *Genetics* **129,** 991–1005.

Singer, B., and Kusmierek, J. T. (1982). Chemical mutagenesis. *Annu. Rev. Biochem.* **52,** 655–693.

Singleton, C. K. (1983). Effects of salts, temperature, and stem length on supercoil-induced formation of cruciforms. *J. Biol. Chem.* **258,** 7661–7668.

Singleton, C. K., Klysik, J., Stirdivant, S. M., and Wells, R. D. (1982). Left-handed Z-DNA is induced by supercoiling in physiological ionic conditions. *Nature (London)* **299,** 312–316.

Singleton, C. K., Klysik, J., and Wells, R. D. (1983). Conformational flexibility of junctions between contiguous B- and Z-DNAs in supercoiled plasmids. *Proc. Natl. Acad. Sci. USA* **80,** 2447–2451.

Smith, F. W., and Feigon, J. (1992). Quadruplex structure of *Oxytricha* telomeric oligonucleotides. *Nature (London)* **356,** 164–168.

Snyder, M., Buchman, A. R., and Davis, R. W. (1986). Bent DNA at a yeast autonomously replicating sequence. *Nature (London)* **324,** 87–89.

Snyder, U. K., Thompson, J. F., and Landy, A. (1989). Phasing of protein-induced DNA bends in a recombination complex. *Nature (London)* **341,** 255–257.

Sowers, L. C., Fazakerley, G. V., Eritja, R., Kaplan, B. E., and Goodman, M. F. (1986). Base pairing and mutagenesis: Observations of a protonated base pair between 2-aminopurine and cytosine in an oligonucleotide by proton NMR. *Proc. Natl. Acad. Sci. USA* **83,** 5434–5438.

Sowers, L. C., Goodman, M. F., Eritja, R., Kaplan, B. E., and Fazakerley, G. V. (1989). Ionized and wobble base-pairing for bromouracil-guanine in equilibrium under physiological conditions: A nuclear magnetic resonance study on an oligonucleotide containing a bromouracil–guanine base-pair as a function of pH. *J. Mol. Biol.* **205,** 437–447.

Spengler, S. J., Stasiak, A., and Cozzarelli, N. R. (1985). The stereostructure of knots and catenanes produced by phage lambda integrative recombination: Implications for mechanism and DNA structure. *Cell* **42,** 325–334.

Stasiak, A. (1992). Three-stranded DNA structure: Is this the secret of DNA homologous recognition? *Mol. Microbiol.* **6,** 3267–3276.

Steck, T. R., Pruss, G. J., Manes, S. H., Burg, L., and Drlica, K. (1984). DNA supercoiling in gyrase mutants. *J. Bacteriol.* **158,** 397–403.

Stettler, U. H., Weber, H., Koller, Th., and Weissmann, Ch. (1979). Preparation and characterization of form V DNA, the duplex DNA resulting from association of complementary, circular single-stranded DNA. *J. Mol. Biol.* **131,** 21–40.

Stirdivant, S. M., Klysik, J., and Wells, R. D. (1982). Energetic and structural inter-relationship between DNA supercoiling and the right- to left-handed Z helix transitions in recombinant plasmids. *J. Biol. Chem.* **257,** 10159–10165.

Stonington, O. G., and Pettijohn, D. E. (1971). The folded genome of *Escherichia coli* isolated in a protein–DNA–RNA complex. *Proc. Natl. Acad. Sci. USA* **68,** 6–9.

Streisinger, G., Okada, Y., Emrich, J., Newton, J., Tsugita, A., Terzaghi, E., and Inouye, M. (1966). Frameshift mutations and the genetic code. *Cold Spring Harbor Symp. Quant. Biol.* **31,** 77–84.

Strobel, S. A., and Dervan, P. B. (1990). Site-specific cleavage of a yeast chromosome by oligonucleotide-directed triple-helix formation. *Science* **249,** 73–75.

Suck, D., Lahm, A., and Oefner, C. (1988). Structure refined to 2 Å of a nicked DNA octanucleotide complex with DNase I. *Nature (London)* **332,** 464–468.

Sullivan, K. M., and Lilley, D. M. (1986). A dominant influence of flanking sequences on a local structural transition. *Cell* **47,** 817–827.

Sullivan, K. M., and Lilley, D. M. (1987). Influence of cation size and charge on the extrusion of a salt-dependent cruciform. *J. Mol. Biol.* **193,** 397–404.

Sullivan, K. M., Murchie, A. I., and Lilley, D. M. (1988). Long range structural communication between sequences in supercoiled DNA. Sequence dependence of contextual influence on cruciform extrusion mechanism. *J. Biol. Chem.* **263,** 13074–13082.

Sutherland, J. C., and Griffen, K. P. (1983). Vacuum ultraviolet circular dichroism of poly(dI-dC) · poly(dI-dC): No evidence for a left-handed double helix. *Biopolymers* **22,** 1445–1448.

Symington, L., and Kolodner, R. (1985). Partial purification of an endonuclease from *Saccharomyces cerevisiae* that cleave Holliday junctions. *Proc. Natl. Acad. Sci. USA* **82,** 7247–7251.

Takasugi, M., Guendouz, A., Chassignol, M., Decout, J. L., Lhomme, J., Thuong, N. T., and Hélène, C. (1991). Sequence-specific photo-induced cross-linking of the two strands of double-helical DNA by a psoralen covalently linked to a triple helix-forming oligonucleotide. *Proc. Natl. Acad. Sci. USA* **88,** 5602–5606.

Tanaka, I., Appelt, K., Dijk, J., White, S. W., and Wilson, K. S. (1984). 3-Å Resolution structure of a protein with histone-like properties in prokaryotes. *Nature* (*London*) **310,** 376–381.

Tang, M., Htun, H., Cheng, Y., and Dahlberg, J. E. (1991). Suppression of cyclobutane and <6–4> dipyrimidines formation in triple-stranded H-DNA. *Biochemistry* **30,** 7021–7026.

Taylor, A. F., and Smith, G. R. (1990). Action of RecBCD enzyme on cruciform DNA. *J. Mol. Biol.* **211,** 117–134.

Taylor, W. H., and Hagerman, P. J. (1990). Application of the method of T4 phage DNA ligase-catalyzed ring-closure to the study of DNA structure. *J. Mol. Biol.* **212,** 363–376.

Thoma, F. (1991). Structural changes in nucleosomes during transcription: Strip, split or flip? *Trends Genet.* **7,** 175–177.

Toose, J. (1982). "Molecular Biology of Tumor Viruses." Cold Spring Harbor Laboratory Press, Cold Spring Harbor, New York.

Topal, M. D., and Fresco, J. R. (1976). Complementary base pairing and the origin of substitution mutations. *Nature* (*London*) **263,** 285–289.

Trifonov, E. N. (1985). Curved DNA. *CRC Crit. Rev. Biochem.* **19,** 89–106.

Trifonov, E. N., and Sussman, J. L. (1980). The pitch of chromatin DNA is reflected in its nucleotide sequence. *Proc. Natl. Acad. Sci. USA* **77,** 3816–3820.

Trifonov, E. N., and Ulanovsky, L. E. (1989). Inherently curved DNA and its structural elements. *In* "Unusual DNA Structures" (R. D. Wells and S. C. Harvey, eds.), pp. 173–187. Springer-Verlag, New York.

Trinh, T. Q., and Sinden, R. R. (1991). Preferential DNA secondary structure mutagenesis in the lagging strand of replication in *E. coli. Nature* (*London*) **352,** 544–547.

Trinh, T. Q., and Sinden, R. R. (1993). The influence of primary and secondary DNA structure in deletion and duplication between direct repeats in *Escherichia coli. Genetics* **134,** 409–422.

Tse-Dinh, Y. C., and Beran, R. K. (1988). Multiple promoters for transcription of the *Escherichia coli* DNA topoisomerase I gene and their regulation by DNA supercoiling. *J. Mol. Biol.* **202,** 735–742.

Ulanovsky, L. E., and Trifonov, E. N. (1987). Estimation of wedge components in curved DNA. *Nature (London)* **326,** 720–722.

Ulanovsky, L., Bodner, M., Trifonov, E. N., and Choder, M. (1986). Curved DNA: Design, synthesis, and circularization. *Proc. Natl. Acad. Sci. USA* **83,** 862–866.

Ulrich, M. J., Gray, W. J., and Ley, T. J. (1992). An intramolecular DNA triplex is disrupted by point mutations associated with hereditary persistence of fetal hemoglobin. *J. Biol. Chem.* **267,** 18649–18658.

Umek, R. M., and Kowalski, D. (1987). Yeast regulatory sequences preferentially adopt a non-B conformation in supercoiled DNA. *Nucleic Acids Res.* **15,** 4467–4480.

Umek, R. M., and Kowalski, D. (1988). The ease of DNA unwinding as a determinant of initiation at yeast replication origins. *Cell* **52,** 559–567.

Ussery, D. W., and Sinden, R. R. (1993). Environmental influences on the *in vivo* level of intramolecular triplex DNA in *Escherichia coli. Biochemistry* **32,** 6206–6213.

Ussery, D. W., Hoepfner, R. W., and Sinden, R. R. (1992). Probing DNA structure with psoralen *in vitro. Meth. Enzymol.* **212,** 242–262.

Vallee, B. L., Coleman, J. E., and Auld, D. S. (1991). Zinc fingers, zinc clusters, and zinc twists in DNA-binding protein domains. *Proc. Natl. Acad. Sci. USA* **88,** 999–1003.

van de Sande, J. H., McIntosh, L. P., and Jovin, T. M. (1982). Mn^{2+} and other transition metals at low concentration induce the right-to-left helical transformation of poly[d(G-C)]. *EMBO J.* **1,** 777–782.

van de Sande, J. H., Ramsing, N. B., Germann, M. W., Elhorst, W., Kalisch, B. W., von Kitzing, E., Pon, R. T., Clegg, R. C., and Jovin, T. M. (1988). Parallel stranded DNA. *Science* **241,** 551–557.

van Holde, K. E. (1989). "Chromatin." Springer-Verlag, New York.

van Holde, K. E., Lohr, D. E., and Robert, C. (1992). What happens to nucleosomes during transcription? *J. Biol. Chem.* **267,** 2837–2840.

Vardimon, L., and Rich, A. (1984). In Z-DNA the sequence G-C-G-C is neither methylated by *Hha* I methyltransferase nor cleaved by *Hha* I restriction endonuclease. *Proc. Natl. Acad. Sci. USA* **81,** 3268–3272.

Veselkov, A. G., Malkov, V. A., Frank-Kamenetskii, M. D., and Dobrynin, V. N. (1993). Triplex model of chromosome ends. *Nature (London)* **364,** 496.

Vojtiskova, M., and Palecek, E. (1987). Unusual protonated structure in the homopurine · homopyrimidine tract of supercoiled and linearized plasmids recognized by chemical probes. *J. Biomol. Struct. Dynam.* **5,** 283–296.

Vologodskii, A. V., and Frank-Kamenetskii, M. D. (1982). Theoretical study of cruciform states in superhelical DNA. *FEBS Lett.* **143,** 257–260.

Wahls, W. P., Wallace, L. J., and Moore, P. D. (1990). The Z-DNA motif $d(TG)_{30}$ promotes reception of information during gene conversion events while stimulating homologous recombination in human cells in culture. *Mol. Cell. Biol.* **10,** 785–793.

Wang, A. H. J., Quigley, G. J., Kolpak, F. J., Crawford, J. L., van Boom, J. H., Van der Marel, G., and Rich, A. (1979). Molecular structure of a left-handed double helical DNA fragment at atomic resolution. *Nature (London)* **282,** 680–686.

Wang, C. I., and Taylor, J. S. (1991). Site-specific effect of thymine dimer formation on $dA_n \cdot dT_n$ tract bending and its biological implications. *Proc. Natl. Acad. Sci. USA* **88,** 9072–9076.

Wang, J. C. (1985). DNA topoisomerases. *Annu. Rev. Biochem.* **54,** 665–697.

Ward, G. K., McKenzie, R., Zannis-Hadjopoulos, M., and Price, G. B. (1990). The dynamic distribution and quantification of DNA cruciforms in eukaryotic nuclei. *Exp. Cell Res.* **188,** 235–246.

Ward, G. K., Shihab-el-Deen, A., Zannis-Hadjopoulos, M., and Price, G. B. (1991). DNA cruciforms and the nuclear supporting structure. *Exp. Cell Res.* **195,** 92–98.

Waring, M. J. (1981). DNA modification and cancer. *Annu. Rev. Biochem.* **50,** 159–192.

Warren, G. L., and Green, R. L. (1985). Comparison of the physical and genetic properties of palindromic DNA sequences. *J. Bacteriol.* **161,** 1103–1111.

Wasserman, S. A., and Cozzarelli, N. R. (1986). Biochemical topology: Applications to DNA recombination and replication. *Science* **232,** 951–960.

Watson, J. D., and Crick, F. H. C. (1953a). The structure of DNA. *Cold Spring Harbor Symp. Quant. Biol.* **18,** 123–131.

Watson, J. D., and Crick, F. H. C. (1953b). Molecular structure of nucleic acids—A structure for deoxyribose nucleic acid. *Nature (London)* **171,** 737–738.

Weaver, D. T., and DePamphilis, M. L. (1984). Role of palindromic and non-palindromic sequences in arresting DNA synthesis *in vitro* and *in vivo*. *J. Mol. Biol.* **180,** 961–986.

Weinreb, A., Collier, D. A., Birshtein, B. K., and Wells, R. D. (1990). Left-handed Z-DNA and intramolecular triplex formation at the site of an unequal sister chromatid exchange. *J. Biol. Chem.* **265,** 1352–1359.

Weinreb, A., Katzenberg, D. R., Gilmore, G. L., and Birshtein, B. K. (1988). Site of unequal sister chromatid exchange contains a potential Z-DNA-forming tract. *Proc. Natl. Acad. Sci. USA* **85,** 529–533.

Weintraub, H., Worcel, A., and Alberts, B. (1976). A model for chromatin based upon two symmetrically paired half-nucleosomes. *Cell* **9,** 409–417.

Weller, S. K., Spadaro, A., Schaffer, J. E., Murray, A. W., Maxam, A. M., and Schaffer, P. A. (1985). Cloning, sequencing, and functional analysis of *ori*L, a herpes simplex virus type 1 origin of DNA synthesis. *Mol. Cell. Biol.* **5**, 930–942.

Wells, R. D. (1988). Unusual DNA structures. *J Biol. Chem.* **263**, 1095–1098.

Wells, R. D., and Sinden, R. R. (1993). Defined ordered sequence DNA, DNA structure, and DNA-directed mutation. *In* "Genome Analysis" (K. E. Davies and S. T. Warren, eds.), Vol. 7, pp. 107–138. Cold Spring Harbor Laboratory Press, Cold Spring Harbor, New York.

Wells, R. D., Collier, D. A., Hanvey, J. C., Shimizu, M., and Wohlrab, F. (1988). The chemistry and biology of unusual DNA structures adopted by oligopurine · oligopyrimidine sequences. *FASEB J.* **2**, 2939–2949.

West, S. C. (1992). Enzymes and molecular mechanisms of genetic recombination. *Annu. Rev. Biochem.* **61**, 603–640.

West, S. C., and Korner, A. (1985). Cleavage of cruciform DNA structures by an activity from *Saccharomyces cerevisae. Proc. Natl. Acad. Sci. USA* **82**, 6445–6449.

West, S. C., Parsons, C. A., and Picksley, S. M. (1987). Purification and properties of a nuclease from *Saccharomyces cerevisiae* that cleaves DNA at cruciform junctions. *J. Biol. Chem.* **262**, 12752–12758.

Weston-Hafer, K., and Berg, D. E. (1989). Palindromy and the location of deletion endpoints in *Escherichia coli. Genetics* **121**, 651–658.

Weston-Hafer, K., and Berg, D. E. (1991). Deletions in plasmid pBR322: Replication slippage involving leading and lagging strands. *Genetics* **127**, 649–655.

White, J. H., and Bauer, W. R. (1987). Superhelical DNA with local substructures: A generalization of the topological constraint in terms of the intersection number and the ladder-like correspondence surface. *J. Mol. Biol.* **195**, 205–213.

White, S.W., Appelt, K., Wilson, K. S., and Tanaka, I. (1989). A protein structural motif that bends DNA. *Protein Struct. Funct. Genet.* **5**, 281–288.

Wilkins, M. H. F., and Randall, J. T. (1953). Crystallinity in sperm heads: Molecular structure of nucleoprotein *in vivo. Biochim. Biophys. Acta* **10**, 192.

Wilkins, M. H. F., Stokes, A. R., and Wilson, H. R. (1953). Molecular structure of deoxypentose nucleic acids. *Nature (London)* **171**, 738–740.

Williams, W. L., and Muller, U. R. (1987). Effects of palindrome size and sequence on genetic stability in the bacteriophage ϕX174 genome. *J. Mol. Biol.* **196**, 743–755.

Williamson, J. R., Raghuraman, M. K., and Cech, T. R. (1989). Monovalent cation-induced structure of telomeric DNA: The G-quartet model. *Cell* **59,** 871–880.

Wittig, B., Dorbic, T., and Rich, A. (1989). The level of Z-DNA in metabolically active, permeabilized mammalian cell nuclei is regulated by torsional strain. *J. Cell Biol.* **108,** 755–764.

Wittig, B., Dorbic, T., and Rich, A. (1991). Transcription is associated with Z-DNA formation in metabolically active permeabilized mammalian cell nuclei. *Proc. Natl. Acad. Sci. USA* **88,** 2259–2263.

Wohlrab, F., and Wells, R. D. (1989). Slight changes in conditions influence the family of non-B-DNA conformations of the herpes simplex virus type 1 DR2 repeats. *J. Biol. Chem.* **264,** 8207–8213.

Wohlrab, F., McLean, M. J., and Wells, R. D. (1987). The segment inversion site of herpes simplex virus type 1 adopts a novel DNA structure. *J. Biol. Chem.* **262,** 6407–6416.

Wolberger, C. (1993). Transcription factor structure and DNA binding. *Curr. Opin. Struct. Biol.* **3,** 3–10.

Wolffe, A. (1992). "Chromatin Structure and Function." Academic Press, San Diego.

Woodcock, C. L. F., Frado, L. L., and Rattner, J. B. (1984). The higher-order structure of chromatin: Evidence for a helical ribbon arrangement. *J. Cell Biol.* **99,** 42.

Woodworth-Gutai, M., and Lebowitz, J. (1976). Introduction of interrupted secondary structure in supercoiled DNA as a function of superhelix density: Consideration of haripin structures in superhelical DNA. *J. Virol.* **18,** 195–204.

Worcel, A., and Benyajati, C. (1977). Higher order coiling of DNA in chromatin. *Cell* **12,** 83–100.

Worcel, A., and Strogatz, S., and Riley, D. (1981). Structure of chromatin and the linking number of DNA. *Proc. Natl. Acad. Sci. USA* **78,** 461.

Workman, J. L., and Roeder, R. G. (1987). Binding of transcription factor TFIID to the major late promoter during *in vitro* nucleosome assembly potentiates subsequent initiation by RNA polymerase II. *Cell* **51,** 613–622.

Wu, H. M., and Crothers, D. M. (1984). The locus of sequence-directed and protein-induced DNA bending. *Nature* (*London*) **308,** 509–513.

Yu, H., Eritja, R., Bloom, L. B., and Goodman, M. F. (1993). Ionization of bromouracil and fluorouracil stimulates base mispairing frequencies with guanine. *J. Biol. Chem.* **268,** 15935–15943.

Zacharias, W., Larson, J. E., Kilpatrick, M. W., and Wells, R. D. (1984). *Hha*I methylase and restriction endonuclease as probes for B to Z DNA conformational changes in d(GCGC) sequences. *Nucleic Acids Res.* **12,** 7677–7692.

Zacharias, W., Jaworski, A., Larson, J. E., and Wells, R. D. (1988). The B- to

Z-DNA equilibrium *in vivo* is perturbed by biological processes. *Proc. Natl. Acad. Sci. USA* **85,** 7069–7073.

Zahn, K., and Blattner, F. R. (1985). Sequence-induced DNA curvature at the bacteriophage lambda origin of replication. *Nature (London)* **317,** 451–453.

Zahn, K., and Blattner, F. R. (1987). Direct evidence for DNA bending at the lambda replication origin. *Science* **236,** 416–442.

Zamenhof, S., Brawerman, G., and Chargaff, E. (1952). On the deoxypentose nucleic acids from several microorganisms. *Biochim. Biophys. Acta* **9,** 402–405.

Zhang, S., Lockshin, C., Herbert, A., Winter, E., and Rich, A. (1992). Zuotin, a putative Z-DNA binding protein in *Saccharomyces cerevisiae.* EMBO J. **11,** 3787–3796.

Zheng, G., and Sinden, R. R. (1988). Effect of base composition at the center of inverted repeated DNA sequences on cruciform transitions in DNA. *J. Biol. Chem.* **263,** 5356–6461.

Zheng, G., Kochel, T., Hoepfner, R. W., Timmons, S. E., and Sinden, R. R. (1991). Torsionally tuned cruciform and Z-DNA probes for measuring unrestrained supercoiling at specific sites in DNA of living cells. *J. Mol. Biol.* **221,** 107–129.

Zhurkin, V. B., Raghunathan, G., Ulyanov, N. B., Camerini-Otero, R. D., and Jernigan, R. L. (1994). A Parallel DNA Triplex as a model for the intermediate in homologous recombination. *J. Mol. Biol.* In press.

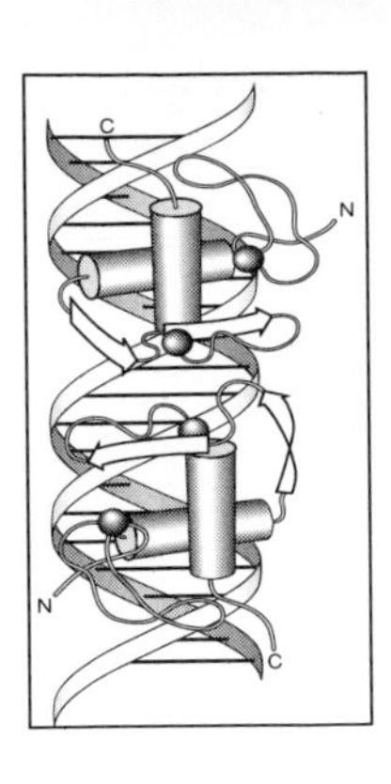

Author Index

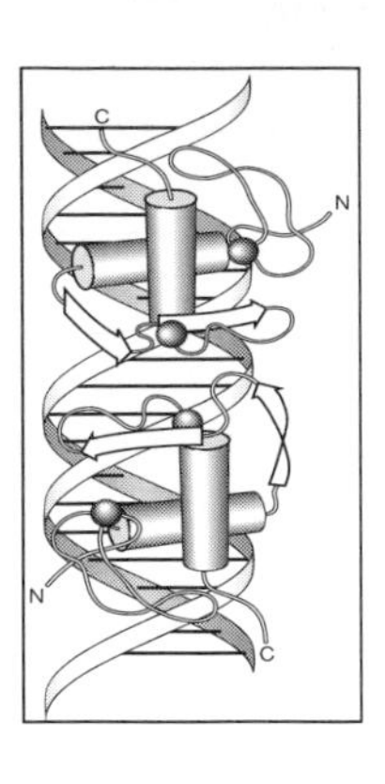

Subject Index